BIOLOGY

A course to 16 +

Geoff Jones

Health Science Section Leader
Oxford College of Further Education

Mary Jones

Head of Biology
Magdalen College School, Brackley

Cambridge University Press
Cambridge
London New York New Rochelle
Melbourne Sydney

Acknowledgements

The authors and publisher are grateful to the following for permission to reproduce their photographs.

A–Z Collection 14.12, 14.31; Dr T Allen, Christie Hospital and Holt Radium Institute, Manchester 11.14; All-Sport 5.20;American Museum of Natural History Neg no 36194, Photo by A E Anderson 16.3; Heather Angel 8.27b and c; Ardea London 14.14, 14.27, 13.12 (John Mason), 14.15b (Gert Behrens), 14.30 (Bob Gibbons), 16.7 (Pat Morris); Mrs J H Bebb (ARC Weed Research Organization) 8.35, 8.36a and b;Biophoto Associates 6.14, 6.20, 6.28, 14.25; Tony & Liz Bomford/Survival Anglia 14.18; Centre for Cell and Tissue Research, York 2.21, 3.3, 3.5, 6.19, 10.15, 11.10, 11.13; Dr J A Chapman 11.1; W H Cousins, Oxford College of Further Education 16.11; Dr Richard Foote, Systematic Entomology Laboratory, USDA and Nigel Luckhurst 16.18; Forestry Commission 7.13, 17.16; E Frei and R D Preston, F.R.S. 1.6; Al Giddings/Survival Anglia 2.9; Dr R H F Hunter, University of Edinburgh 8.17; Alan Hutchinson Library 15.16; Geoff Jones 11.3, 13.3, 13.4, 16.8; Dr H B D Kettlewell 16.4a and b; Frank Lane Agency 14.1 (Steve McCutcheon), 14.5 (Ronald Thompson), 16.10b Frants Hartmann; Nigel Luckhurst 6.7;National Portrait Gallery, London 16.9; Natural History Photographic Agency 7.16, 8.27a, 9.3, 14.4; Nature Conservancy Council/P.Wakely 14.8; Dr A J Pontin, Royal Holloway College 14.28; Press-tige Pictures 7.12, 14.32; Roy Shaw 8.26; Topham 9.7, 9.13, 14.3, 14.7, 17.15; C James Webb 14.19; Dr Paul Wheater 1.4, 6.13, 11.4, 15.1; Bill Whitfield 14.15a; D P Wilson/Eric & David Hosking 17.24

The authors and publisher are also grateful to Guinness Superlatives Ltd., for permission to extract many of the 'fact!' items of interest from their publications, *The Guinness Book of Records*, and *The Guinness Book of Animal Facts and Feats*.

Published by the Press Syndicate of the University of Cambridge
The Pitt Building, Trumpington Street, Cambridge CB2 1RP
32 East 57th Street, New York, NY 10022, USA
10 Stamford Road, Oakleigh, Melbourne 3166, Australia

© Cambridge University Press 1984
The line illustrations in this book are © Geoff Jones 1984

First published 1984
Fourth printing 1986

Printed in Great Britain by Ebenezer Baylis & Son,
The Trinity Press, Worcester and London

British Library cataloguing in publication data

Jones, G.H.
Biology–A course to 16+.

1. Biology
I. Title II. Jones, M.R.
574 QH308.7

ISBN 0 521 28532 1

MX

Preface

This book is intended for students of a wide ability range, working towards 'O'-level, CSE or 16+ examinations in biology. We hope that the readability of the text and the clear, fully-labelled diagrams will enable most CSE candidates to understand important facts and principles, yet there is ample material here to provide more able students with the information and ideas that they need to obtain an A grade at 'O'-level.

Many people have helped in the production of this book. In particular, our very warmest thanks go to Dr. John Gillman, who encouraged us to take on such a project and gave us unwavering support. The help and encouragement of our parents has also been much appreciated. Mrs. Betty Cameron's assistance in particular has been invaluable. Not only has she critically read the text and helped with indexing, but has also taken two small children off our hands for many long hours – without that, nothing would ever have been written!

Geoff and Mary Jones January 1984

The authors

Geoff Jones studied botany at Oxford, graduating in 1970, and then microbial biochemistry at Imperial College, London, where he obtained an MSc. After teaching biology and chemistry for some years at St. Edward's School, Oxford, he moved to Oxford College of Further Education where he is now Health Science Section Leader.

Mary Jones graduated in zoology at Oxford in 1971, and then took a PGCE in 1972. Since then she has taught at Headington School and Milham Ford School, both in Oxford, and at Banbury School. She is now Head of Biology at Magdalen College School, Brackley. She has examined in 'O'- and 'A'-level biology examinations for Oxford Local, London and Cambridge Local Examination Boards.

Contents

Key to organisms shown in the cover picture

1 Hawthorn (*Crataegus monogyna*)
2 Puffin (*Fratercula arctica*)
3 Grey seal (*Halichoerus grypus*)
4 Gannet (*Sula bassana*)
5 Herring gull (*Larus argentatus*)
6 Common mussel (*Mytilus edulis*)
7 Beadlet anemone (*Actinia equina*)
8 Common limpet (*Patella vulgata*)
9 Common periwinkle (*Littorina littorea*)
10 Parasitic anenome (*Calliactis parasitica*)
11 Hermit crab (*Pagurus bernhardus*) in a Whelk shell (*Buccinum undatum*)
12 Dahlia anemone (*Tealia felina*)
13 Common blenny (*Blennius pholis*)
14 Serrated wrack (*Fucus serratus*)
15 Jellyfish (*Chrysaora isosceles*)
16 Snakelocks anemone (*Anemonia sulcata*)
17 Bladder wrack (*Fucus vesiculosus*)
18 Shore crab (*Carcinus maenas*)
19 Common starfish (*Asterias rubens*)
20 Barnacle (*Balanus balanoides*)

1 Cells

1.1 All living things have certain characteristics.

Biology is the study of living things, which are often called **organisms.** Living organisms have several features or **characteristics** which make them different from objects which are not alive.

1 They **reproduce.**
2 They **feed.**
3 They **respire**—that is, they release energy from their food, often by combining it with oxygen.
4 They **grow.**
5 They **excrete**—that is, they get rid of substances which they do not want. These have been made by some of the chemical reactions going on inside them.
6 They **move.**
7 They are **sensitive**—that is, they can sense and respond to changes in their surroundings.
8 They are made of **cells.**

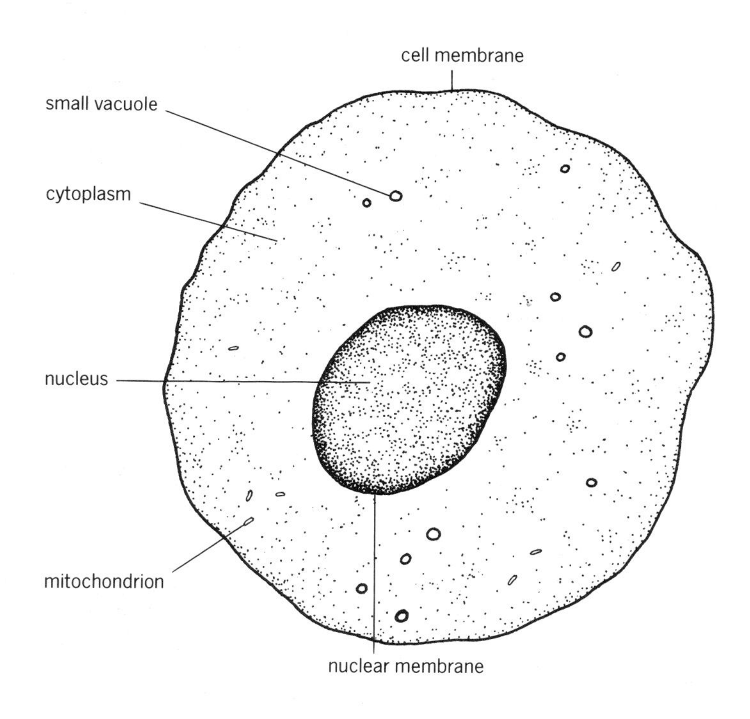

1.1 A typical animal cell as seen with a light microscope

Cell structure

1.2 Microscopes are used to study cells.

All living things are made of cells. Cells are very small, so large organisms contain millions of cells.

To see cells clearly, you need to use a microscope. The kind of microscope used in a school laboratory is called a **light microscope** because it shines light through the piece of animal or plant you are looking at. It uses glass lenses to magnify and focus the image. A very good light microscope can magnify about 1500 times, and can show all the structures in Figs 1.1 and 1.2.

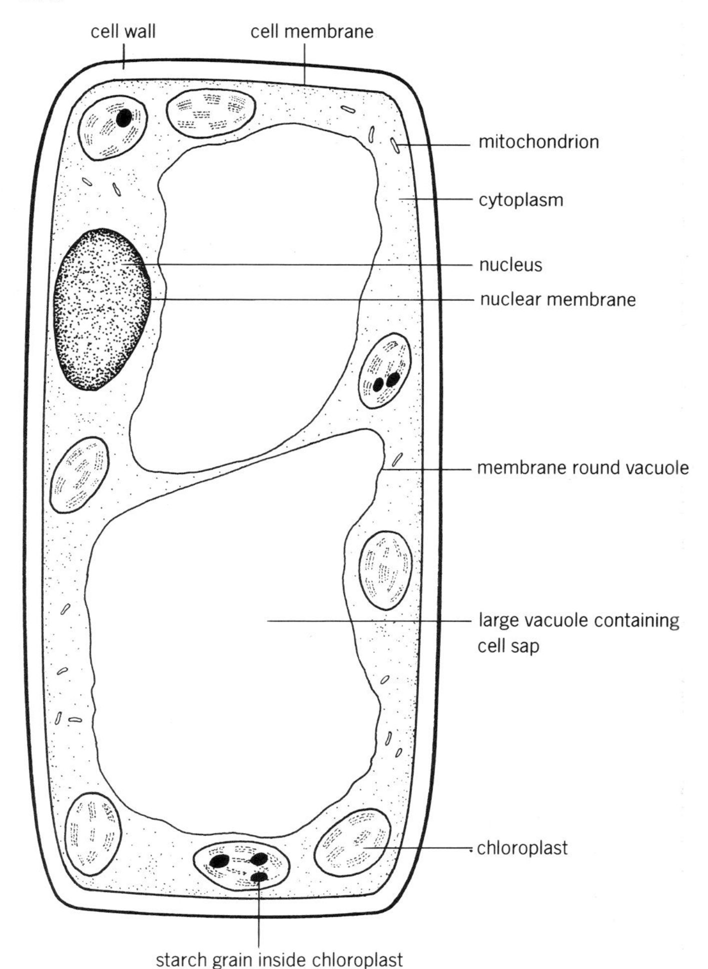

1.2 A typical plant cell as seen with a light microscope

1.3 Equipment used in looking at biological material

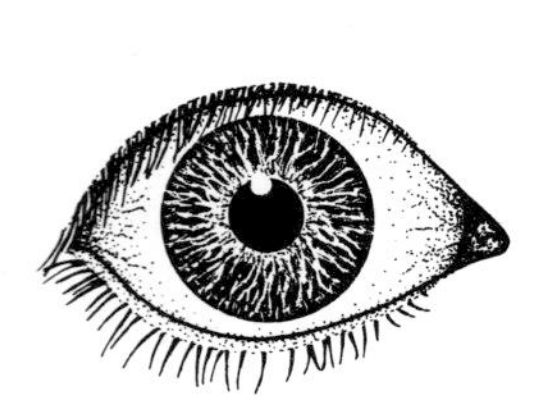

Eye

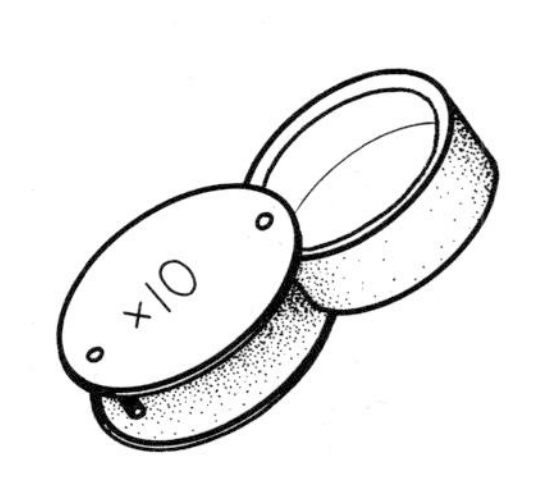

Hand lens magnifies x10

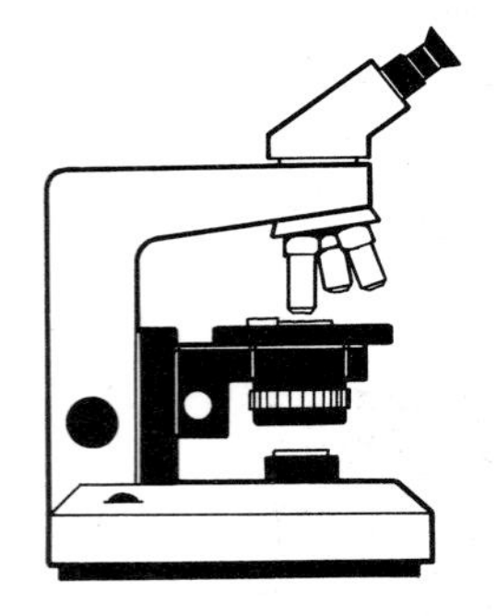

light microscope magnifies x400

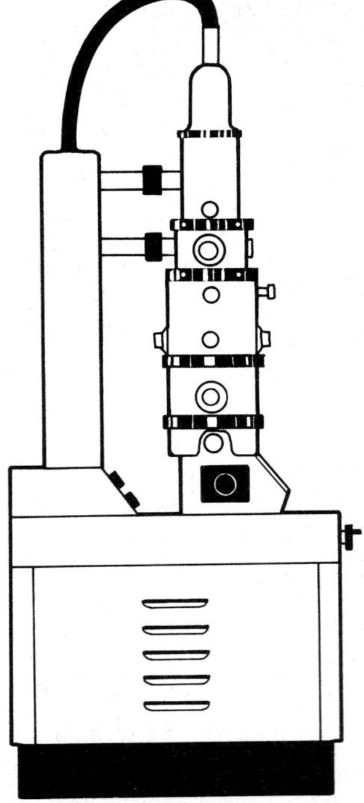

electron microscope magnifies x40,000

The human eye cannot see most cells

Using a hand lens, a cell can be seen as a dot.

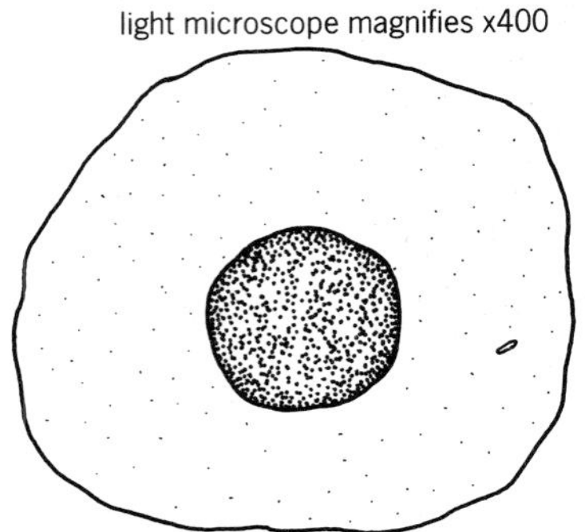

With a light microscope, you can see some of the structures inside a cell, such as a mitochondrion.

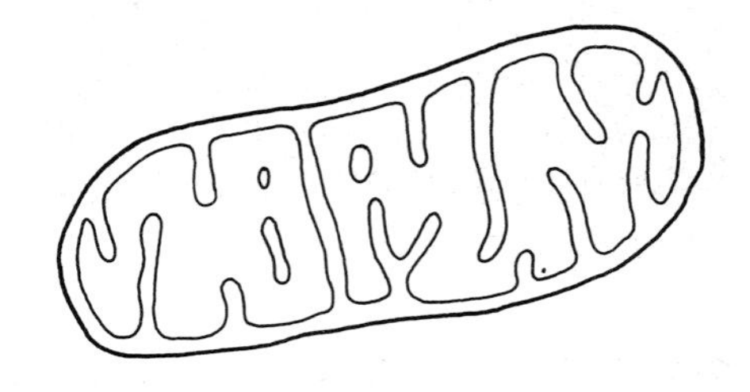

With an electron microscope, the internal structures of a mitochondrion can be seen.

To see even smaller things inside a cell, an **electron microscope** is used. This uses a beam of electrons instead of light, and can magnify up to 500 000 times. This means that a lot more detail can be seen inside a cell, as shown in Fig 1.5.

Questions

1 List the characteristics of living things.
2 How many times can a good light microscope magnify?
3 How many times can an electron microscope magnify?

1.3 All cells have a cell membrane.

Whatever sort of animal or plant they come from, all cells have a **cell membrane** around the outside. Inside it is a jelly-like substance called **cytoplasm,** in which are found many small structures called **organelles.** The most obvious of these organelles is usually the **nucleus.** The whole content of the cell is called **protoplasm.**

1.4 Plant cells have a cell wall.

All plant cells are surrounded by a **cell wall** made of **cellulose.** Paper, which is made from cell walls, is also made of cellulose. Animal cells never have cell walls made of cellulose. Cellulose belongs to a group of substances called **polysaccharides,** which are described in section 2.8. It forms fibres, which criss-cross over one another to form a very strong covering to the cell (see Fig 1.6). This helps to protect and support the cell.

Because of the spaces between the fibres, even very large molecules are able to go through the cellulose cell wall. It is therefore said to be **fully permeable.**

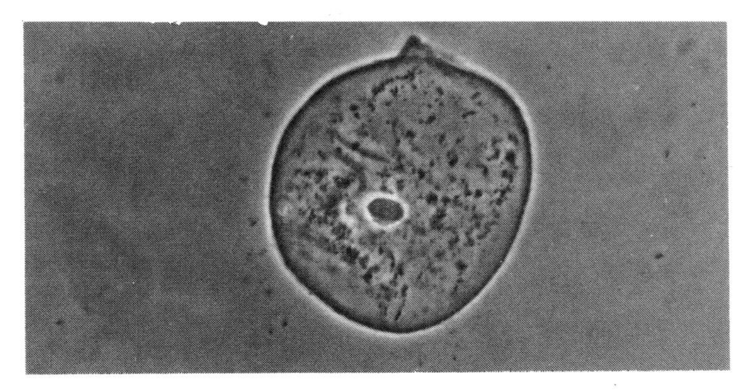

1.4 A human cheek cell as it appears in a light microscope

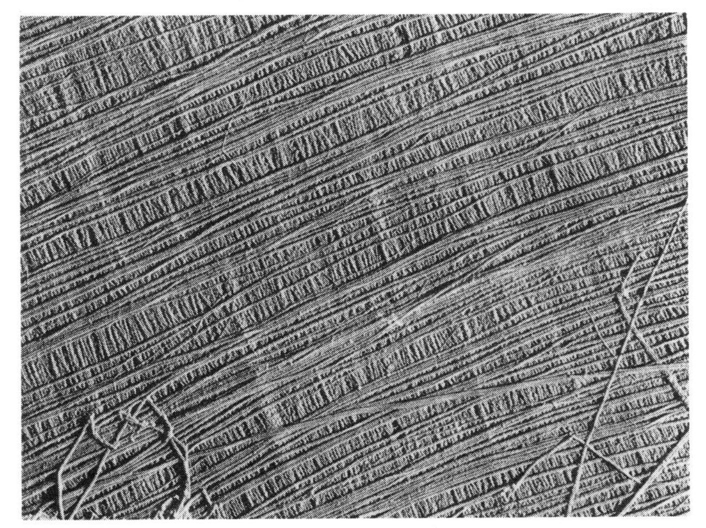

1.6 Cellulose fibres in a plant cell wall

1.5 Cell membranes are selectively permeable.

All cells have a membrane surrounding the cell. It is called the cell membrane or **plasma membrane.** In a plant cell, it is very difficult to see, because it is right against the cell wall.

The cell membrane is a very thin layer of protein and fat. It is very important to the cell because it controls what goes in and out of it. It is said to be **selectively permeable,** which means that it will let some substances through but not others.

1.6 Cytoplasm is a complex solution.

Cytoplasm is a clear jelly. It is nearly all water; about 70% is water in many cells. It contains many substances dissolved in it, especially proteins.

1.7 Most cells contain vacuoles.

A vacuole is a space in a cell, surrounded by a membrane, and containing a solution. Plant cells have very large vacuoles, which contain a solution of sugars and other substances called **cell sap.**

Animal cells have much smaller vacuoles, which may contain food or water.

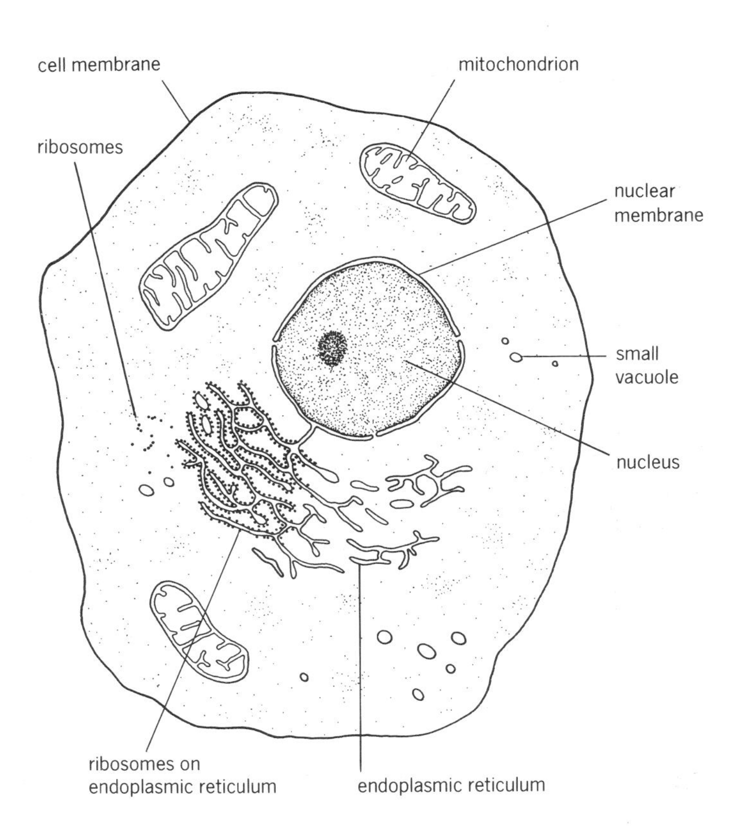

1.5 A typical animal cell, showing structures that can be seen with an electron microscope

1.8 Endoplasmic reticulum is a membrane network.

Endoplasmic reticulum is a maze of membranes which runs all through the cytoplasm. They are assembly lines for making fats and proteins out of smaller molecules in the cell. Endoplasmic reticulum is found in all cells.

1.9 Chloroplasts trap the energy of sunlight.

Chloroplasts are never found in animal cells, but most of the cells in the green parts of plants have them. They contain the green colouring or pigment called **chlorophyll.** Chlorophyll absorbs sunlight, and the energy of sunlight is then used for making food for the plant by **photosynthesis** (see Chapter 3).

Chloroplasts often contain starch grains, which have been made by photosynthesis. Animal cells never contain starch grains.

1.10 Mitochondria release energy from food.

Every cell has mitochondria, because it is here that the cell releases energy from food. The energy is needed to help it move and grow. Mitochondria are sometimes called the 'powerhouses' of the cell. The energy is released by combining food with oxygen, in a process called **respiration.** The more active a cell, the more mitochondria it has.

Investigation 1.1 Looking at animal cells

The easiest place to find animal cells is on yourself. If you colour or **stain** the cells they are quite easy to see using a light microscope (see Fig 1.7).

1. Using a clean fingernail, or section lifter, gently rub off a little of the lining from the inside of your cheek.
2. Put your cells on to the middle of a clean microscope slide, and gently spread them out. You will probably not be able to see anything at all at this stage.
3. Put on a few drops of **methylene blue.**
4. Gently lower a coverslip over the stained cells, trying not to trap any air bubbles.
5. Use filter paper or blotting paper to clean up the slide, and then look at it under the low power of a microscope.
6. Make a labelled drawing of a few cells.

Questions

1. Which part of the cell stained the darkest blue?
2. Name two structures which you could not see in your cells, but which you would have been able to see if you were using an electron microscope.

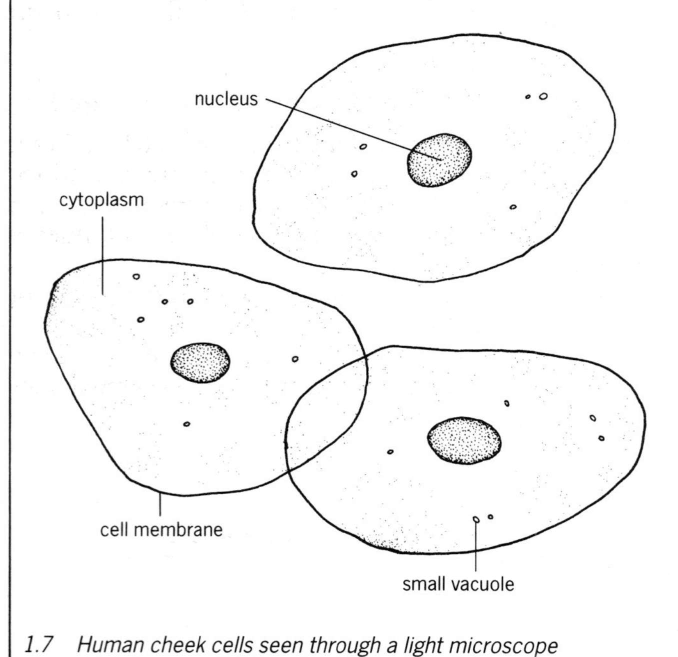

1.7 Human cheek cells seen through a light microscope

1.11 Ribosomes make proteins.

Ribosomes are very tiny, round objects, often attached to the endoplasmic reticulum. It is here that proteins are made, by joining together smaller molecules called **amino acids.** To get the amino acids in the right order, the ribosomes follow instructions from the nucleus. This process is described in Chapter 15.

1.12 The nucleus stores inherited information.

The nucleus is where the information is stored which helps the ribosomes to make the right sort of proteins. The information is kept on the **chromosomes,** which are inherited from the organism's parents.

Chromosomes are very long, but so thin that they cannot easily be seen even by the electron microscope. However, when the cell is dividing, they become short and thick, and can be seen with a good light microscope.

Investigation 1.2 Looking at plant cells

To be able to see cells clearly under a microscope, you need a very thin layer. It is best if it is only one cell thick. An easy place to find such a layer is inside an onion bulb.

1. Cut a small piece from an onion bulb, and use forceps to gently peel a small piece of thin skin, called epidermis, from the inside of it. Do not let it get dry.
2. Put a drop or two of water on to the centre of a clean microscope slide. Put the piece of epidermis into it, and spread it flat.
3. Gently lower a coverslip on to it.
4. Use filter paper or blotting paper to clean up the slide, and then look at it under the low power of a microscope.
5. Make a labelled drawing of a few cells.

Questions

1. Name two structures which you can see in these cells, but which you could not see on the cheek cells.
2. Most plant cells have chloroplasts, but these do not. Suggest a reason for this.

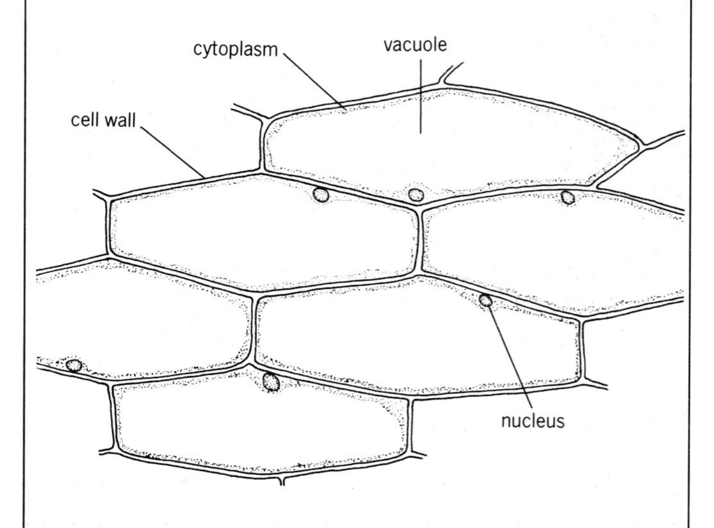

1.8 Onion epidermis cells seen through a light microscope

Table 1.1 A comparison between animal and plant cells

Similarities

1 Both have a cell membrane surrounding the cell.
2 Both have cytoplasm.
3 Both contain a nucleus.
4 Both contain mitochondria.
5 Both contain endoplasmic reticulum.
6 Both contain ribosomes.

Differences

	Plant cells	Animal cells
1	Have a cellulose cell wall outside cell membrane	No cell wall
2	Often have chloroplasts containing chlorophyll	No chloroplasts
3	Often have very large vacuoles, containing cell sap	Only have small vacuoles
4	Often have starch granules	Never have starch granules; sometimes have glycogen granules
5	Often regular in shape	Often irregular in shape

Animal and plant cells obtain their food in different ways. Plants make their own food, so their cells contain chloroplasts. Starch granules store some of the food they make. Animals often have to move to find their food. This is made easier if their cells do not have a rigid wall.

Questions

1 What sort of cells are surrounded by a cell membrane?
2 What are plant cell walls made of?
3 What does fully permeable mean?
4 What does selectively permeable mean?
5 What is the main constituent of cytoplasm?
6 What is a vacuole?
7 What is cell sap?
8 Chloroplasts contain chlorophyll. What does chlorophyll do?
9 What happens inside mitochondria?
10 Where are proteins made?
11 What is stored in the nucleus?
12 Why can chromosomes be seen only when a cell is dividing?

Cells and organisms

1.13 There is division of labour between cells.

A large organism such as yourself may contain many millions of cells, but not all the cells are alike. Almost all of them can carry out the activities which are characteristic of living things (see section 1.1), but many of them specialise in doing some of these better than other cells do. Muscle cells, for example, are specially adapted for movement. Cells in the leaf of a plant are specially adapted for making food by photosynthesis.

1.14 Similar cells are grouped to form tissues.

Often, cells which specialise in the same activity will be found together. A group of cells like this is called a **tissue.** An example of a tissue is the layer of cells lining your stomach. These cells make enzymes to help to digest your food (see Fig 1.9).

The stomach also contains other tissues. For example, there is a layer of muscle in the stomach wall, made of cells which can move. This muscle tissue makes the wall of the stomach move in and out, churning the food and mixing it up with the enzymes.

1.15 An organ contains tissues working together.

All the tissues in the stomach work together, although they each have their own job to do. A group of tissues like this makes up an **organ.** The stomach is an organ. Other organs include the heart, the kidneys and the lungs.

1.16 A system contains organs working together.

The stomach is only one of the organs which help in the digestion of food. The mouth, the intestines and the stomach are all part of the **digestive system.** The heart is part of the circulatory system, while each kidney is part of the excretory system.

The way in which organisms are built up can be summarised like this:–

Organelles make up **cells** which make up **tissues** which make up **organs** which make up **systems** which make up **organisms.**

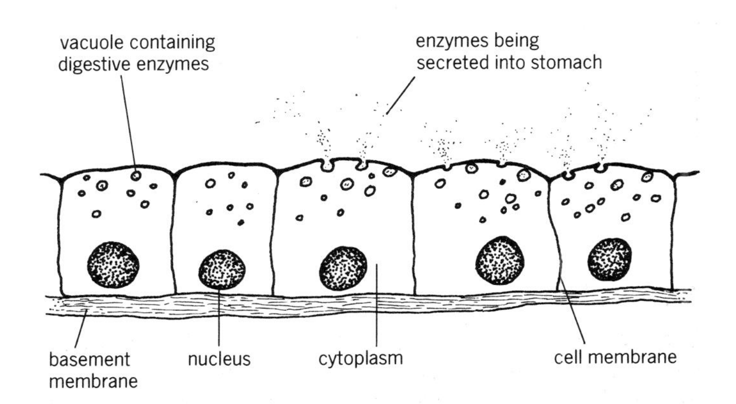

1.9 Cells lining the stomach; an example of a tissue

Chapter revision questions

1 Make a list of the parts of a cell which can be seen:
(a) in the light microscope as well as the electron microscope.
(b) only in the electron microscope.

2 Arrange these structures in order of size, beginning with the smallest.
stomach mitochondrion starch grain cheek cell nucleus ribosome

3 For each of the following, state whether it is an organelle, a cell, a tissue, an organ, a system, or an organism:
(a) heart, (b) chloroplast, (c) nucleus, (d) cheek cell, (e) onion epidermis, (f) onion bulb, (g) onion plant, (h) mitochondrion, (i) human being, (j) lung.

4 State which part of a plant cell:
(a) makes food by photosynthesis,
(b) releases energy from food,
(c) controls what goes in and out of the cell,
(d) stores information about making proteins,
(e) contains cell sap,
(f) protects the outside of the cell.

5 A tree is a living organism, but a bicycle is not. Using these as examples, explain the differences between living and non-living things.

6 With the aid of large labelled diagrams, make a comparison of a typical plant cell and a typical animal cell. Whenever possible, explain the reasons for their differences and similarities.

2 How animals feed

Nutrition

2.1 All organisms feed.

All living organisms need to take many different substances into their bodies. Some of these may be used to make new parts, or repair old parts. Others may be used to release energy.

Taking in useful substances is called feeding, or **nutrition.**

2.2 Green plants turn simple chemicals into complex ones.

Green plants take in simple substances, which they get from the air and soil. They use carbon dioxide, water and minerals, which are simple **inorganic** substances. The plant builds them into more complex materials, such as sugars. Sugars are **organic** materials. Organic substances are ones which have been made by living things.

The way in which plants feed is called **autotrophic nutrition.** 'Auto' means self, and 'trophic' means feeding. Therefore 'autotrophic' means that the plant feeds itself, and does not rely on other living things to make its food for it. Autotrophic nutrition is described in Chapter 3.

2.3 Animals take complex substances from plants.

Animals cannot make their own food. They feed on organic substances which have originally been made by plants. This is called **heterotrophic nutrition.** 'Hetero' means other, so 'heterotrophic' means that an animal feeds on substances made by other organisms. Some animals eat other animals, but all the substances passing from one animal to another were first made by plants.

2.4 Your diet is the food you eat each day.

The food which an animal eats every day is called its **diet.** Most animals need seven kinds of food in their diet. These are:

Carbohydrates,
Proteins,
Fats,
Vitamins,
Minerals,
Water,
Roughage.

A diet which contains all of these things, in the correct amounts and proportions, is called a balanced diet (see section 2.19).

Questions

1 Give two examples of inorganic substances.
2 Give one example of an organic substance.
3 Explain the difference between autotrophic and heterotrophic nutrition.

Carbohydrates

2.5 Starch and sugars are carbohydrates:

Carbohydrates include starches and sugars. Their molecules contain three kinds of atom–carbon (C), hydrogen (H), and oxygen (O). A carbohydrate molecule has about twice as many hydrogen atoms as carbon or oxygen atoms.

2.6 Glucose is a simple sugar.

The simplest kinds of carbohydrate are the simple sugars or **monosaccharides. Glucose** is a simple sugar. A glucose molecule is made of six carbon atoms joined in a ring, with the hydrogen and oxygen atoms pointing out from and into the ring (Fig. 2.2).

The molecule contains six carbon atoms, twelve hydrogen atoms, and six oxygen atoms. To show this, its **molecular formula** can be written $C_6H_{12}O_6$. This formula stands for one molecule of a simple sugar, and tells you which atoms it contains, and how many of each kind.

Although they contain many atoms, simple sugar molecules are very small. They are soluble, and they taste sweet.

2.7 Sucrose is a complex sugar.

If two simple sugar molecules join together, a large molecule called a complex sugar or **disaccharide** is made (Fig 2.1b). Two examples of complex sugars are **sucrose** (the sugar you use on the table) and **maltose** (malt sugar). Like simple sugars, they are soluble and they taste sweet.

2.8 Starch is a polysaccharide.

If many simple sugars join together, a very large molecule called a **polysaccharide** is made. Some polysaccharide molecules contain thousands of sugar molecules joined together in a long line. The **cellulose** of plant cell walls is a polysaccharide and so is **starch**, which is often found inside plant cells (Fig 2.1c).

Most polysaccharides are insoluble, and they do not taste sweet.

Investigation 2.1 Testing foods for carbohydrates

Whenever you are doing any kind of food tests, there are certain procedures you should always follow.

Always do a standard test first. For example, if you are testing foods for simple sugar, begin by testing a known simple sugar, such as glucose. Keep the result of this test, so that you can compare your other results with it.

Keep foods completely separate from each other. This means using clean tubes, spatulas and knives for each kind of food.

Use the same amount of reagents for each test.

To test for simple sugars

1 Copy the results chart, ready to fill it in.
2 Cut or grind a little of the test food into very small pieces. Put these into a test tube. Add some water, and shake it up to try to dissolve it.
3 Add some **Benedict's solution.** Benedict's solution is blue, because it contains copper salts.
4 Boil it. If there is any simple sugar in the food, the Benedict's solution will turn orange red.

Results table

Food	Colour with Benedict's solution	Simple sugar present

This test works because the simple sugar reduces the blue copper salts to a red compound. Sugars which do this are called **reducing sugars.** All simple sugars are reducing sugars, and so are some complex sugars.

To test for non-reducing sugars

Some complex sugars, such as sucrose, are not reducing sugars, so they will not turn Benedict's solution red. To test for them, you first have to break them down to simple

continued

sugars. Then you must do the Benedict's test.

1 Draw a results chart.
2 Make a solution of the food to be tested.
3 Do the simple sugar test, to check that there is no reducing sugar in the food.
4 Boil a fresh tube of food solution with hydrochloric acid. This breaks apart each complex sugar molecule into two simple sugar molecules.
5 Add sodium hydrogen carbonate solution until the contents stop fizzing to neutralise any left-over hydrochloric acid in the tube.
6 Now add Benedict's solution and boil. A red colour shows that there is now reducing sugar in the food, which was produced from non-reducing sugar.

To test for starch

There is no need to dissolve the food for this test.

1 Draw a results chart.
2 Put a small piece of the food onto a white tile.
3 Add a drop or two of **iodine solution.** Iodine solution is brown, but it turns bluish black if there is starch in the food.

Questions

1 Why should you cut food into small pieces and try to dissolve it before you do a food test?
2 If you test a food, and find that it contains a simple sugar, there is no point in doing the non-reducing sugar test. Why not?
3 How could you test a solution to see if it contained iodine?

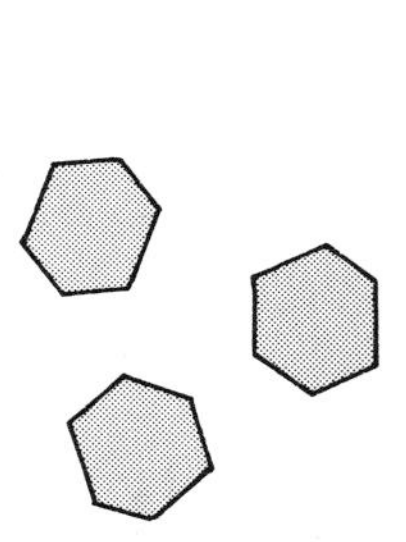

(a) Simple sugar molecules, e.g. glucose

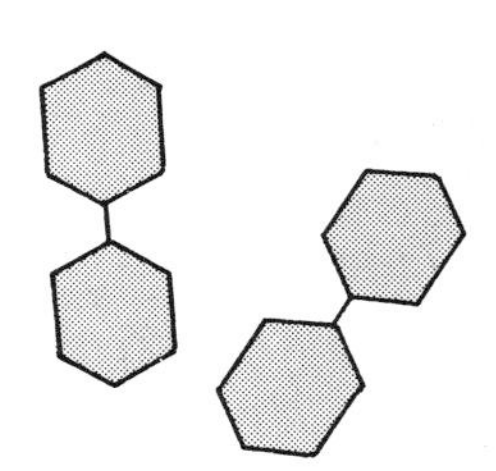

(b) Complex sugar molecules, e.g. maltose

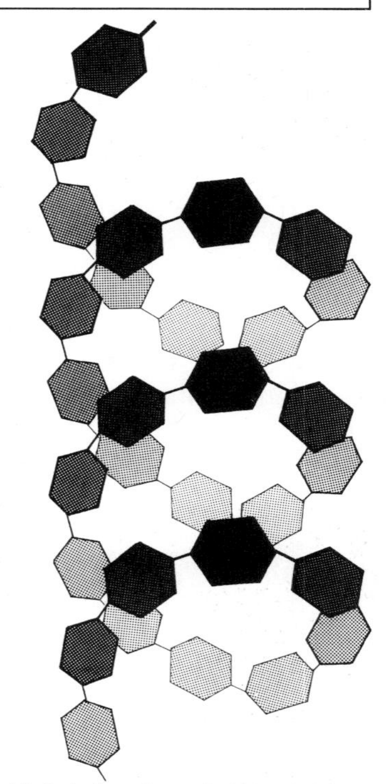

(c) Part of a polysaccharide molecule, e.g. starch.

2.1 Carbohydrate molecules

2.9 Animals get energy from carbohydrates.

Carbohydrates are needed for energy. One gram of carbohydrate releases 17 kJ (kilojoules) of energy in the body. The energy is released by respiration (see Chapter 5).

Questions

1. Which three elements are contained in all carbohydrates?
2. The molecular formula for glucose is $C_6H_{12}O_6$. What does this tell you about a glucose molecule?
3. To which group of carbohydrates do each of these substances belong: (a) glucose, (b) starch, (c) sucrose?
4. Why do animals need carbohydrates?

2.2 *The structure of a glucose molecule*

2.3 *Carbohydrate foods*

Fact! The longest time that anyone has ever lived without food or water is 18 days. An 18-year old man in Austria was put into a police cell on 1st April, 1979–and forgotten. When he was discovered on 18th April, he was very close to death.

Proteins

2.10 Proteins are long chains of amino acids.

Protein molecules contain some kinds of atoms which carbohydrates do not. As well as carbon, hydrogen and oxygen, they also contain nitrogen (N) and small amounts of sulphur (S).

Like polysaccharides, protein molecules are made of long chains of smaller molecules joined end to end. These smaller molecules are called **amino acids.** There are about twenty different kinds of amino acid. Any of these twenty can be joined together in any order to make a protein molecule (see Fig 2.5). Each protein is made of molecules with amino acids in a precise order. Even a small difference in the order of amino acids makes a different protein, so there are millions of different proteins which could be made.

Some proteins are soluble, such as haemoglobin, the red pigment in blood. Others are insoluble, such as keratin. Hair and fingernails are made of keratin.

Investigation 2.2 Testing food for proteins

The Biuret test

1. Draw a results chart.
2. Put the food into a test tube, and add a little water.
3. Add some potassium hydroxide solution.
4. Add two drops of copper sulphate solution.
5. Shake the tube gently. If a purple colour appears, then there is protein present.

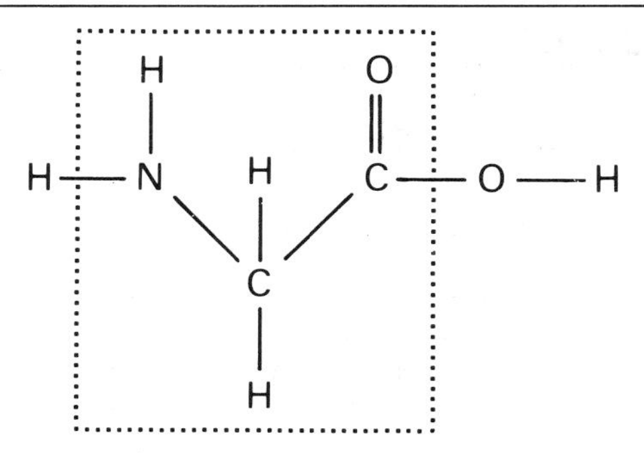

2.4 *The structure of an amino acid molecule*

2.11 Proteins are used for growth and repair.

Unlike carbohydrates, proteins are not normally used to provide energy. Many of the proteins in the food you eat are used for making new cells. New cells are needed for growing, and for repairing damaged parts of the body. In particular, **cell membranes** and **cytoplasm** contain a lot of protein.

Proteins are also needed to make **antibodies.** These fight bacteria and viruses inside the body. **Enzymes** are also proteins.

Questions

1. Name two elements found in proteins which are not found in carbohydrates.
2. How many different amino acids are there?
3. In what way are protein molecules similar to polysaccharides?
4. Give two examples of proteins.
5. List three reasons why animals need proteins.

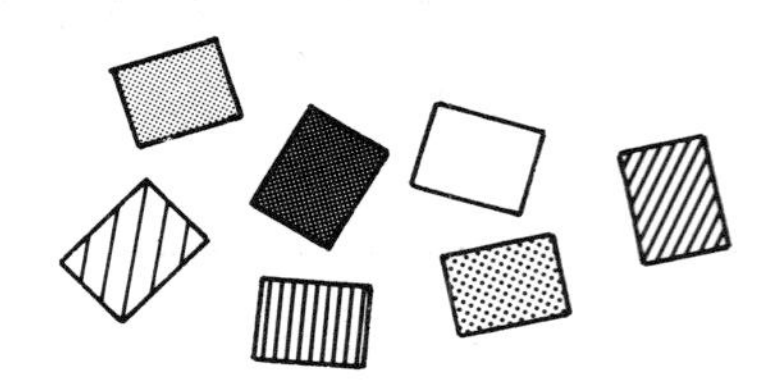

Amino acid molecules.

Part of a protein molecule.

2.5 Protein and amino acid molecules

2.6 Protein foods

Fats

2.12 Fats are made of glycerol and fatty acids.

Like carbohydrates, fats contain only three kinds of atom—carbon, hydrogen and oxygen. A fat molecule is made of four molecules joined together. One of these is **glycerol.** Attached to the glycerol are three long molecules called **fatty acids** (see Fig 2.7).

Fats are insoluble in water.

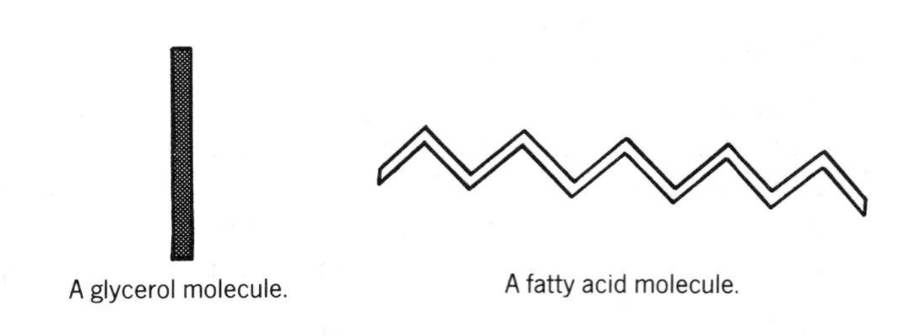

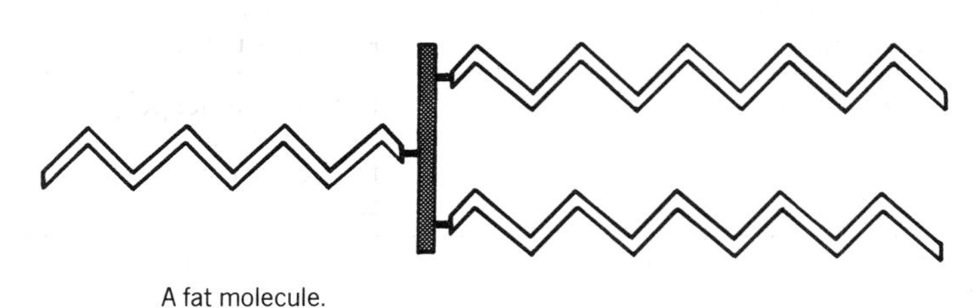

2.7 Fat molecule

Investigation 2.3 Testing food for fats

The emulsion test

Fats will not dissolve in water, but they will dissolve in alcohol. If a solution of fat in alcohol is added to water, the fat forms tiny globules which float in the water. This is called an **emulsion.** The globules of fat make the water look milky.

1. Draw a results chart.
2. Chop or grind a small amount of food, and put some into a very clean, dry test tube. Add some **absolute** (pure) **alcohol** (ethanol). Shake it thoroughly.
3. Put some distilled water in another tube.
4. Pour some of the liquid part, but not any solid, from the first tube into the water. A milky appearance shows that there is fat in the food.

The grease-spot test

1. Draw a results chart.
2. Rub some of the food onto some filter paper.

continued

3 A translucent mark suggests that there is fat in the food. To check that it is a greasy mark, and not a wet one, dry the paper gently. If the mark goes, it was made by water. If it stays, it was made by fat.

Questions

1 What is an emulsion?
2 Sometimes, a food gives a positive result with the emulsion test, but not with the grease-spot test. Can you suggest why?

2.13 Animals store energy in fat.

Like carbohydrates, fats can be used in a cell to release energy. A gram of fat gives about 39 kJ of energy. This is more than twice as much energy as that released by a gram of carbohydrate. However, the chemical reactions which have to take place to release the energy

2.8 Fatty foods

Table 2.1 Carbohydrates, fats and proteins

	Carbohydrates	*Proteins*	*Fats*
Elements which they contain	C, H, O	C, H, O, S, N	C, H, O
Smaller molecules of which they are made	Simple sugars, (monosaccharides)	Amino acids	Fatty acids and glycerol
Solubility in water	Sugars are soluble; polysaccharides are not very soluble, often completely insoluble	Some are soluble, some insoluble	Insoluble
Why animals need them	Easily available energy (17kJ/g)	Making cells, antibodies, enzymes; only used for energy when all other stores have run out (17kJ/g)	Storage of energy (39kJ/g), insulation
Some foods which contain them	Bread, cakes, potatoes, rice	Meat, fish, eggs, milk, cheese, peas, beans	Butter, lard, margarine, oil, fat meat, peanuts

2.9 Whales such as this humpback in Glacier Bay, Alaska, have thick layers of blubber under their skin to keep in body heat

Table 2.2 Vitamins

Vitamin	*Foods which contain them*	*Why they are needed*	*Deficiency disease*
A	Butter, egg yolk, cod liver oil, carrots	To keep the cells lining the respiratory system healthy; to make a pigment in the rod cells in the retina of the eye, needed for seeing in dim light	Infections of cells lining respiratory system, night blindness
B (There are at least twelve different B vitamins, but they usually occur together.)	Wholemeal bread, yeast extract, liver, brown rice	Involved in many chemical reactions in the body, for example respiration	**Beri-beri,** which is common in S.E. Asia, where polished (not brown) rice is the staple diet; this disease causes muscular weakness and paralysis
C	Oranges, lemons, black-currants, raw vegetables, potatoes	Keeps tissues in good repair	**Scurvy,** which causes pains in joints and muscles, and bleeding from gums and other places; this used to be a common disease of sailors, who had no fresh vegetables on long voyages.
D	Butter, egg yolk; can be made by the skin when sunlight shines on it	Helps calcium and phosphate to be used for making bones	**Rickets,** which causes bones to become soft and deformed; this disease was common among young children in industrial areas, who rarely got out into sunshine. It also occurs in some immigrant children in the UK, not used to the limited amounts of sunshine.

from the fat are quite long and complicated. This means that a cell tends to use carbohydrates first when it needs energy, and only uses fats when all the available carbohydrates have been used.

The extra energy which they contain makes fats very useful for storing energy. Some cells, particularly ones underneath the skin, become filled with large drops of fat or oils. These stores can be used to release energy when needed. This layer of cells is called **adipose tissue.** It also helps to keep heat inside the body—that is, it insulates the body. Animals, such as seals and some whales which live in very cold places, often have especially thick layers of adipose tissue, called blubber.

Questions

1 Which three elements are contained in all fats?
2 List two reasons why animals need fats in their diet.

2.14 Vitamins are needed in small amounts.

Vitamins are organic substances, which are only needed in tiny amounts. If you do not have enough of a vitamin, you may get a deficiency disease. Table 2.2 shows some of the most important ones.

2.15 Minerals are needed in small amounts.

Minerals are inorganic substances. Only small amounts of them are needed in a diet. Table 2.3 shows some of the most important ones.

2.16 Water dissolves substances in cells.

Inside every living organism, chemical reactions are going on all the time. These reactions are called metabolism. Metabolic reactions can only take place if the chemicals which are reacting are dissolved in water. This is one reason why water is so important to living organisms. If their cells dry out, the reactions stop, and the organism dies.

Water is also needed for other reasons. For example, plasma, the liquid part of blood, must contain a lot of water, so that substances like glucose can dissolve in it. Dissolved substances are transported around the body.

2.17 Roughage keeps the alimentary canal moving.

Roughage, or fibre, is food which cannot be digested. It goes right through the digestive system from one end to the other, and is egested in the faeces (see section 2.39).

Table 2.3 Minerals

Mineral element	*Foods which contain it*	*Why it is needed*	*Deficiency disease*
Calcium (Ca)	Milk, cheese, bread	For bones and teeth	Brittle bones and teeth
Phosphorus (P)	Milk	For bones and teeth	Brittle bones and teeth
Fluorine (F)	Fluoride toothpaste, fluoridated water	Makes tooth enamel resistant to decay	Bad teeth
Iodine (I)	Seafood, table salt	For making hormone thyroxine	**Goitre,** a swelling in the neck; slow metabolic rate
Iron (Fe)	Liver, egg yolk	For making haemoglobin, the red pigment in blood which carries oxygen	**Anaemia**–not enough red blood cells, so the tissues are short of oxygen and cannot release energy

In food the minerals are present in the form of ions. For example, phosphorus is not present as the element (which burns in air!), but as phosphate ions.

Roughage helps to keep the alimentary canal working properly. Food moves through the alimentary canal (see section 2.31) because the muscles contract and relax to squeeze it along. This is called **peristalsis** (see Fig 2.18). The muscles are stimulated to do this when there is food in the alimentary canal. Soft foods do not stimulate the muscles very much. The muscles work more when there is harder, less digestible food, like roughage, in the alimentary canal. Roughage keeps the digestive system in good working order, and helps to prevent constipation.

All plant food, such as fruit and vegetables, contains roughage. This is because the plant cells have cellulose cell walls. Humans cannot digest cellulose.

One common form of roughage is the outer husk of cereal grains, such as oats, wheat and barley. This is called bran. Some of this husk is also found in wholemeal bread. Brown or unpolished rice, is also a good source of roughage.

Questions

1. Which vitamin prevents rickets?
2. Why do you need iron in your diet?
3. What is metabolism?
4. Why do organisms die if they do not have enough water?
5. Describe how food is moved along the alimentary canal.
6. Plant foods contain a lot of roughage. Explain.

Fact! The largest animal in the world, the Blue whale, is a filter feeder. It feeds on tiny shrimp-like animals called krill.

Balanced diet

2.18 Diets should provide the right amount of energy.

Every day, a person uses up energy. The amount you use partly depends on how old you are, which sex you are and what job you do. A few examples are shown in Fig 2.10.

The energy you use each day comes from the food

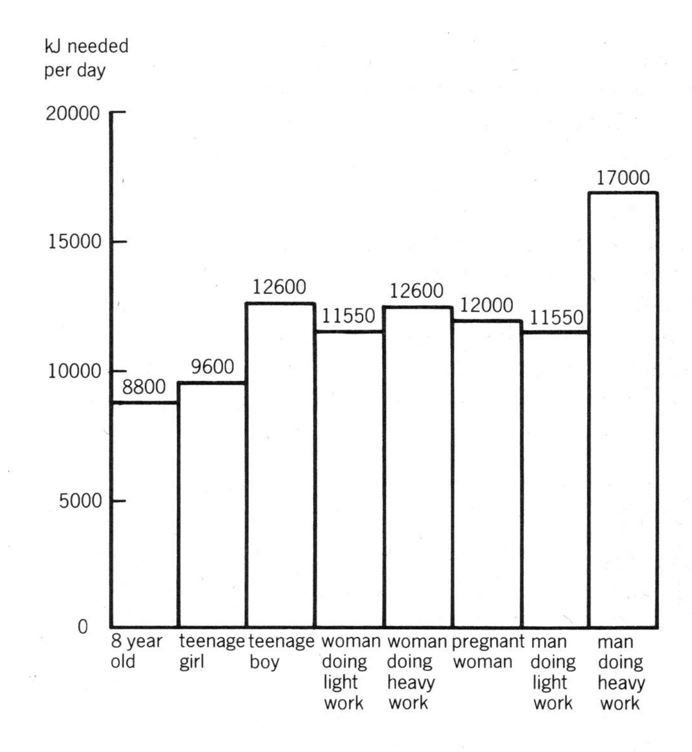

2.10 Daily energy requirements

you eat. If you eat too much food, some of the extra will probably be stored as fat. If you eat too little, you may not be able to obtain as much energy as you need. This will make you feel tired.

All food contains some energy. Scientists have worked out how much energy there is in a particular kind of food. You can look up this information in reference books. A few examples are given in Table 2.4. One gram of fat contains twice as much energy as one gram of protein or carbohydrate. This is why fried foods should be avoided if you are worried about putting on weight.

Table 2.4 Energy content of some different kinds of food

Food	*kJ/100g*	*Food*	*kJ/100g*
Breakfast foods		tomatoes	60
cornflakes	1567	rice	1536
oatmeal	1698	spaghetti	1612
boiled egg	612	**Desserts**	
brown bread	948	apples	196
white bread	991	bananas	326
milk	272	melon	96
sugar	1680	oranges	150
butter	3041	canned peaches	373
low fat spread	1506	strawberries	109
marmalade	315	double cream	1841
Main meals		apple pie	1179
stewed steak	932	currant bun	1385
roast chicken	599	trifle	674
ham	1119	ice cream	698
roast lamb	1209	custard	496
fried liver	1016	**Snacks**	
grilled chop	1380	chocolate	2214
grilled sausage	1500	fruit yoghurt	405
steak and kidney pie	1195	crisps	2224
fish fingers	749	plain biscuits	1925
sardines	906	chocolate biscuits	2197
cheddar cheese	1682	roast peanuts	2364
cottage cheese	402	**Drinks**	
baked beans	270	unsweetened fruit juice	143
cabbage	66	coffee	12
carrots	98	tea	0
lettuce	36	Coca-cola	168
peas	161	wine	284
boiled potatoes	339	orange squash	15
chips	1065		

2.19 Diets should contain a variety of food.

As well as providing you with energy, food is needed for many other reasons. To make sure that you eat a balanced diet you must eat food containing carbohydrate, fat and protein. You also need each kind of vitamin and mineral, roughage and water. If you miss out any of these things, your body will not be able to work properly.

Many people eat too many carbohydrates. This is partly because they are cheap, and partly because they are 'satisfying' foods; that is, they make you feel full. The staple foods of many countries are carbohydrate foods. Rice, the staple food of China, is largely a carbohydrate food. So are potatoes, the staple food for much of Europe.

Digestion

2.20 Digestion makes food easier to absorb.

An animal's alimentary canal is a long tube running from one end of its body to the other (see Fig 2.11). Before food can be of any use to the animal, it has to get out of the alimentary canal and into the bloodstream. This is called **absorption.** To be absorbed, molecules of food have to get through the walls of the alimentary canal. They need to be quite small to be able to do this.

The food you eat usually contains some large molecules (see section 2.8). Before they can be absorbed, they must be broken down into small ones. This is called **digestion.**

2.21 Not all foods need digesting.

Large carbohydrate molecules, such as polysaccharides have to be broken down to simple sugars. Proteins are broken down to amino acids. Fats are broken down to fatty acids and glycerol (see Fig 2.13).

Simple sugars, water, vitamins and minerals are small molecules, and can be absorbed just as they are.

2.22 Digestion may be mechanical and chemical.

Often the food an animal eats is in quite large pieces. These need to be broken up by teeth, and by the churning movements of the alimentary canal. This is called **mechanical digestion.**

Once any pieces of food have been ground up, the large molecules present are then broken down into small ones. This is called **chemical digestion.** It involves a chemical change from one sort of molecule to another. Enzymes are involved in this process.

Questions

1. What is digestion?
2. Name two groups of food which do not need to be digested.
3. What does digestion change each of these kinds of food into: (a) polysaccharides, (b) proteins, (c) fats?
4. What is meant by chemical digestion?

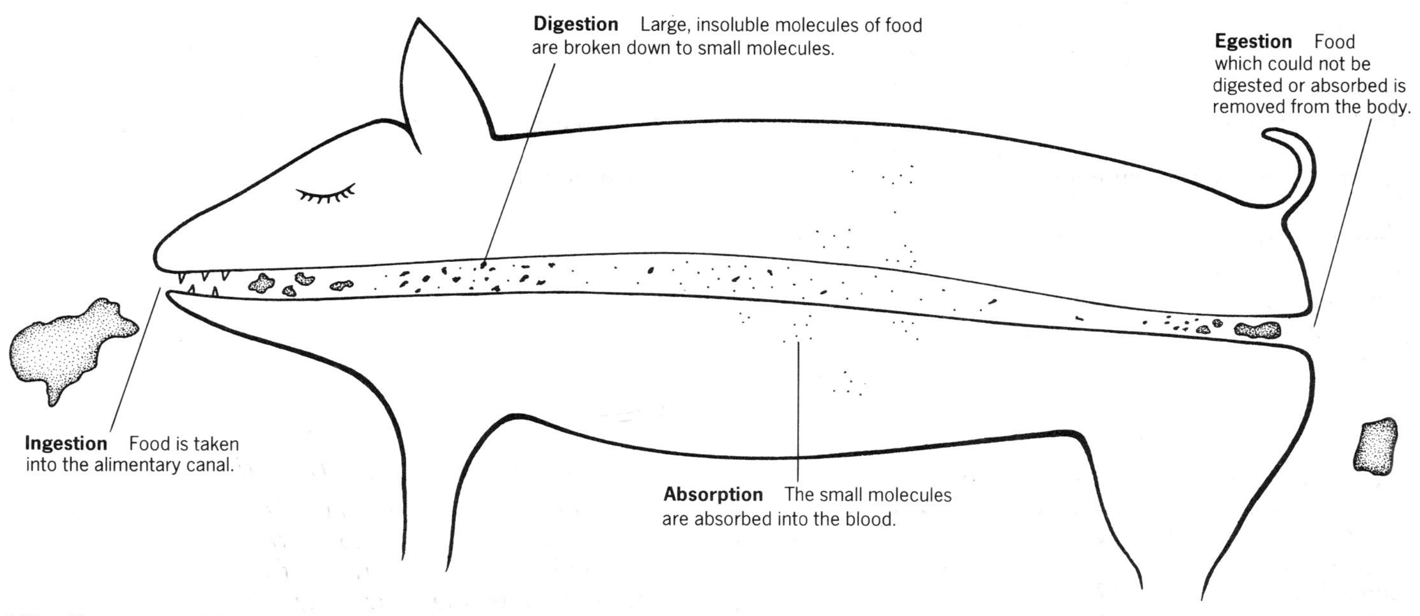

2.11 How an animal deals with food

Enzymes

2.23 Enzymes can speed up reactions.

If you made a solution of starch in water, and kept it at the temperature of your body, it would be a very long time before any of the starch molecules changed into glucose molecules. Yet this reaction happens quite quickly inside your alimentary canal.

Many chemical reactions can be speeded up by substances called **catalysts.** A catalyst alters the rate of a chemical reaction, without being changed itself.

Almost all the chemical reactions which go on inside the body are controlled by catalysts. The catalysts which are found in living organisms are called **enzymes.**

In several parts of the alimentary canal, digestive juices are secreted. These digestive juices contain enzymes which speed up the breakdown of large molecules to small ones. They are called **digestive enzymes.** This breakdown is called **hydrolysis** because water is also involved,('lysis' means splitting, and 'hydro' means water).

2.24 Enzymes are given special names.

Digestive enzymes are often named according to the substances whose breakdown they are catalysing. Enzymes which catalyse the breakdown of carbohydrates are called **carbohydrases.** If they break down proteins, they are **proteases.** If they break down fats (lipids) they are **lipases.**

Sometimes, they are given more specific names than this. For example, a carbohydrase which breaks down starch is called **amylase.** One which breaks down maltose is called **maltase.** One which breaks down sucrose is called **sucrase.**

2.25 Enzymes change substrates to products.

A chemical reaction always involves one substance changing into another. The substance which is present at the beginning of the reaction is called the **substrate.** The substance which is made by the reaction is called the **product.**

For example, in saliva there is an enzyme called amylase. It catalyses the breakdown of starch to the complex sugar maltose.

$$\text{starch} \xrightarrow{\text{amylase}} \text{maltose}$$

In this reaction, starch is the substrate, and maltose is the product.

2.26 All enzymes have certain properties.

1 All enzymes are proteins This may seem rather odd, because some enzymes actually digest proteins.

2 Enzymes are made inactive by high temperature This is because they are protein molecules, which are damaged by heat (see section 9.2).

3 Enzymes work best at a particular temperature The digestive enzymes which are found in the alimentary canal of a human usually work best at about 37 °C. This is called the **optimum** temperature.

4 Enzymes work best at a particular pH pH is a measure of how acid or alkaline a solution is. Some enzymes work best in acid conditions (low pH). Others work best in alkaline conditions (high pH).

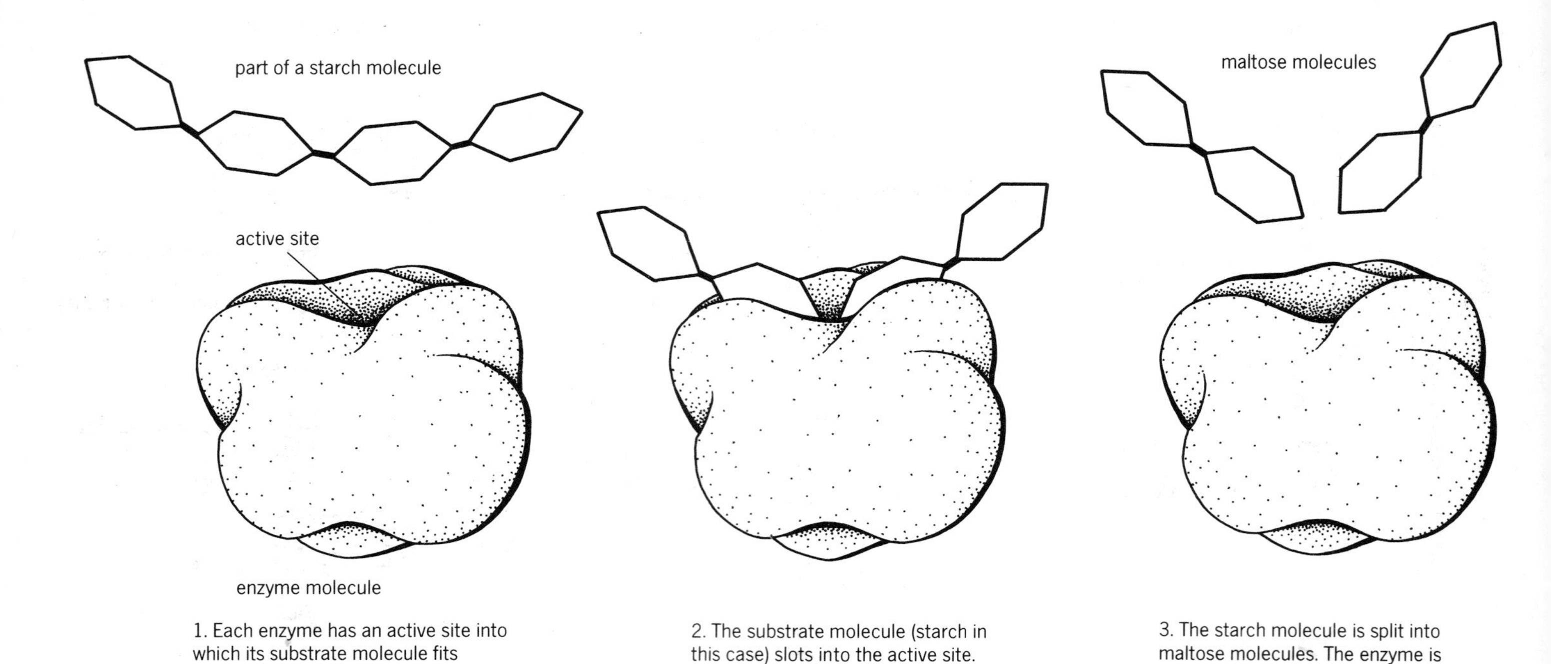

2.12 How an enzyme works

5 Enzymes are catalysts They are not changed in the chemical reactions which they control. They can be used over and over again, so a small amount of enzyme can change a lot of substrate to a lot of product.

6 Enzymes are specific This means that each kind of enzyme will only catalyse one kind of chemical reaction.

Questions

1 What is a catalyst?
2 What are the catalysts inside a living organism called?
3 Which kinds of reaction inside a living organism are controlled by enzymes?
4 What is meant by a carbohydrase?
5 Give one example of a carbohydrase.
6 Name the substrate and product of a reaction involving a carbohydrase.
7 Why are enzymes damaged by high temperatures?
8 What is meant by an optimum temperature?

Digestion in humans–teeth

2.27 Human beings feed holozoically.

Like all animals, mammals eat organic food. This may be plants, or it may be other animals which themselves have eaten plants. This is called heterotrophic nutrition (see section 2.3). Mammals feed by taking in, or ingesting, pieces of food which they digest inside their alimentary canal. This is a particular sort of heterotrophic nutrition, called **holozoic nutrition.** Humans are mammals. They feed holozoically.

2.28 The structure of a tooth

Teeth help with the ingestion and mechanical digestion of food. They can be used to bite off pieces of food. They then chop, crush or grind them into smaller pieces. This gives the food a larger surface area, which makes it easier for the enzymes to work. It also helps to dissolve soluble parts of the food.

The structure of a tooth is shown in Fig 2.15. The part of the tooth which is embedded in the gum is called the **root.** The part which can be seen is the **crown.** The crown is covered with **enamel.** Enamel is the hardest substance made by animals. It is very difficult to break or chip it. However, it can be dissolved by acids. Bacteria will feed on sweet foods left on the teeth. This makes acids, which dissolve the enamel and decay then sets in.

Under the enamel is a layer of **dentine,** which is rather like bone. This is also quite hard, but not as hard as enamel. It has channels in it which contain living cytoplasm.

In the middle of the tooth is the **pulp cavity.** It contains nerves and blood vessels. These supply the cytoplasm in the dentine with food and oxygen.

The root of the tooth is covered with **cement.** This has fibres growing out of it. These attach the tooth to the jawbone, but allow it to move slightly when biting or chewing.

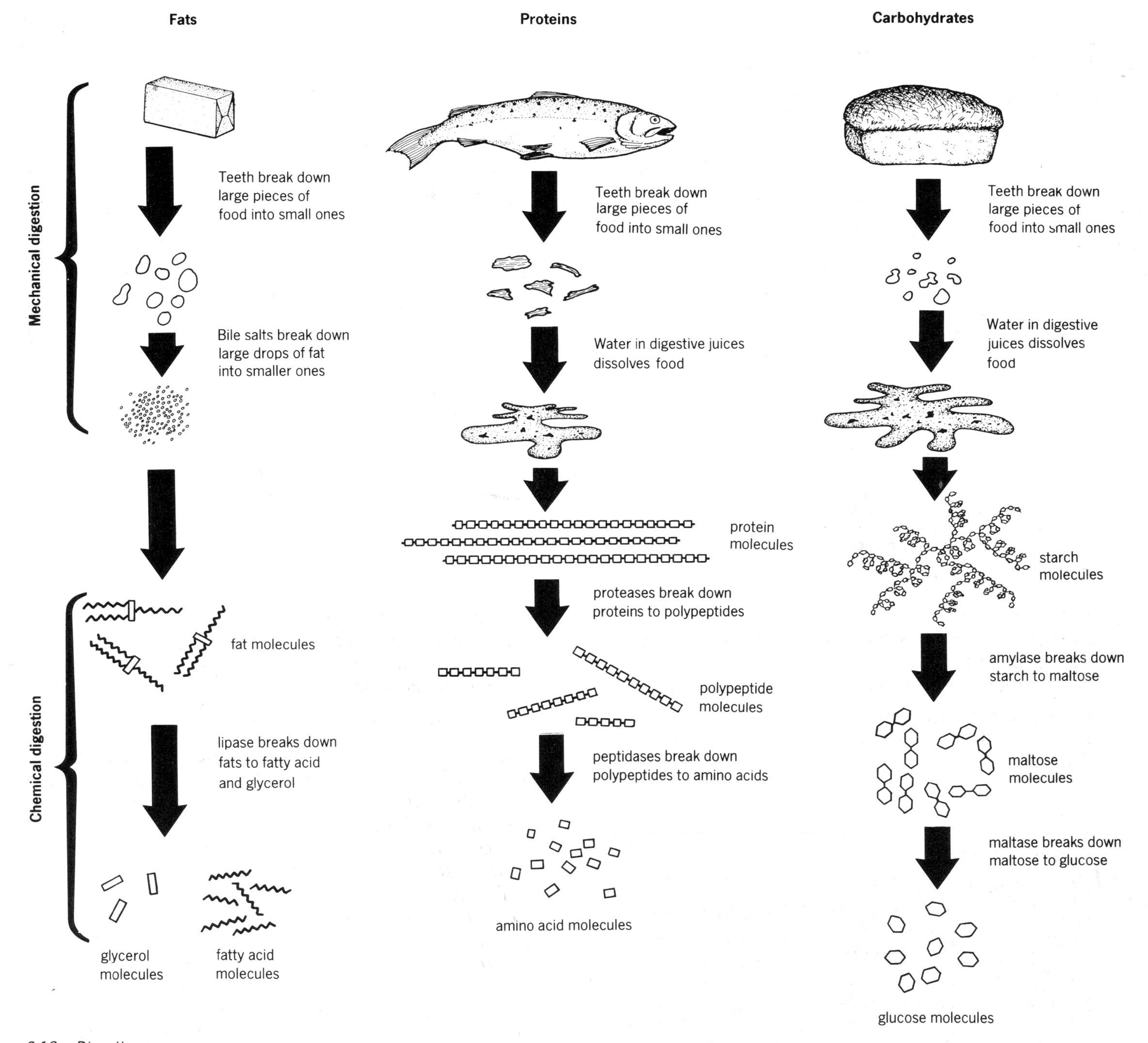

2.13 Digestion

2.29 Mammals have different types of teeth.

One of the ways in which mammals differ from other animals is that they have different kinds of teeth. Most mammals have four kinds (see Fig 2.16). **Incisors** are the sharp-edged, chisel-shaped teeth at the front of the mouth. They are used for biting off pieces of food. **Canines** are the more pointed teeth at either side of the incisors. **Premolars** and **molars** are the large teeth towards the back of the mouth. They are used for chewing food. The ones right at the back are sometimes called wisdom teeth. They do not grow until much later than the others.

2.30 Mammals have two sets of teeth in their life.

Mammals also differ from other animals in having two sets of teeth. The first set is called the **milk teeth** or deciduous teeth. In humans, these start to grow through the gum, one or two at a time, when a child is about five months old. By the age of eighteen or twenty months, most children have a set of 20 teeth.

This first set of teeth are milk teeth, and they begin to fall out when the child is about seven years old. They are all replaced by new ones, and twelve new teeth also grow, making up the complete set of **permanent teeth.** There are 32 altogether. Most people have all their permanent teeth by about seventeen years of age.

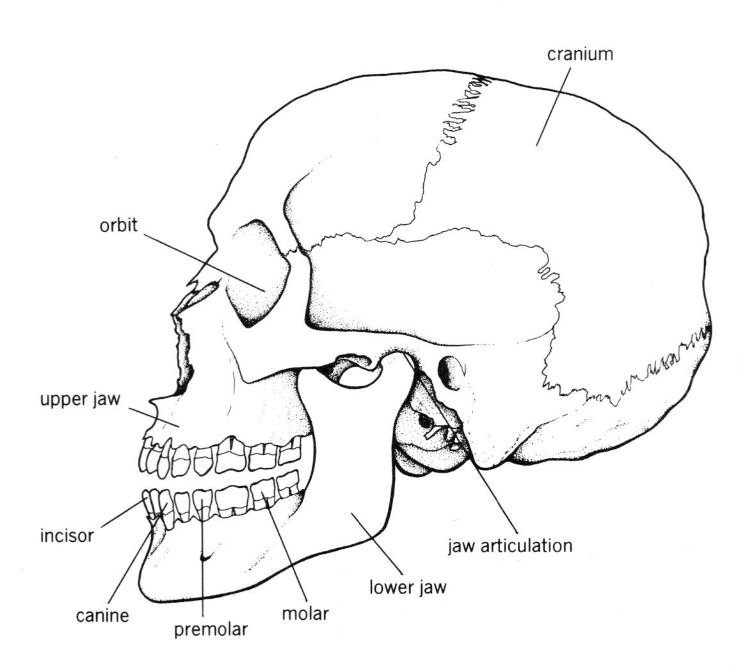

2.14 A human skull

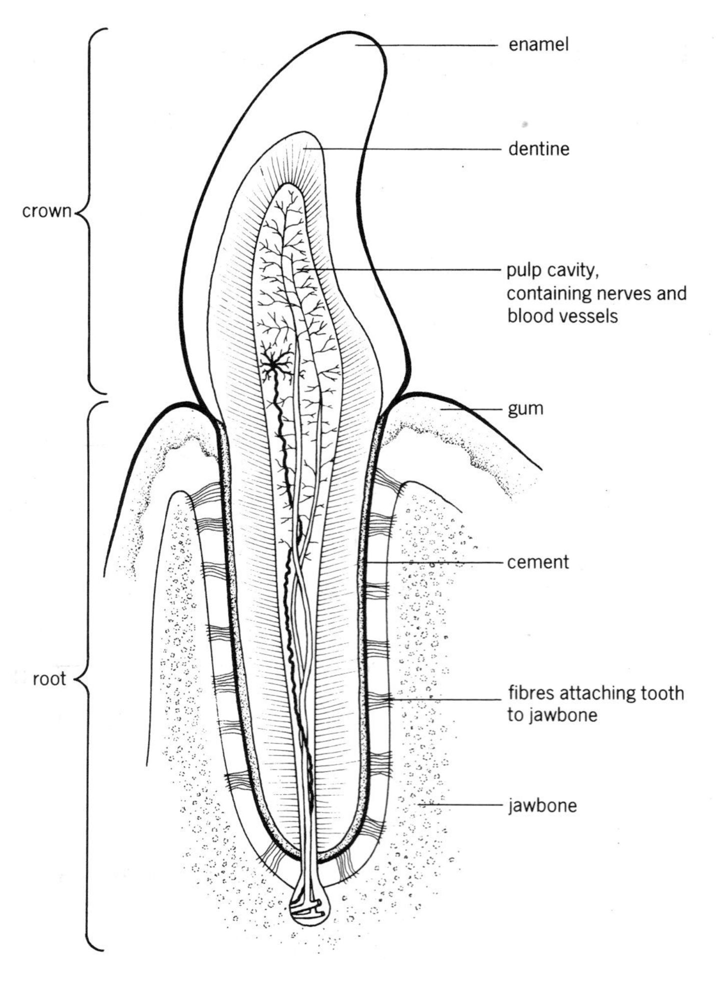

2.15 Longitudinal section through an incisor tooth

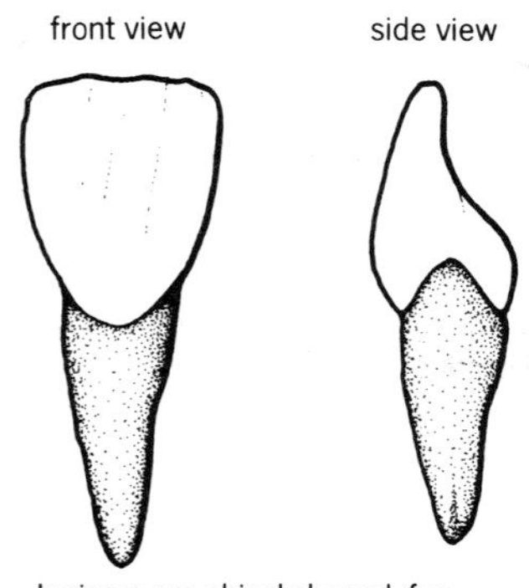

Incisors are chisel shaped, for biting off pieces of food.

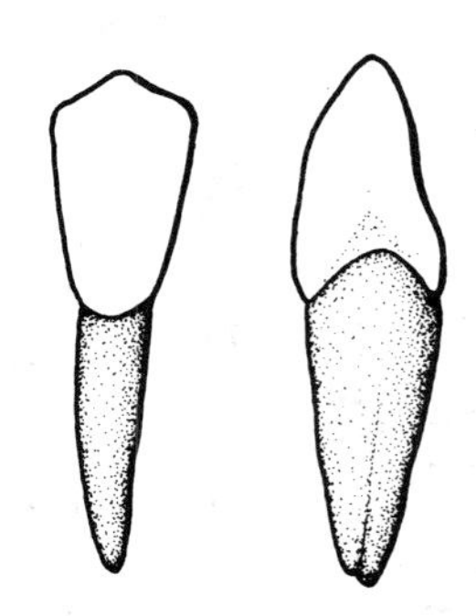

Canines are very similar to incisors in humans.

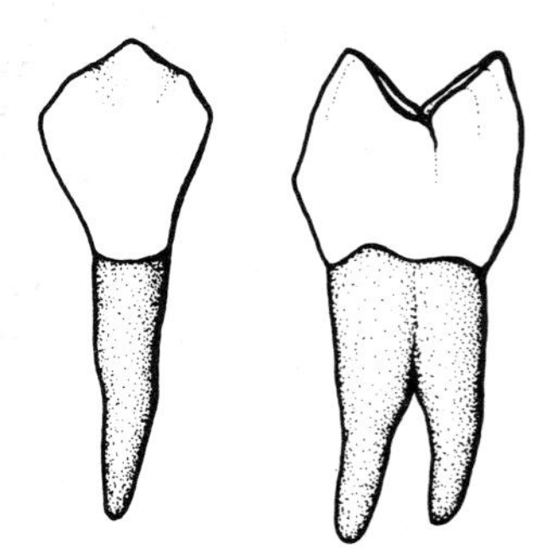

Premolars have wide surfaces, for grinding food.

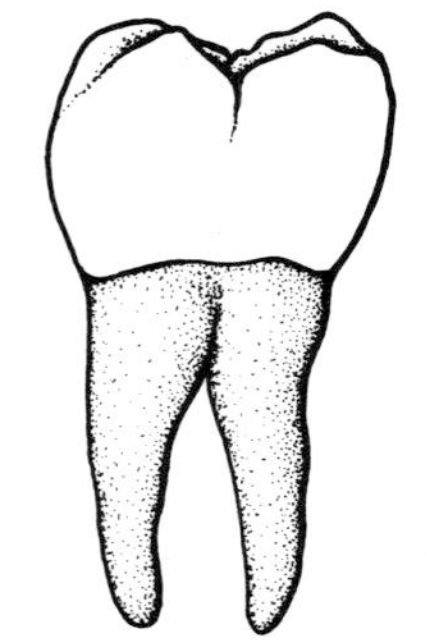

Molars, like premolars, are used for grinding food.

2.16 Types of human teeth

Questions

1 What is holozoic nutrition?
2 Explain how sugary foods can cause tooth decay.
3 What are incisors, and what are they used for?
4 Describe two ways in which mammals' teeth differ from those of other animals.

Fact!

The heaviest man who ever lived was Jon Brower Minnoch, of Washington, USA. At his greatest weight, he was estimated to be about 635 kg. He died in September 1983.

Digestion in humans—the alimentary canal

2.31 The alimentary canal is a muscular tube.

The alimentary canal (see Fig 2.17) is a long tube which runs from the mouth to the anus. The wall of the tube contains muscles, which contract and relax to make food move along. This movement is called **peristalsis** (Fig 2.18).

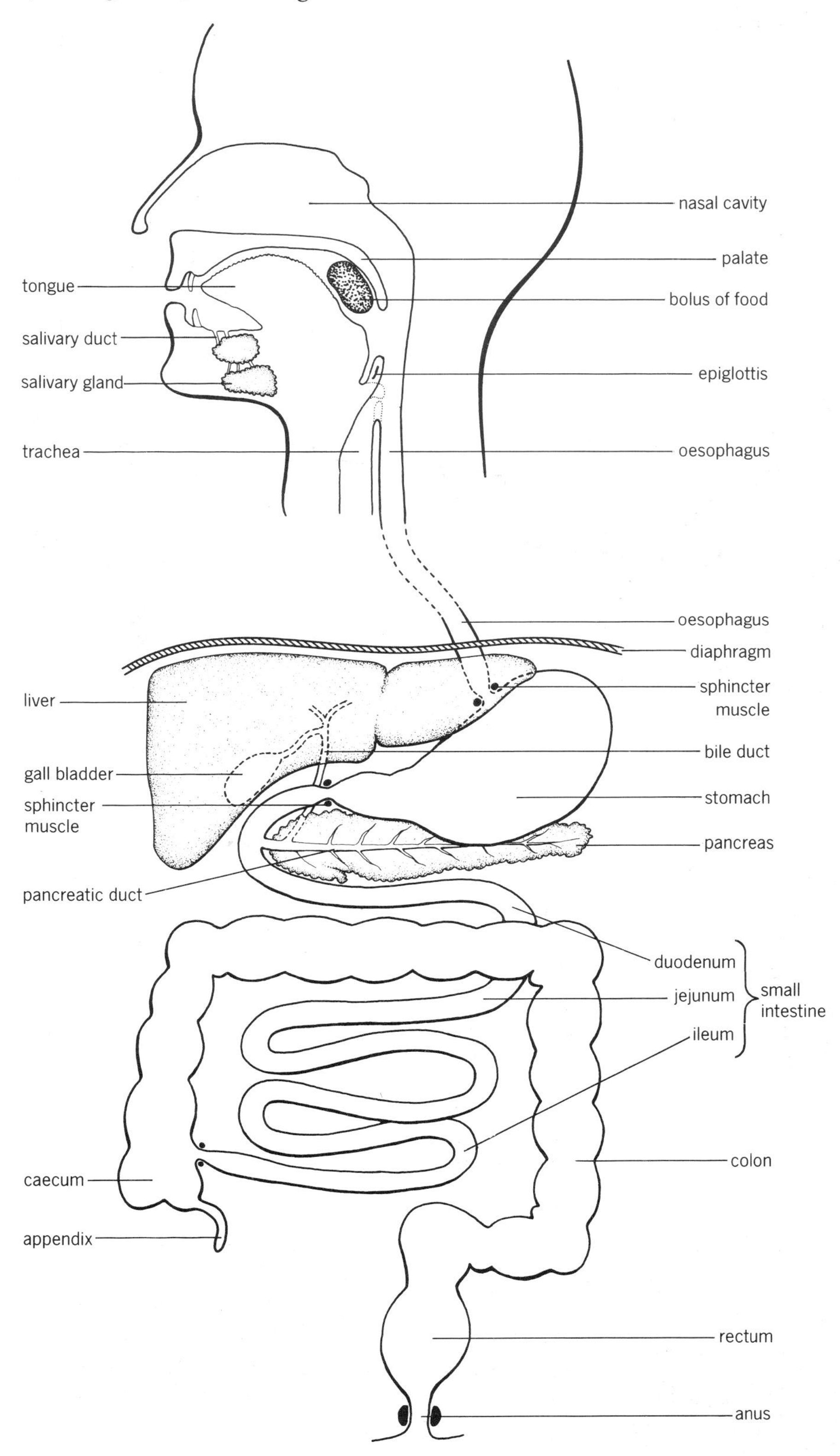

2.17 The human alimentary canal

Sometimes, it is necessary to keep the food in one part of the alimentary canal for a while, before it is allowed to move into the next part. Special muscles can close the tube completely in certain places. They are called **sphincter muscles.**

To help the food to slide easily through the alimentary canal, it is lubricated with **mucus.** Mucus is made in goblet cells which occur all along the alimentary canal.

Each part of the alimentary canal has its own part to play in the digestion, absorption, and egestion of food.

2.32 In the mouth, food is mixed with saliva.

Food is ingested using the teeth, lips and tongue. The teeth then bite or grind the food into smaller pieces. The tongue mixes the food with saliva, and forms it into a **bolus.** The bolus is then swallowed.

Saliva is made in the **salivary glands.** It is a mixture of water, mucus and the enzyme amylase. Amylase begins to digest starch in the food to maltose. Usually, it does not have time to finish this because the food is not kept in the mouth for very long.

2.33 The oesophagus carries food to the stomach.

There are two tubes leading down from the back of the mouth. The one in front is the **trachea** or windpipe, which takes air down to the lungs. Behind the trachea is the **oesophagus,** which takes food down to the stomach.

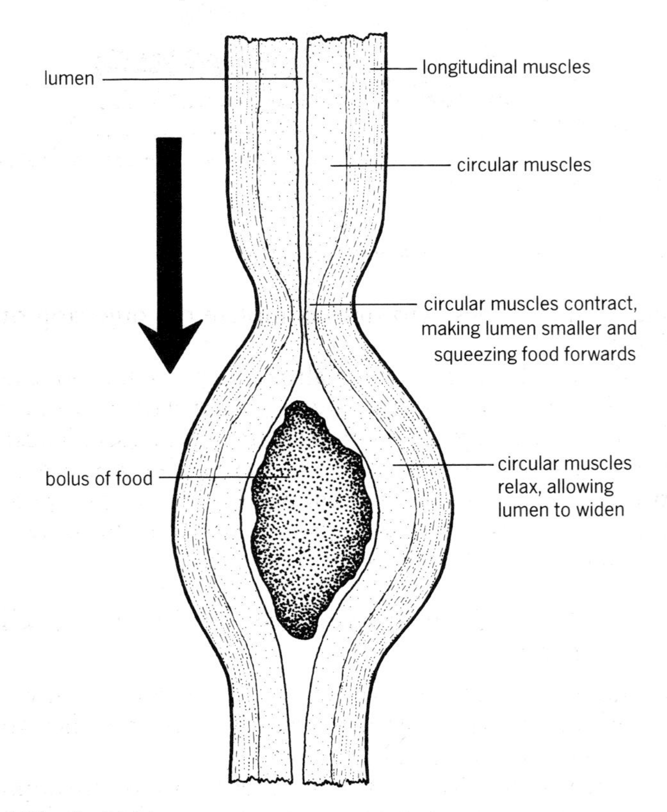

2.18 Peristalsis

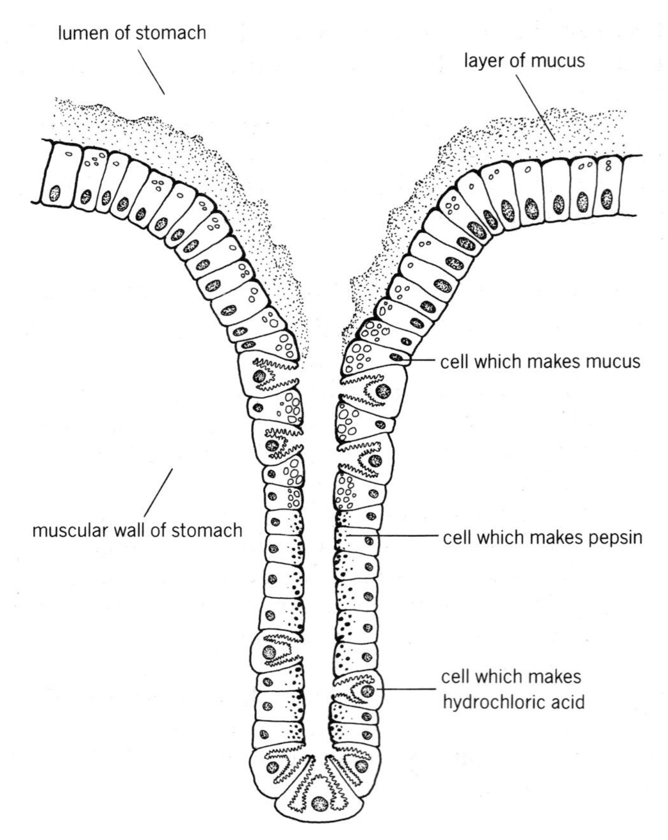

2.19 A gastric pit

When you swallow, a piece of cartilage covers the entrance to the trachea. It is called the **epiglottis,** and it stops food from going down into the lungs.

A sphincter muscle at the bottom of the oesophagus opens to let the food into the stomach.

2.34 The stomach stores food and digests proteins.

The stomach has strong, muscular walls. The muscles contract and relax to churn the food and mix it with the enzymes and mucus. The mixture is called **chyme.**

Like all parts of the alimentary canal, the stomach wall contains goblet cells which secrete mucus. It also contains other cells which produce an enzyme called **pepsin,** and others which make **hydrochloric acid.** These are situated in pits in the stomach wall (see Fig 2.19).

Pepsin is a protease. It begins to digest proteins by breaking them down into polypeptides. Pepsin works best in acid conditions. The acid also helps to kill any bacteria in the food.

The stomach can store food for quite a long time. After one or two hours, the sphincter at the bottom of the stomach opens and lets the chyme into the duodenum.

2.35 The small intestine is very long.

The small intestine is the part of the alimentary canal

between the stomach and the colon. It is about 5 m long. It is called the small intestine because it is quite narrow.

Different parts of the small intestine have different names. The first part, nearest to the stomach, is the **duodenum.** The middle section is the **jejunum.** The last part, nearest to the colon, is the **ileum.**

2.36 Pancreatic juice flows into the duodenum.

Several enzymes are secreted into the duodenum. They are made in the **pancreas,** which is a cream coloured gland, lying just underneath the stomach. A tube called the pancreatic duct leads from the pancreas into the duodenum. Pancreatic juice, which is a fluid made by the pancreas, flows along this tube.

This fluid contains many enzymes. One is **amylase,** which breaks down starch to maltose. Another is **trypsin,** which is a protease and breaks down proteins and polypeptides to amino acids. Another is **lipase,** which breaks down fats to fatty acids and glycerol.

These enzymes do not work well in acid environments, but the chyme which has come from the stomach contains hydrochloric acid. Pancreatic juice contains **sodium hydrogen carbonate** which neutralises the acid.

2.37 Bile helps to digest fats.

As well as pancreatic juice, another fluid flows into the duodenum. It is called **bile.** Bile is a yellowish green, watery liquid. It is made in the liver, and then stored in the gall bladder. It flows to the duodenum along the bile duct.

Bile does not contain any enzymes. It does, however, help to digest fats. It does this by breaking up the large drops of fat into very small ones, making it easier for the lipase in pancreatic juice to digest them. This is called **emulsification,** and it is done by salts in the bile called bile salts.

Bile also contains yellowish pigments. These are made by the liver when it breaks down old red blood cells. They are made from the haemoglobin. The pigments are not needed by the body, so they are eventually excreted in the faeces.

2.38 Digestion is completed in the small intestine.

As well as receiving enzymes made in the pancreas, the small intestine makes some enzymes itself. They are made by cells in its walls.

The inner wall of all three parts of the small intestine–the duodenum, jejunum and ileum–is covered with millions of tiny projections. They are called **villi.** Each villus is about 1 mm long (see Figs 2.20, 2.21, 2.22). It is the cells covering the villi which make the enzymes. The enzymes do not come out into the lumen of the small intestine, but stay close to the cells which make them. These enzymes complete the digestion of food.

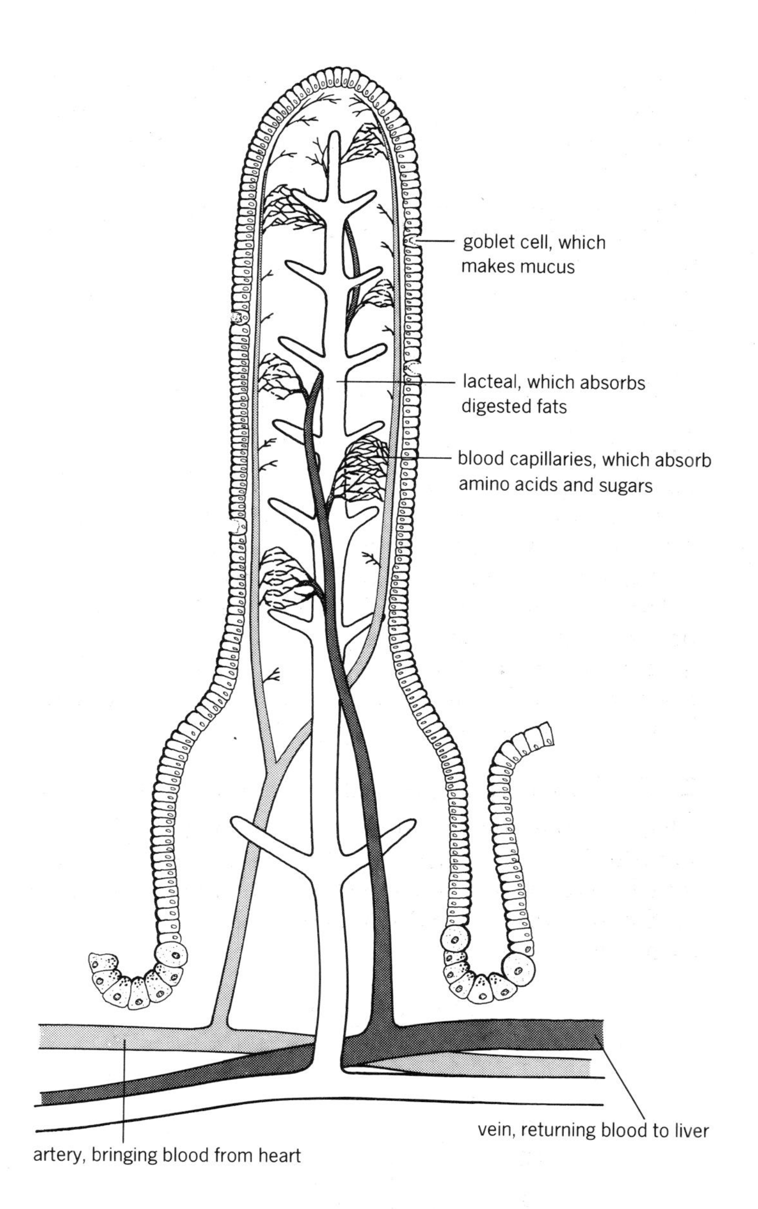

2.20a A longitudinal section through a villus

Maltase breaks down maltose to glucose. **Sucrase** breaks down sucrose to glucose and fructose. **Lactase** breaks down lactose to glucose and galactose. These three enzymes are all carbohydrases. There are also **proteases,** which finish breaking down any polypeptides into amino acids. **Lipase** completes the breakdown of fats to fatty acids and glycerol.

2.39 Digested food is absorbed in the small intestine.

By now, most carbohydrates have been broken down to simple sugars, proteins to amino acids, and fats to fatty acids and glycerol.

These molecules are small enough to pass through the wall of the small intestine and into the blood. This is called absorption. The small intestine is especially adapted to allow absorption to take place very efficiently. Some of its features are listed in Table 2.5.

2.40 The colon absorbs water.

Not all the food that is eaten can be digested, and this undigested food cannot be absorbed in the small intestine. It travels on, through the caecum past the appendix and into the colon. In humans, the caecum and appendix have no function. In the colon, water and salt are absorbed.

The colon and rectum are sometimes called the **large intestine,** because they are wider tubes than the duodenum and ileum.

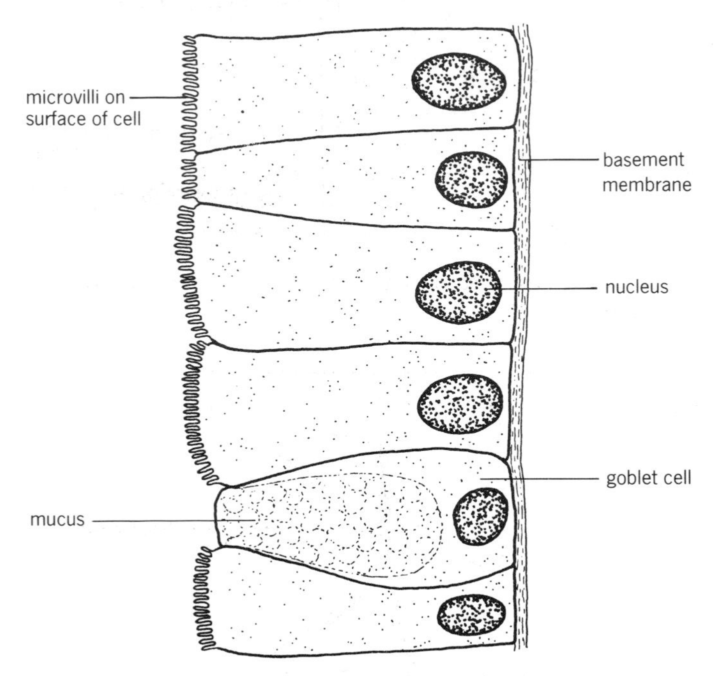

2.20b Detail of villus surface

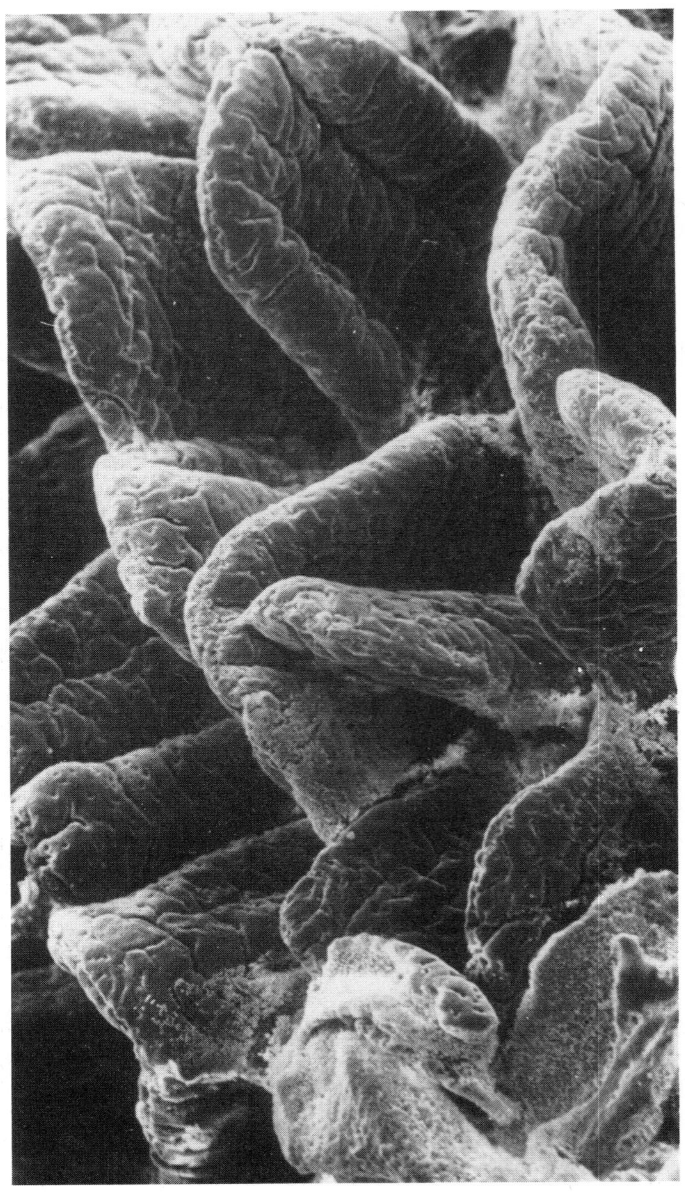

2.21 The inner surface of the ileum. Each projection is a villus, about 1 mm long

Table 2.5 How the small intestine is adapted for absorbing digested food

	Feature	*How this helps absorption to take place*
1	It is very long, about 5 m in an adult.	This gives plenty of time for digestion to be completed, and for digested food to be absorbed as it passes through.
2	It has villi. Each villus is covered with cells which have even smaller projections on them, called microvilli.	This gives the inner surface of the small intestine a very large surface area. The larger the surface area, the faster food can be absorbed.
3	Villi contain blood capillaries.	Digested food passes into the blood, to be taken to the liver and then round the body.
4	Villi contain lacteals, which are part of the lymphatic system.	Fats are absorbed into the lacteals.
5	Villi have walls only one cell thick.	The digested food can easily cross the wall to reach the blood capillaries and lacteals.

2.41 The rectum temporarily stores undigested food.

By the time the food reaches the rectum, most of the substances which can be absorbed have gone into the blood. All that remains is undigestible food (roughage), bacteria, and some dead cells from the inside of the alimentary canal. This mixture forms the **faeces,** which are passed out at intervals through the anus.

Questions

1 What is a sphincter muscle?
2 Name two places in the alimentary canal where sphincter muscles are found.
3 In which parts of the alimentary canal is mucus secreted? Explain why.
4 Name two parts of the alimentary canal where amylase is secreted. What does it do?
5 What is the epiglottis?
6 Why do the walls of the stomach secrete hydrochloric acid?
7 Which three parts of the alimentary canal make up the small intestine?
8 Which two digestive juices are secreted into the duodenum?
9 How do bile salts help in digestion?
10 Name three enzymes made by the cells covering the villi in the small intestine and explain what they do.
11 (a) In which part of the alimentary canal is digested food absorbed?
(b) Describe three ways in which this part is adapted for absorption.
12 In which part of the alimentary canal is water absorbed?
13 What do faeces contain?

2.42 All absorbed food goes straight to the liver.

After it has been absorbed into the blood, the food is taken to the liver, in the **hepatic portal vein** (see Fig 2.23). The liver processes some of it, before it goes any further (refer to Table 10.2). Some of the food can be broken down, some converted into other substances, some stored and the remainder left unchanged.

The food dissolved in the blood plasma, is then taken to other parts of the body where it may become **assimilated** as part of the cells.

Fact!

Dried twisted gut is still used for stringing tennis rackets.

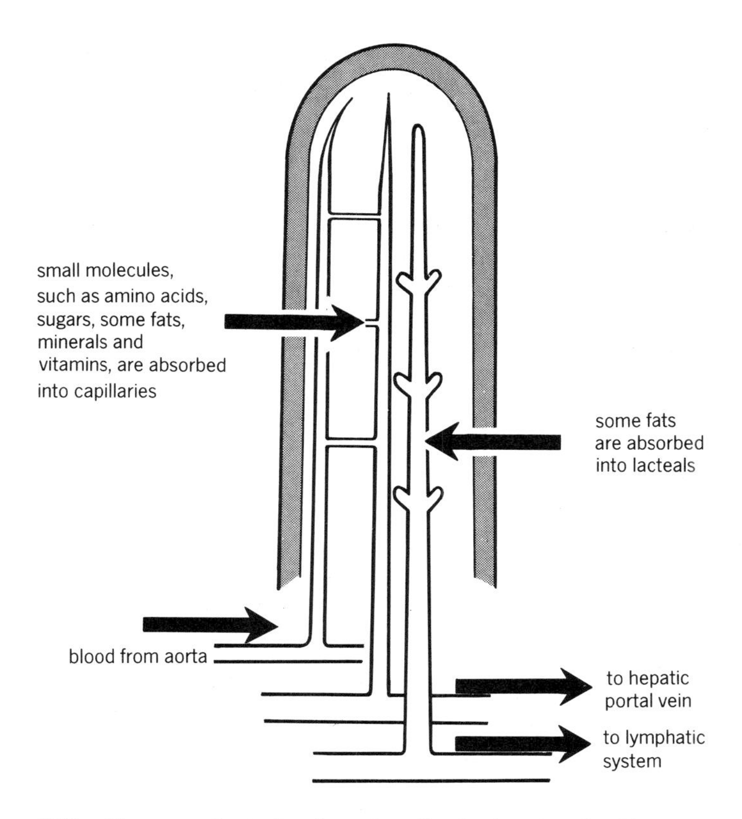

2.22 Diagrammatic section through a villus to show how food is absorbed

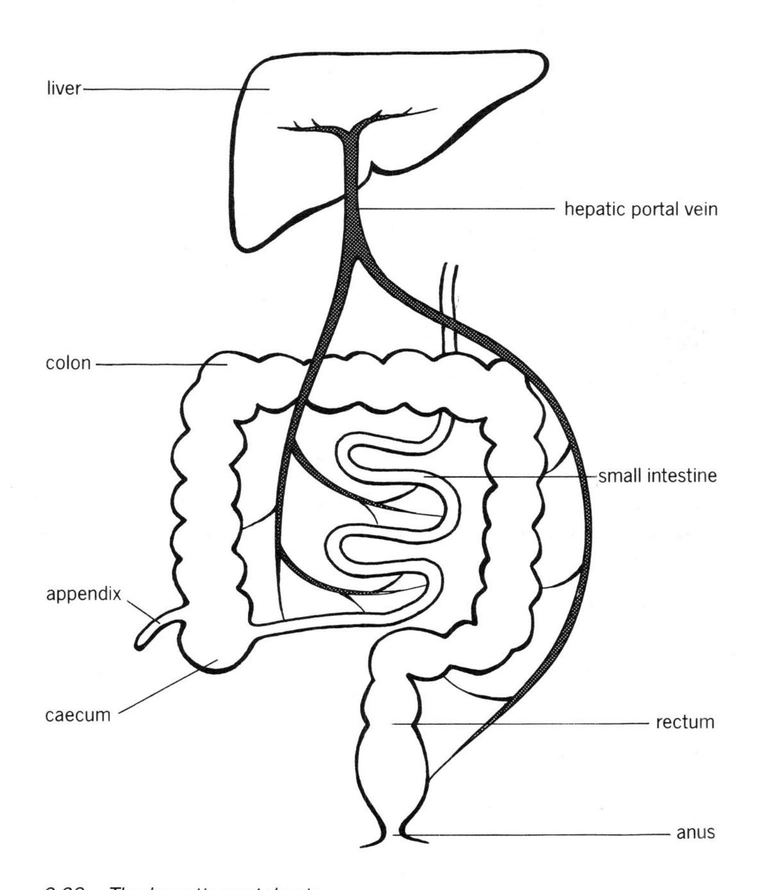

2.23 The hepatic portal vein

Table 2.6 Summary of chemical digestion in the human alimentary canal

Part of canal	*Juices secreted*	*Where made*	*Enzymes in juice*	*Substrate*	*Product*	*Other substances in juice*	*Function of other substances in juice*
Mouth	Saliva	Salivary glands	Amylase	Starch	Maltose		
Oesophagus	None						
Stomach	Gastric juice	In pits in wall of stomach	Protease (pepsin)	Proteins	Polypeptides	Hydrochloric acid	Acid environment for pepsin; kills bacteria.
Duodenum and jejunum	Pancreatic juice	Pancreas	Amylase	Starch	Maltose	Sodium hydrogen carbonate ($NaHCO_3$)	Neutralizes acidity of chyme, to make an alkaline environment for enzymes
			Protease (trypsin)	Proteins and polypeptides	Amino acids		
			Lipase	Emulsified fats	Fatty acids and glycerol		
	Bile	Liver, stored in gall bladder	None			Bile salts	Emulsify fats
						Bile pigments	Excretory products
Ileum	No juice secreted; enzymes remain in or on the cells covering the villi	By cells covering the villi	Maltase	Maltose	Glucose		
			Sucrase	Sucrose	Glucose and fructose		
			Lactase	Lactose	Glucose and galactose		
			Peptidase	Polypeptides	Amino acids		
			Lipase	Emulsified fats	Fatty acids and glycerol		

All of the digestive juices also contain **water.** Water is used in splitting the large food molecules (see section 2.23). It also acts as a solvent for the enzymes, substrates and products. The juices also contain **mucus,** which is a lubricant, and helps to protect the walls of the alimentary canal from being digested by the enzymes. The colon and rectum play no part in digestion. They are concerned with absorption and egestion. They are not included in this table.

Digestion in other mammals

2.43 Herbivores eat plants; carnivores eat animals.

Humans eat a very wide variety of food. They are **omnivores.** The human alimentary canal is designed to cope with a diet containing both plant and animal material. However many animals have a much more restricted diet than this. Some, such as rabbits, eat only plants. They are called **herbivores.** Others, such as cats, eat only animal material. They are called **carnivores.**

Plant material is much more difficult to digest than animal material. This is because each plant cell is surrounded by a tough cellulose cell wall (see section 1.4). This makes it difficult for the digestive enzymes to reach the food material inside plant cells. Very few types of animal can make an enzyme which can digest cellulose. Animal cells do not contain cellulose, so they are much easier to digest.

The alimentary canals of herbivores and carnivores, therefore, are very different from one another.

2.44 Different diets need different kinds of teeth.

Rabbits use their teeth for cropping grass. They then grind it as thoroughly as possible, to break down the cellulose cell walls. Cats use their teeth for killing their prey, and cutting meat into pieces small enough to swallow.

Incisors The incisors of a rabbit are long and chisel-shaped. The upper and lower ones cut together to chop off pieces of grass (see Fig 2.24). A cat, however, has much smaller incisors. They are peg-shaped, and are used for gripping food (see Fig 2.25).

Canines A rabbit has no canines. A cat has long pointed canines, which it uses for killing its prey.

A rabbit has a long, toothless gap between its incisors and its premolars. This is called a **diastema.** The diastema is used for manipulating the grass in the rabbit's mouth, as it is repeatedly turned around to be chewed from different angles.

Premolars and molars A rabbit's premolars and molars have broad, ridged surfaces. The ridges on the top teeth fit into the grooves on the bottom ones. They are moved in a circular, sideways motion, grinding the grass between them (see Figs 2.26 and 2.27).

A cat's premolars and molars have sharp, cutting edges. The largest ones are called **carnassial** teeth. As the jaw moves up and down, the top and bottom teeth cut past one another like scissor blades, chopping the food into pieces (see Fig 2.28).

Pulp cavity Rabbits chew their food for a long time, and so their teeth are always being worn away by grinding the cellulose in the grass they eat. To make up for this wear, their teeth continue to grow all their lives. The pulp cavity is large, and has a wide opening at the bottom which does not restrict the blood supply. This helps to provide the substances needed to make the tooth grow.

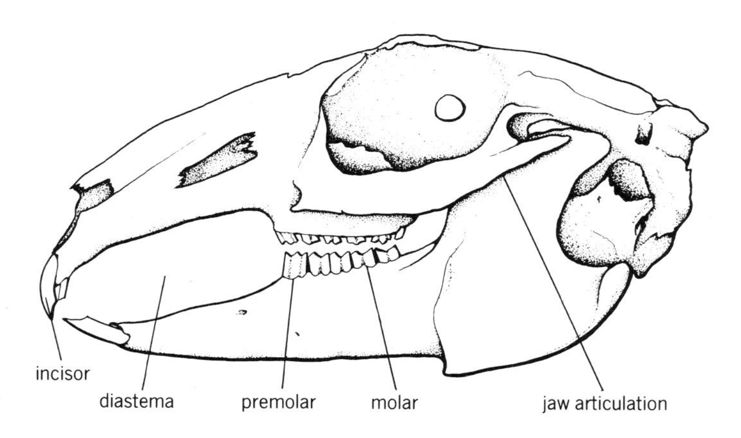

2.24 *A rabbit skull*

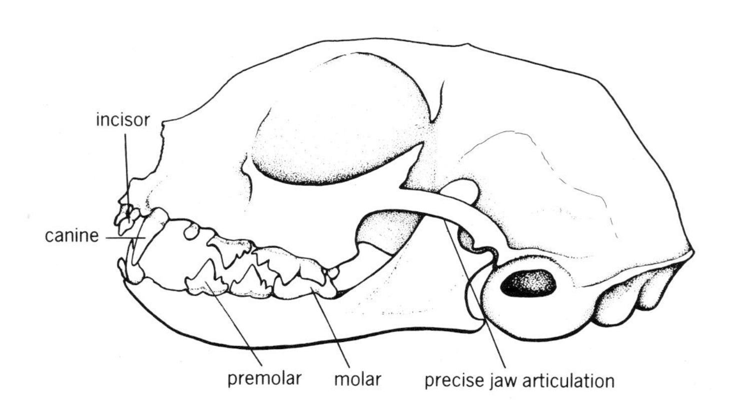

2.25 *A cat skull*

Cats, however, keep their food in their mouths for only a short time. Their teeth do not get worn down so fast, and there is no need for them to grow continuously. The pulp cavities of a cat's teeth are smaller than those of a rabbit, allowing in only enough food and oxygen to keep the tooth alive.

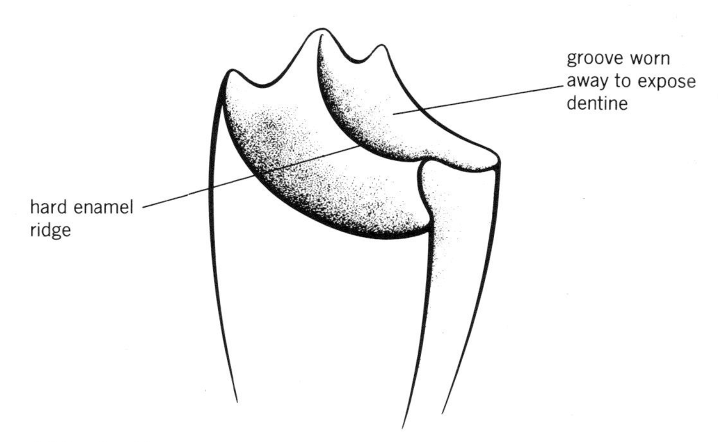

2.26 *A rabbit's molar tooth showing the surface ridges*

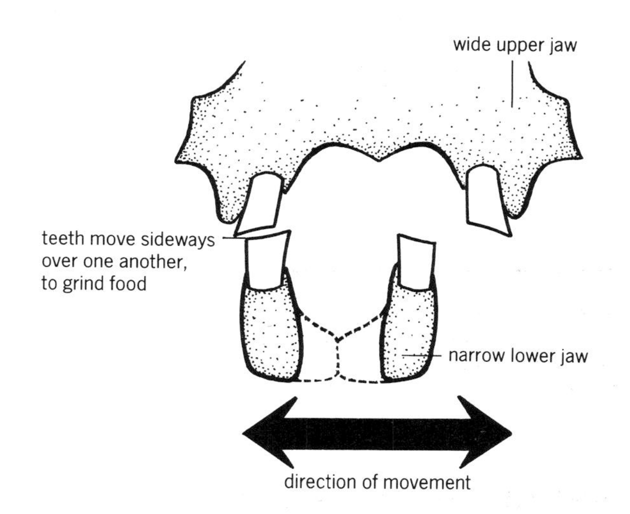

2.27 *Vertical section through a rabbit's jaws, looking from the front*

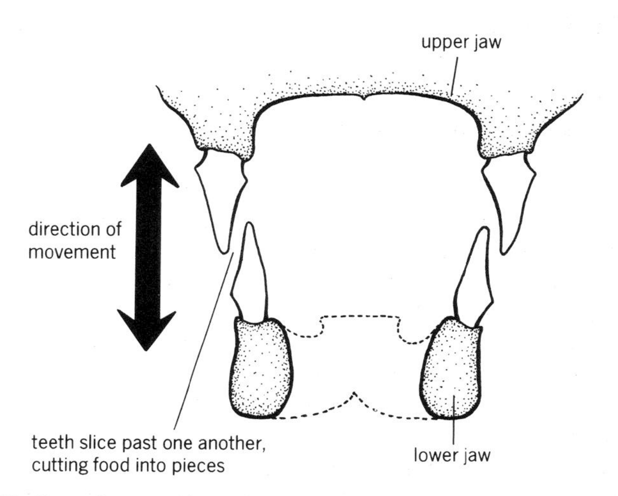

2.28 *Vertical section through a cat's jaws, looking from the front*

2.45 Herbivores have loosely linked jaws.

Rabbits move their jaws from side to side, as well as up and down in order to grind their food as thoroughly as possible. This means that the lower jaw must be quite loosely linked, or **articulated,** with the upper jaw. Cats, on the other hand, need a crisp, sharp chopping movement to deal with their food. The lower jaw is articulated precisely with the upper jaw, allowing only a very slight side to side movement.

2.46 Bacteria help herbivores to digest cellulose.

Caecum and appendix The thorough chewing of its food by a rabbit crushes the cellulose cell walls of the grass. This allows the rabbit's digestive enzymes to work on the contents of the cells. The cellulose itself, however, has not been broken into smaller molecules.

Cellulose is a polysaccharide. It is made of many glucose molecules joined together. To break it up into these glucose molecules, an enzyme called cellulase is needed. Rabbits cannot make cellulase.

In the caecum of a rabbit there are large numbers of bacteria which can make cellulase. So cellulose is broken down in the caecum to small molecules, which can be absorbed. Both the rabbit and the bacteria benefit from this arrangement. The rabbit gets useful food material from the cellulose which would otherwise be wasted. The bacteria have a warm place to live, and constant food supply. This is an example of **symbiosis** or mutualism (see section 14.30).

There is one problem, however. By the time the cellulose is digested, it has already gone past the small intestine, which is the only place where digested food can be absorbed. The food therefore continues onwards, through the colon and rectum, and is passed out as soft, mucus-coated pellets. The rabbit eats these immediately, so that the food can pass through the small intestine and be absorbed.

Because the caecum has such an important function in the rabbit, it is quite large (see Fig 2.29). A cat, however, has no cellulose in its diet and, although it still has a caecum, it is not used. It is said to be a vestigial structure.

Length of intestine A rabbit has a longer intestine than a cat. This is because its food is more difficult to digest, and needs to be in the alimentary canal for a longer time.

Table 2.7 Summary of the differences between the alimentary canals of a rabbit and a cat

Rabbit, a herbivore	*Cat, a carnivore*
Lower jaw loosely articulated, allowing side to side and up and down movement	Jaw precisely articulated, allowing only up and down movement
Incisors long, chisel-shaped, forward-pointing for cropping grass	Incisors short, peg-shaped, backward-pointing, for holding prey
No canines	Canines long and pointed for killing prey
Diastema for manipulating grass	No diastema
Broad, ridged, premolars and molars for grinding grass	Sharp-edged premolars and molars for cutting meat
Open roots allow teeth to grow continuously	Constricted openings in roots do not allow growth of teeth
Long intestine gives time for digestion	Shorter intestine, because food does not take so long to digest
Large caecum and appendix, containing bacteria which secrete cellulase	Small vestigial caecum

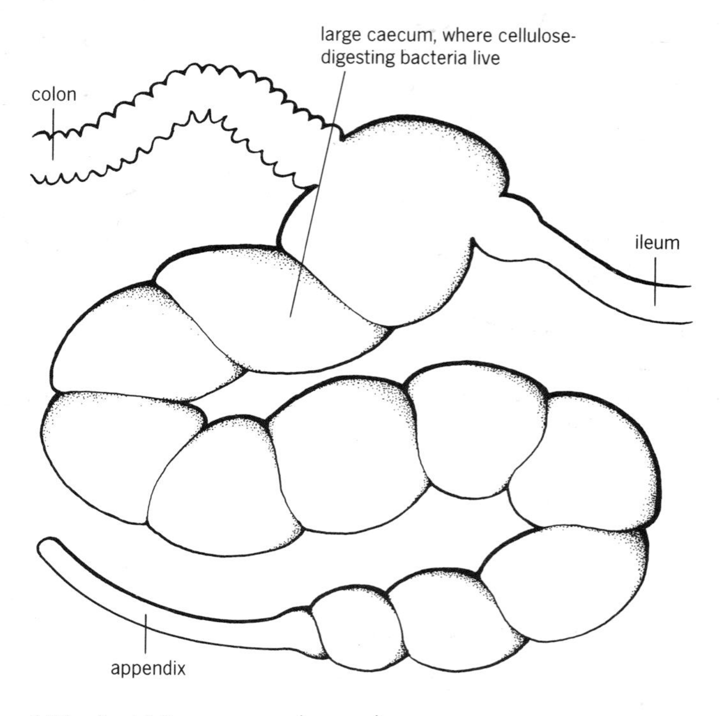

2.29 A rabbit's caecum and appendix

Questions

1 Why is plant material more difficult to digest than animal material?
2 What is a diastema?
3 Explain how (a) a rabbit, and (b) a cat, use their premolars and molars.
4 Rabbits have large pulp cavities, but cats have only small ones. Explain why.
5 What happens in a rabbit's caecum and appendix?

Heterotrophic nutrition–*Amoeba*

2.47 *Amoeba* feeds by phagocytosis.

Amoeba (Fig 2.30) is a microscopic organism which consists of only one cell–it is unicellular. It belongs to the kingdom Protista (see Chapter 17).

Amoeba lives in ponds and slow-moving streams. It moves slowly over the bottom, or over the surface of dead leaves, using its **pseudopodia** (false feet). There are many other protistans in the water which are smaller than the *Amoeba*. *Amoeba* feeds on these.

It has no special sense organs, but the surface of the cell is sensitive to chemicals in the water. This is rather like the human sense of smell. All living organisms release chemical substances and *Amoeba* can therefore sense where its potential prey is. It slowly moves towards it, travelling up a chemical gradient–that is, moving towards the place where the chemical is most concentrated.

When a small protistan has been found, *Amoeba* puts out pseudopodia around it. The pseudopodia completely surround the protistan, and join up around it. It is now enclosed in a food vacuole inside the cell, in a drop of water. This process is called **phagocytosis** ('cell feeding').

The protistan must now be digested. Enzymes are secreted into the food vacuole, which digest the protistan in a similar way to the digestion of food in your alimentary canal. The digested, soluble food is then absorbed into the cytoplasm of the *Amoeba*.

Some of the protistan may not have been digested. These undigestible remains stay in the *Amoeba* for some time, but periodically they are passed out.

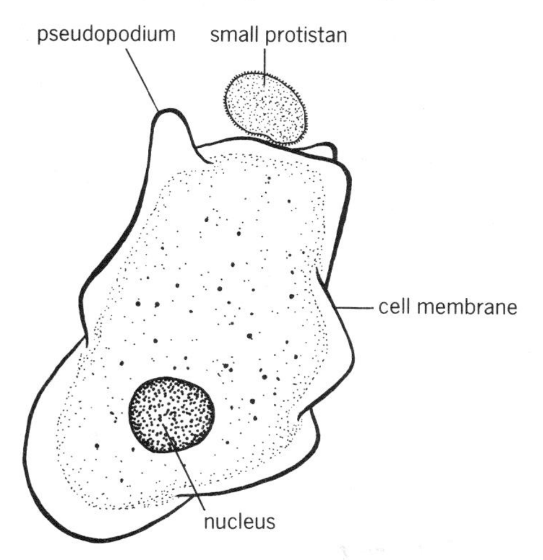

Amoeba finds prey, by moving along a chemical gradient.

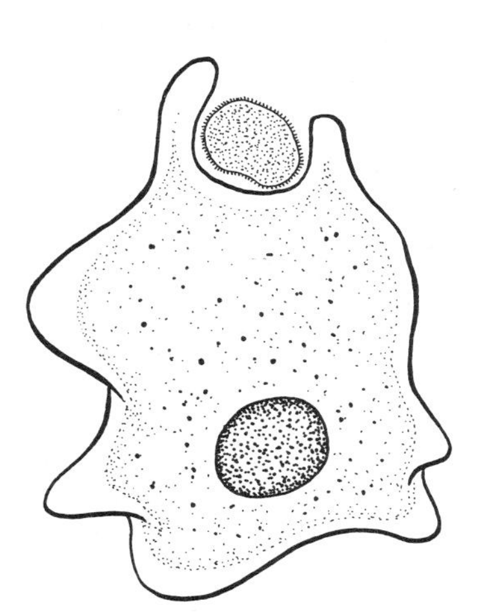
Pseudopodia surround prey

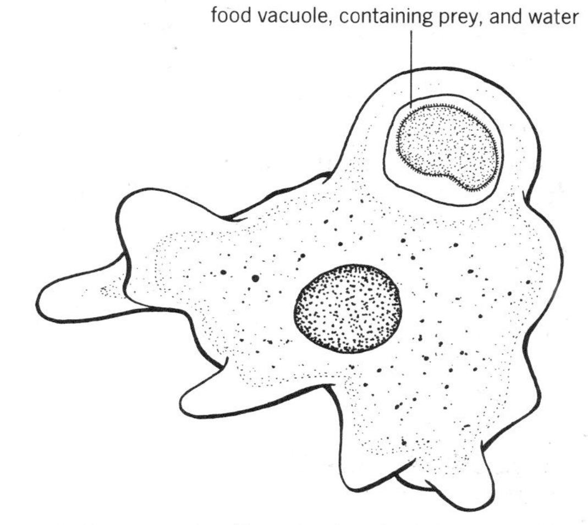

Prey is enclosed in a vacuole.

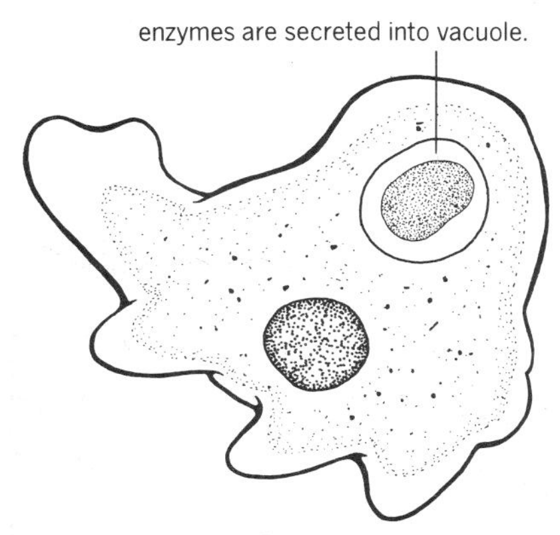

Digestive enzymes are secreted into the vacuole.

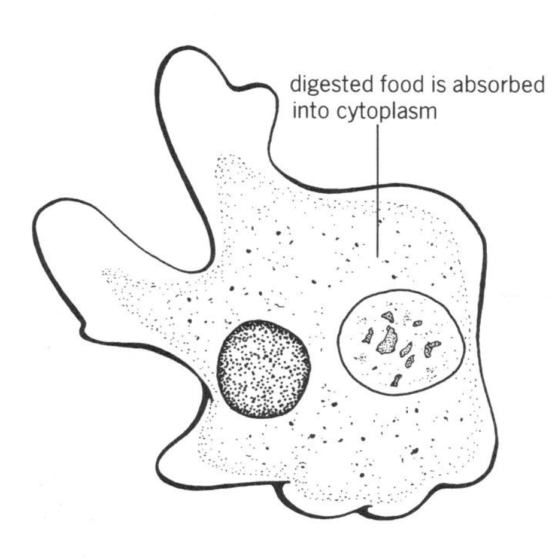

The enzymes break down the digestible parts of the prey, so that they become soluble and are absorbed.

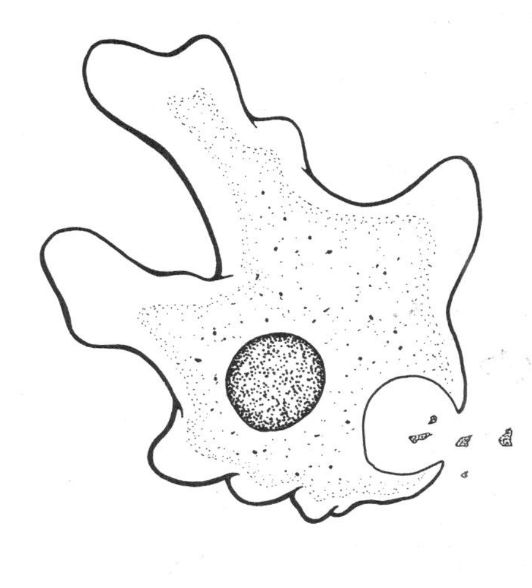
Indigestible parts of the prey are egested.

2.30 Feeding in Amoeba

2.48 *Amoeba* digests food intracellularly.

Amoeba takes its food into its cell, and then secretes enzymes onto it to digest it. This is called **intracellular** digestion, because it takes place inside a cell ('intra' means inside).

Mammals digest their food inside their alimentary canal. The food is not inside the cells, and so this is called **extracellular** digestion, because it takes place outside a cell ('extra' means outside).

Heterotrophic nutrition—insects

2.49 Houseflies cover their food with saliva.

The housefly feeds on almost any kind of organic material. Saliva flows along a **proboscis** (see Fig 2.31), onto the food. Enzymes in the saliva digest the food, making a solution. The dissolved food is then sucked up through the proboscis, using powerful muscles.

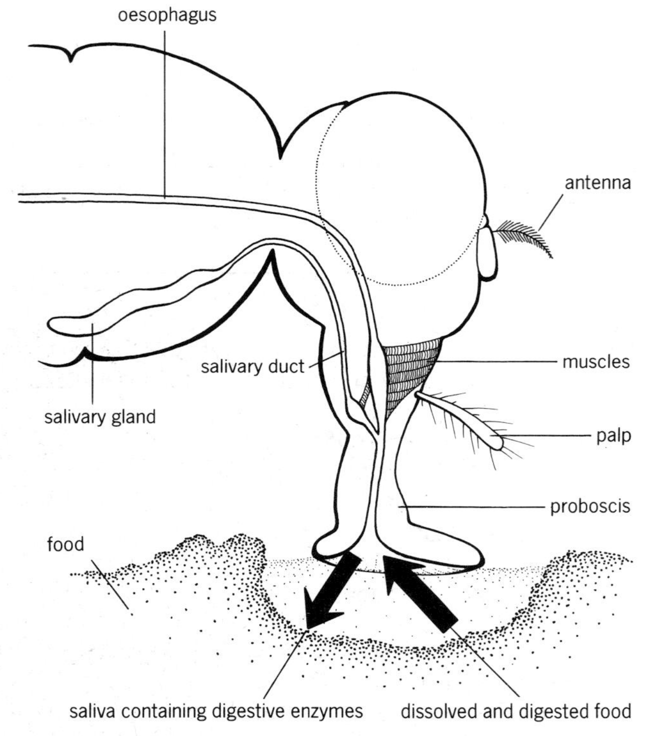

2.31 *Feeding in a housefly*

2.50 Locusts have jaws to bite their food.

Locusts feed in a similar way to mammals in that they bite off pieces of food using hard jaws made of **chitin** (see Fig 2.32). The food is then digested inside their alimentary canal.

> **Fact!** The desert locust is the most destructive insect in the world. A large swarm of locusts can consume 20 000 tonnes of grain and vegetables in a day.

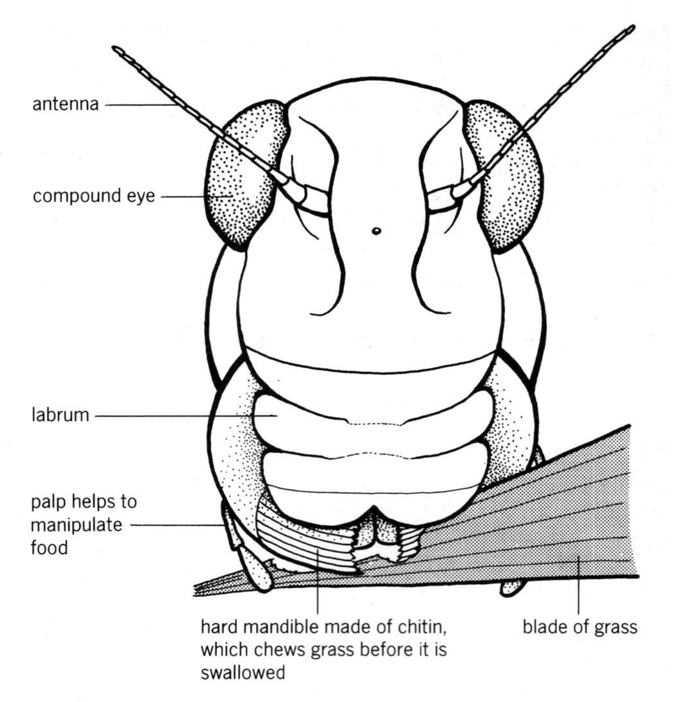

2.32 *Feeding in a locust*

Heterotrophic nutrition—fish

2.51 Some animals filter food from water.

Some fish, such as sharks, feed by biting. Others, such as young herrings, feed by straining out microscopic organisms called **plankton** from the current of water which passes over their gills (see Figs 5.21 and 5.22). The plankton is then swallowed. This is called **filter feeding.**

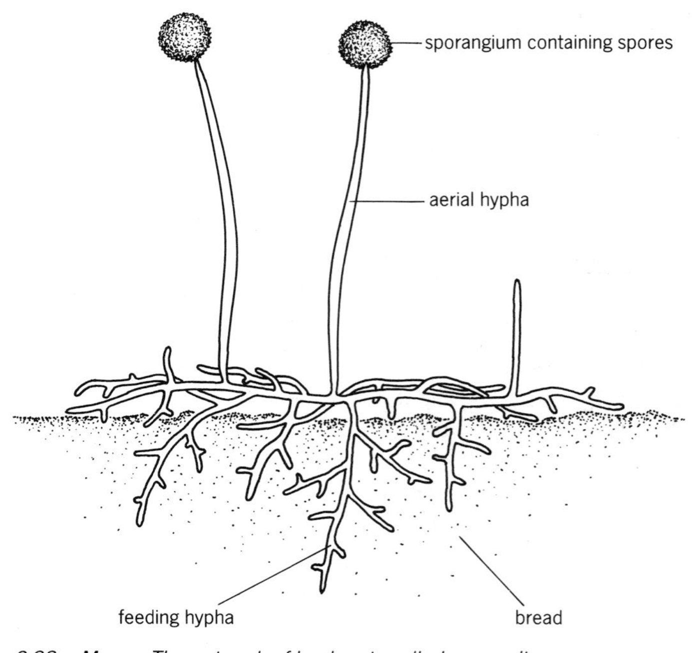

2.33 *Mucor. The network of hyphae is called a mycelium.*

Questions

1 What does unicellular mean?
2 How does *Amoeba* find its food?
3 What is phagocytosis?
4 Explain the difference between intracellular and extracellular digestion.
5 What is saprophytic nutrition?

Heterotrophic nutrition–fungi

2.52 Fungi feed saprophytically.

Moulds and mushrooms belong to the kingdom Fungi (see Chapter 17). A common mould is the bread mould *Mucor* (see Fig 2.33). *Mucor* grows on many kinds of non-living organic material, such as bread. It consists of threads or **hyphae,** which make up a **mycelium.**

Mucor feeds on the substances on which it grows. The tips of the hyphae secrete enzymes, which digest the bread. The starch is broken down to glucose, which is soluble, and diffuses into the hyphae. Proteins in the bread are broken down to amino acids, and fats to fatty acids and glycerol. All of these are absorbed by the hyphae. The hyphae then grow forward into the space made by the dissolved bread.

This type of nutrition is called **saprophytism.** All fungi feed in this way, and so do many bacteria. It is the way in which dead organisms are decayed, and is very important because it helps to release nutrients from them which would not otherwise be available.

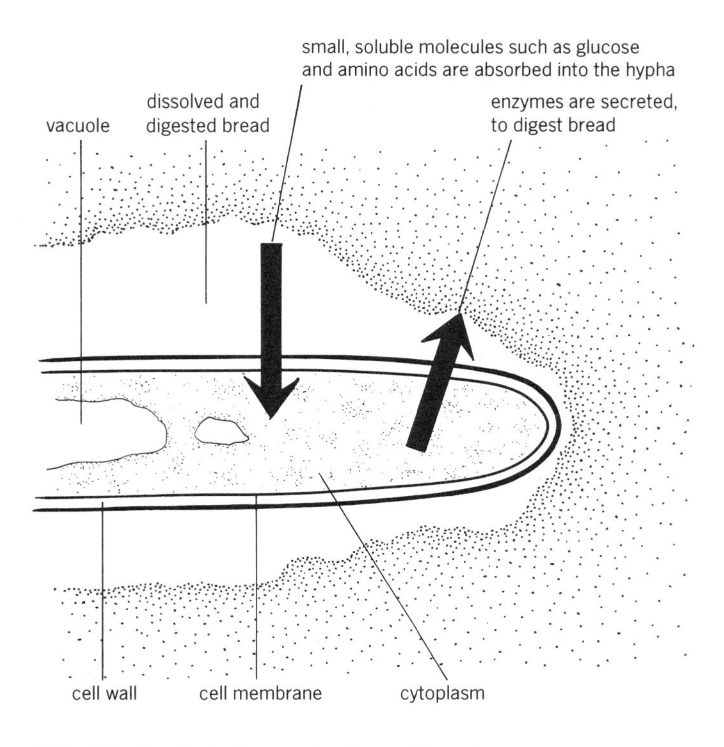

2.34 The tip of a hypha growing through bread

Table 2.8 Methods of feeding on organic substances originally made by green plants—heterotrophic nutrition

holozoic nutrition
feeding on quite large pieces of food, which are taken into the alimentary canal and digested extracellularly eg: human beings

filter feeding
feeding on small organisms and food particles in water which are trapped by a filtration system, and then digested as for holozoic nutrition eg: young herrings

parasitism
feeding on and living in close association with, another living organism called the host eg: tapeworm

saprophytism
feeding on dead organic material, which is digested extracellularly and then absorbed eg: *Mucor*

Parasites

2.53 Parasites feed on other living organisms.

A parasite is an organism which lives in or on another organism, and feeds from it. An example of a parasite is the tapeworm, which is described in Fig 14.18.

Chapter revision questions

1 With the aid of examples wherever possible, explain the differences between each of the following pairs of terms.
(a) autotrophic; heterotrophic,
(b) monosaccharide; disaccharide,
(c) inorganic; organic,
(d) enamel; dentine,
(e) digestion; absorption,
(f) herbivore; carnivore,
(g) symbiosis; saprophytism,
(h) intracellular digestion; extracellular digestion,
(i) substrate; product,
(j) cellulose; cellulase
2 (a) What is meant by a balanced diet?
(b) Using Table 2.4 and Fig 2.10, plan menus for one day which would provide a balanced diet for (i) a teenage boy, and (ii) a pregnant woman. For each food you include, state how much energy, and which types of nutrients it contains.
3 All of these organisms feed heterotrophically:
(a) horse, (b) herring, (c) mushroom, (d) tapeworm, (e) *Amoeba,* (f) mouse, (g) sparrow, (h) flea, (i) mildew, (j) locust.
For each one, choose the appropriate method of nutrition from these four: parasitism, holozoic nutrition, saprophytism, and filter feeding.

continued

4 For each of these digestive juices, state (a) where it is made, (b) where it works, and (c) what it does.

(i) pancreatic juice
(ii) saliva
(iii) bile
(iv) gastric juice

5 Describe what would happen to a piece of steak, containing only protein; and a chip, containing starch and fat, as they passed through your alimentary canal.

6 The function of the stomach is to store food, and to begin to digest it. The small intestine, however, first completes the digestion of food, and then absorbs it. In the form of a table, keeping equivalent points opposite one another, make a comparison between these two organs, bearing in mind their different functions.

3 How green plants feed

3.1 Green plants feed autotrophically.

Green plants make their food by using inorganic substances to build up organic substances. This is called **autotrophic nutrition** (see section 2.2). The inorganic substances which a green plant uses are carbon dioxide, water and a variety of minerals. Using these simple substances, the plant makes all the carbohydrates, fats, proteins and vitamins which it needs.

3.2 Plants make glucose from carbon dioxide and water.

To make carbohydrates, green plants combine carbon dioxide with water. The carbohydrate which is made is **glucose.** At the same time, oxygen molecules are produced.

carbon dioxide + water → glucose + oxygen

However, if you mixed together carbon dioxide and water, they would not combine to make glucose and oxygen. They have to be given energy to make them combine. The energy which green plants use for this is **sunlight** energy. The reaction is therefore called **photosynthesis** ('photo' means light, and 'synthesis' means manufacture).

3.3 Chlorophyll absorbs sunlight.

However, sunlight shining onto water and carbon dioxide still will not make them react together in this way. The sunlight energy has to be trapped, and then used in the reaction. Green plants have a substance which does this. It is called **chlorophyll.**

Chlorophyll is the pigment which makes plants look green. It is kept inside the chloroplasts of plant cells, arranged on a series of membranes (see Fig 3.9). Spread out like this, it can trap the maximum amount of sunlight.

When sunlight falls on a chlorophyll molecule, the energy is absorbed. The chlorophyll molecule then releases the energy. The energy makes carbon dioxide combine with water, with the help of enzymes inside the chloroplast.

3.4 Photosynthesis is a chemical process.

The full equation for photosynthesis is written like this

$$\text{carbon dioxide} + \text{water} \xrightarrow[\text{chlorophyll}]{\text{sunlight}} \text{glucose} + \text{oxygen}$$

To show the number of molecules involved in the reaction, a balanced equation needs to be written. Carbon dioxide contains two atoms of oxygen, and one of carbon, so its molecular formula is CO_2. Water has the formula H_2O. Glucose (see section 2.6) has the formula $C_6H_{12}O_6$. Oxygen molecules contain two atoms of oxygen, and so they are written O_2.

The balanced equation for photosynthesis is this.

$$6\,CO_2 + 6\,H_2O \xrightarrow[\text{chlorophyll}]{\text{sunlight}} C_6H_{12}O_6 + 6\,O_2$$

Questions

1. Which inorganic substances does a plant use to make food?
2. What kind of energy does a plant use to make carbon dioxide combine with water?
3. What is chlorophyll?
4. What happens when sunlight falls on a chlorophyll molecule?
5. What does a balanced equation show?

Leaves

3.5 Plant leaves are food factories.

Photosynthesis happens inside chloroplasts. This is where the enzymes and chlorophyll are which catalyse and supply energy to the reaction. In a typical plant, most chloroplasts are in the cells in the leaves. A leaf is a factory for making carbohydrates.

Leaves are therefore specially adapted to allow photosynthesis to take place as quickly and efficiently as possible.

Fact!

The largest leaves of any plant belong to the raffia palm, which grows on the Mascarene Islands in the Indian Ocean, and also the Amazonian bamboo palm. They both have leaves up to 19.8m long.

3.6 The structure of leaves.

A leaf consists of a broad, flat part called the **lamina** (see Fig 3.1), which is joined to the rest of the plant by a leaf stalk or **petiole.** Running through the petiole are vascular bundles (see section 6.29), which then form the **veins** in the leaf. These contain tubes which carry substances to and from the leaf.

Although a leaf looks thin, it is in fact made up of several layers of cells. You can see these if you look at a transverse section (TS) of a leaf under a microscope (see Fig 3.2).

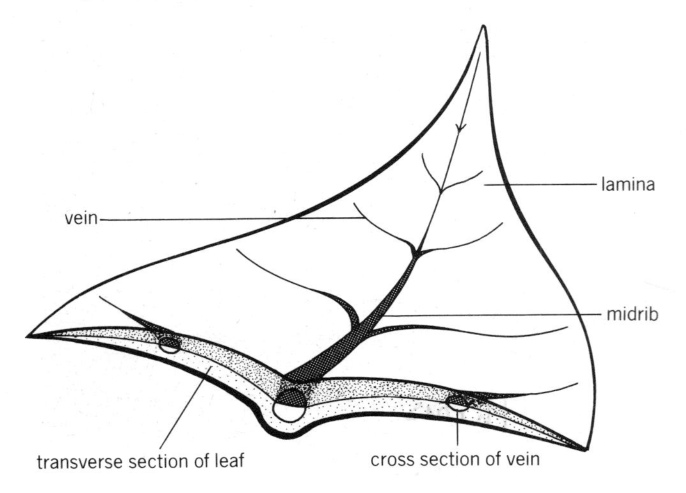

3.1 The structure of a leaf

The top and bottom of the leaf are covered with a layer of closely fitting cells called the **epidermis** (see Fig 3.4). These cells do not contain chloroplasts. Their function is to protect the inner layers of cells in the leaf. The cells of the upper epidermis often secrete a waxy substance, which lies on top of them. It is called the **cuticle,** and it helps to stop water evaporating from the leaf. There is sometimes a cuticle on the underside of a leaf as well.

In the lower epidermis, there are small holes called **stomata** (singular: stoma). Each stoma is surrounded by a pair of sausage shaped **guard cells** (see Fig 3.4) which can open or close the hole (see Fig 10.14). Guard cells, unlike the other cells in the epidermis, do contain chloroplasts.

The middle layers of the leaf are called the **mesophyll** ('meso' means middle, and 'phyll' means leaf). These cells all contain chloroplasts. The cells nearer to the top of the leaf are arranged like a fence or palisade, and they form the **palisade layer.** The cells beneath them are rounder, and arranged quite loosely, with large air spaces between them. They form the **spongy layer.**

Running through the mesophyll are veins. Each vein contains large, thick-walled **xylem vessels** (see section 6.27) for carrying water, and smaller, thin-walled **phloem tubes** (see section 6.28) for carrying away food which the leaf has made.

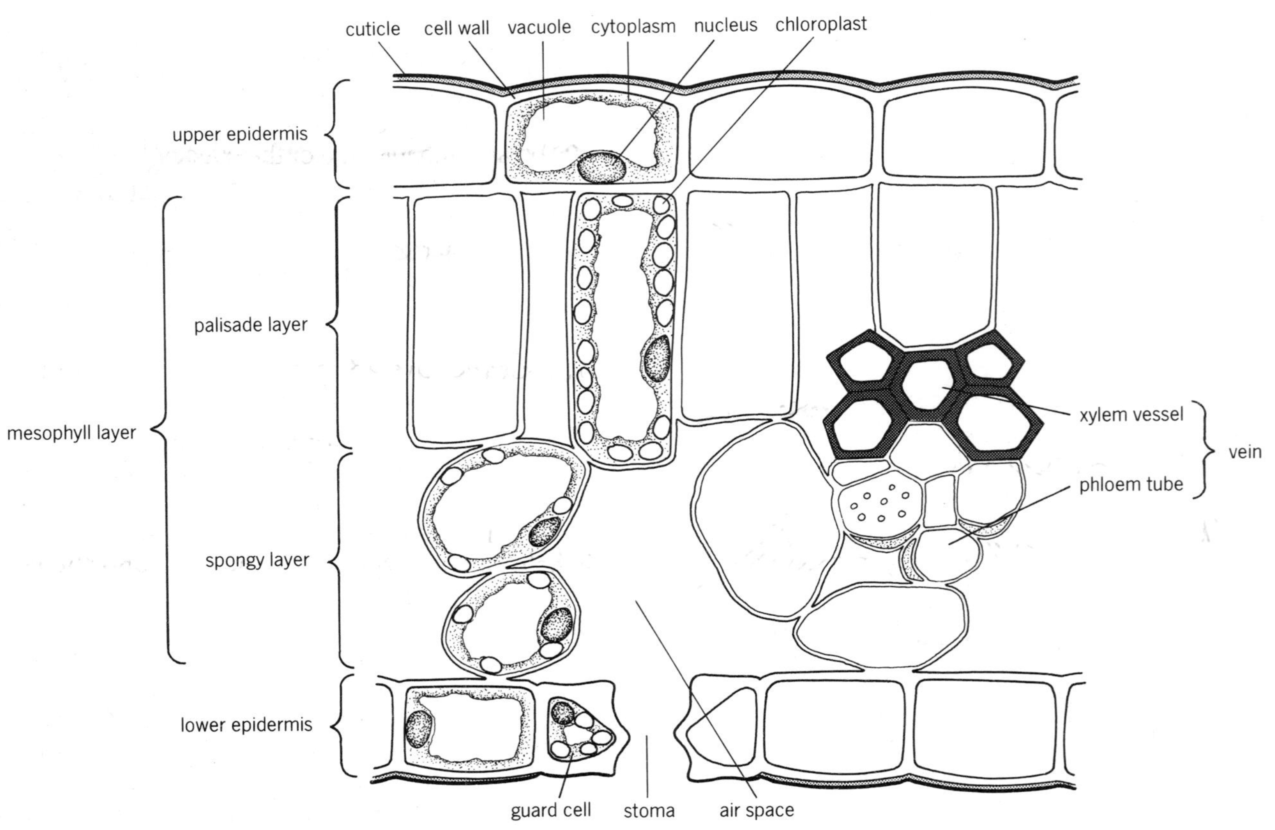

3.2 Transverse section through part of a leaf

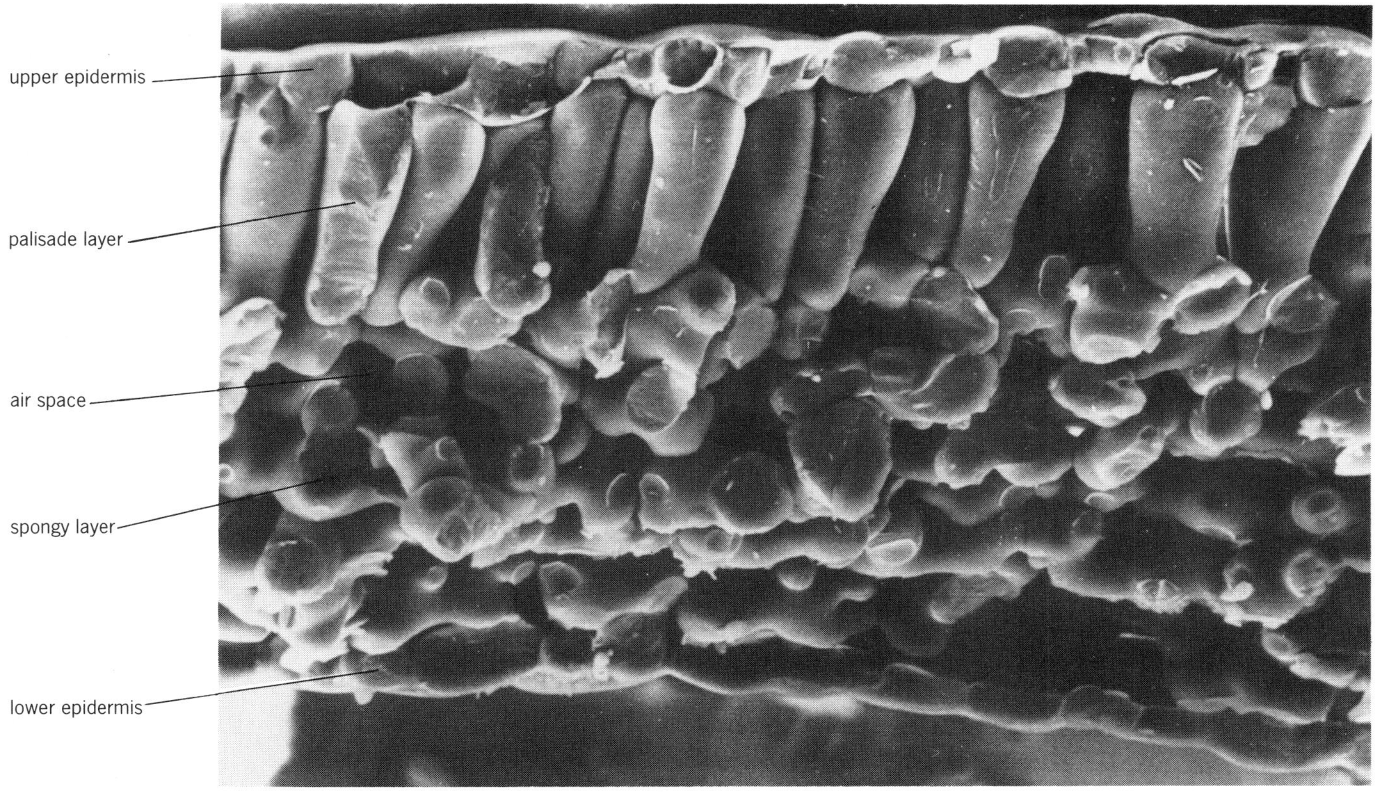

3.3 A scanning electron micrograph showing the cells inside a leaf; notice the many air spaces between the cells

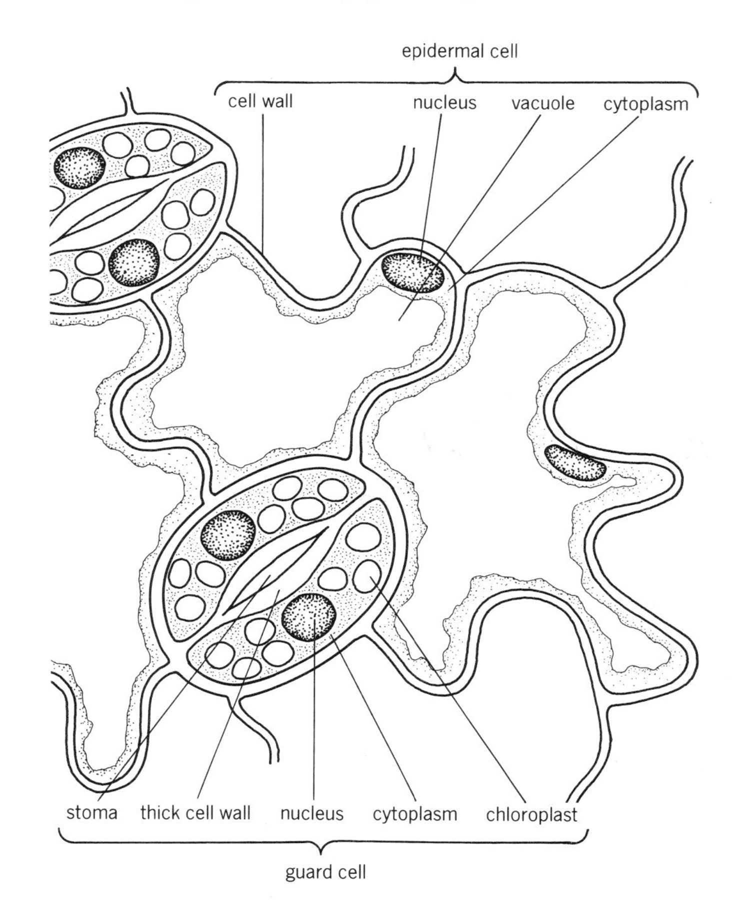

3.4 Surface view of the lower epidermis of a leaf

Questions

1. What is another name for a leaf stalk?
2. Which kind of cells make the cuticle on a leaf?
3. What is the function of the cuticle?
4. What are stomata?
5. What are guard cells?
6. List three kinds of cells in a leaf which contain chloroplasts, and one kind which does not.

3.7 Leaves are adapted to obtain carbon dioxide, water and sunlight.

Carbon dioxide Carbon dioxide is obtained from the air. There is not very much available, because only about 0.03% of the air is carbon dioxide. Therefore the leaf must be very efficient at absorbing it. The leaf is held out into the air by the stem and the leaf stalk, and its large surface area helps to expose it to as much air as possible.

The cells which need the carbon dioxide are the mesophyll cells, inside the leaf. The carbon dioxide can get into the leaf through the stomata. It does this by diffusion, which is described in Chapter 4. Behind each stoma is an air space (see Fig 3.2) which connects up with other air spaces between the spongy mesophyll cells. The carbon dioxide can therefore diffuse to all the cells in the leaf. It can then diffuse through the cell

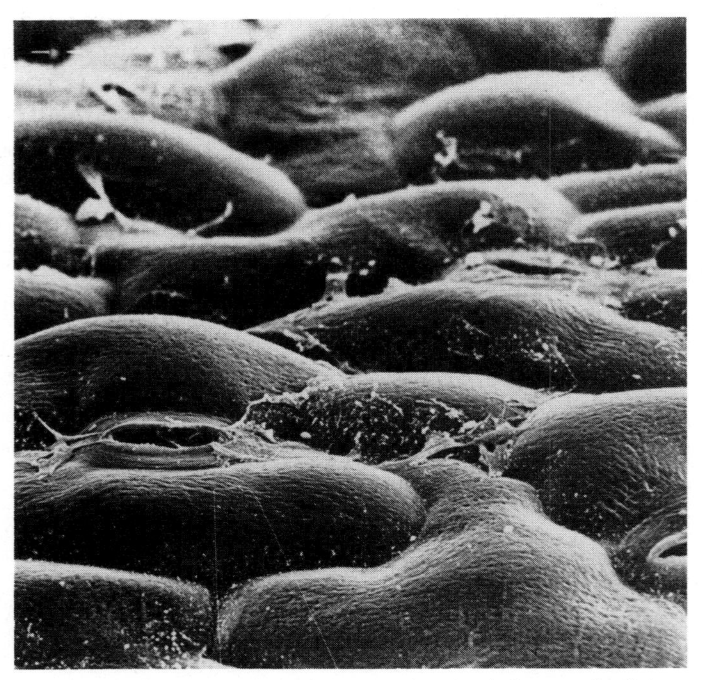

3.5 The lower surface of a leaf, showing the closely-fitting cells of the epidermis; the oval holes are stomata

wall and cell membrane of each cell, and into the chloroplasts.

Water Water is obtained from the soil. It is absorbed by the root hairs (see section 6.38), and carried up to the leaf in the xylem vessels. It then travels from the xylem vessels to the mesophyll cells by osmosis, which is described in Chapter 4. The path it takes is shown in Figs 3.6 and 3.7.

Sunlight The position of a leaf and its broad, flat, surface help it to obtain as much sunlight as possible. If you look up through the branches of a tree, you will see that the leaves are arranged so that they do not cut off light from one another more than necessary. Plants which live in shady places often have particularly big leaves.

The cells which need the sunlight are the mesophyll cells. The thinness of the leaf allows the sunlight to penetrate right through it, and reach all the cells. To help this the epidermal cells are transparent, with no chloroplasts.

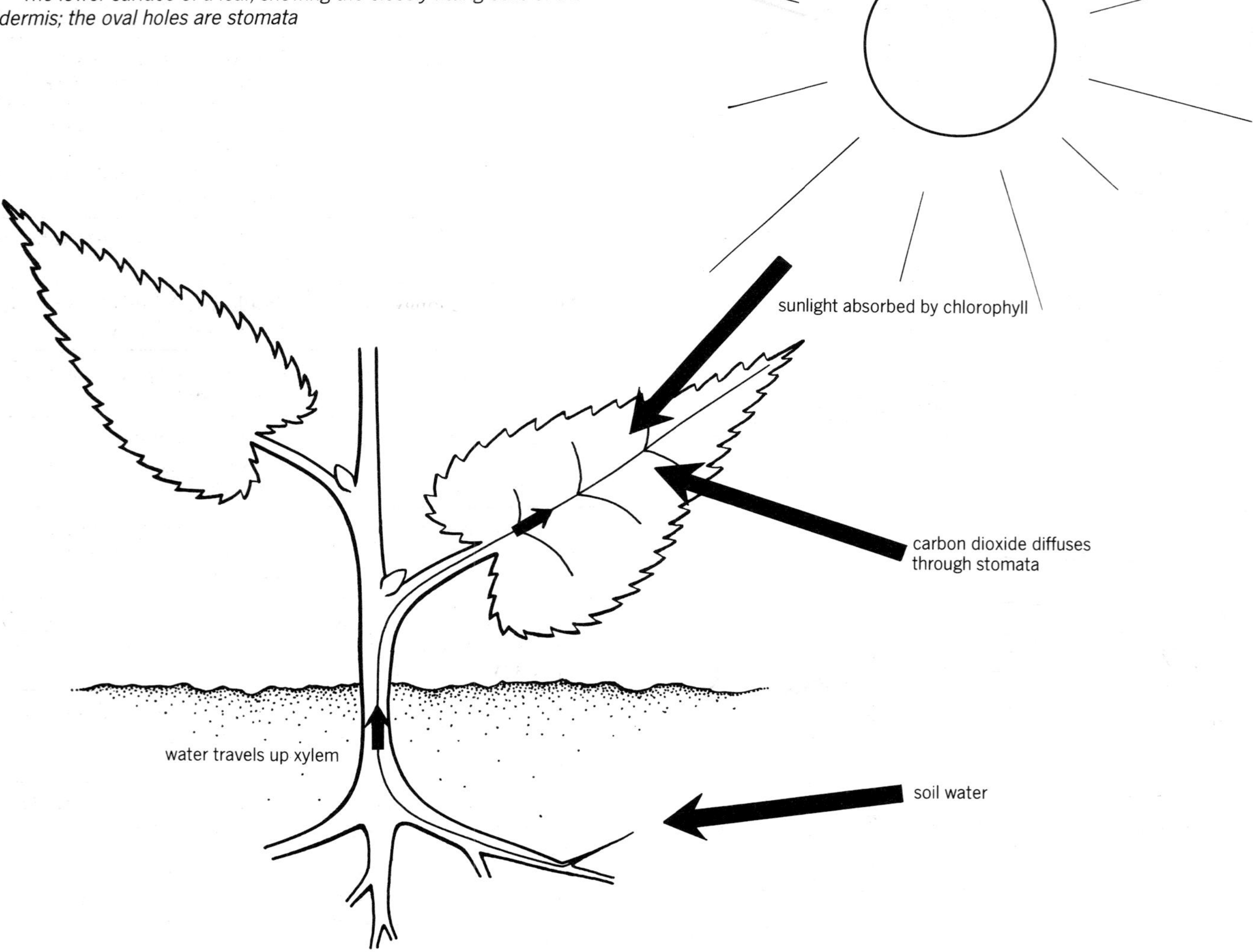

3.6 How the materials for photosynthesis get into a leaf

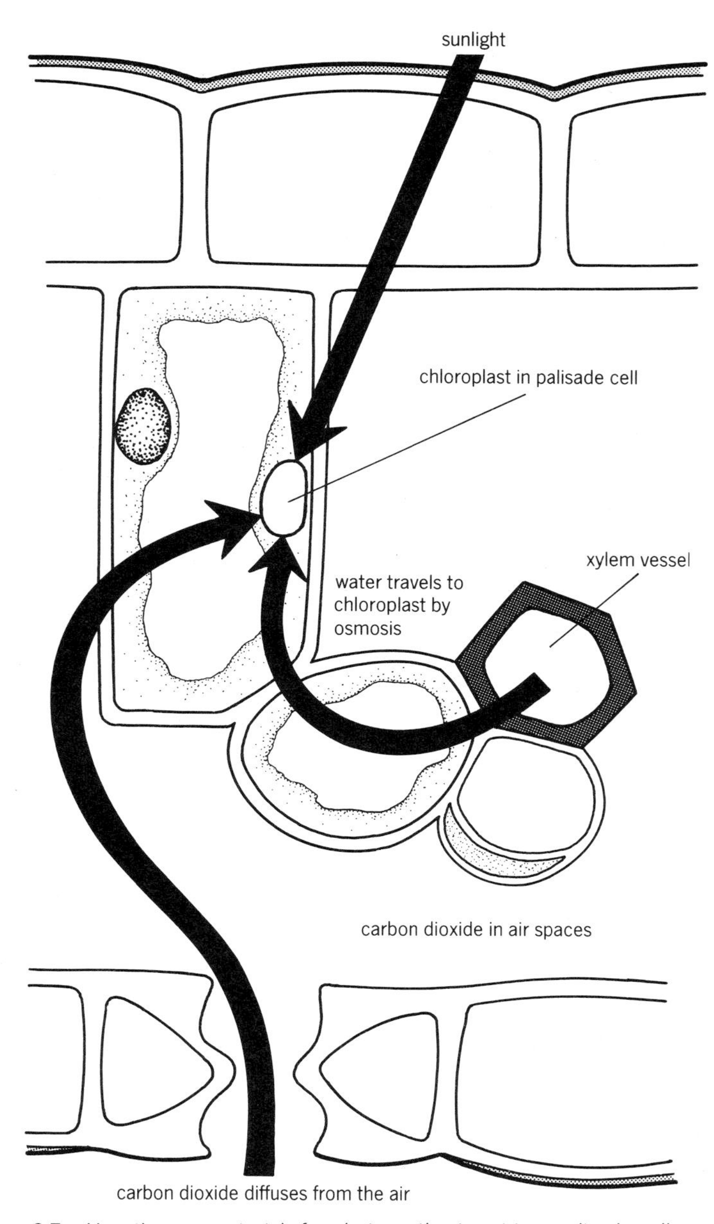

3.7 How the raw materials for photosynthesis get to a palisade cell

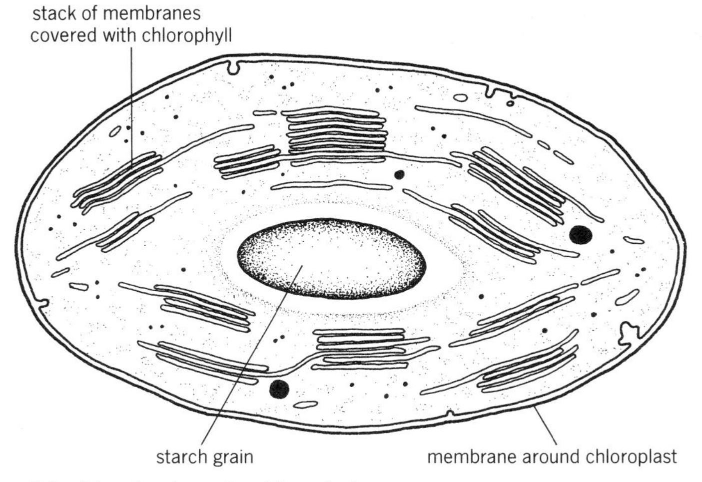

3.8 The structure of a chloroplast

In the mesophyll cells, the chloroplasts are arranged to get as much sunlight as possible, particularly those in the palisade cells. They can lie broadside on to do this, but in strong sunlight, they often arrange themselves end on. This reduces the amount of light absorbed. Inside them, the chlorophyll is arranged on flat membranes (see Fig 3.8) to expose as much as possible to the sunlight.

Questions

1 What are the raw materials needed for photosynthesis?
2 What percentage of the air is carbon dioxide?
3 How does carbon dioxide get into a leaf?
4 How does a leaf obtain its water?
5 Give two reasons why leaves need to have a large surface area.
6 Why are leaves thin?

Table 3.1 Adaptations of leaves for photosynthesis

Adaptation	*Function*
Supported by stem and petiole	To expose as much of it as possible to sunlight and air
Large surface area	To expose as much of it as possible to sunlight and air
Thin	To allow sunlight to penetrate to all cells; to allow CO_2 to diffuse in and O_2 to diffuse out as quickly as possible
Stomata in lower epidermis	To allow $CO_2$and O_2 to diffuse in and out
Air spaces in spongy mesophyll	To allow CO_2 and O_2 to diffuse to and from all cells
No chloroplasts in epidermis	To allow sunlight to penetrate to mesophyll layer
Chloroplasts containing chlorophyll present in mesophyll layer	To absorb sunlight, to provide energy to combine CO_2 and H_2O
Palisade cells arranged end on	To keep as few cell walls as possible between sunlight and chloroplasts
Chloroplasts in palisade cells arranged broadside on, especially in dim light	To expose as much chlorophyll as possible to sunlight
Chlorophyll arranged on flat membranes inside chloroplast	To expose as much chlorophyll as possible to sunlight
Xylem vessels within short distance of every mesophyll cell	To supply water to chloroplasts for photosynthesis
Phloem tubes within short distance of every mesophyll cell	To take away organic products of photosynthesis

Investigation 3.1 Looking at the epidermis of a leaf

Using a piece of epidermis

1 Using forceps, carefully peel a small piece of epidermis from the underside of a leaf.
2 Put the piece of epidermis into a drop of water on a microscope slide.
3 Spread it out carefully, trying not to let any part of it fold over. Cover it with a cover slip.
4 Look at your slide under the microscope, and make a labelled drawing of a few cells.

Making a nail varnish impression

1 Paint the underside of a leaf with transparent nail varnish. Leave to dry thoroughly.
2 Peel off part of the nail varnish, and mount it in a drop of water on a microscope slide.
3 Spread it out carefully, and cover with a coverslip.
4 Look at your slide under the microscope, and make a labelled drawing of the impressions made by a few cells.
5 Repeat with the upper surface of a leaf.

Questions

1 On which surface of the leaf did you find most stomata?
2 Which of these two techniques for examining the epidermis of a leaf do you consider (a) is easiest, and (b) gives you the best results?
3 There are two kinds of cell in the lower epidermis of a leaf. What are they, and what are their functions?

3.8 Glucose is used in different ways.

One of the first carbohydrates to be made in photosynthesis is **glucose**. There are several things which may then happen to it.

Energy may be released from glucose in the leaf All cells need energy, which they obtain by the process of respiration (see section 5.4). Some of the glucose which a leaf makes will be broken down by respiration, to release energy.

Glucose may be turned into starch and stored in the leaf Glucose is a simple sugar (see section 2.6). It is soluble, and quite a reactive substance. It is not, therefore, a very good storage molecule. Firstly, being reactive, it might get involved in chemical reactions where it was not wanted. Secondly, it would dissolve in the water in and around the plant cells, and might be lost from the cell. Thirdly, when dissolved, it would increase the strength of the solution in the cell, which could damage the cell.

The glucose is therefore converted into **starch** to be stored. Starch is a polysaccharide, made of many glucose molecules joined together. Being such a large molecule, it is not very reactive, and not very soluble. It can be made into granules which can be easily stored inside the chloroplasts.

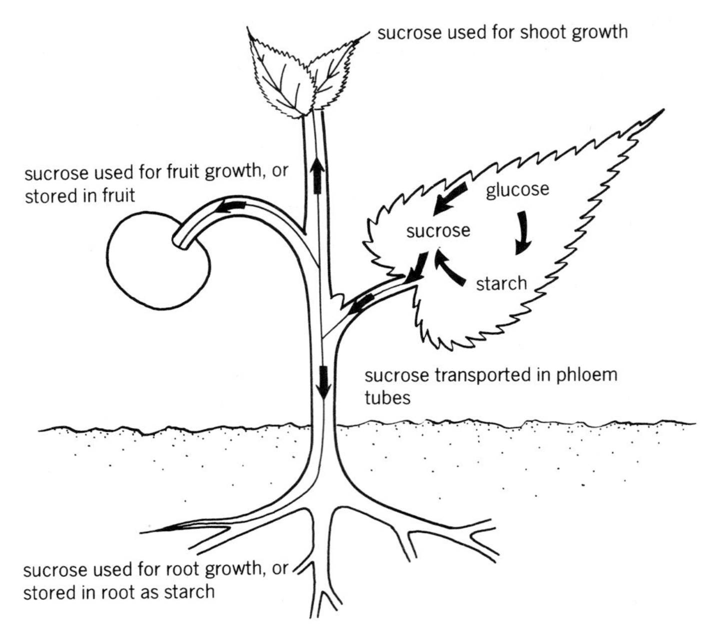

3.9 The products of photosynthesis

Glucose may be used to make other organic substances The plant can use glucose as a starting point for making all the other organic substances it needs. These include the carbohydrates sucrose and cellulose. Plants also make **oils** (liquid fats).

With the addition of minerals containing nitrogen and sulphur, amino acids can be made. These are then joined together to make **proteins**. The plant also makes other substances such as chlorophyll and vitamins.

The mineral substances required are all obtained from the soil. They are absorbed through the root hairs. Some of the most important ones are listed in Table 3.2. Water culture experiments (see Fig 3.10) can show which minerals a plant needs, and what happens to it if it does not have them.

Glucose may be transported to other parts of the plant A molecule has to be small and soluble to be transported easily. Glucose has both of these properties, but it is also rather reactive. It is therefore converted to the complex sugar sucrose to be transported to other parts of the plant. Sucrose molecules are also quite small and soluble, but less reactive than glucose. They dissolve in the sap in the phloem vessels, and can be distributed to whichever parts of the plant need them (see Fig 3.9).

The sucrose may later be turned back into glucose again, to be broken down to release energy, or turned into starch and stored, or used to make other substances which are needed for growth.

Table 3.2 Mineral salts required by plants

Element	*Mineral salt*	*Why it is needed*	*Deficiency disease*
Nitrogen	Nitrates or as organic compounds from nitrogen-fixing bacteria	To make proteins	Poor growth, yellow leaves
Sulphur	Sulphates	To make proteins	Poor growth, yellow leaves
Phosphorus	Phosphates	To make ATP (see section 5.5)	Poor growth, especially of roots
Magnesium	Magnesium salts	To make chlorophyll	Yellowing between veins of leaves
Iron	Iron salts	To make chlorophyll; iron is not contained in chlorophyll, but is needed for its manufacture	Yellowing in young leaves
Potassium	Potassium salts	To keep correct salt balance for cells	Mottled leaves

All of these minerals are obtained from the soil. They are absorbed in solution through the root hairs.

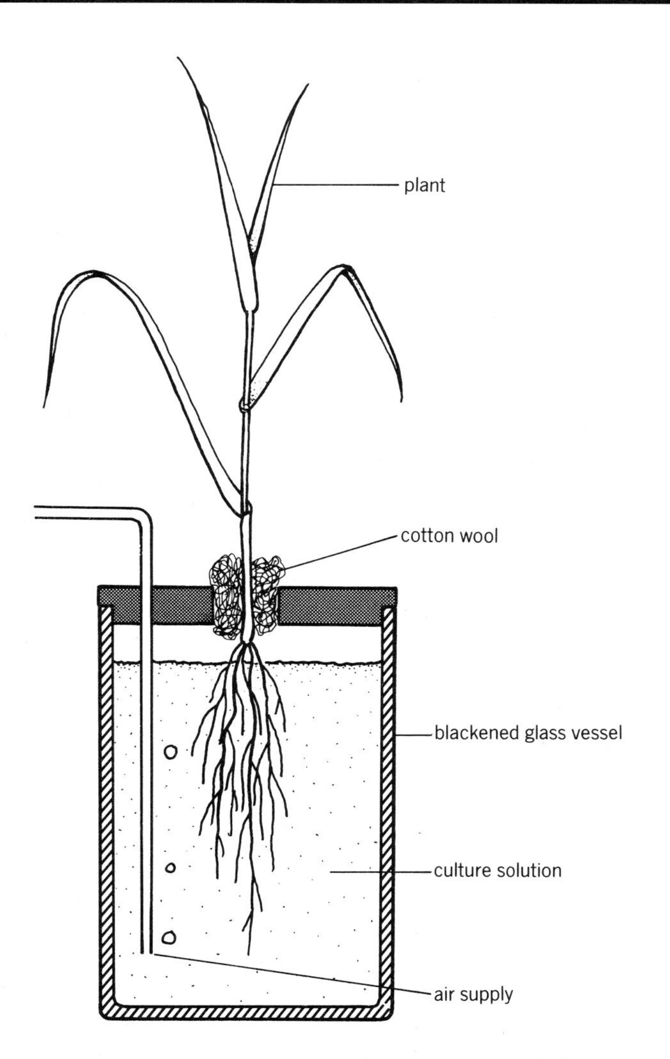

3.10 Apparatus for water culture experiments.
Several plants are grown like this, in identical conditions. Some, the controls, are given a culture solution containing all the minerals needed by plants. Others have a solution from which one mineral is missing. By comparing each plant with the control plants, the effects of a lack of each mineral can be seen.

Questions

1 Why is glucose not very good for storing in a leaf?
2 What substances does a plant need to be able to convert glucose into proteins?
3 Why do plants need iron?
4 How do parts of the plant such as the roots which cannot photosynthesise obtain food?

Photosynthesis experiments

3.9 Experiments need controls.

If you do Investigations 3.3, 3.4 and 3.5, you can find out for yourself which substances a plant needs for photosynthesis. In each experiment, the plant is given everything it needs, except for one substance. Another plant is used at the same time. This is a **control**. The control plant is given everything it needs, including the substance being tested for.

Both plants are then treated in exactly the same way. Any differences between them at the end of the experiment, therefore, must be because of the substance being tested.

At the end of the experiment, test a leaf from your experimental plant and your control to see if they have made starch. By comparing them, you can find out which substances are necessary for photosynthesis.

3.10 Plants for photosynthesis experiments must be destarched.

It is very important that the leaves you are testing should not have any starch in them at the beginning of the experiment. If they did, and you found that the leaves contained starch at the end of the experiment, you could not be sure that they had been photosynthesising. The starch might have been made before the experiment began.

So, before doing any of these experiments, you must destarch the plants. The easiest way to do this is to leave them in a dark cupboard for at least 24 hours. The plants cannot photosynthesise while they are in the cupboard because there is no light. Therefore they use up their stores of starch. To be certain that they are thoroughly destarched, test a leaf for starch before you begin your experiment.

3.11 Iodine solution can stain starch in leaves.

Iodine solution is used to test for starch. A bluish black colour shows that starch is present. However, if you put iodine solution onto a leaf which contains starch, it will not immediately turn black. This is because the starch is right inside the cells, inside the chloroplasts (see Fig 1.2). The iodine solution cannot get through the cell membranes to reach the starch and react with it. Another difficulty is that the green colour of the leaf and the brown iodine solution can look black together.

Therefore before testing a leaf for starch, you must break down the cell membranes, and get rid of the green colour (chlorophyll). The way this is done is described in Investigation 3.2. The cell membranes are first broken down by boiling water, and then the chlorophyll is removed by dissolving it out with alcohol.

Investigation 3.2 Testing a leaf for starch

1 Take a leaf from a healthy plant, and drop it into boiling water in a water bath. Leave it for about 30 seconds. (The length of time needed varies for different types of leaves. 30s is about right for geranium leaves.)
2 Remove the leaf, which will be very soft, and drop it into a tube of alcohol in the water bath (Fig 3.11). Leave it until all the chlorophyll has been dissolved out of the leaf.

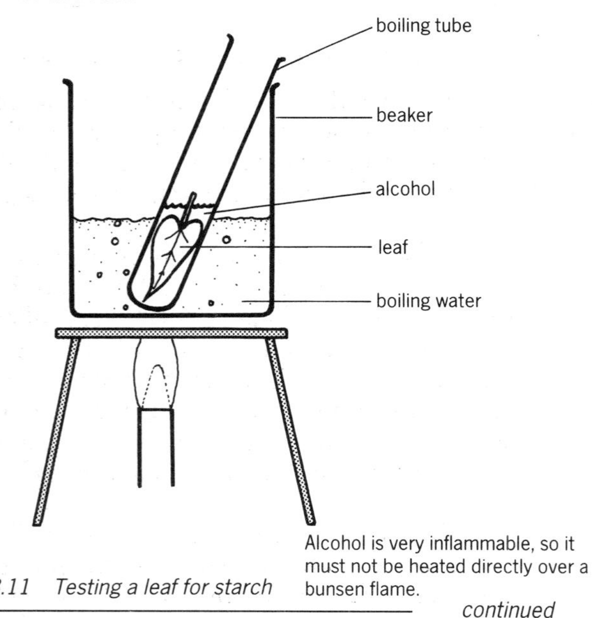

3.11 Testing a leaf for starch

continued

3 The leaf will now be brittle. Remove it from the alcohol, and dip it into water again to soften it.
4 Spread out the leaf on a white tile, and cover it with iodine solution. A black colour shows that the leaf contains starch.

Questions

1 Why was the leaf put into boiling water?
2 Why did the alcohol become green?
3 Why was the leaf put into alcohol after being put into boiling water?

Investigation 3.3 To see if light is necessary for photosynthesis

1 Take a healthy geranium plant, growing in a pot. Leave it in a cupboard for a few days, to destarch it.
2 Test one of its leaves for starch, to check that it does not contain any.
3 Using a folded piece of black paper or aluminium foil, a little larger than a leaf, cut out a shape (see Fig 3.12). Fasten the paper or foil firmly over both sides of a leaf on your plant, making sure that the edges are held firmly together. Don't take the leaf off the plant!

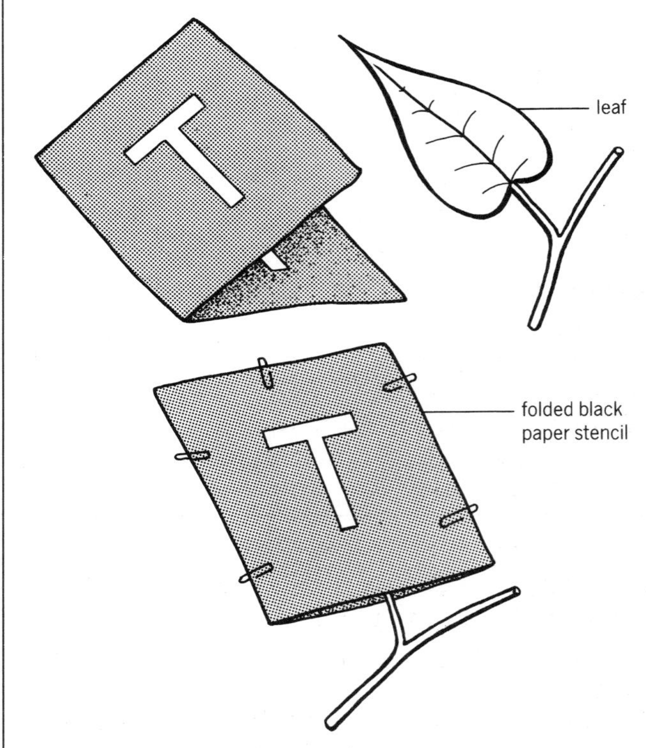

3.12 To see if light is necessary for photosynthesis

4 Leave the plant near a warm, sunny window for a few days.
5 Remove the cover from your leaf, and test it for starch.
6 Make a labelled drawing of the appearance of your leaf after testing for starch.

continued

Questions

1 Why was the plant destarched before the beginning of the experiment?
2 Why was part of the leaf left uncovered?
3 What do your results tell you about light and photosynthesis?

Investigation 3.4 To see if carbon dioxide is necessary for photosynthesis

1 Destarch a plant.
2 Set up your apparatus as shown in Fig 3.13. Take special care that no air can get into the flasks. Leave the plant in a warm sunny window for a few days.
3 Test each treated leaf for starch.

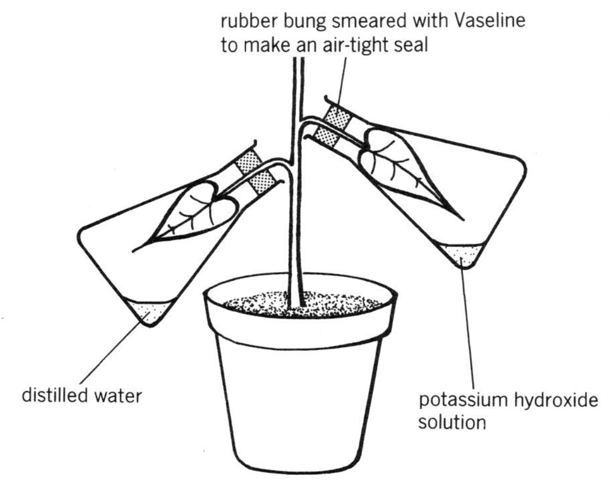

3.13 To see if carbon dioxide is necessary for photosynthesis

Questions

1 Why was potassium hydroxide put in with one leaf, and water with the other?
2 Which was the control?
3 Why was Vaseline put around the tops of the flasks?
4 What do your results suggest about carbon dioxide and photosynthesis?

Investigation 3.5 To see if chlorophyll is necessary for photosynthesis

1 Destarch a plant with variegated (green and white) leaves.
2 Leave your plant in a warm, sunny spot for a few days.
3 Test one of the leaves for starch.
4 Make a drawing of your leaf before and after testing

Questions

1 What was the control in this experiment?
2 What do your results tell you about chlorophyll and photosynthesis?

Investigation 3.6 To show that oxygen is produced during photosynthesis

1 Set up the apparatus as shown in Fig 3.14. Make sure that the test tube is completely full of water.

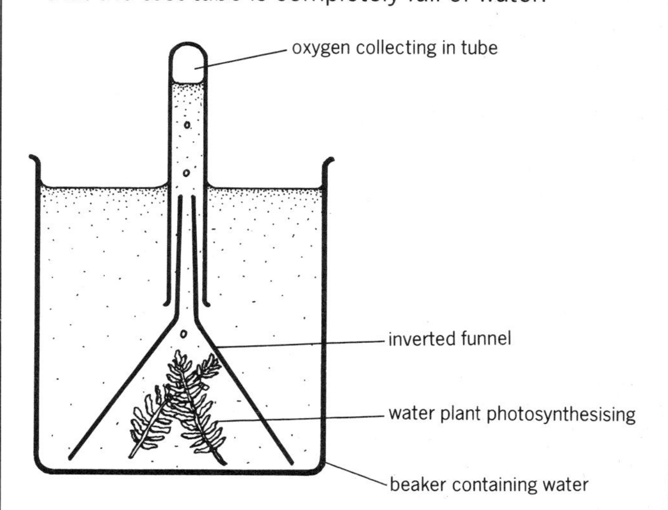

3.14 To show that oxygen is produced during photosynthesis

2 Leave the apparatus near a warm, sunny window for a few days.
3 Carefully remove the test tube from the top of the funnel, allowing the water to run out, but not allowing the gas to escape.
4 Light a wooden splint, and then blow it out so that it is just glowing. Carefully put it into the gas in the test tube. If it bursts into flame, then the gas is oxygen.

Questions

1 Why was this experiment done under water?
2 This experiment has no control. Try to design one.

3.12 Radioactive chemicals show CO_2 uptake.

Investigation 3.4 shows that carbon dioxide is necessary for photosynthesis, but it does not prove that the carbon dioxide is actually absorbed by the leaf. To show that this is what happens, a **radioactive isotope** can be used.

The atoms of some elements can occur in more than one form. The different forms are called isotopes. Some isotopes produce radiation, and are called radioactive. The most common isotope of carbon is carbon twelve, written ^{12}C. It is not radioactive. There is, however, another isotope, ^{14}C, which is radioactive. In this experiment, a leaf is supplied with carbon dioxide containing ^{14}C, and then tested for radioactivity.

The apparatus for this experiment is shown in Fig 3.15. The solution of sodium hydrogen carbonate gives off carbon dioxide containing ^{14}C. The apparatus is left in a warm, sunny place, to allow the leaf to photosynthesise. A second identical piece of apparatus is also set up, but left in darkness.

After a few hours, both leaves are tested to see if they have absorbed any of the ^{14}C. A simple way to do this is to press the leaf onto a piece of X-ray photographic film (see Fig 3.16). This must be done in darkness, so that no light falls onto the film. After the leaf has been in contact with the film for some time, the film is developed.

The developed film onto which the leaf in the sunlight was pressed will have a dark area, in the shape of the leaf. This happens because the ^{14}C in the leaf emitted radiation, which affected the X-ray film in just the same way that X-rays do. This shows that the leaf has absorbed the radioactive carbon dioxide.

The other leaf, which had not been able to photosynthesise, makes no mark–it has not absorbed the ^{14}C.

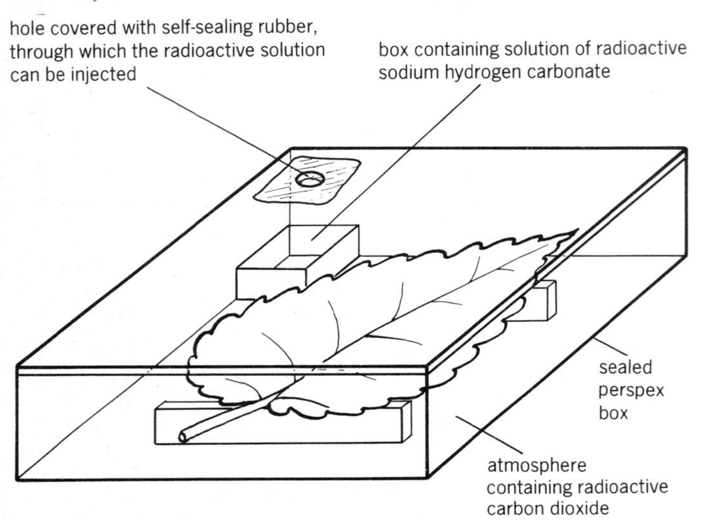

3.15 Apparatus for supplying a leaf with $^{14}CO_2$

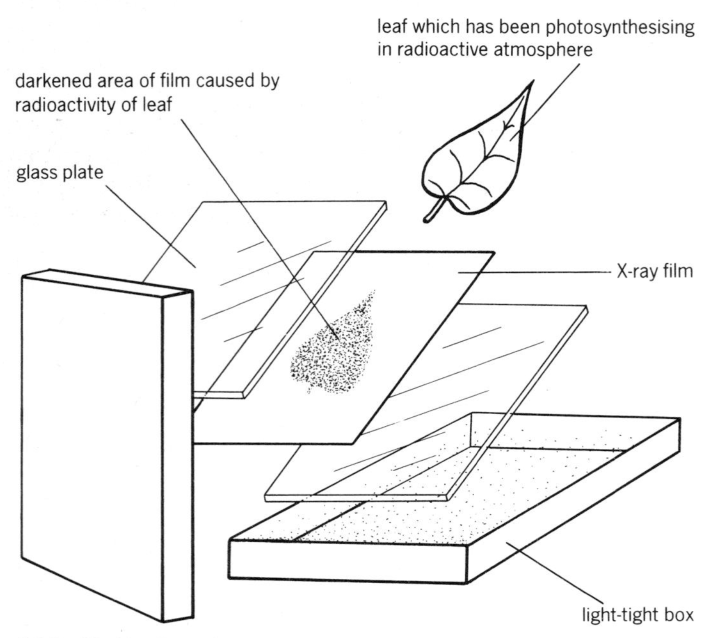

3.16 Testing for radioactivity in a leaf

3.13 Many factors affect rate of photosynthesis.

If a plant is given plenty of sunlight, carbon dioxide and water, the limit on the rate at which it can photosynthesise is its own ability to absorb these materials, and make them react. However, quite often plants do not have unlimited supplies of these materials, and so their rate of photosynthesis is not as high as it might be.

Sunlight In the dark, a plant cannot photosynthesise at all. In dim light, it can photosynthesise slowly. As light intensity increases, the rate of photosynthesis will increase, until the plant is photosynthesising as fast as it can. At this point, even if the light becomes brighter, the plant cannot photosynthesise any faster (see Fig 3.17a).

Over the first part of the curve, in Fig 3.17a, between A and B, light is a **limiting factor.** The plant is limited in how fast it can photosynthesise because it does not have enough light. You can show this because when the plant is given more light it photosynthesises faster. Between B and C, however, light is not a limiting factor. You can show this because, even if more light is shone on the plant, it still cannot photosynthesise any faster. It already has as much light as it can use.

Carbon dioxide Carbon dioxide can also be a limiting factor (see Fig 3.17b). The more carbon dioxide a plant is given, the faster it can photosynthesise up to a point, but then a maximum is reached.

Temperature The chemical reactions of photosynthesis will only take place very slowly at low temperatures (see section 9.1), so a plant can photosynthesise faster on a warm day than a cold one.

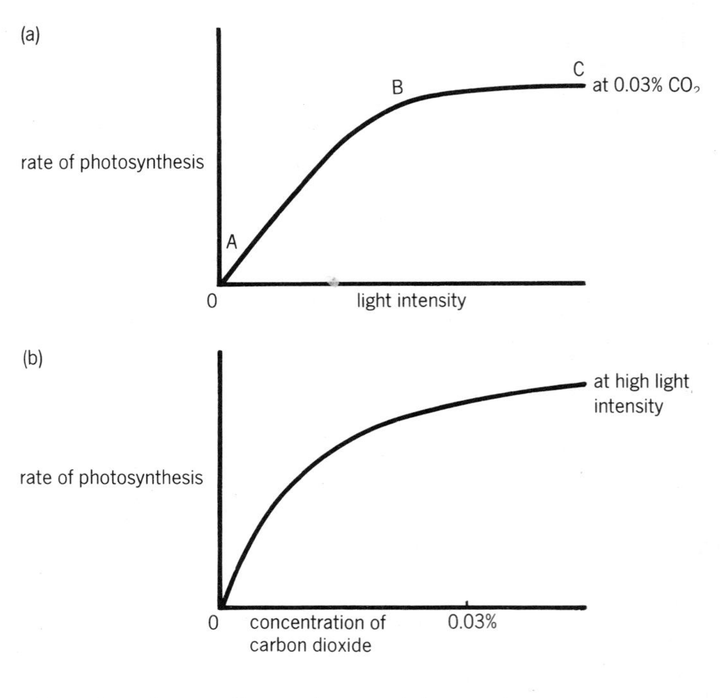

3.17 Some factors affecting the rate of photosynthesis

Stomata The carbon dioxide which a plant uses passes into the leaf through the stomata. If the stomata are closed, then photosynthesis cannot take place. Stomata often close if the weather is very hot and sunny, to prevent too much water being lost (see Fig 10.14). On a really hot day, therefore, photosynthesis may slow down or stop for a time.

Questions

1 What is meant by a radioactive isotope?
2 How does radioactivity affect an X-ray photographic film?
3 What is meant by a limiting factor?
4 Name two factors which may limit the rate of photosynthesis of a healthy plant.
5 Why do plants sometimes stop photosynthesising on a very hot, dry day?

3.14 Photosynthesis takes place in stages.

The photosynthesis equation given in section 3.4 is actually a summary of a whole series of chemical reactions which take place inside chloroplasts. In fact, carbon dioxide and water are combined to make glucose in two stages.

Firstly, energy is absorbed from sunlight by the chlorophyll. This energy is used to split water molecules into hydrogen and oxygen. This is called **photolysis** ('light splitting'). This stage requires light, and so it is called the **light reaction.** The oxygen from the water is given off as a gas.

Secondly the hydrogen is combined with carbon dioxide to produce glucose. This stage does not need light, so it is called the **dark reaction.** This is a little misleading, because it does take place in the daytime.

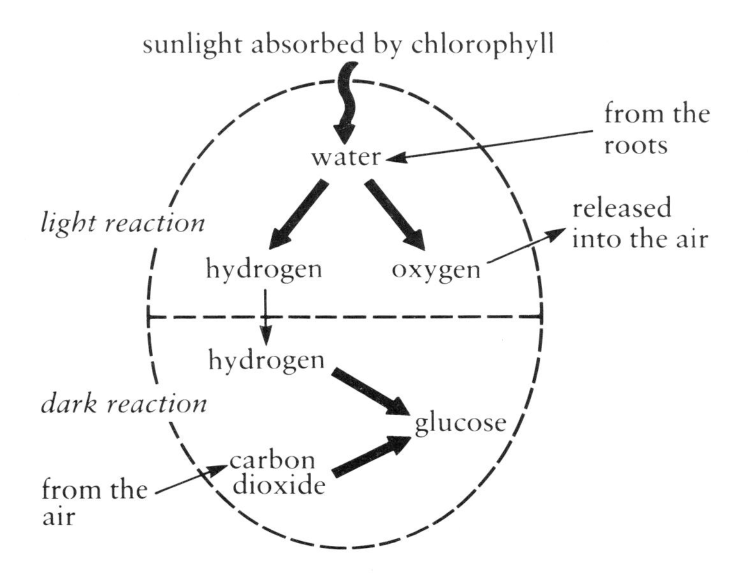

3.18 *The reactions of photosynthesis*

Chapter revision questions

1 Give one use of each of these substances when performing experiments to investigate photosynthesis:
(a) ^{14}C,
(b) wooden splint,
(c) variegated leaf,
(d) potassium hydroxide solution,
(e) alcohol.

2 Copy and complete this table.

	obtained from	used for
nitrates	soil, through root hairs	making proteins
water		
magnesium		
carbon dioxide		

3 Explain the following.
(a) There is an air space behind each stoma.
(b) The epidermal cells of a leaf do not have chloroplasts.
(c) Leaves have a large surface area.
(d) The veins in a leaf branch repeatedly.
(e) A leaf containing starch does not turn black immediately when you put iodine solution onto it.
(f) Chloroplasts have many membranes in them.

4 Which carbohydrate does a plant use for each of these purposes? Explain why.
(a) transport
(b) storage

5 Describe how a carbon atom in a carbon dioxide molecule in the air could become part of a starch molecule in a carrot root. Mention all the structures it would pass through, and what would happen to it at each stage.

continued

6 Read the following passage carefully, then answer the questions, using both the information in the passage and your own knowledge.

White light is made up of all the colours of the rainbow. Sea water acts as a light filter which screens off some of the light energy, starting at the red end of the spectrum. As sunlight travels downwards through the water, first the red light is lost, then green and yellow and finally blue. In very clear water, the blue light can penetrate to a maximum of 1 000 m. Below this, all is dark.

The upper layers of the sea contain a large community of microscopic floating organisms called plankton, many of which are tiny plants known as phytoplankton. These act as a gigantic solar cell, which feeds all the animals of the sea and supplies both them and the atmosphere above with oxygen.

Nearer the shore, larger plants are found. Seaweeds grow on rocky shores, brown and green ones high on the shore, and red ones lower down, where they are covered with deep water when the tide is in. The colours of the seaweeds are due to their light-absorbing pigments, not all of which are chlorophyll.

(a) Why are no green plants found below the upper few hundred metres of the sea?
(b) Some living organisms are found in the permanently dark depths of the oceans. What might they feed on?
(c) What are phytoplankton?
(d) Explain as fully as you can the last sentence of paragraph 2, "These act as .... with oxygen."
(e) Chlorophyll is a green pigment. Which colours of light does it (i) absorb, and (ii) reflect?
(f) What colour light would you expect the pigment of red seaweeds to absorb?
(g) Why are red seaweeds normally found lower down the shore than green ones?

7 An experiment was performed to find out how fast a plant photosynthesised as the concentration of CO_2 in the air around it was varied. The results were as follows.

CO_2 concentration % by volume in air	*Rate of photosynthesis in arbitrary units*	
	low light intensity	*high light intensity*
0	0	0
0.02	20	33
0.04	29	53
0.06	35	68
0.08	39	79
0.10	42	86
0.12	45	89
0.14	46	90
0.16	46	90
0.18	46	90
0.20	46	90

(a) Plot these results on a graph, drawing one line for low and one for high light intensity, both on the same pair of axes.
(b) What is the CO_2 concentration of normal air?
(c) What is the rate of photosynthesis at this CO_2 concentration in a high light intensity?
(d) Market gardeners often add carbon dioxide to the air in greenhouses. What is the advantage of doing this?
(e) Up to what values does CO_2 concentration act as a limiting factor at high light intensities?

4 Diffusion and osmosis

4.1 Diffusion results from random movement.

Atoms, molecules and ions are always moving. The higher the temperature, the faster they move. In a solid substance the molecules cannot move very far, because they are held together by forces between them. In a liquid they can move more freely, knocking into one another and rebounding. In a gas they are freer still. Molecules and ions can also move freely when they are in solution.

When they can move freely, molecules tend to spread themselves out as evenly as they can. This happens with gases, solutions, and mixtures of liquids. Imagine, for example, a rotten egg in one corner of a room, giving off hydrogen sulphide gas. To begin with, there will be a very high concentration of the gas near the egg, but none in the rest of the room. However, before long the hydrogen sulphide molecules have spread throughout the air in the room. Soon, you will not be able to tell where the smell first came from—the whole room will smell of hydrogen sulphide.

The hydrogen sulphide molecules have spread out or **diffused** through the air. Diffusion is the movement of particles from a place where they are in a high concentration, to a place where they are in a lower concentration. Diffusion evens out the molecules.

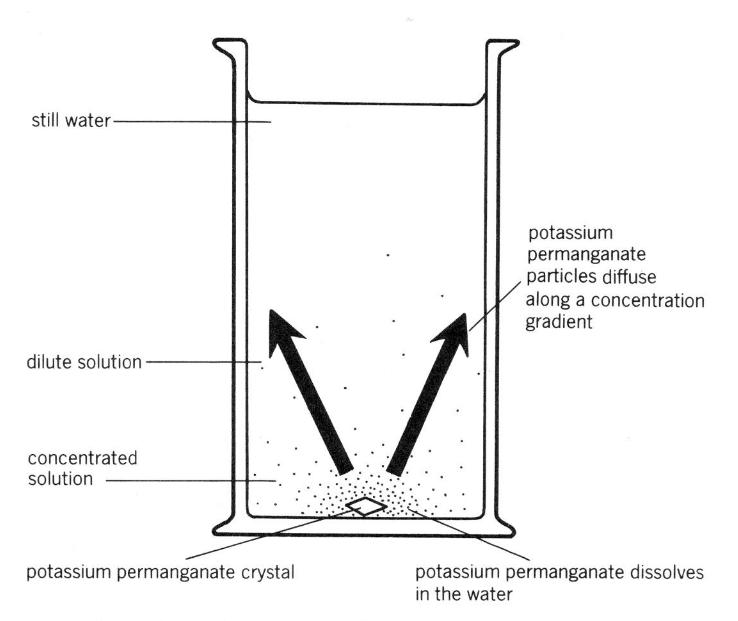

4.1 Apparatus to demonstrate diffusion

Investigation 4.1 To show diffusion in a solution

1. Fill a gas jar with water. Leave for several hours to let it become completely still.
2. Drop a small crystal of potassium permanganate into the water.
3. Make a labelled drawing of the gas jar to show how the colour is distributed.
4. Leave the gas jar completely undisturbed for several days.
5. Make a second drawing to show how the colour is distributed.
6. You can try this with other salts as well, such as copper sulphate or potassium dichromate.

Questions

1. Why was it important to leave the water to become completely still before the crystal was put in?
2. Why had the colour spread through the water at the end of your experiment?

4.2 Diffusion is important to living organisms.

Living organisms obtain many of their requirements by diffusion. They also get rid of many of their waste products in this way. For example, plants need carbon dioxide for photosynthesis. This diffuses from the air into the leaves, through the stomata. It does this because there is a lower concentration of carbon dioxide inside the leaf. This is because the cells are using it up. Outside the leaf in the air, there is a higher concentration. Carbon dioxide molecules therefore diffuse into the leaf, along this concentration gradient.

Oxygen, which is a waste product of photosynthesis, diffuses out in the same way. There is a higher concentration of oxygen inside the leaf, because it is being made there. Oxygen therefore diffuses out through the stomata into the air.

Diffusion is also important in gas exchange for respiration in animals and plants (see section 5.11). Some of the products of digestion are absorbed from the ileum of mammals by diffusion (see section 2.39).

> **Questions**
> 1 What effect does temperature have on the movement of atoms, molecules and ions?
> 2 What is diffusion?
> 3 List three examples of diffusion which occur in living organisms.

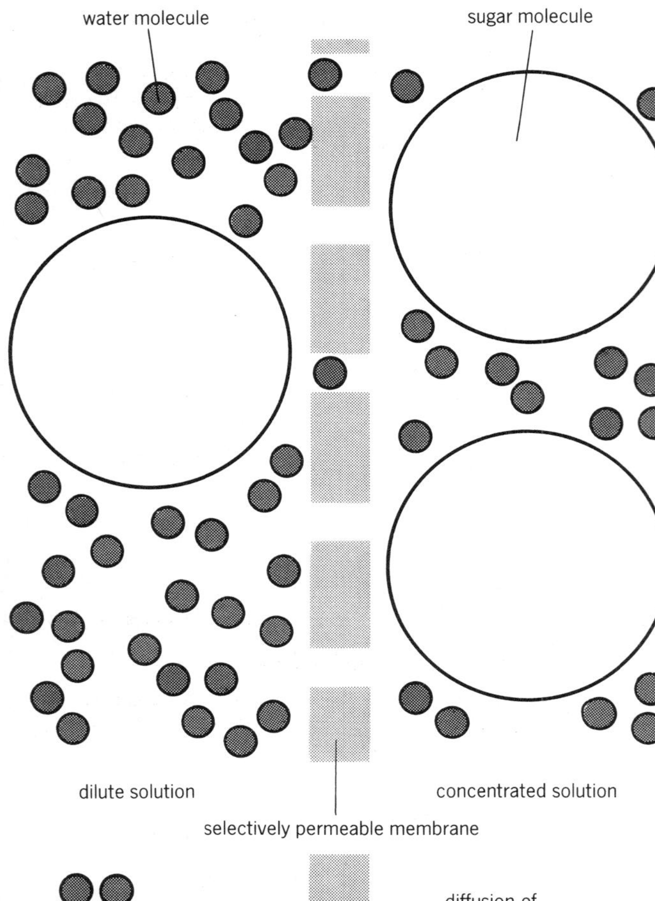

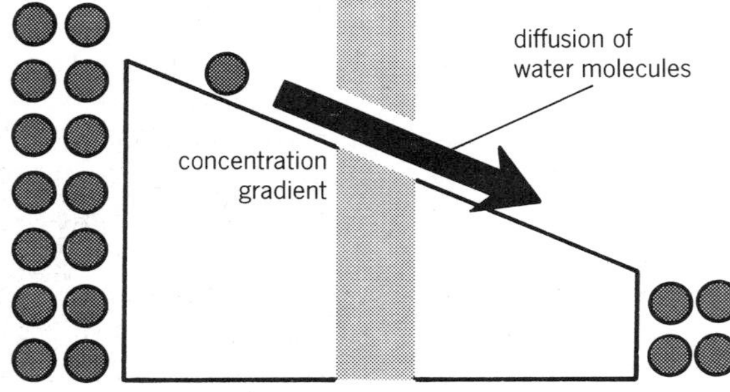

4.2 Osmosis.
There are more water molecules on the left hand side of the membrane than the right, so there is a concentration gradient for water molecules. They travel down the gradient, from left to right.
Is there a concentration gradient for sugar molecules? Why don't they travel down it?

4.3 In osmosis, water diffuses through a membrane.

Fig 4.2 illustrates a concentrated sugar solution, separated from a dilute sugar solution by a membrane. The membrane has holes or **pores** in it which are very small. An example of a membrane like this is visking tubing.

Water molecules are very small. Each one is made of two hydrogen atoms and one oxygen atom. Sugar molecules are many times larger than this. In visking tubing, the holes are big enough to let the water molecules through, but not the sugar molecules. It is called a **selectively permeable membrane** because it will let some molecules through but not others.

There is a higher concentration of sugar molecules on the right-hand side of the membrane in Fig 4.2, but lower on the left-hand side. If the membrane was not there, the sugar molecules would diffuse from the concentrated solution into the dilute one until they were evenly spread out. However, they cannot do this because the pores in the membrane are too small for them to get through.

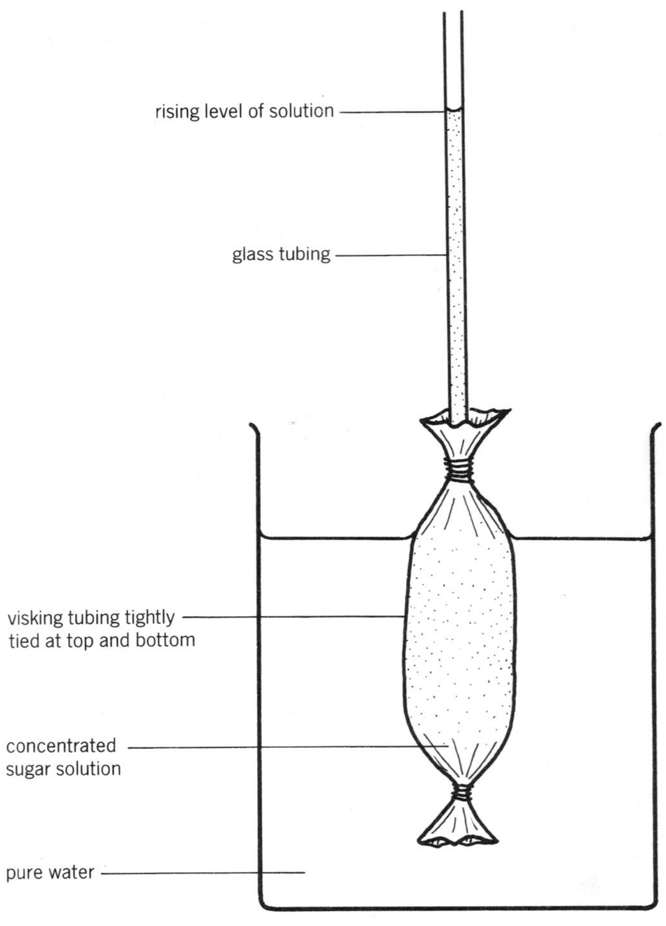

4.3 Apparatus to demonstrate osmosis. Water diffuses from the pure water into the sugar solution, along its concentration gradient. Sugar molecules cannot diffuse out, because the pores in the visking tubing are too small for them. Therefore, the level of the sugar solution rises, as it is diluted by the water diffusing into it by osmosis.

There is also a concentration gradient for the water molecules. On the left-hand side of the membrane, there is a high concentration of water molecules. On the right-hand side, the concentration of water molecules is lower because a lot of the space is taken up by sugar molecules. The water molecules therefore diffuse from the left-hand side into the right-hand side. They can do this because the pores in the membrane are large enough for them to get through.

What is the result of this? Water has diffused from the dilute solution, through the selectively permeable membrane, into the concentrated solution. The concentrated solution will become more dilute, because of the extra water molecules coming into it.

This process is called **osmosis.** Osmosis is the diffusion of water molecules from a place where they are in a higher concentration (such as a dilute sugar solution), to a place where the water molecules are in a lower concentration (such as a concentrated sugar solution) through a selectively permeable membrane.

Questions

1. Which is larger – a water molecule or a sugar molecule?
2. What is meant by a selectively permeable membrane?
3. Give an example of an artificial selectively permeable membrane.
4. How would you describe a solution which has a high concentration of water molecules?
5. What is osmosis?

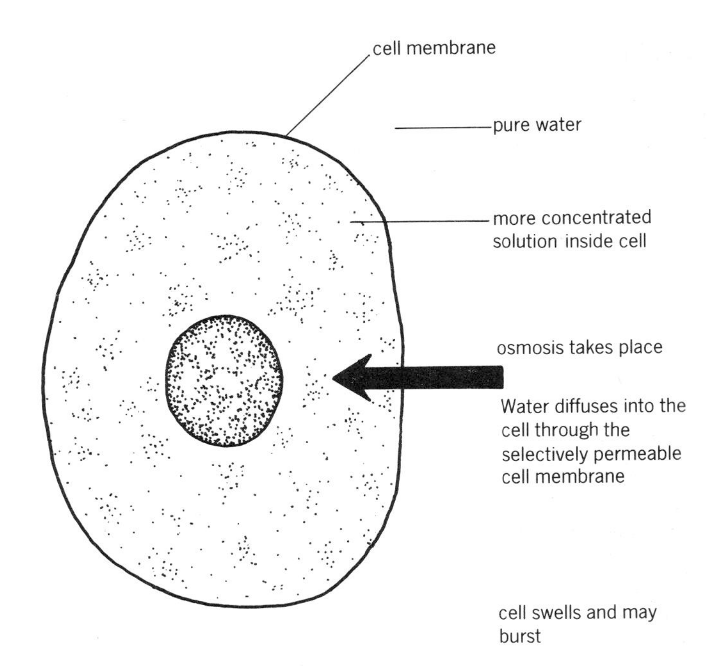

4.4 An animal cell placed in pure water

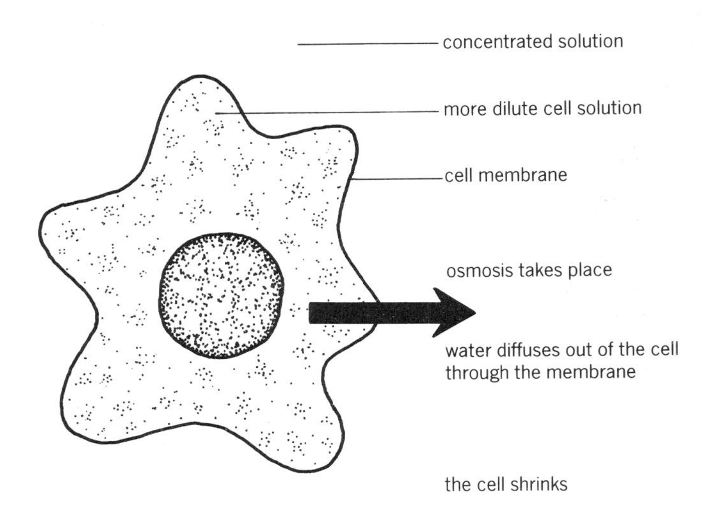

4.5 An animal cell placed in a concentrated solution

4.4 Cell membranes are selectively permeable.

Cell membranes behave very like visking tubing. They will let some substances pass through them, but not others. They are selectively permeable membranes.

There is always cytoplasm on one side of any cell membrane. Cytoplasm is a solution of proteins and other substances in water. There is usually a solution on the other side of the membrane, too. Inside large animals, cells are surrounded by tissue fluid (see section 6.22). In the soil, the roots of plants are often surrounded by a film of water. Single-celled organisms such as *Amoeba* are also surrounded by water.

So, cell membranes often separate two different solutions – the cytoplasm, and the solution around the cell. If the solutions are of different concentrations, then osmosis will occur.

4.5 Osmosis and animal cells.

Animal cells burst in pure water Fig 4.4 illustrates an animal cell in pure water. The cytoplasm inside the cell is a fairly concentrated solution. The proteins and many other substances dissolved in it are too large to get through the cell membrane. Water molecules, though, can get through.

If you compare this situation with Fig 4.2, you will see that they are similar. The dilute solution in Fig 4.2 and the pure water in Fig 4.4 are each separated from a concentrated solution by a selectively permeable membrane. In Fig 4.4 the concentrated solution is the cytoplasm and the selectively permeable membrane is the cell membrane. Therefore osmosis will occur.

Water molecules will diffuse from the dilute solution into the concentrated solution. What happens to the cell? As more and more water enters it, it swells. The cell membrane has to stretch as the cell gets bigger, until eventually the strain is too much, and the cell bursts. You can try this with blood cells, in Investigation 4.2.

Animal cells shrink in concentrated solutions Fig 4.5 illustrates an animal cell in a concentrated solution. If this solution is more concentrated than the cytoplasm, then the water molecules will diffuse out of the cell. Look at Fig 4.2 to see why.

As the water molecules go out through the cell membrane, the cytoplasm shrinks. The cell shrivels up. You can see this if you put blood cells into a concentrated solution, in Investigation 4.2.

Investigation 4.2 To find the effects of different solutions on animal cells

Always take care when using blood. If you have a cut, even a small one, cover it with a plaster before doing this experiment. Wear an overall and do not get any blood onto your skin. Wipe up any spills straight away with disinfectant.

1 Set up a microscope.
2 Take three clean microscope slides. Label them A, B and C.
3 Put a drop of distilled water onto the centre of slide A.
4 Put a drop of medium strength salt solution onto slide B.
5 Put a drop of concentrated salt solution onto slide C.
6 Put a drop of blood into the solutions on each of your three slides.
7 Carefully cover each with a coverslip. Use filter paper to remove any excess liquid.
8 Look at each of your slides under the microscope. Make a labelled drawing of a small part of each one.

Questions

1 What happened to the cells which were put into distilled water? Explain your answer.
2 What happened to the cells which were put into medium strength salt solution? Explain your answer.
3 What happened to the cells which were put into concentrated salt solution? Explain your answer.

4.6 Osmosis and plant cells.

Plant cells do not burst in pure water Fig 4.6 illustrates a plant cell in pure water. Plant cells are surrounded by a cell wall. This is fully permeable, which means it will let any molecules go through it, so osmosis will not occur across it.

Although it is not easy to see, a plant cell also has a cell membrane just like an animal cell. The cell membrane is selectively permeable. A plant cell in pure water will take in water by osmosis through its selectively permeable cell membrane in the same way as an animal cell. As the water goes in, the cytoplasm and vacuole will swell.

However, the plant cell has a very strong cell wall around it. The cell wall is much stronger than the cell membrane and it stops the plant cell from bursting. The cytoplasm presses out against the cell wall, but the cell wall resists and presses back on the contents.

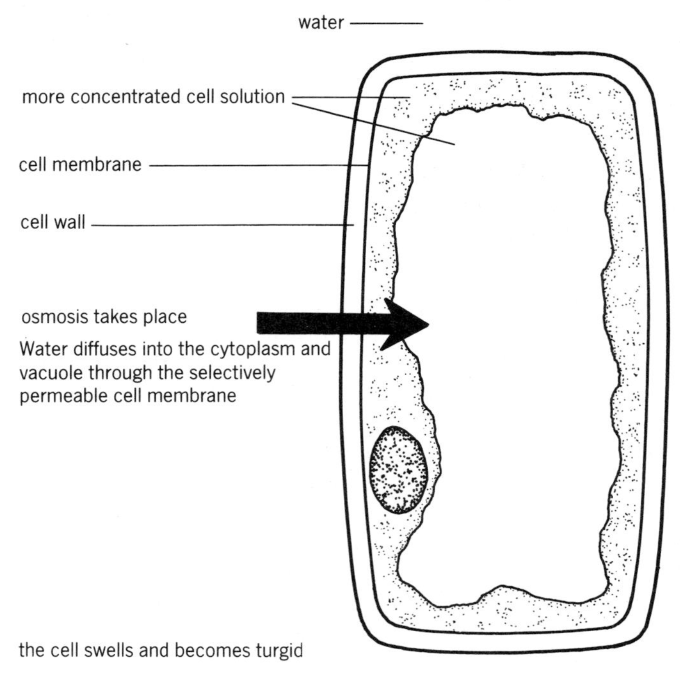

4.6 A plant cell placed in pure water

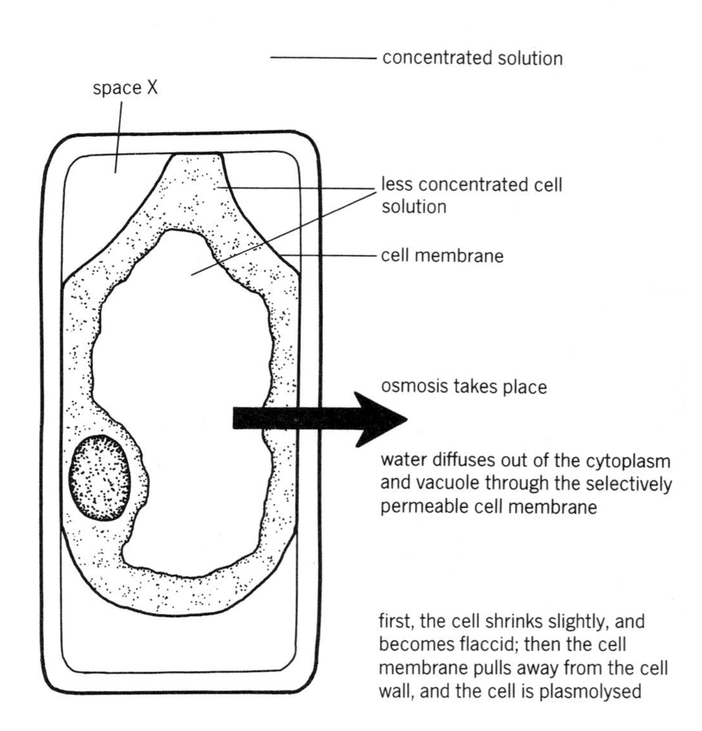

4.7 A plant cell in a concentrated solution

A plant cell in this state is rather like a blown-up tyre – tight and firm. It is said to be **turgid.** The turgidity of its cells helps a plant that has no wood in it to stay upright, and keeps the leaves firm. Plant cells are usually turgid.

Plant cells plasmolyse in concentrated solutions Fig 4.7 illustrates a plant cell in a concentrated solution. Like the animal cell in Fig 4.5, it will lose water by osmosis. The cytoplasm and the vacuole will shrink. As the cytoplasm shrinks, it stops pushing outwards on the cell wall. Like a tyre when some of the air has leaked out, the cell becomes floppy. It is said to be **flaccid.** If the cells in a plant become flaccid, then the plant loses its firmness and begins to wilt.

If the solution is very concentrated, then a lot of water will diffuse out of the cell. The cytoplasm and vacuole go on shrinking. The cell wall, though, is too stiff to be able to shrink much. As the cytoplasm shrinks further and further into the centre of the cell, the cell wall gets left behind. The cell membrane, surrounding the cytoplasm, tears away from the cell wall (see Fig 4.7).

A cell like this is said to be **plasmolysed.** This does not normally happen because plant cells are not usually surrounded by very strong solutions. However, you can make cells become plasmolysed if you do Investigation 4.3. Plasmolysis usually kills a plant cell because the cell membrane is damaged as it tears away from the cell wall.

Questions

1 What happens to an animal cell in pure water?
2 Explain why this does not happen to a plant cell in pure water.
3 Which part of a plant cell is (a) fully permeable and (b) selectively permeable?
4 What is meant by a turgid cell?
5 What is plasmolysis?
6 How can plasmolysis be brought about?
7 In Fig 4.7, what fills space X? Explain your answer.

Fact!

Pumpkin–growers can make their prize specimens even larger by injecting a strong sugar solution into the pumpkin. The increase in size can be dramatic. Can you explain why?

Investigation 4.3 To find the effects of different solutions on plant cells

1 Set up a microscope.
2 Take three clean microscope slides. Label them A, B and C.
3 Put a drop of distilled water onto the centre of slide A.
4 Put a drop of medium strength sugar solution onto slide B.
5 Put a drop of concentrated sugar solution onto slide C.
6 Peel off a very thin layer of the red epidermis from a rhubarb petiole (leaf stalk). To get good results, it should be as thin as possible (only one cell thick).
7 Cut three squares of this epidermis, each with sides about 5 mm long.
8 Put one square into the drop of solution on each of your three slides.
9 Carefully cover each one with a coverslip. Clean excess liquid from your slides with filter paper.
10 Look at each of your slides under the microscope. Make a labelled drawing of a few cells from each one.

Questions

1 Which part of the cell is coloured red?
2 What has happened to the cells in pure water? Explain your answer.
3 What has happened to the cells in medium strength sugar solution? Explain your answer.
4 What has happened to the cells in concentrated sugar solution? Explain your answer.

Investigation 4.4 To demonstrate osmosis using eggs

Beneath the shell of a hen's egg is a thin, selectively permeable membrane. If you can dissolve away the shell, you will expose this membrane.

1 Take two fresh hen's eggs. Put each into a beaker and cover them with dilute hydrochloric acid. Leave them for several hours, until all the shell has dissolved.
2 Take the eggs out of the acid, and wash them gently.
3 Label two clean beakers A and B. Put an egg into each.
4 Cover egg A with distilled water. Cover egg B with concentrated salt solution. Leave the eggs in these solutions for a few days.
5 Make a labelled drawing of each egg in its solution.
6 Remove the eggs from their solutions, and examine them. Note down any differences between them.

Questions

1 What happened to egg A? Explain your answer.
2 What happened to egg B? Explain your answer.
3 Hen's eggs are not usually immersed in salt solution or water. What substances do you think normally diffuse through the membrane around the egg?

Investigation 4.5 To demonstrate osmosis using potatoes

1 Label four petri dishes A, B, C and D.
2 Cook a small potato, with its peel still on, in boiling water for five minutes.
3 Meanwhile, cut a raw potato in half. Peel part of each half (see Fig 4.8) and cut out a hollow in the centre. Do *not* cut right through it.

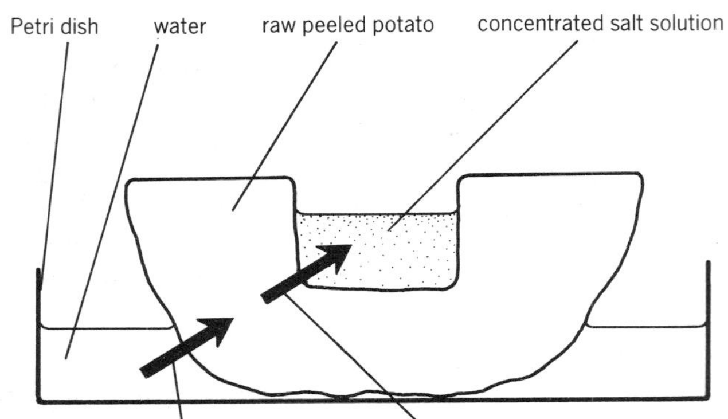

4.8 Experiment to demonstrate osmosis using potatoes

4 Put one half into dish A and the other into dish B.
5 Do the same with the boiled potato. Put one half into dish C and the other into dish D.
6 Pour distilled water into dishes A and C and into the hollows in potatoes B and D.
7 Pour concentrated salt solution into dishes B and D and into the hollows in potatoes A and C.
8 Make labelled drawings of all your potatoes and dishes. Leave them for a day or so.
9 Examine your potatoes and make labelled drawings of your results.

Questions

1 What has happened in potato A? Explain your answer.
2 What has happened in potato B? Explain your answer.
3 When potatoes C and D were boiled, the cell membranes were destroyed. Explain your results for these two potatoes.
4 Why is it very important not to cut right through the potatoes when you are making the hollows?

Chapter revision questions

1 Which of these is an example of (a) diffusion, (b) osmosis, or (c) neither? Explain your answer in each case.
(i) Water moves from a dilute solution in the soil into the cells in a plant's roots.
(ii) Saliva flows out of the salivary glands into your mouth.
(iii) A spot of blue ink dropped into a glass of still water quickly colours all the water blue.
(iv) Carbon dioxide goes into a plant's leaves when it is photosynthesising.

2 An experiment is set up as shown in Fig 4.9. Starch molecules are too big to go through visking tubing. Water molecules and iodine molecules can go through.

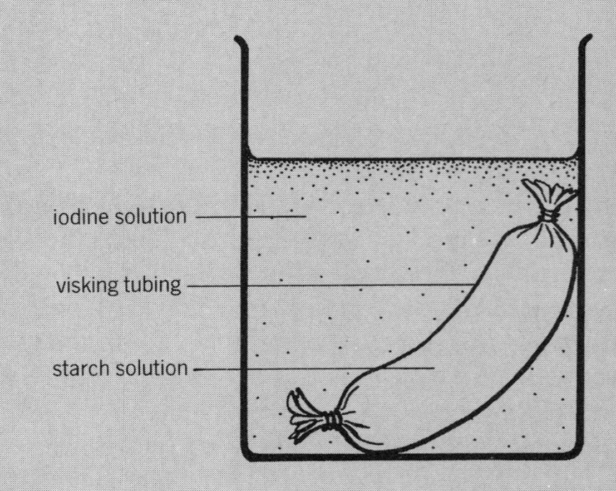

4.9

(a) In which direction will the iodine molecules diffuse? Explain your answer.
(b) In which direction will the water molecules diffuse? Explain your answer.
(c) Draw and label a diagram to show what the apparatus would look like after an hour. Label the colours of the two solutions.

5 Respiration

5.1 Respiration releases energy from food.

Every cell in every living organism needs energy. Cells get their energy from food. The energy is released from the food by a process called **respiration.**

5.2 Sugars release energy when oxidised.

A food which is often used for obtaining energy is sugar. Energy can be released from sugar, by combining it with oxygen. This is called **oxidation.** Fig 5.1 illustrates apparatus in which sugar and oxygen can be made to react together. Oxygen is fed into the space around the sugar. A small electric current is passed into the sugar. When it gets hot enough, the sugar suddenly begins to combine with the oxygen. As it does so, energy is released from it. The energy is heat energy. The sugar burns and the water around it heats up.

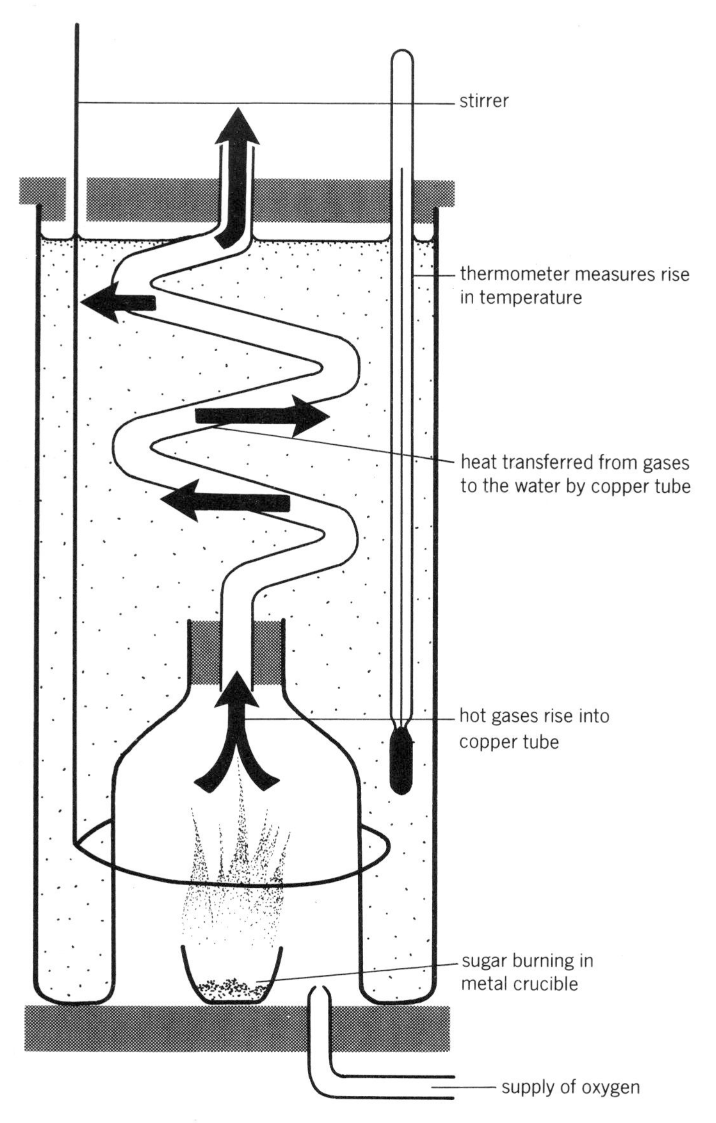

5.1 Apparatus for oxidising sugar

5.3 Respiration oxidises sugars in stages.

Obviously, this does not happen in living cells. If sugar was oxidised like this, the cells would get so hot that they would be killed. In the apparatus in Fig 5.1, the sugar is oxidised quickly in one violent reaction. In a cell, during respiration, the sugar is oxidised very gradually in a series of small, controlled reactions. The reactions are controlled by enzymes. The result, though, is similar. Sugar is combined with oxygen, releasing energy. Carbon dioxide and water are formed.

5.4 The respiration equation.

The process of respiration can be summarised like this.

sugar + oxygen → carbon dioxide + water + energy

The sugar which is normally used is glucose, $C_6H_{12}O_6$. The balanced equation for respiration is this.

$$C_6H_{12}O_6 + 6O_2 \rightarrow 6CO_2 + 6H_2O + \text{energy}$$

The carbon dioxide and water are by-products. The reaction produces energy for the cell.

5.5 ATP stores the energy released from food.

Respiration is going on in all living cells all the time. What happens to the energy which is released?

In the cell are molecules of a substance called **ADP.** ADP stands for adenosine diphosphate. The energy which is released by respiration is used to join a phosphate group onto an ADP molecule. The molecule which is made is called **ATP** (adenosine triphosphate, see Fig 5.4).

ADP + phosphate + energy → ATP

ADP has a low energy content, but ATP has a high energy content. ATP stores energy for short periods. When the cell needs energy, the ATP can be broken down again, releasing its energy.

ATP → ADP + phosphate + energy

Investigation 5.1 To show that carbon dioxide is produced in respiration

1 Set up the apparatus as in Fig 5.2.
2 Note the colour of the bicarbonate indicator solution or lime water in each flask.
3 Turn on the pump, to draw air through the apparatus. Leave it running until one of the solutions has changed colour.

Bicarbonate indicator solution changes from red to yellow when carbon dioxide is bubbled through it. Lime water changes from clear to milky white.

Questions

1 Potassium hydroxide solution absorbs carbon dioxide. Why was the air bubbled through this solution before it reached the mouse?
2 Why was the air bubbled through lime water or bicarbonate indicator solution before reaching the mouse?
3 Which solution changed colour? What does this show?

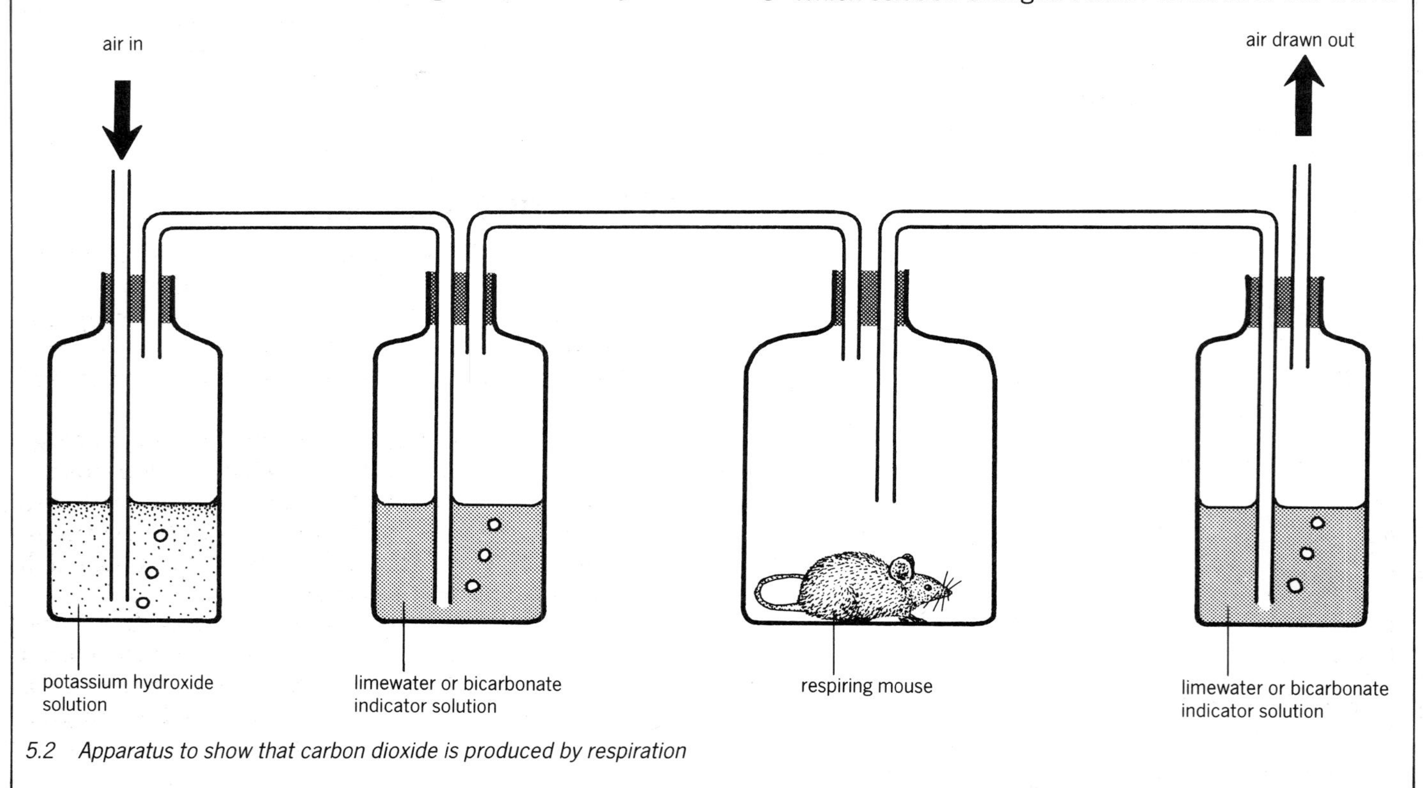

5.2 Apparatus to show that carbon dioxide is produced by respiration

Investigation 5.2 To show the uptake of oxygen during respiration

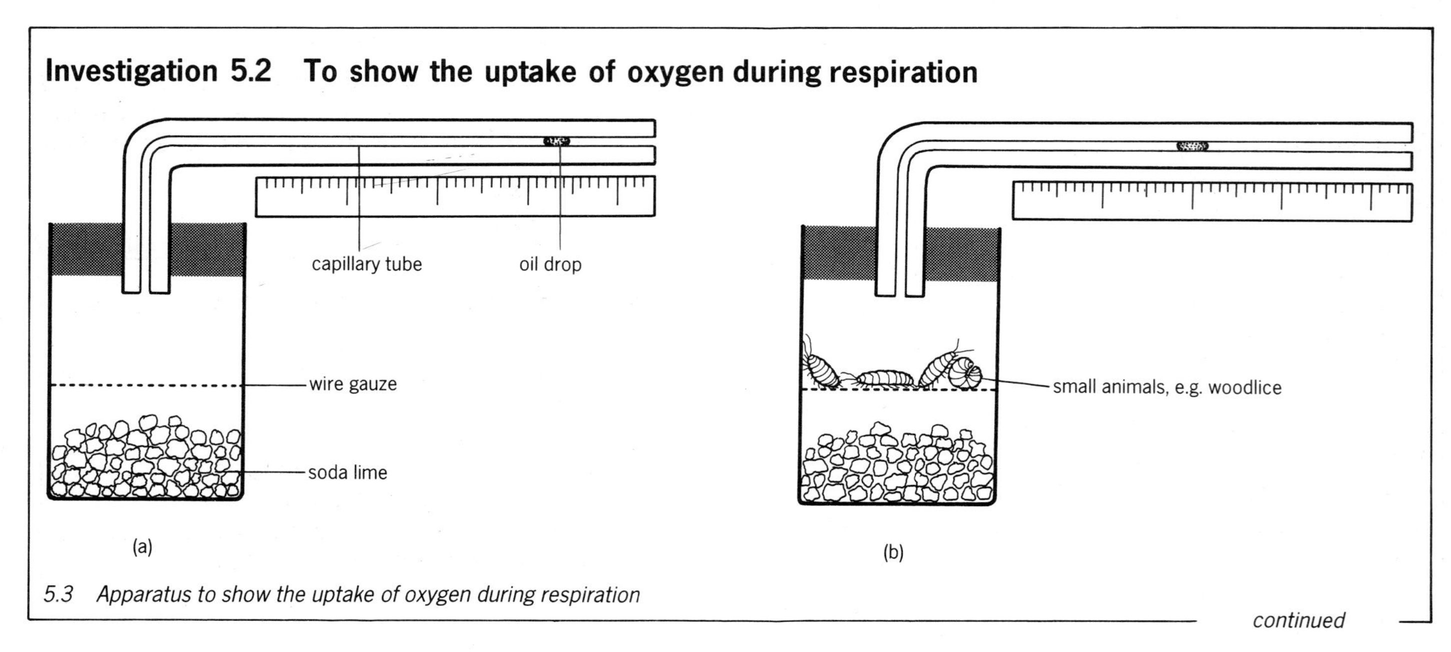

5.3 Apparatus to show the uptake of oxygen during respiration

continued

1 Copy out the results table, ready to fill it in.
2 Set up both pieces of apparatus shown in Fig 5.3(a) and (b). Make sure that the connections between the capillary tube, bung and chamber are airtight. Vaseline will help to seal them.
3 Dip the end of the capillary tube of each apparatus into oil, so that a drop is introduced into it.
4 Watch the movement of the oil drop in apparatus (a). This should move quite quickly at first, as the soda-lime absorbs any carbon dioxide already in the apparatus.
5 When the oil drop in apparatus (a) slows down or stops, set your stop clock to time 0, and record the position of the oil drop in *both* pieces of apparatus.
6 At suitable time intervals, note both time and distance travelled by both oil drops. Record these results in your results table.
7 Plot a graph of the distance in (b) minus distance in (a), against time. Put time on the bottom axis.

Results table

Time in minutes	0	1	2	3	4	etc.
Distance travelled by oil drop in (b) in cm						
Distance travelled by oil drop in (a) in cm						
Distance (b)-(a)						

Questions

1 What gas is absorbed by soda lime?
2 Why does the oil drop in apparatus (a) move quickly at first?
3 Why does the oil drop in (a) slow down or stop?
4 Why does the oil drop carry on moving in apparatus (b), when the oil drop stops in (a)?
5 Why does the CO_2 breathed out by the woodlice not cause the oil drop in (b) to move to the right?
6 What is the purpose of apparatus (a)?
7 What might cause the oil drop to move in apparatus (a) after its first rapid movement?
8 How could another similar apparatus be used to work out the amount of CO_2 breathed out by the woodlice?

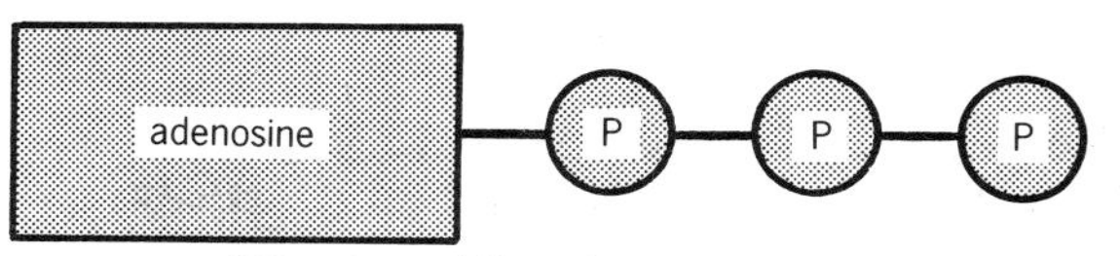

ATP molecule *When the end phosphate group is removed, energy is released.*

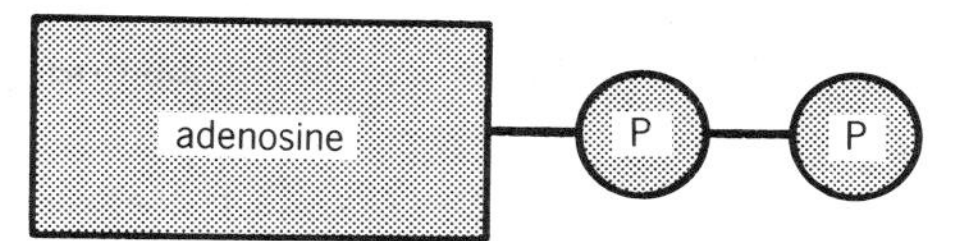

ADP molecule *This can be made into an ATP molecule by adding another phosphate group.*

5.4 ATP and ADP

5.6 The advantages of ATP.

Why have ATP? Why not just use the energy as it is released, when sugar is oxidised in respiration?

There are three very important reasons. Firstly, the cell might not need the energy straight away. By storing it in ATP, the cell can release energy when and where it needs it.

Secondly, the cell might not want very much energy at once. The amount of energy released by oxidising one glucose molecule is quite large. It is large enough to make many ATP molecules. By storing the energy in small packages (ATP molecules), the cell can use small quantities as required. This avoids waste.

Thirdly, ATP stores the energy in a form that is useful to the cell. The chemical energy in ATP can be used to drive many different chemical reactions which would otherwise not be able to take place fast enough. This process is very efficient and can be carefully controlled by the cell.

5.7 Respiration produces heat.

The energy released by respiration is not all stored as ATP. Some of it does escape as heat. In fact, many animals use the heat from respiration to keep their bodies warm (see section 9.5).

Investigation 5.3 To show that heat is produced in respiration

1 Soak some peas in water for a day, so that they begin to germinate.
2 Boil a second set of peas, to kill them.
3 Wash both sets of peas in dilute disinfectant, so that any bacteria and fungi on them are killed.
4 Put each set of peas into a Thermos flask as shown in Fig 5.5. Do not fill the flasks completely.

continued

5 Note the temperature of each flask.
6 Support each flask upside down, and leave them for a few days.
7 Note the temperature of each flask at the end of your experiment.

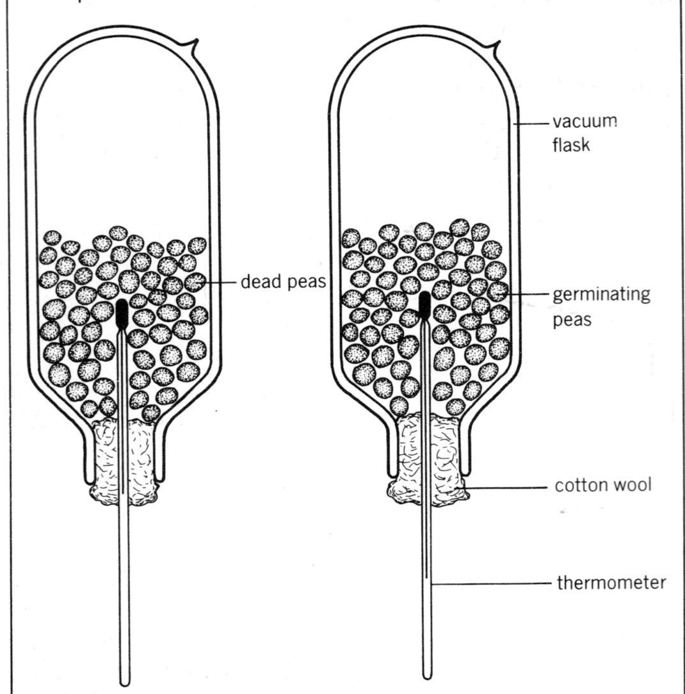

5.5 *Apparatus to show that heat is produced during respiration*

Questions

1 Which flask showed the higher temperature at the end of the experiment? Explain your answer.
2 Why is it important to kill any bacteria and fungi on the peas?
3 Why should the flasks not be completely filled with peas?
4 Carbon dioxide is a heavy gas. Why were the flasks left upside down, with porous cotton wool plugs in them?
5 Not all of the energy produced by the respiring peas will be given off as heat. What happens to the rest of it?

Fact! A resting human uses about 40 kg of ATP in one day. During strenuous exercise, a person may use as much as 0.5 kg of ATP a minute.

5.8 ATP is made inside mitochondria.

Respiration happens inside every living cell, whether plant or animal. Inside these cells are **mitochondria.** Mitochondria are called the 'power-houses' of a cell, because it is there that ATP is made. The more energy a cell needs, the more mitochondria it has. Muscle cells, for example, contain large numbers of mitochondria.

Questions

1 In respiration, sugar is oxidised. What does this mean?
2 What is special about the way that oxidation happens inside cells?
3 What is the purpose of respiration?
4 What is the energy released in respiration used for?
5 In which part of a cell does respiration take place?

5.9 Respiration sometimes occurs without oxygen.

The process described so far in this chapter releases energy from sugar by combining it with oxygen. It is called **aerobic respiration,** because it uses air (which contains oxygen).

It is possible, though, to release energy from sugar without using oxygen. It is not such an efficient process and not much energy is released, but the process is used by some organisms. It is called **anaerobic respiration** ('an' means without).

Yeast, a single-celled fungus, can respire anaerobically. It breaks down sugar to alcohol.

$$\text{sugar} \rightarrow \text{alcohol} + \text{carbon dioxide} + \text{energy}$$

$$C_6H_{12}O_6 \rightarrow 2C_2H_5OH + 2CO_2 + \text{energy}$$

As in aerobic respiration, carbon dioxide is made and the energy is used to make ATP.

Some of the cells in your body, particularly muscle cells, can also respire anaerobically for a short time. They make lactic acid instead of alcohol. This is described in section 5.25.

Table 5.1 A comparison of aerobic and anaerobic respiration

Similarities		
1	Energy released by breakdown of sugar	
2	ATP made	
3	Some energy lost as heat	
Differences		
	aerobic respiration	*anaerobic respiration*
1	Uses oxygen gas	Does not use oxygen gas
2	No alcohol or lactic acid made	Alcohol or lactic acid made
3	Large amount of energy released	Small amount of energy released
4	CO_2 always made	CO_2 sometimes made

5.10 Yeast is used for baking and brewing.

Breaking down sugar to alcohol and carbon dioxide, which is the way in which yeast respires anaerobically, is called **fermentation.** Fermentation is used to make drinks such as beer and wine.

Investigation 5.4 To show that carbon dioxide is produced when yeast respires anaerobically

1 Boil some water, to drive off any dissolved air.
2 Dissolve a small amount of sugar in the boiled water, and allow it to cool.
3 When it is cool, add yeast and stir with a glass rod.
4 Set up the apparatus as in Fig 5.6. Add the liquid paraffin by trickling it gently down the side of the tube, using a pipette.
5 Set up an identical piece of apparatus, but use boiled yeast instead of living yeast.
6 Leave your apparatus in a warm place.
7 Observe what happens to the bicarbonate indicator solution after half an hour.

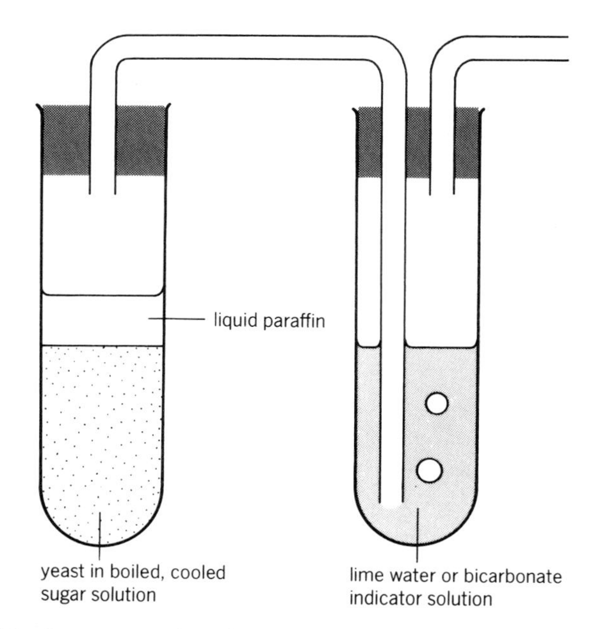

5.6 Apparatus to show that carbon dioxide is produced when yeast respires anaerobically

Questions

1 Why is it important to boil the water?
2 Why must the sugar solution be cooled before adding the yeast?
3 What is the liquid paraffin for?
4 What happened to the lime water or bicarbonate indicator solution in each of your pieces of apparatus? What does this show?
5 What new substance would you expect to find in the sugar solution containing living yeast at the end of the experiment?

Brewing and wine making To make beer, yeast is dissolved in a warm liquid containing the sugar maltose. The maltose comes from germinating barley seeds. The yeast respires, breaking down the maltose and making alcohol and carbon dioxide. The carbon dioxide makes the beer fizzy.

Wine is made in a similar way, but the sugar comes from grapes.

Bread making When making bread, flour is mixed with water to make a dough. Flour contains starch and some of this breaks down to the sugar maltose when the flour is moistened. Yeast is added to the dough and breaks down the sugar as it respires.

There is air in the dough, so the yeast respires aerobically at first, until the oxygen is used up. It makes carbon dioxide, and bubbles of this gas get caught in the dough, making it rise. The yeast is killed when the bread is cooked.

Questions

1 What is anaerobic respiration?
2 Name one organism which can respire anaerobically.
3 What is fermentation?
4 Why does bread not taste of alcohol after baking?

Gas exchange

5.11 Gas exchange occurs at respiratory surfaces.

If you look back at the respiration equation in section 5.4, you will see that two substances are needed. They are glucose and oxygen. The way in which cells obtain glucose is described in Chapters 2 and 3. Animals get sugar from carbohydrates which they eat. Plants make their's, by photosynthesis.

Oxygen is obtained in a different way. Animals and plants get their oxygen directly from their surroundings. The part of the organism through which oxygen enters the body is called the **respiratory surface.**

If you look again at the respiration equation you can see that carbon dioxide is made. This is a waste product and it must be removed from the organism. It leaves across the respiratory surface.

So, at a respiratory surface, oxygen comes in and carbon dioxide goes out. This is called **gas exchange.**

5.12 *Amoeba* exchanges gases through its membrane.

Amoeba is a single-celled organism which lives in water. Its respiratory surface is its **cell membrane** (see Fig 5.7).

Oxygen dissolves in water, so there are usually oxygen molecules in the water around the *Amoeba.* Inside the cell, however, oxygen is being used up in respiration. Therefore, there is a higher concentration of oxygen molecules outside the cell, but a lower concentration inside. So oxygen diffuses into the cell, across the cell membrane, down a concentration gradient. Carbon dioxide diffuses in the opposite direction.

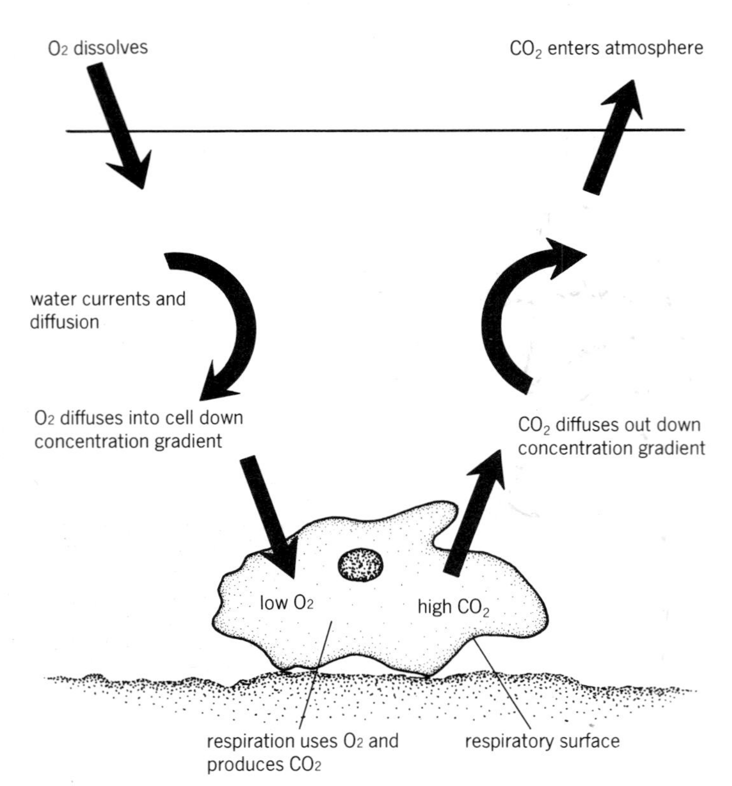

5.7 Gas exchange in Amoeba

5.13 Large organisms need transport systems.

Amoeba is a very small organism made of only one cell. Oxygen can quickly diffuse into the centre of the cell because it does not have far to go.

This would not work, though, for a large organism like a human. It would take too long for oxygen to diffuse from the air to every cell in your body. Some sort of transport system is needed to get the oxygen all around the body as quickly as possible. In humans this is the blood system.

5.14 Surface area to volume ratio.

Large organisms have another problem with gas exchange. The larger their volume, the more oxygen they need. This oxygen has to get through their respiratory surface. The amount that can get into their body depends on the area of this surface.

So, the need for oxygen increases with an organism's **volume,** but the supply of oxygen increases with its **respiratory surface area.** This is a problem because, as an organism gets larger, its surface area and volume increase by different amounts.

Imagine a cube shaped organism (see Fig 5.8), where the outer surface is the respiratory surface. A good way to compare the surface area to the volume is to divide the surface area by the volume. This is called the surface area to volume ratio. For the small cube, this is 6. For the medium sized cube it is 0.6, and for the large cube only 0.06. So, the larger an organism is, the smaller its surface area to volume ratio is.

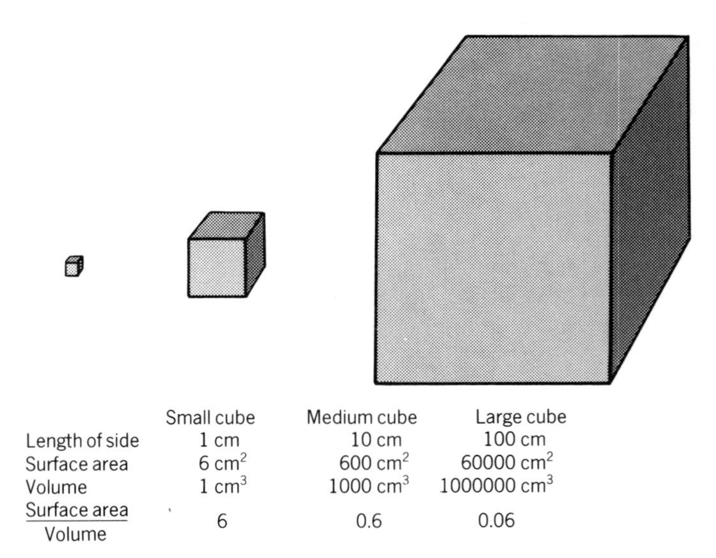

	Small cube	Medium cube	Large cube
Length of side	1 cm	10 cm	100 cm
Surface area	6 cm^2	600 cm^2	60000 cm^2
Volume	1 cm^3	1000 cm^3	1000000 cm^3
Surface area / Volume	6	0.6	0.06

5.8 Surface area to volume ratio

A small organism, which has a large surface area compared to its volume, can use its body surface for gas exchange. But a large organism cannot use its ordinary body surface alone, because it is not large enough. Large organisms need special respiratory surfaces, which provide the large areas required for gas exchange. Specialised parts of their surface are highly divided or folded to provide this extra area for gas exchange (see Fig 5.9). Sometimes, these surfaces are tucked inside the body for protection. This also helps to prevent too much water evaporating from organisms which live on land.

5.15 Respiratory surfaces are thin and moist.

Before it can diffuse across a respiratory surface, oxygen has to be dissolved in water. If an organism is surrounded by water, this is no problem. A fish's gills, for example, are always surrounded by water in which the oxygen is dissolved.

On land, though, the respiratory surface must be kept moist by the organism itself. Your lungs, for example, contain cells which make a liquid to keep the respiratory surface wet (see Fig 5.14).

Table 5.2 Properties of respiratory surfaces of large organisms

1. They should be **thin** to allow gases to diffuse across them quickly.
2. They should be close to an efficient **transport system** to take gases to and from the cells which need them.
3. They should be kept **moist** as oxygen will not diffuse across them unless it dissolves in water.
4. They should have a **large surface area** so that a lot of oxygen can diffuse across at the same time.
5. They should have a good **supply of oxygen.**

A microscopic organism like an *Amoeba* has a large surface area in comparison to its volume. The cell membrane has a large enough area to supply all the oxygen it needs.

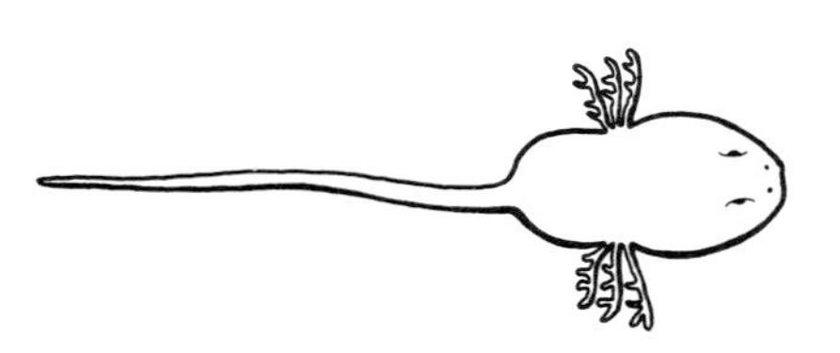

The body surface of a larger organism, such as a tadpole, is not big enough to supply all its oxygen. It increases its respiratory surface area by means of gills.

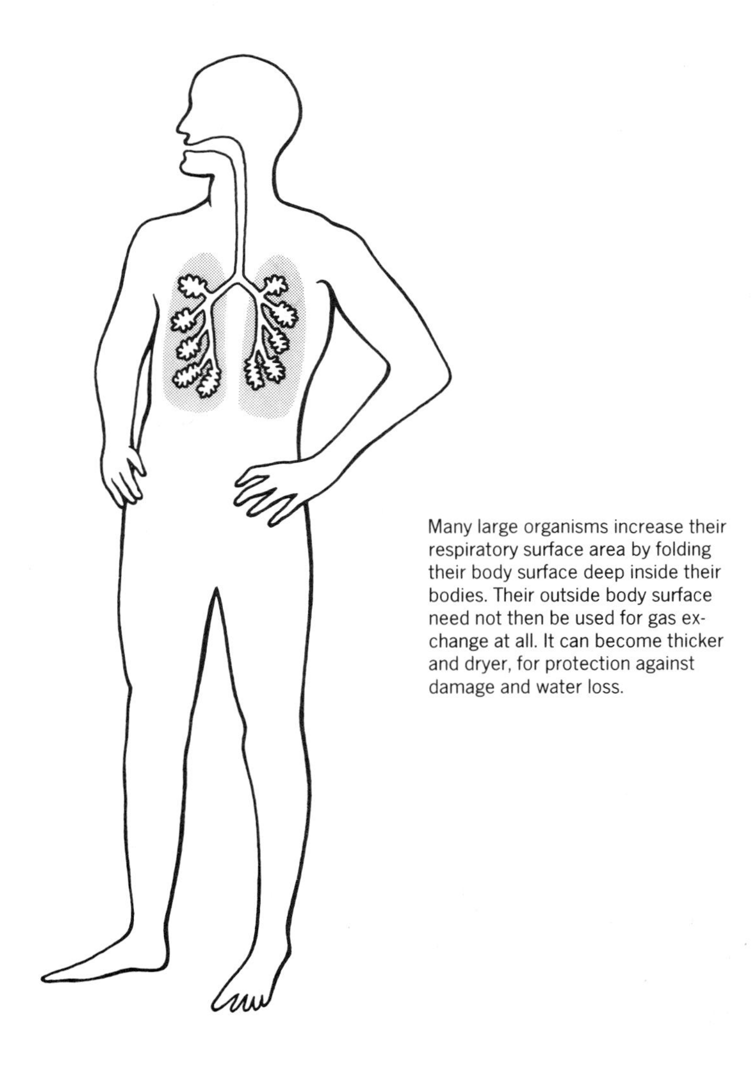

Many large organisms increase their respiratory surface area by folding their body surface deep inside their bodies. Their outside body surface need not then be used for gas exchange at all. It can become thicker and dryer, for protection against damage and water loss.

5.9 Respiratory surfaces

A respiratory surface should also be as thin as possible. The thinner it is, the more quickly oxygen can diffuse across it. This means that respiratory surfaces are often very delicate.

5.16 Earthworms exchange gases through the skin.

An earthworm's respiratory surface is its skin. The skin is kept moist in three ways. Firstly worms usually live in damp places. Secondly tiny pores in the skin allow fluid from inside the worm, called **coelomic fluid,** to leak out onto the skin. Thirdly, mucous glands in the worm's skin secrete mucus onto the surface which helps prevent drying out. If the worm dries out, then it will not be able to absorb oxygen, and will die.

The skin is quite thin and has a good blood supply. The blood transports oxygen and carbon dioxide to and from all the cells in the earthworm's body. The blood contains **haemoglobin** (see section 6.19) which helps to carry oxygen.

Questions

1 What is a respiratory surface?
2 What is the respiratory surface of an *Amoeba*?
3 How does gas exchange take place across an *Amoeba*'s respiratory surface?
4 Explain, as briefly as you can, why large organisms need special respiratory surfaces.
5 Why are the respiratory surfaces of most terrestrial (land living) animals inside their bodies?
6 Why must respiratory surfaces be kept moist?
7 Explain how an earthworm's respiratory surface is kept moist.

Fact!

A resting human breathes out about 500 litres of carbon dioxide every 24 hours.

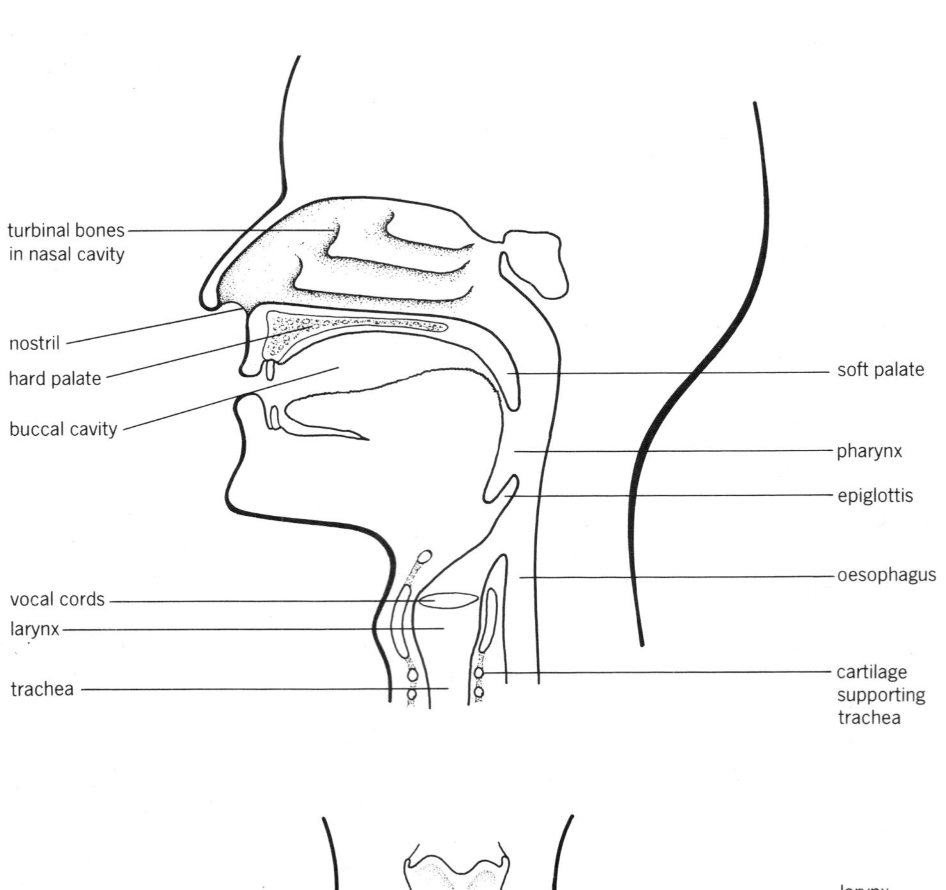

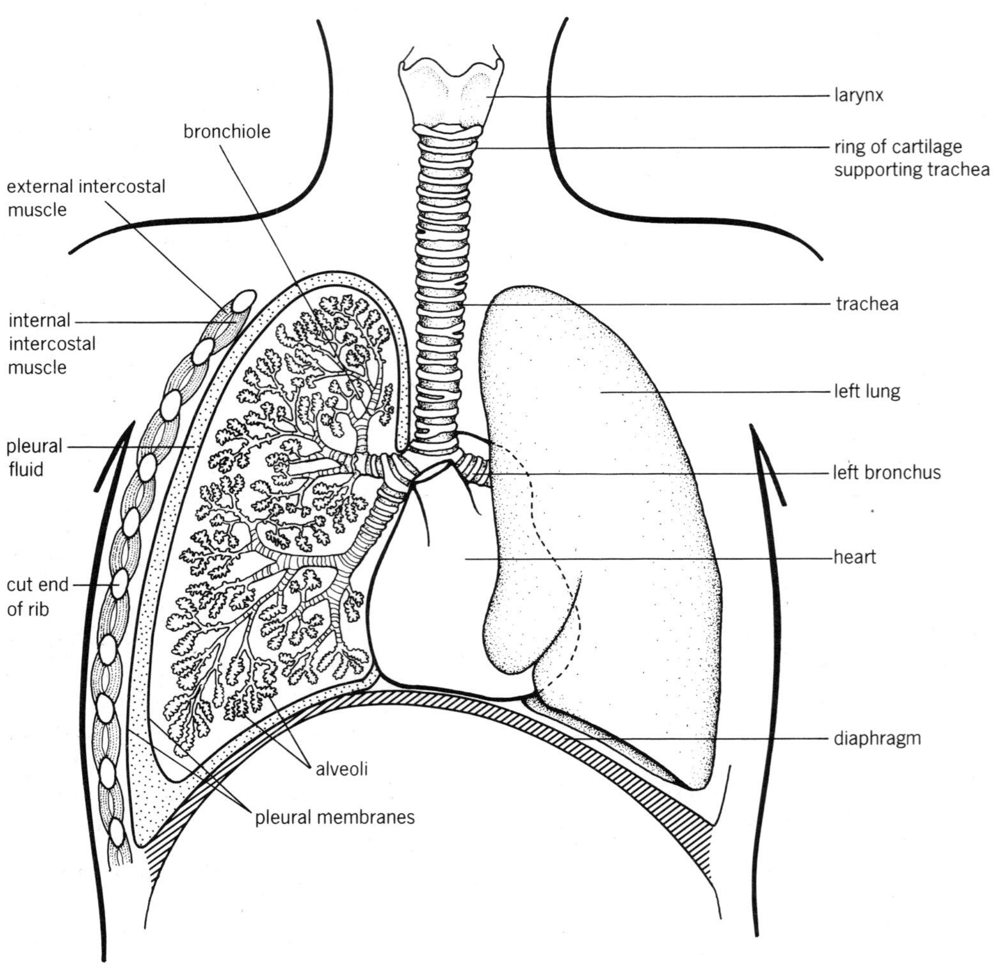

5.10 The human respiratory system

Gas exchange in humans

5.17 The structure of the respiratory system.

Fig 5.10 shows the structures which are involved in gas exchange in a human. The most important are the two lungs. Each lung is filled with many tiny air spaces called air sacs or **alveoli.** It is here that oxygen diffuses into the blood. Because they are so full of spaces, lungs feel very light and spongy to touch. The lungs are supplied with air through the windpipe or **trachea.**

5.18 The path taken by air to the lungs.

The nose and mouth Air can enter the body either through the nose or mouth. The nose and mouth are separated by the **palate** (see Fig 5.10), so you can breathe through your nose even when you are eating.

It is better to breathe through your nose, because the structure of the nose allows the air to become warm, moist and filtered before it gets to the lungs. Inside the nose are some thin bones called **turbinal bones** which are covered with a thin layer of cells. Some of these cells make a liquid containing water and mucus which evaporates into the air in the nose and moistens it (see Fig 5.11).

Other cells have very tiny hair like projections called cilia. The cilia are always moving and bacteria or particles of dust get trapped in them and in the mucus. Cilia are found all along the trachea and bronchi, too. They waft the mucus, containing bacteria and dust, up to the back of the throat, so that it does not block up the lungs. Cilia in the nose also waft mucus into the oesophagus where it is swallowed.

The trachea The air then passes into the windpipe or **trachea.** At the top of the trachea is a piece of cartilage called the **epiglottis.** This closes the trachea and stops food going down the trachea when you swallow. This is a reflex action which happens automatically when a bolus of food touches the soft palate.

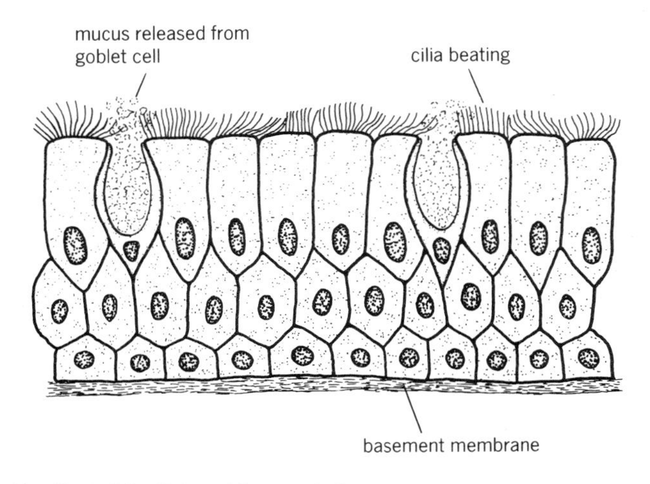

5.11 Part of the lining of the respiratory passages

Investigation 5.5 Examining lungs

1 Examine some sheep lungs obtained from a butcher's shop or abattoir.

Questions

1 What colour are the lungs? Why are they this colour?
2 Push them gently with your finger. What do they feel like? Why do they feel like this?
3 What is covering the surface of the lungs? What is its name, and why is it there?
4 Find the two tubes leading down to the lungs. Which one is the oesophagus? Follow it along, and notice that it goes right past the lungs. Where is it going to?
5 The other tube is the trachea. What does it feel like? Why does it feel like this?
6 What is the name of the wide part at the top of the trachea? What is its function?
7 If the lungs have not been badly cut, take a long glass tube (such as a burette tube) and push it down through the trachea. Hold the trachea tightly against it, and blow down it. What happens?

Just below the epiglottis is the voice box or **larynx.** This contains the vocal cords. The vocal cords can be tightened by muscles so that they make sounds when air passes over them. The trachea has rings of cartilage around it which keep it open.

The bronchi The trachea goes down through the neck and into the **thorax.** The thorax is the upper part of your body from the neck down to the bottom of the ribs and diaphragm. In the thorax the trachea divides into two. The two branches are called the right and left **bronchi.** One bronchus goes to each lung and then branches out into many smaller tubes called **bronchioles.**

The alveoli At the end of each bronchiole are many tiny air sacs or **alveoli** (see Fig 5.12). This is where gas exchange takes place.

5.19 Alveolar walls form the respiratory surface.

The walls of the alveoli are the respiratory surface. Tiny blood vessels, called **capillaries,** are closely wrapped around the the outside of the alveoli (see Fig 5.12). Oxygen diffuses across the walls of the alveoli into the blood (see Fig 5.14). Carbon dioxide diffuses the other way.

The walls of the alveoli have several features which make them an efficient respiratory surface.

They are very thin They are only one cell thick. The capillary walls are also only one cell thick. An oxygen molecule only has to diffuse across this small thickness to get into the blood.

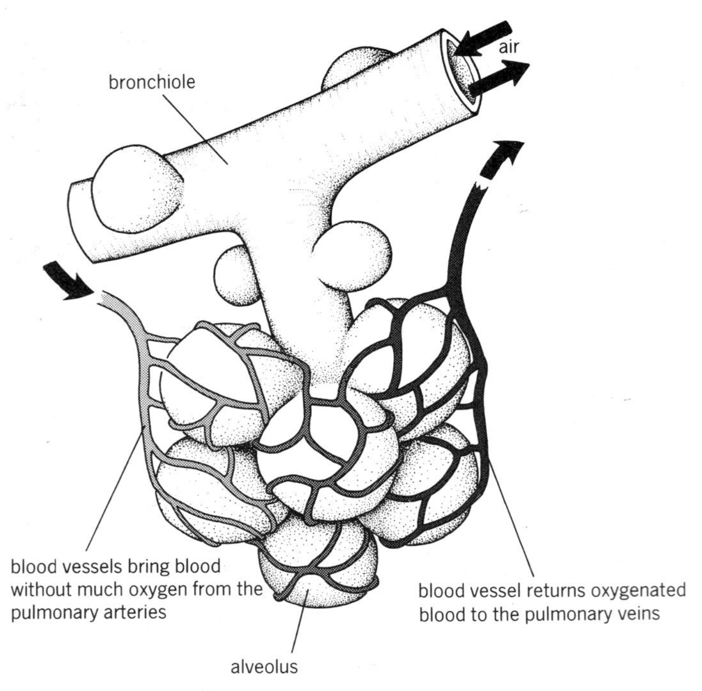

5.12 Alveoli

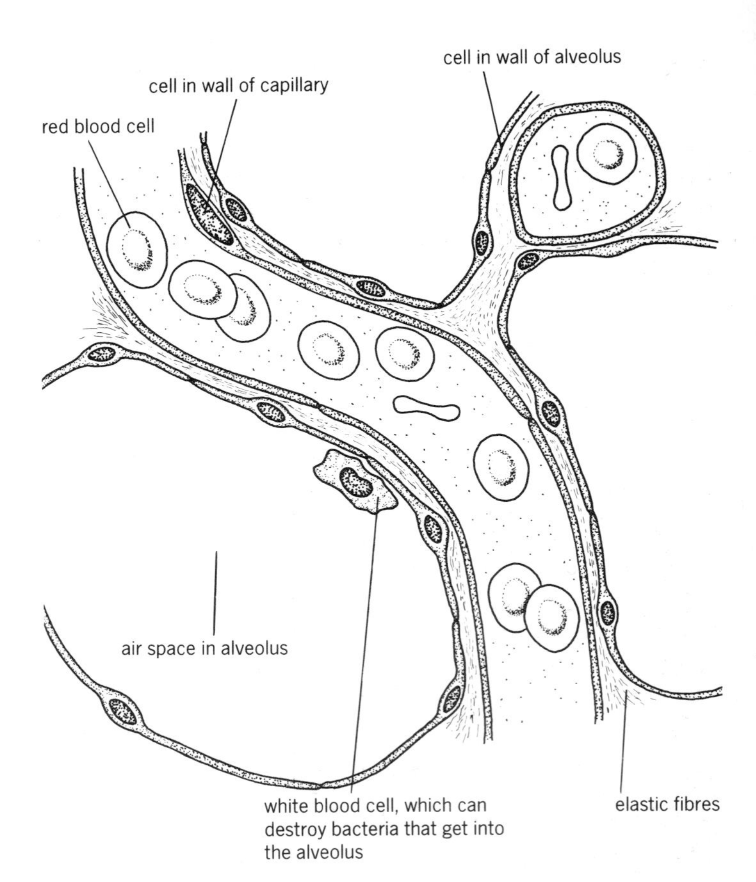

5.13 Section through part of a lung, magnified

They have an excellent transport system Blood is constantly pumped to the lungs along the pulmonary artery. This branches into thousands of capillaries which take blood to all parts of the lungs. Carbon dioxide in the blood can diffuse out into the air spaces in the alveoli and oxygen can diffuse into the blood. The blood is then taken back to the heart in the pulmonary vein, ready to be pumped to the rest of the body.

The way in which the blood carries oxygen and carbon dioxide is explained in section 6.19.

They are moist Special cells in the alveoli secrete a watery liquid. Oxygen dissolves in this liquid before diffusing across the wall of the alveolus.

They have a large surface area In fact, the surface area is enormous! The total surface area of all the alveoli in your lungs is over 100 m^2.

They have a good supply of oxygen Your breathing movements keep your lungs well supplied with oxygen.

Questions

1. Why is it better to breathe through your nose than through your mouth?
2. What is the function of the cilia in the respiratory passages?
3. What is the larynx?
4. What is the respiratory surface of a human?
5. How many cells does an oxygen molecule have to pass through, to get from an alveolus into the blood?

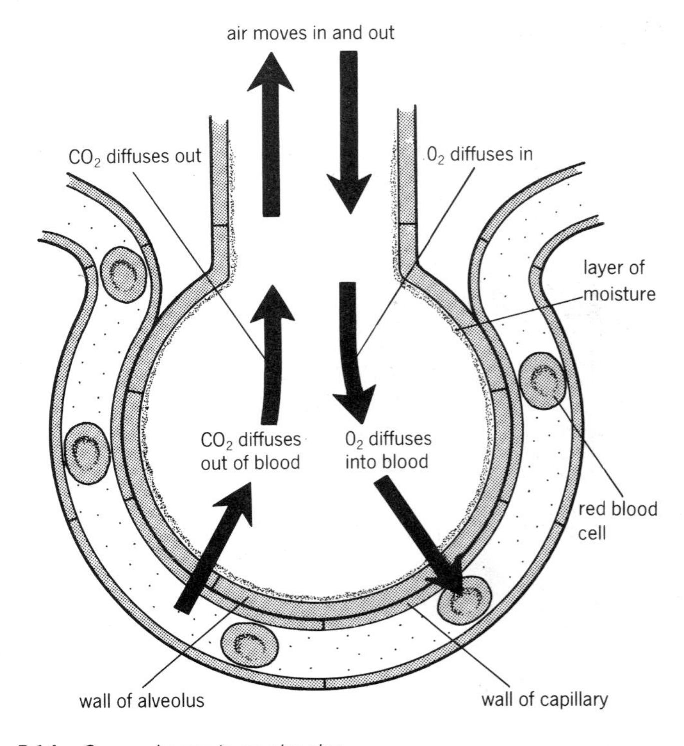

5.14 Gas exchange in an alveolus

5.20 Ribs and diaphragm move during breathing.

To make air move in and out of the lungs, you must keep changing the volume of your thorax. First, you make it large so that air is sucked in. Then you make it smaller again so that air is squeezed out. This is called breathing.

There are two sets of muscles which help you to breathe. One set is in between the ribs. This set is called the **intercostal muscles** (see Fig 5.15). The other set is in the **diaphragm.** The diaphragm is a large sheet of muscle and elastic tissue which stretches across your body, underneath the lungs and heart.

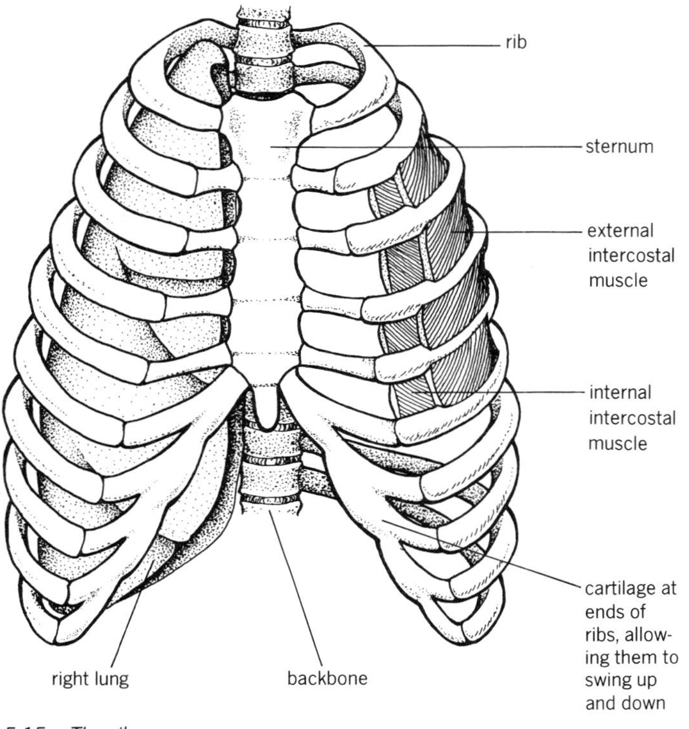

5.15 The rib cage

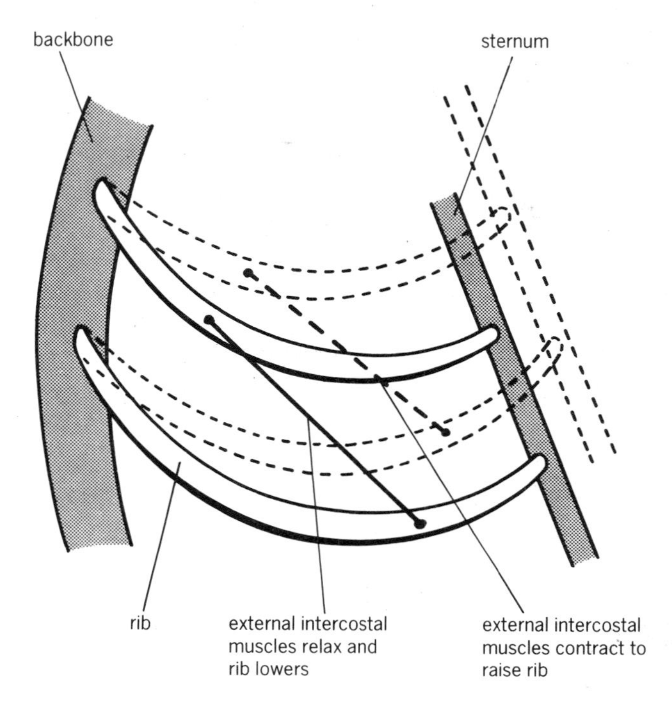

5.17 The external intercostal muscles raise and lower the ribs

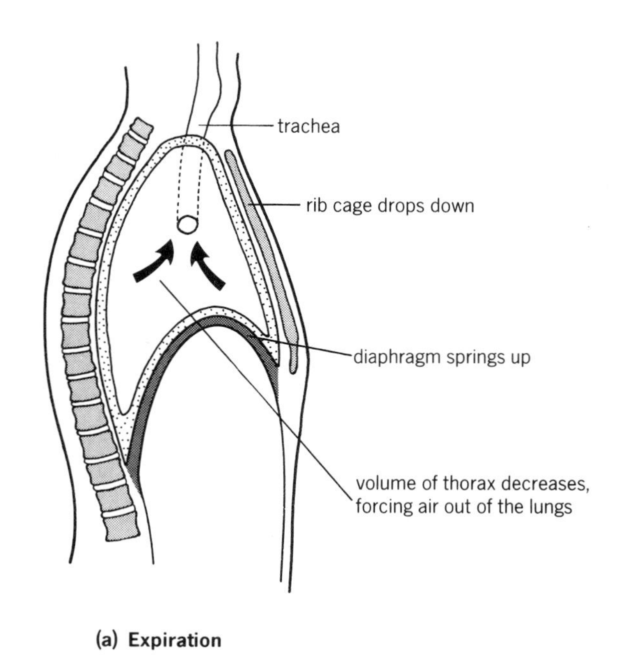

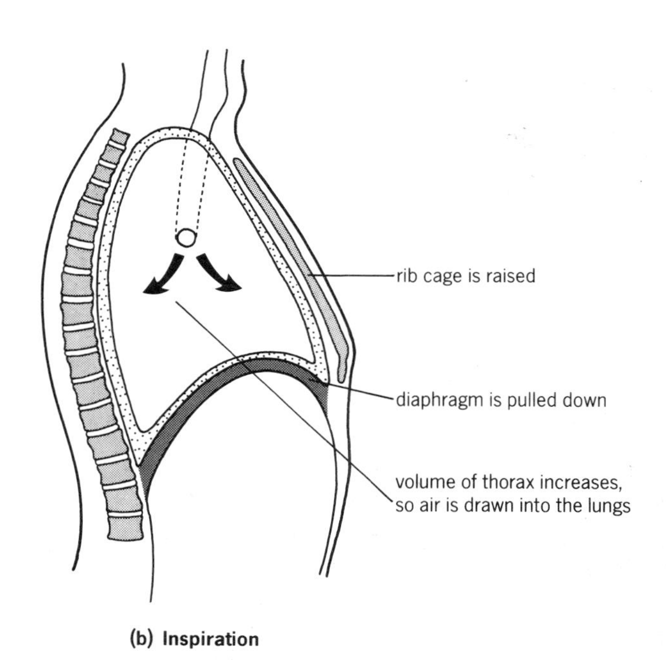

5.16 How the thorax changes shape during breathing

5.21 Breathing in is called inspiration.

When breathing in the muscles of the diaphragm contract. This pulls the diaphragm downwards, which increases the volume in the thorax (see Fig 5.16(b)). At the same time, the external intercostal muscles contract. This pulls the rib cage upwards and outwards (see Fig 5.17). Together, these movements increase the volume of the thorax.

As the volume of the thorax increases, the pressure inside it falls below atmospheric pressure. Extra space has been made and something must come in to fill it up. Air therefore rushes in along the trachea and bronchi into the lungs.

5.22 Breathing out is called expiration.

When breathing out the muscles of the diaphragm relax. The diaphragm springs back up into its domed shape because it is made of elastic tissue. This decreases the volume in the thorax. The external intercostal muscles also relax. The rib cage drops down again into its normal position. This also decreases the volume of the thorax (see Fig 5.16(a)).

As the volume of the thorax decreases, the pressure inside it increases. Air is squeezed out through the trachea into the nose and mouth, and on out of the body.

Investigation 5.6 Using a model to show the action of the diaphragm

1 Use the apparatus in Fig 5.18. Begin with the plunger in as far as it will go. Put your thumb tightly over the hole, and pull the plunger outwards.
2 Repeat this, but this time do not cover the hole with your thumb.

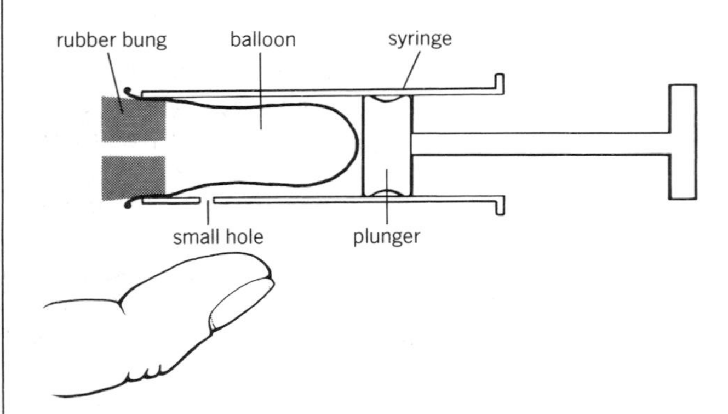

5.18 A model to show the action of the diaphragm

Questions

1 What does the balloon represent?
2 What does the plunger represent?
3 What happens to the balloon when you cover the hole and pull the plunger? Explain why.
4 What happens to the balloon when you pull the plunger without covering the hole? Explain why.
5 In what ways is this a misleading demonstration of the mechanism of breathing?

Investigation 5.7 Comparing the carbon dioxide content of inspired and expired air

You can use either lime water or bicarbonate indicator solution for this experiment. Lime water changes from clear to cloudy when carbon dioxide dissolves in it. Bicarbonate indicator solution changes from red to yellow when carbon dioxide dissolves in it.

1 Set up the apparatus as in Fig 5.19.
2 Breathe in and out gently through the rubber tubing. Do not breathe too hard. Keep doing this until the liquid in one of the flasks changes colour.

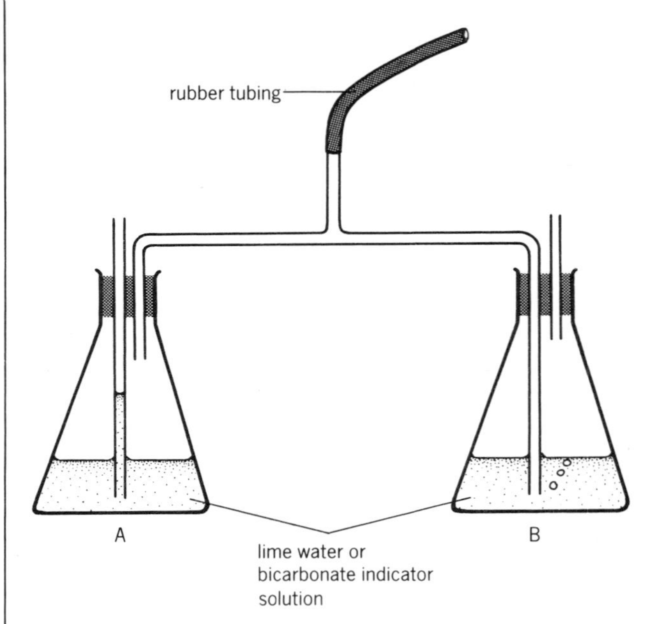

5.19 Comparing the carbon dioxide content of inspired and expired air

Questions

1 In which flask did bubbles appear when you breathed out? Explain why.
2 In which flask did bubbles appear when you breathed in? Explain why.
3 What happened to the liquid in flask A?
4 What happened to the liquid in flask B?
5 What do your results tell you about the amount of carbon dioxide in inspired air and expired air?

5.23 Internal intercostal muscles can force air out.

Usually, you breathe out by relaxing the external intercostal muscles and the muscles of the diaphragm, as explained in section 5.22. Sometimes, though, you breathe out more forcefully – when coughing, for example. Then the internal intercostal muscles contract strongly, making the rib cage drop down even further. The muscles of the abdomen wall also contract, helping to squeeze extra air out of the thorax.

Table 5.3 Comparison of inspired and expired air. Inspired air is the air you breathe in. Expired air is the air you breathe out.

	Inspired air	*Expired air*	*Reason for difference*
Oxygen	21%	16%	Oxygen is absorbed across respiratory surface, then used by cells in respiration
Carbon dioxide	0.03%	4%	Carbon dioxide is made by cells as a waste product of respiration, and is released across the respiratory surface
Argon and other noble gases	1%	1%	
Nitrogen	78%	78%	Nitrogen gas is not used by cells
Water content (humidity)	variable	always higher	Respiratory surface must be kept moist. Some of this moisture evaporates and is lost as air is breathed out
Temperature	variable	always higher	Air is warmed as it passes through the respiratory passages

5.24 Pleural membranes help with breathing.

Each lung is covered with a thin, smooth membrane. Another similar membrane lines the inside of the rib cage. These are called the **pleural membranes** (see Fig 5.10).

The pleural membranes make a liquid called pleural fluid. This fills the space between the two membranes. As the lungs inflate and deflate, the pleural fluid helps to lubricate them so that they do not rub against the rib cage too much. Also, as the rib cage expands, the pleural fluid ensures that the lungs adhere closely to the moving ribs.

Sometimes, in a car accident for example, the thorax becomes punctured so that air gets in between the pleural membranes. If this happens, then the lungs will not work and they collapse.

The pleural membranes also help to keep the two lungs separate from one another. If one lung is punctured in an accident so that it collapses, the other pleural cavity will still be airtight, so the other lung can work normally.

Table 5.4 The differences between respiration, gas exchange and breathing

Respiration is chemical reactions which happen in all living cells, in which food is broken down to release energy, usually by combining it with oxygen.

Gas exchange is the exchange of gases across a respiratory surface. For example, oxygen is taken into the body, and carbon dioxide is removed from it. Gas exchange also takes place during photosynthesis and respiration in plants.

Breathing is muscular movements which keep the respiratory surface supplied with oxygen.

Questions

1. What is breathing?
2. Which muscles help in breathing?
3. Where are the pleural membranes?
4. Give two functions of pleural fluid.

5.25 Exercise can create an oxygen debt.

All the cells in your body need oxygen for respiration and all of this oxygen is supplied by the lungs. The oxygen is carried by the blood to every part of the body.

Sometimes, cells may need a lot of oxygen very quickly. Imagine you are running in a race. The muscles in your legs are using up a lot of energy. To produce this energy, the mitochondria in the muscles will be combining oxygen with glucose as fast as they can, to make ATP which will provide the energy for the muscles.

A lot of oxygen is needed to work as hard as this. You breathe deeper and faster to get more oxygen into your blood. Your heart beats faster to get the oxygen to the leg muscles as quickly as possible. Eventually a limit is reached. The heart and lungs cannot supply oxygen to the muscles any faster. But more energy is still needed for the race. How can that extra energy be found?

Extra energy can be produced by anaerobic respiration. Some glucose is broken down without combining it with oxygen.

glucose → lactic acid + energy

As explained in section 5.9, this does not release very much energy, but a little extra might make all the difference.

When you stop running, you will have quite a lot of lactic acid in your muscles and your blood. This lactic acid must be broken down by combining it with oxygen. So, even though you do not need the energy any more, you go on breathing hard. You are taking in extra oxygen to break down the lactic acid.

While you were running, you built up an **oxygen debt.** You 'borrowed' some extra energy, without 'paying' for it with oxygen. Now, as the lactic acid is combined with oxygen, you are paying off the debt. Not until all the lactic acid has been used up, does your breathing rate and rate of heart beat return to normal.

5.20 Sebastian Coe pays back his oxygen debt after a race

Investigation 5.8 Investigating how breathing rate changes with exercise

1. Copy the table, ready to fill in your results.
2. Sit quietly for two minutes, to make sure you are completely relaxed.
3. Count how many breaths you take in one minute. Record it in your table.
4. Wait one minute, then count breaths again, and record.
5. Now do some vigorous exercise, such as stepping up and down onto a chair, for exactly two minutes. At the end of this time, sit down. Immediately count your breaths in the next minute, and record.
6. Continue to record your breaths per minute every other minute, until they have returned to near the level before you started to exercise.
7. Draw a graph of your results, putting time on the bottom (x) axis.

Questions

1. Why does your breathing rate rise so quickly during exercise?
2. Why did your breathing rate not go back to normal as soon as you finished exercising?

continued

3. Work out how many minutes it took your breathing rate to return to normal after exercise. Collect everyone's results for this, and work out the class average.

   Now compare individual results with the average one. Do fit people who regularly play a lot of sport, come above or below this average result? Try to explain this?

Results table

Time	Number of breaths per minute
1st minute	
3rd minute	
6th minute	
8th minute	
10th minute	

Gas exchange in fish

5.26 Fish absorb oxygen through gills.

Fish use oxygen which is dissolved in water, just as *Amoebas* do. Their respiratory surface is the surface of their **gills.**

Fig 5.21 shows the structure of a fish's gills. A fish has several gills. These have spaces in between them called **gill pouches.** The gill pouches open to the outside. In bony fish, the opening is covered by a piece of skin and bone called the **operculum.**

Each gill is supported by a piece of bone called a **gill bar.** On the outer surface of the gill bar are many thin, soft flaps of tissue. These are the **gill lamellae** and it is here where gas exchange takes place. Because they are so finely divided, they have a very large surface area. They are thin and have a good blood supply.

On the other side of each gill bar are the **gill rakers.** These trap particles of dirt and stop them from clogging up the lamellae. Some fish, such as the herring, use their gill rakers for filter feeding (see section 2.51).

5.27 Some fish make breathing movements.

Oxygen is brought to the gills as water flows over them. Water flows in through the mouth, over the gills and out through the gill slits.

Some fish, such as the herring, swim fast with their mouths open. This makes water pass over their gills.

Others, such as trout, lie in fast running water, facing upstream. Fish which cannot swim fast, or which live in still water, have to make breathing movements. These are explained in Fig 5.22.

Investigation 5.9 Investigating the structure of gills

1 Examine a dead fish. Find the operculum. Using a seeker, lift it up gently. Notice the gills lying underneath. What colour are they? Why are they this colour?
2 Gently push a seeker into the fish's mouth, and out under the operculum. What normally travels along this path?
3 Using scissors, cut the operculum neatly where it joins the body. Using forceps and scissors, cut out *one* gill, as close to its ends as possible.
4 Put the gill into a small dish containing water. Make a drawing of it. Label gill bar, gill lamellae and gill rakers. What is the function of each of these?
5 Take the gill out of the water, and lay it on a dry tile. What difference is there between the position of the gill lamellae now, and when they were floating in the water? Can you think of one reason why a fish cannot breathe out of the water?

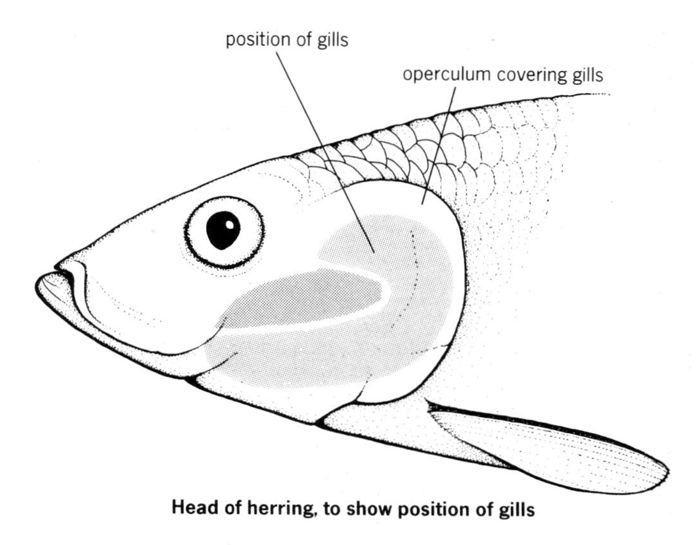

Head of herring, to show position of gills

A single gill

5.21 A herring's gills

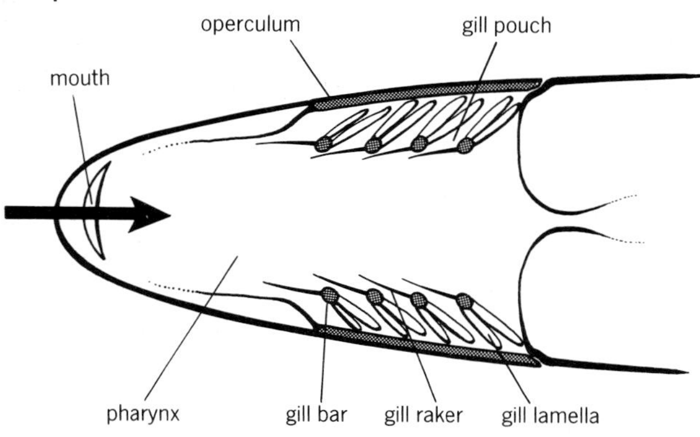

In inspiration the mouth opens. The floor of the pharynx is pulled down, increasing the space inside it. Water therefore flows into the mouth and pharynx.

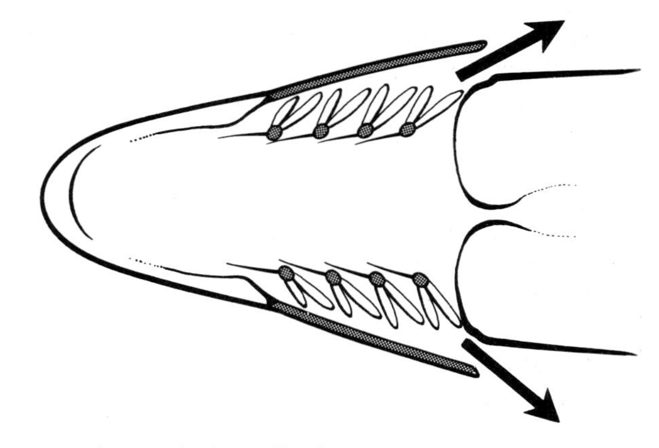

On expiration the mouth closes. The floor of the pharynx goes up, squeezing the water out past the gills.

5.22 Breathing movements in a fish

Gas exchange in insects

5.28 Insects have a tracheal system for gas exchange.

Although the lungs of a mammal and the gills of a fish appear at first sight to be very different from one another, they enable quite similar methods of gas exchange to take place. In both, the blood collects oxygen at the respiratory surface and then transports it to the rest of the body.

Insects, however, have a completely different arrangement. The blood is not involved at all. Instead, oxygen is supplied directly to every part of the body. The insect's body contains a network of tubes called **tracheae** (see Fig 5.23). Tracheae open to the outside by means of **spiracles.** These can be opened and closed (see Fig 5.25). Air, containing oxygen, diffuses through the spiracles into the tracheae.

Insects which fly often have **air sacs.** As they fly, the movement of the flight muscles makes the air sacs inflate and deflate. This speeds up the movement of air through the tracheal system.

Like the trachea of a mammal, an insect's tracheae have a strong support around them to keep them open. In a mammal, the trachea is supported by rings of cartilage. In an insect, the tracheae are supported by a spiral of **chitin.** Chitin is the tough material which also forms the hard outer covering, or **exoskeleton,** of the insect.

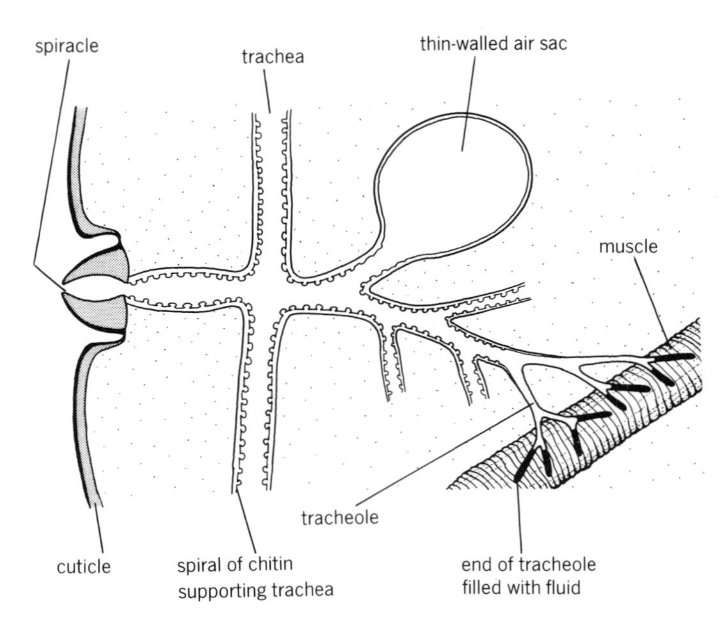

5.24 How oxygen is supplied to a muscle in an insect

5.29 Tracheoles form the respiratory surface.

As the tracheae get deeper into the insect's body, they divide. Eventually, the branches get very small and are called **tracheoles** (see Fig 5.24). The tracheoles go into the muscles and other tissues of the insect.

The tracheoles do not have chitin in their walls. Instead, they have a thin membrane. This is the insect's respiratory surface. The ends of the tracheoles are filled with fluid in which oxygen dissolves. The oxygen then diffuses across the membrane into the tissues of the insect and carbon dioxide diffuses out.

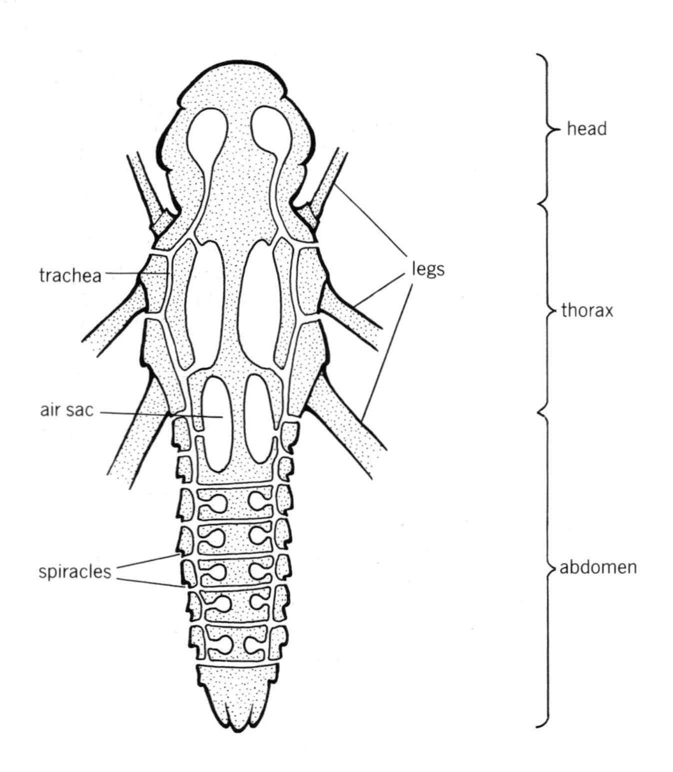

5.23 The tracheal system of an insect, looking down from above

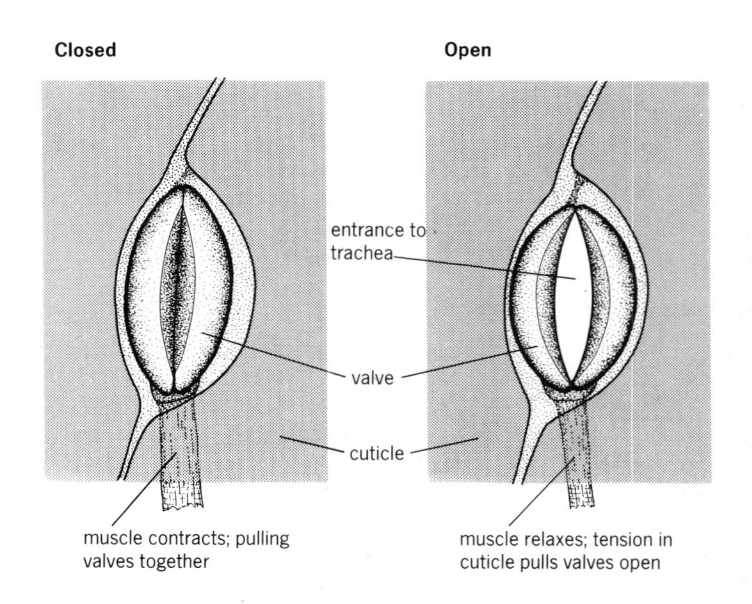

5.25 How an insect's spiracles open and close

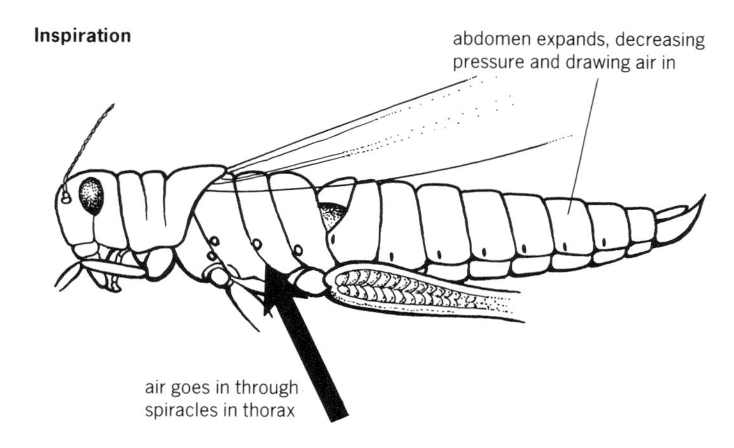

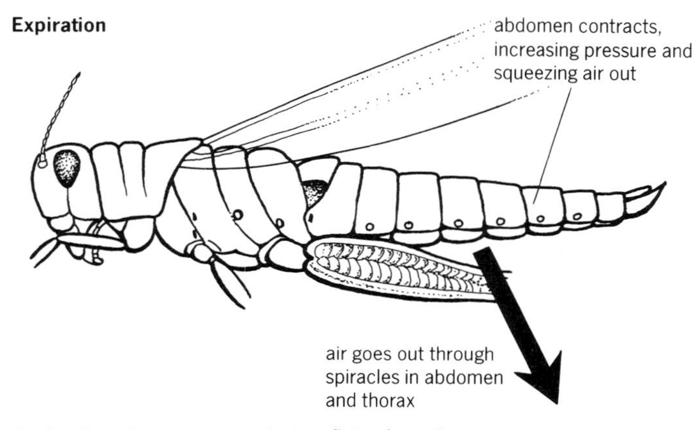

5.26 Breathing movements in a flying locust

5.30 Large insects make breathing movements.

Small insects, particularly ones which are not very active, do not make breathing movements. Because they do not need large amounts of oxygen and because the oxygen does not have far to go, diffusion alone can provide them with all that they need.

Larger, active insects such as locusts need to make breathing movements to get enough oxygen to their muscles. The way a locust breathes is shown in Fig 5.26.

Questions

1. What are the tracheae of an insect?
2. How do tracheae connect to the outside?
3. How are an insect's tracheae similar to those of a mammal?
4. How do the walls of tracheoles differ from the walls of tracheae?
5. What kind of insects make breathing movements?

Gas exchange in flowering plants

5.31 Plants respire and photosynthesise.

Green plants photosynthesise. They make glucose by combining water and carbon dioxide.

$$\text{carbon dioxide} + \text{water} \xrightarrow[\text{chlorophyll}]{\text{sunlight}} \text{glucose} + \text{oxygen}$$

$$6CO_2 + 6H_2O \xrightarrow[\text{chlorophyll}]{\text{sunlight}} C_6H_{12}O_6 + 6O_2$$

This needs energy which comes from sunlight. The energy is trapped by chlorophyll. The glucose which is made contains some of this energy.

When the plant needs energy, it releases it from the glucose in the same way that an animal does – that is, by respiration.

$$\text{glucose} + \text{oxygen} \rightarrow \text{carbon dioxide} + \text{water} + \text{energy}$$

$$C_6H_{12}O_6 + 6O_2 \rightarrow 6CO_2 + 6H_2O + \text{energy}$$

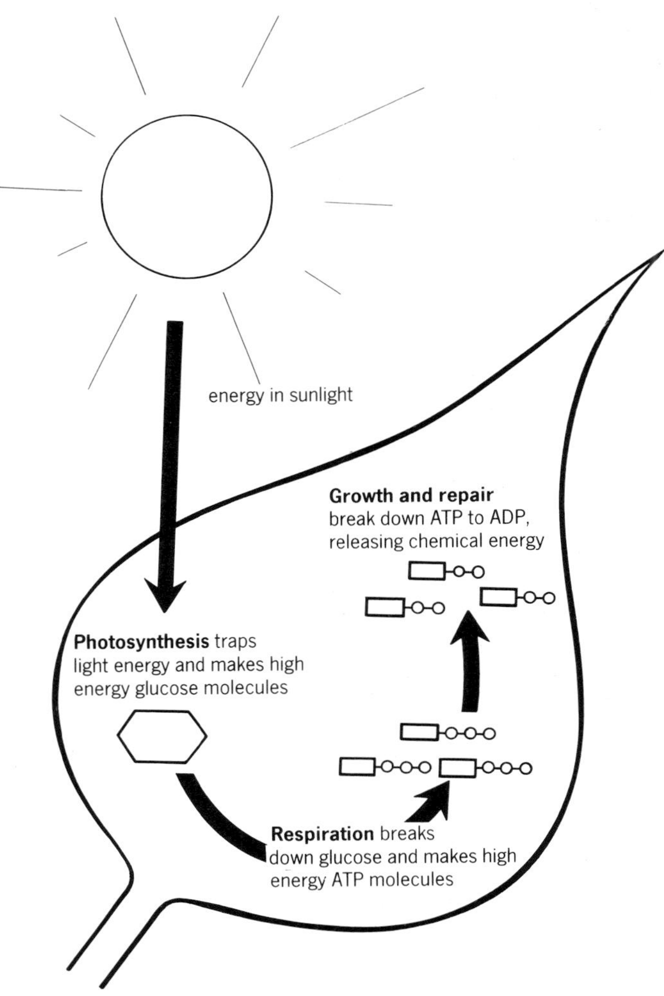

5.27 How the energy in sunlight is changed to useful energy by a green plant

At first sight, this reaction looks like photosynthesis going 'backwards'. In some ways it is. The photosynthesis reaction makes glucose and the respiration reaction breaks it down. However, the reactions are really very different. In photosynthesis, the energy which goes into the reaction is light energy which is trapped by the chloroplast. In respiration, the energy which comes out is chemical energy. This process occurs in the mitochondrion.

As in animals, the energy which is released during respiration is used to make ATP. The ATP can then be used whenever the plant needs energy (see Fig 5.27).

5.32 Plants, like animals, need energy.

Plants do not need as much energy as animals. They are not so active, partly because they do not have to move to find their food. However, all living cells need some energy. Plant cells need energy for growth, reproduction, for transporting food material between cells and inside cells and many other reasons. They need energy all the time. So all living plant cells, like animal cells, are always respiring.

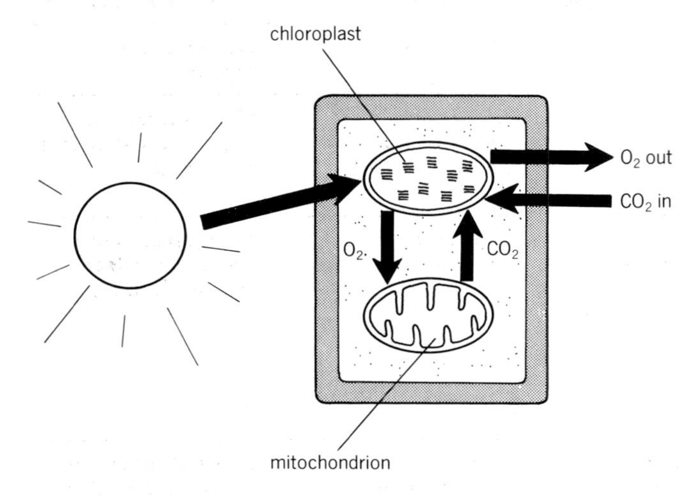

Day; photosynthesis and respiration

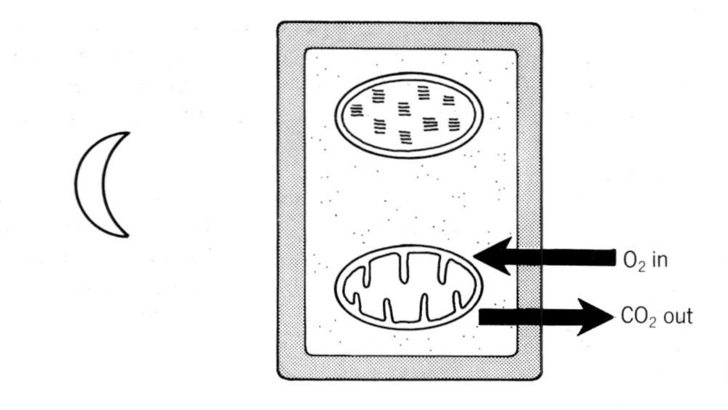

Night; respiration only

5.28 The balance between photosynthesis and respiration in a green plant

Investigation 5.10 To investigate the effect that plants and animals have on the carbon dioxide concentration of water

1 Copy the results chart, ready to fill it in.
2 Take four clean tubes, and put an equal quantity of bicarbonate indicator solution into each. Note the colour of the solution, and fill it in on your results chart.

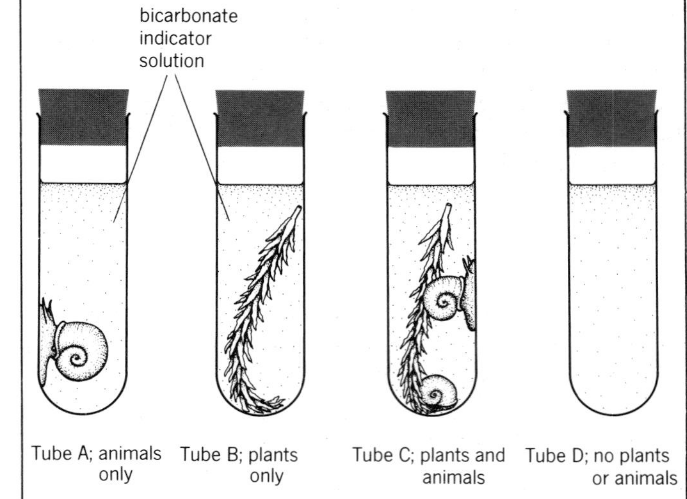

5.29 To investigate the effect that plants and animals have on the carbon dioxide concentration of water

3 Put watersnails and water plants into each tube as shown in Fig 5.29.
4 Stopper each tube firmly with a rubber bung.
5 Leave all four tubes in a light place for several hours.

Results table	A	B	C	D
Contents of tube				
Colour of indicator solution at beginning				
Colour of indicator solution at end				

Questions

1 Remembering that animals and plants respire, and that plants also photosynthesise in the light, explain the differences in the colour of the bicarbonate indicator solution in each tube.
2 What would you expect to happen to the animals and/or the plants in each tube if you left them for a long time?
3 What would happen to the colour of the solution in each of your four tubes if you left them in the dark?

5.33 The balance between photosynthesis and respiration.

Some plant cells, however, also photosynthesise. The cells in a leaf have chloroplasts and they use carbon dioxide and release oxygen during the daytime. At the same time, respiration is happening inside the mitochondria (see Fig 5.28).

In the daytime, photosynthesis is going on much faster than respiration. All of the carbon dioxide that the plant makes by respiration is used up by the chloroplasts in photosynthesis. Even this is not enough, and the plant takes in extra carbon dioxide from the air.

Some of the oxygen which is made by photosynthesis is used up for respiration. There is a lot left over, however, and this diffuses out of the cell.

At night, the chloroplasts stop photosynthesising. The mitochondria, however, continue to respire. Oxygen is used up, and carbon dioxide is released.

Questions

1. Why is respiration not really like 'photosynthesis backwards'?
2. Why do plants need energy?
3. In which parts of a plant cell does (a) respiration, and (b) photosynthesis happen?
4. At what times of day do plant cells respire?
5. At what times of day do plant cells photosynthesise?
6. Why do plants not need a transport system to carry oxygen around their bodies?

Table 5.5 Comparison of gas exchange in living organisms

Organism	*Respiratory surface (RS)*	*How is the RS kept moist?*	*How is the RS provided with oxygen?*	*Transport system in the organism*	*How is the surface area of the RS increased?*
Amoeba	Cell membrane	Surrounded by water	Water currents, diffusion	Diffusion only	Not necessary, as organism is small
Earthworm	Skin	Moisture in soil and on damp vegetation; coelomic fluid leaks through pores	Diffusion	Blood containing haemoglobin	Not necessary, as organism is small and not very active
Mammal	Alveoli in lungs	Cells in alveoli secrete fluid	Breathing movements; muscles of diaphragm and intercostal muscles vary volume of thorax, so that air moves up and down trachea	Blood containing haemoglobin	Many alveoli create very large surface area inside lungs
Fish	Lamellae of gills	Surrounded by water	Water currents and breathing movements; continuous movement of water into mouth and out under operculum	Blood containing haemoglobin	Each gill divided into many thin lamellae and each lamella also subdivided
Insect	Surface of tracheoles	Cells in tracheoles secrete fluid	By diffusion along extensive system of tracheae; breathing movements in active insects such as locust	By diffusion; tracheoles reach every part of body; blood not involved	Large number of very small tracheoles throughout body
Flowering plant	Surface of cells inside leaf	Cells in leaf surrounded by thin film of water brought by xylem vessels	By diffusion through stomata into leaf and stem; from photosynthesising cells; through lenticels in woody stem, by diffusion; from air spaces in soil through epidermis of root, by diffusion	By diffusion	Leaf is very thin, so has large surface area compared with volume

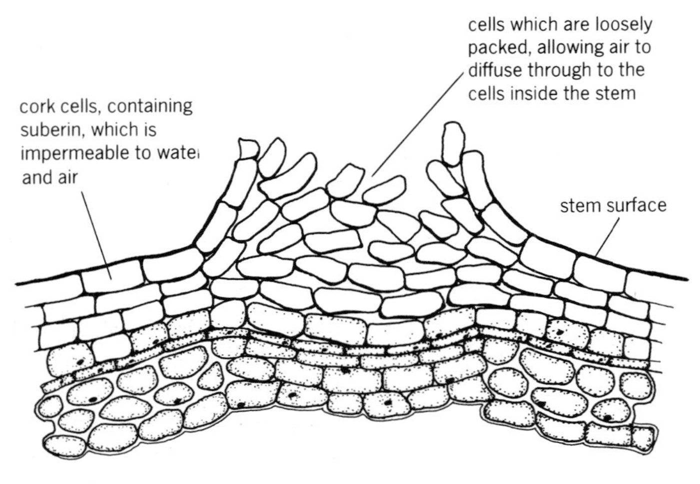

5.30 *Section through a lenticel*

5.34 Plants get oxygen by diffusion.

Plants have a branching shape, so they have quite a large surface area in comparison to their volume. Therefore, diffusion alone can supply all their cells with as much oxygen as they need for respiration. Diffusion occurs in the leaves, stems and roots of plants.

Leaves In the daytime, leaves are photosynthesising. This supplies plenty of oxygen for respiration. At night, oxygen diffuses into the leaves through the stomata (see Fig 3.2). It dissolves in the thin layer of moisture around the cells and diffuses in across their cell walls and membranes.

Stems The stems of herbaceous plants have stomata. Woody stems are covered with a layer of **cork cells** which make up the bark. These cells will not let air through, so the cork cells are packed loosely in places, to let oxygen diffuse in to the cells underneath. These places are called **lenticels** (see Fig 5.30).

Roots Roots get their oxygen from the air spaces in the soil. If the soil is waterlogged for very long, they become short of oxygen. Under these conditions the roots will respire anaerobically, producing alcohol (see section 5.9). This may kill the plant.

Chapter revision questions

1 Match each of these words with its definition listed below: operculum, tracheole, stoma, spiracle, alveolus, lenticel, lamellae, fermentation, mitochondrion, cilia.
(a) A type of anaerobic respiration which makes alcohol and carbon dioxide
(b) A small, thin-walled tube which carries oxygen deep into an insect's body
(c) A bony covering over a fish's gills
(d) An air sac inside a mammal's lungs
(e) Small hair-like projections from a cell
(f) An opening into an insect's tracheal system
(g) A small hole on the surface of a leaf, through which gases diffuse
(h) The part of a cell where sugar is oxidised
(i) Finely divided parts of a fish's gills
(j) Part of a woody stem through which gas exchange takes place

2 Which of these descriptions applies to aerobic respiration, which to anaerobic respiration, and which to both?
(a) lactic acid or alcohol made
(b) carbon dioxide made
(c) energy released from glucose
(d) heat produced
(e) ATP made
(f) glucose oxidised

3 (a) What is meant by a respiratory surface?
(b) Describe, with the aid of diagrams, how the respiratory surfaces of (i) a fish, and (ii) a human are kept supplied with oxygen.
(c) List three properties which these two types of respiratory surface have in common.

4 Describe an experiment you could do to see if germinating seeds give off carbon dioxide. Include a labelled diagram of your apparatus, a control, and explain what you think your results would be.

5 Construct a table to compare the processes of respiration and photosynthesis in a green plant.

6 Transport

6.1 Large organisms need transport systems.

Large organisms need transport systems to supply all their cells with food, oxygen and other materials. This chapter describes the transport systems of mammals and flowering plants, and describes the parts which make up these systems.

Layout of a mammal transport system

6.2 Mammals have double circulatory systems.

The main transport system of a mammal is its blood system. This is sometimes called the vascular system. It is a network of tubes, called **blood vessels.** A pump, the **heart,** keeps blood flowing through the vessels.

Fig 6.1 illustrates the general layout of the blood system. The arrows show the direction of blood flow. If you follow the arrows, beginning at the lungs, you can see that blood flows into the left-hand side of the heart, and then out to the rest of the body. It is brought back to the right-hand side of the heart, before going back to the lungs again.

This is called a **double circulatory system,** because the blood travels through the heart twice on one complete journey around the body.

Lungs

right side of heart

left side of heart

Rest of body

6.1 The general layout of the circulatory system of a human, as seen from the front

6.3 Oxygenated and deoxygenated blood.

The blood in the left-hand side of the heart has come from the lungs. It contains oxygen, which was picked up by the capillaries surrounding the alveoli (see Fig 6.2). It is called **oxygenated blood.**

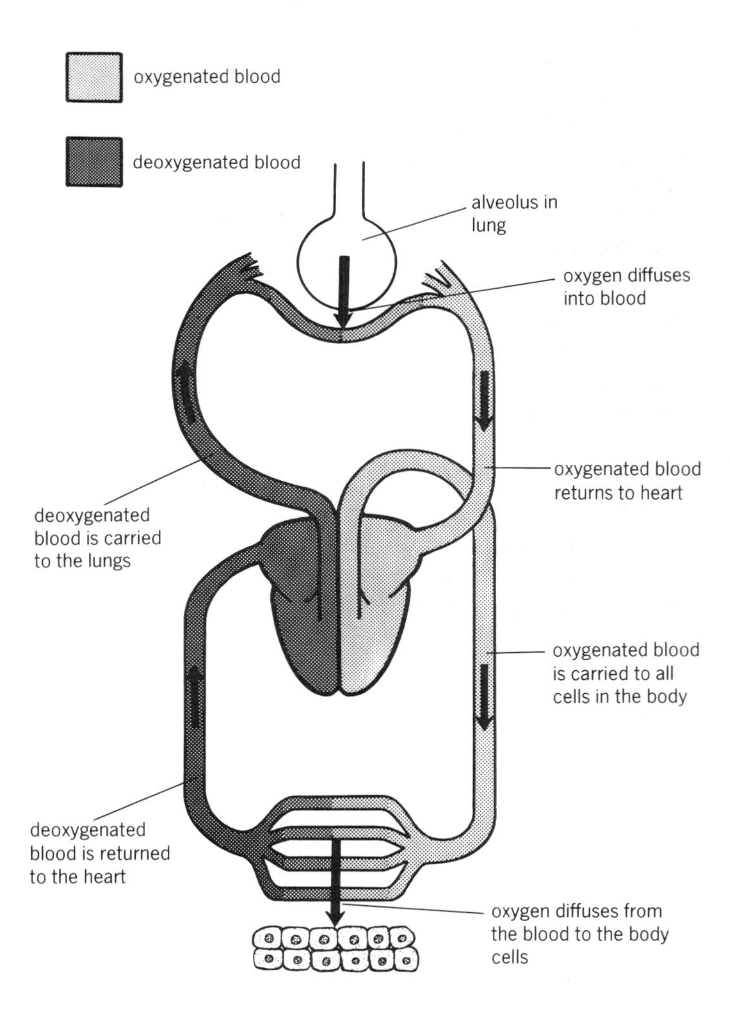

6.2 Oxygenated and deoxygenated blood

This oxygenated blood is then sent around the body. Some of the oxygen in it is taken up by the body cells, which need oxygen for respiration. When this happens the blood becomes **deoxygenated.** The deoxygenated blood is brought back to the right-hand side of the heart. It then goes to the lungs, where it becomes oxygenated once more.

> **Questions**
>
> 1 Why do large organisms need transport systems?
> 2 What is a double circulatory system?
> 3 What is oxygenated blood?
> 4 Where does blood become oxygenated?
> 5 Which side of the heart contains oxygenated blood?

The heart

6.4 The structure of the heart.

The function of the heart is to pump blood around the body. It is made of a special type of muscle called **cardiac muscle.** This muscle contracts and relaxes regularly, throughout life.

Fig 6.3 illustrates a section through a heart. It is divided into four chambers. The two upper chambers are called **atria.** The two lower chambers are **ventricles.** The chambers on the left-hand side are completely separated from the ones on the right-hand side by a septum.

If you look at Fig 6.1, you will see that blood flows into the heart at the top, into the atria. Both of the atria receive blood. The left atrium receives blood from the **pulmonary veins,** which come from the lungs. The right atrium receives blood from the rest of the body, arriving through the **venae cavae.**

From the atria, the blood flows into the ventricles. The ventricles then pump it out of the heart. The blood in the left ventricle is pumped into the **aorta,** which takes the blood around the body. The right ventricle pumps blood into the **pulmonary artery,** which takes it to the lungs.

The job of the ventricles is quite different from the job of the atria. The atria simply receive blood, either from the lungs or the body and supply it to the ventricles. The ventricles pump blood out of the heart and all round the body. To help them to do this, the ventricles have much thicker, more muscular walls than the atria.

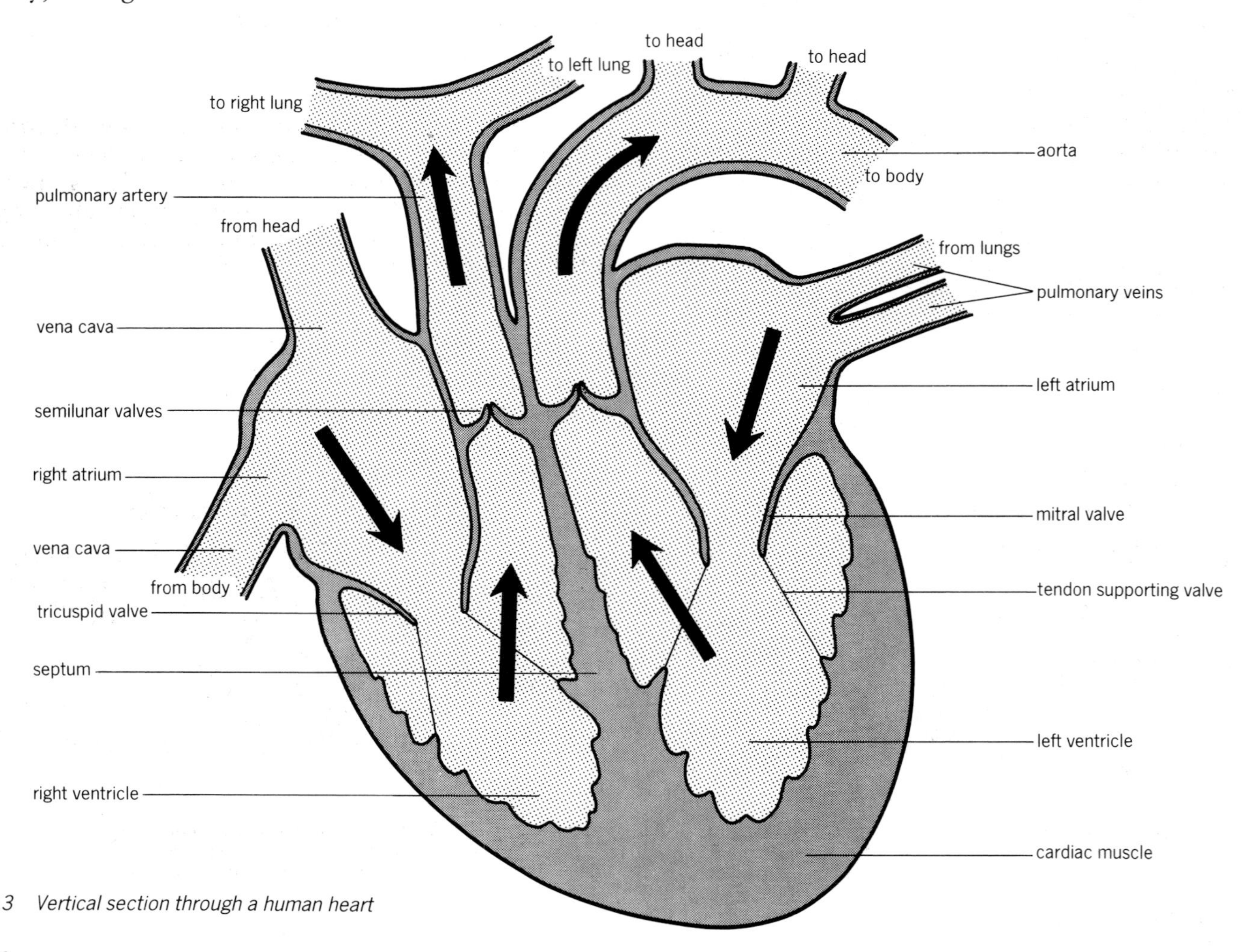

6.3 Vertical section through a human heart

There is also a difference in the thickness of the walls of the right and left ventricles. The right ventricle pumps blood to the lungs, which are very close to the heart. The left ventricle, however, pumps blood all around the body. The left ventricle has an especially thick wall of muscle to enable it to do this.

6.5 Coronary arteries supply heart muscle.

In Fig 6.4, you can see that there are blood vessels on the outside of the heart. They are called the **coronary arteries.** These vessels supply blood to the heart muscles.

It may seem odd that this is necessary, when the heart is full of blood. However, the muscles of the heart are so thick that the food and oxygen in the blood inside the heart would not be able to diffuse to all the muscles quickly enough. The heart muscle needs a constant supply of food and oxygen, so that it can keep contracting and relaxing. The coronary artery supplies this.

If the coronary artery gets blocked, for example by a blood clot, the cardiac muscles run short of oxygen. They cannot contract, so the heart stops beating. This is called a heart attack or **cardiac arrest.**

Questions

1. What kind of muscle is found in the heart?
2. Which parts of the heart receive blood from (a) the lungs, and (b) the body?
3. Which parts of the heart pump blood into (a) the pulmonary artery, and (b) the aorta?
4. Why do the ventricles have thicker walls than the atria?
5. Why does the left ventricle have a thicker wall than the right ventricle?
6. What is the function of the coronary artery?

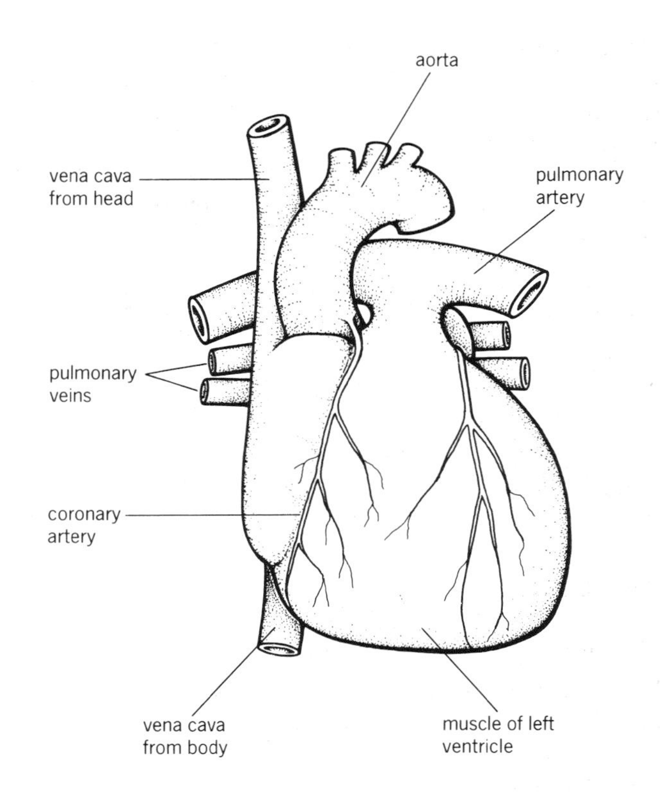

6.4 External appearance of a human heart

6.6 Heart beat.

The heart beats as the cardiac muscles in its walls contract and relax. When they contract, the heart becomes smaller, squeezing blood out. This is called **systole.** When they relax, the heart becomes larger, allowing blood to flow into the atria and ventricles. This is called **diastole.** Fig 6.5 illustrates this.

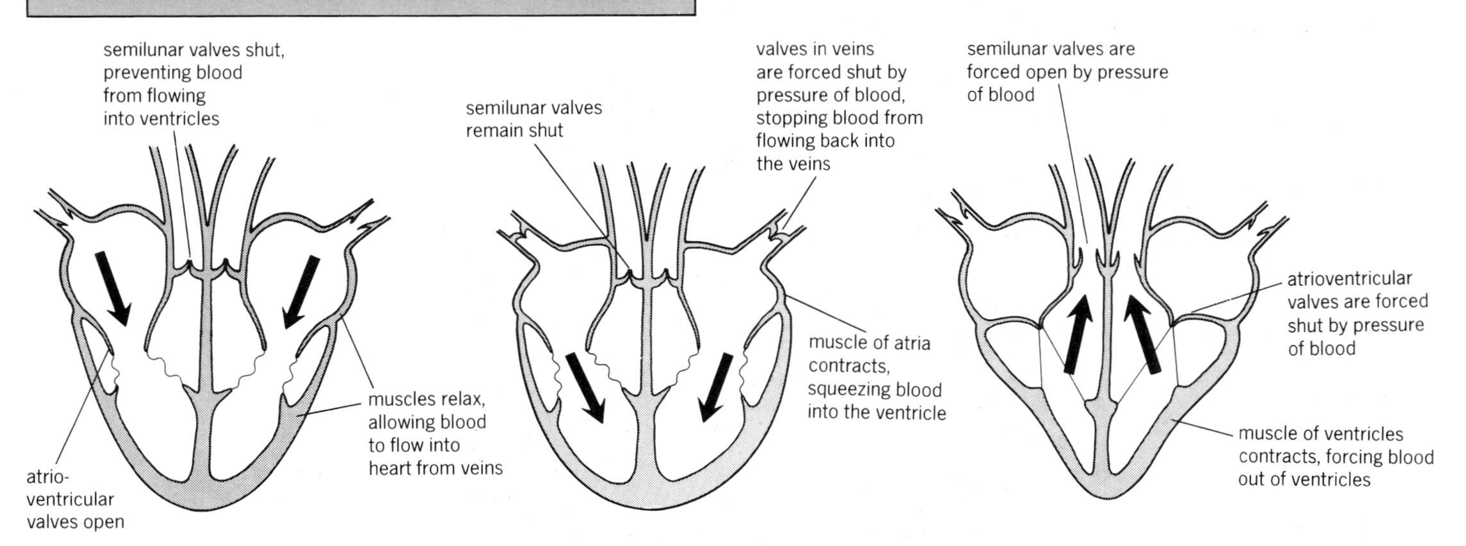

6.5 How the heart pumps blood

Investigation 6.1 To find the effect of exercise on the rate of heart beat

The best way to measure the rate of your heart beat is to take your pulse. Use the first two fingers of your right hand and lie them on the inside of your left wrist. Feel for the tendon near the outside of your wrist. If you rest your fingers lightly just over this tendon, you can feel the artery in your wrist pulsing as your heart pumps blood through it.

Perform the experiment explained in Investigation 5.8.

Questions

1 Why does your heart beat faster during exercise?
2 Why does the heart not return to its normal rate of beating as soon as you finish exercising?

6.7 Blood flows one way through heart valves.

There is a valve between the left atrium and the left ventricle, and another between the right atrium and ventricle. These are called **atrio-ventricular valves** (see Fig 6.5).

The valve on the left-hand side of the heart is made of two parts and is called the **bicuspid valve,** or the **mitral valve.** The valve on the right-hand side has three parts, and is called the **tricuspid valve.**

The function of these valves is to stop blood flowing from the ventricles back to the atria. This is important, so that when the ventricles contract, the blood is pushed up into the arteries, not back into the atria. As the ventricles contract, the pressure of the blood pushes the valves upwards. The tendons attached to them stop them from going up too far.

Questions

1 What is (a) systole, and (b) diastole?
2 Where are the atrio-ventricular valves?
3 What is their function?
4 Why are these valves supported by tendons?

Blood vessels

6.8 There are three kinds of blood vessels.

There are three main kinds of blood vessel: arteries, capillaries and veins. **Arteries** carry blood away from the heart. They divide again and again, and eventually form very tiny vessels called **capillaries.** The capillaries gradually join up with one another to form large vessels called **veins.** Veins carry blood towards the heart.

6.9 Arteries have thick elastic walls.

When blood flows out of the heart, it enters the arteries. The blood is then at very high pressure, because it has been forced out of the heart by the contraction of the muscular ventricles. Arteries therefore need very strong walls to withstand the high pressure of the blood flowing through them.

The blood does not flow smoothly through the arteries. It pulses through, as the ventricles contract and

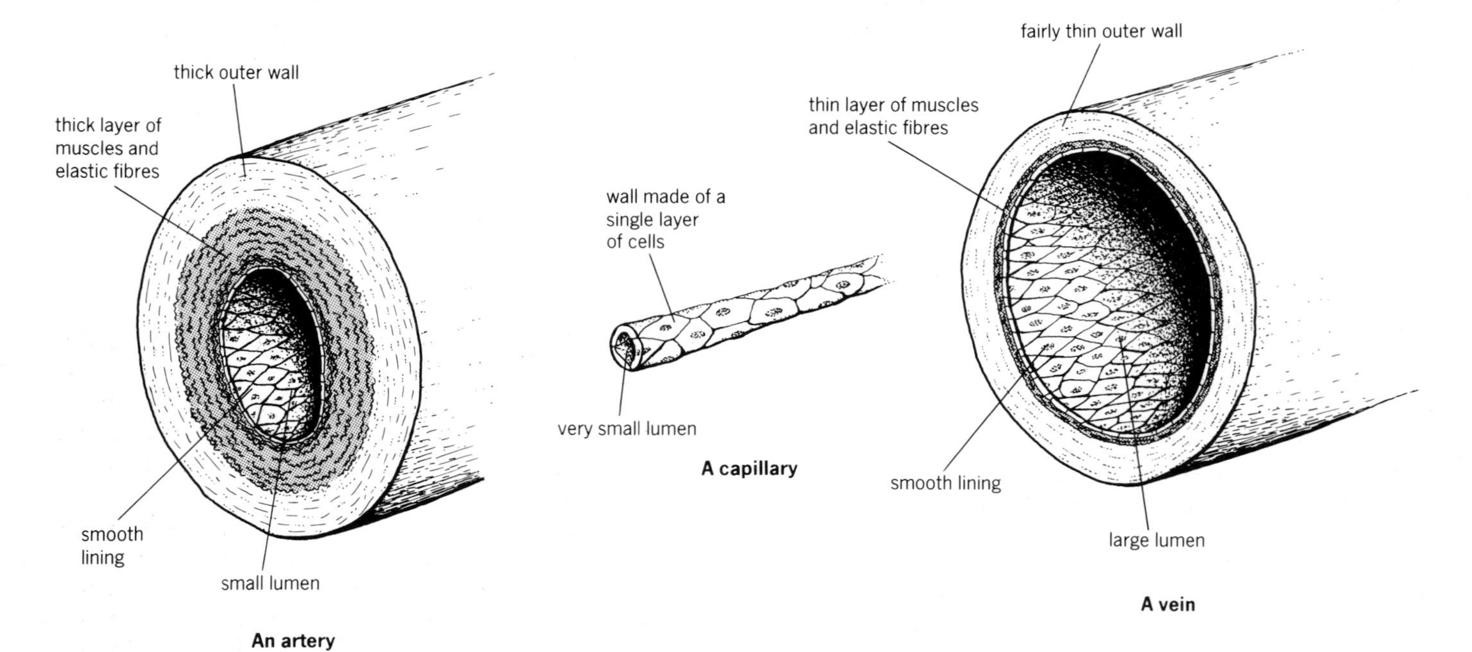

6.6 Sections through the three types of blood vessels

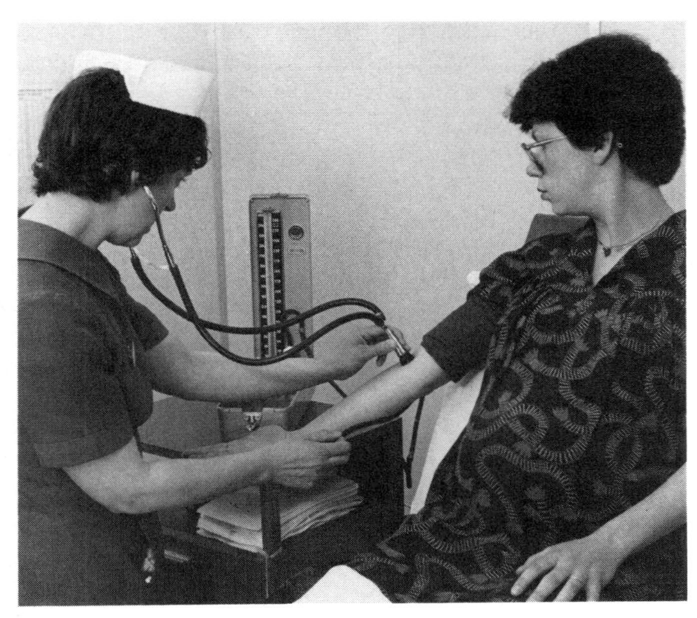

6.7 *A sphygmomanometer being used to measure blood pressure*

relax. The arteries have elastic tissue in their walls which can stretch and recoil with the force of the blood. This helps to make the flow of blood smoother. You can feel your arteries stretch and recoil when you feel your pulse in your wrist.

The blood pressure in the arteries of your arm can be measured using a **sphygmomanometer** (see Fig 6.7).

6.10 Capillaries are very narrow, with thin walls.

The arteries gradually divide to form smaller and smaller vessels (see Fig 6.8). These are the capillaries. The capillaries are very small and penetrate to every part of the body. No cell is very far away from a capillary.

The function of the capillaries is to take food, oxygen and other materials to all the cells in the body, and to take away their waste materials. To do this, their walls must be very thin so that substances can get in and out of them easily. The walls of the smallest capillaries are only one cell thick (see Fig 6.6).

6.11 Veins have one-way valves.

The capillaries gradually join up again to form veins. By the time the blood gets to the veins, it is at a much lower pressure than it was in the arteries. The blood flows more slowly and smoothly now. There is no need for veins to have such thick, strong, elastic walls.

If the veins were narrow, this would slow down the blood even more. To help to keep the blood moving easily through them, the space inside the veins, called the **lumen,** is much wider than the lumen of the arteries.

Veins have valves in them to stop the blood flowing backwards (see Fig 6.9). Valves are not needed in the

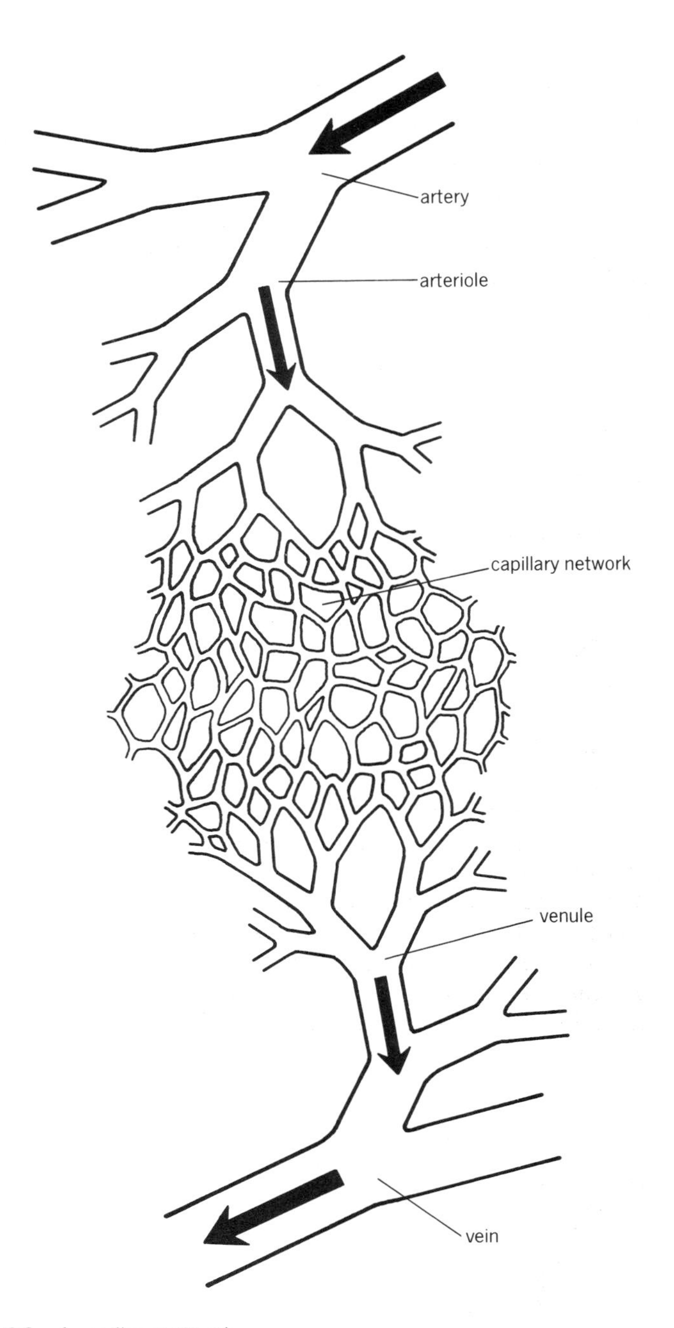

6.8 *A capillary network*

arteries, because the force of the heart beat keeps blood moving forwards through them.

Blood is also kept moving in the veins by the contraction of muscles around them. The large veins in your legs are squeezed by your leg muscles when you walk. This helps to push the blood back up to your heart. If a person is confined to bed for a long time, then there is a danger that the blood in these veins will not be kept moving. A clot may form in them, called a **thrombosis.** If the clot is carried to the lungs, it could get stuck in the arterioles. This is called a **pulmonary embolism,** and it may prevent the circulation reaching part of the lungs. In serious cases this can cause death.

Table 6.1 Arteries, veins, and capillaries

	Function	*Structure of wall*	*Width of lumen*	*Reasons for structure*
Arteries	carry blood away from the heart	thick, strong, containing muscles and elastic fibres	varies, as elastic fibres stretch and recoil	strength needed to resist pulsing of blood as it is pumped by the heart
Capillaries	supply all cells with their requirements, and take away waste products	very thin, often only one cell thick	very narrow	no need for strong walls, as most of force of blood has been lost; thin walls and narrow lumen bring blood into close contact with tissues
Veins	return blood to heart	quite thin	wide, contains valves	no need for strong walls, as most of force of blood has been lost; wide lumens offer less resistance to blood flow; valves prevent backflow

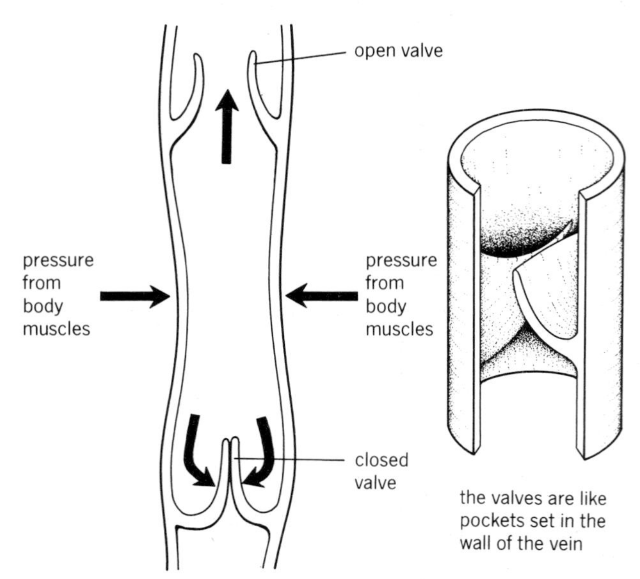

6.9 Valves in a vein

6.12 Each organ has its own blood supply.

Figs 6.10 and 6.11 illustrate the positions of the main arteries and veins in the body.

Each organ of the body, except the lungs, is supplied with oxygenated blood from an artery. Deoxygenated blood is taken away by a vein. The artery and vein are named according to the organ they are connected with. For example, the blood vessels of the kidneys are the renal artery and vein. The liver has the hepatic artery and vein.

All arteries, other than the pulmonary artery, branch from the aorta. All the veins, except the pulmonary veins, join up to one of the two venae cavae.

The liver has two blood vessels supplying it with blood. The first is the **hepatic artery,** which supplies oxygen. The second is the **hepatic portal vein.** This vein brings blood from the digestive system (see Fig 2.23), so that the liver can process the food which has been absorbed, before it travels to other parts of the body. All the blood leaves the liver in the **hepatic vein.**

Questions

1 Which blood vessels carry blood (a) away from, and (b) towards the heart?
2 Why do arteries need strong walls?
3 Why do arteries have elastic walls?
4 What is the function of capillaries?
5 Why do veins have a large lumen?
6 How is blood kept moving in the large veins of the legs?
7 What is unusual about the blood supply to the liver?

Fact!

The body adjusts to changes in blood volume so quickly that there is no detectable change in blood volume when the standard 450 ml – about one tenth of the total blood volume – is taken in a blood donation clinic.

Large quantities of blood are sometimes needed during operations. A 50 year old haemophiliac, who underwent open heart surgery in Chicago in 1970 needed 1080 litres of blood.

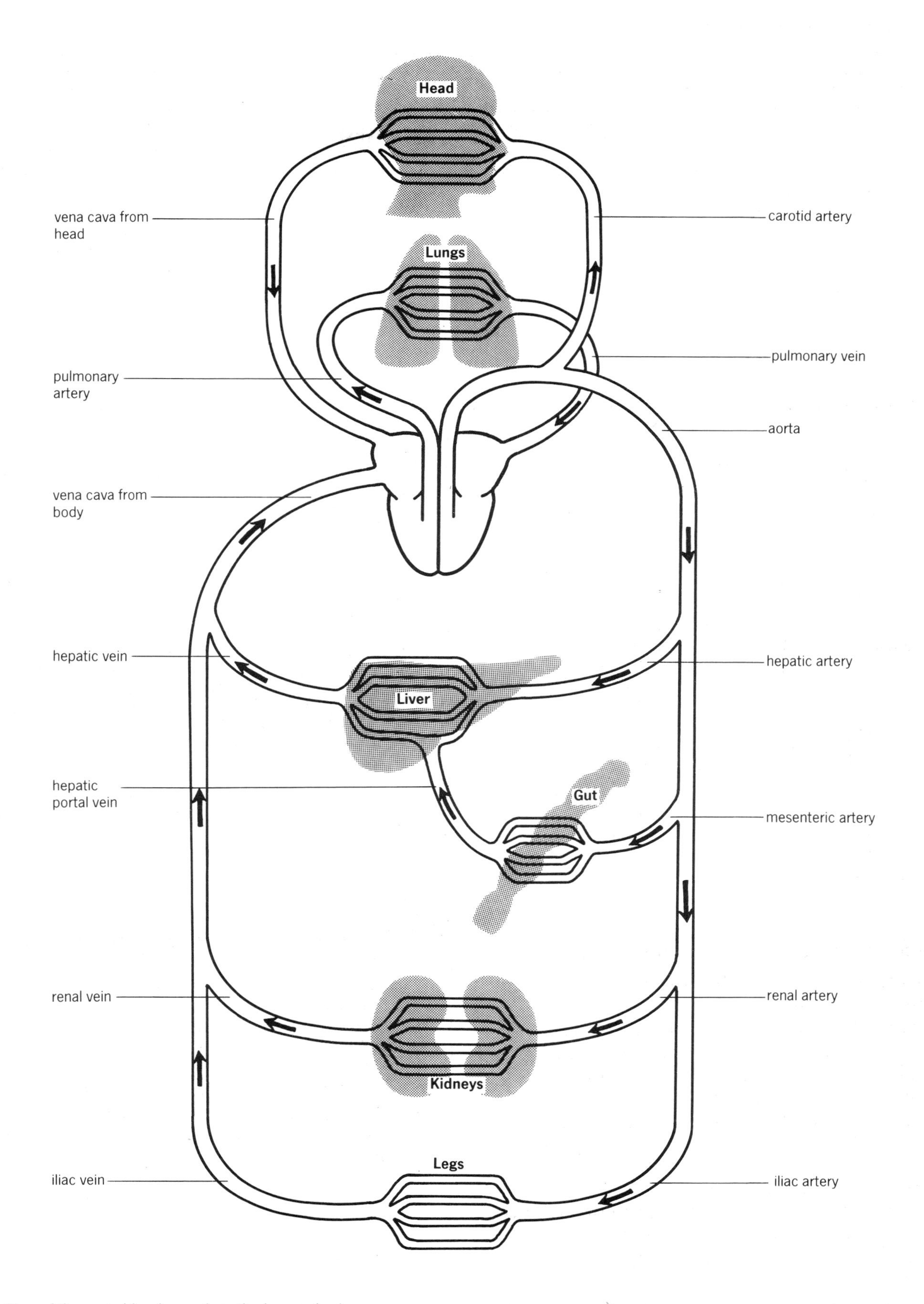

6.10 Plan of the main blood vessels in the human body

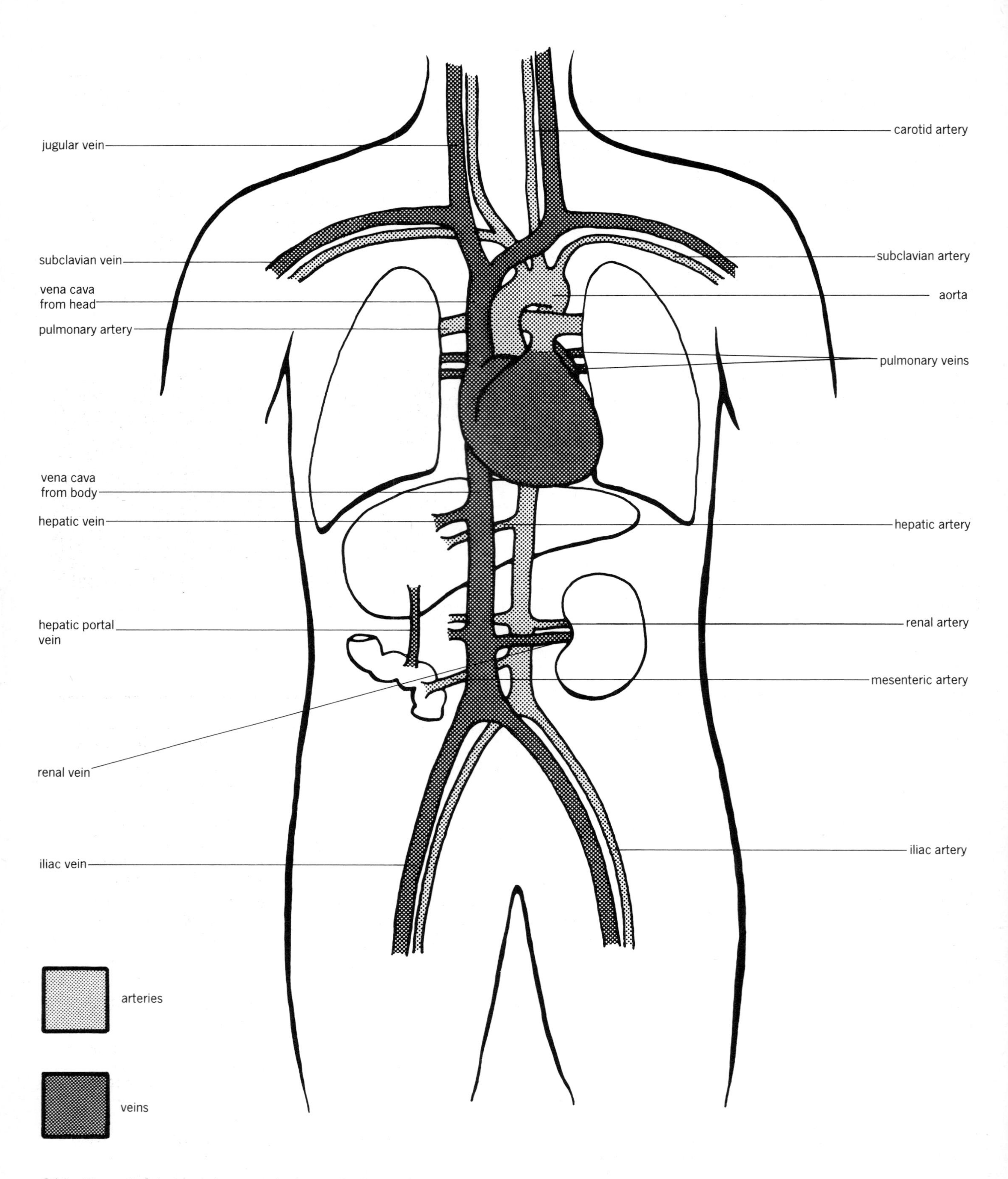

6.11 The main arteries and veins in the human body

Blood

6.13 Blood consists of cells floating in plasma.

The liquid part of blood is called **plasma.** Floating in the plasma are cells. Most of these are **red blood cells.** A much smaller number are **white blood cells.** There are also small fragments formed from special cells in the bone marrow, called **platelets** (see Fig 6.12).

6.14 Plasma is a complex solution.

Plasma is mostly water. Many substances are dissolved in it. Glucose, amino acids, salts, hormones, blood proteins, and antibodies are all dissolved in the plasma. Look at Table 6.2.

6.15 Red blood cells carry oxygen.

Red blood cells are made in the bone marrow of some bones, including the ribs, vertebrae and some limb bones. They are produced at a very fast rate – about 9 000 million per hour!

Red cells have to be made so quickly because they do not live for very long. Each red cell only lives for about four months. One reason for this is that they do not have a nucleus (see Fig 6.12).

Red cells are red because they contain the pigment **haemoglobin.** This carries oxygen. Haemoglobin is a protein, and contains iron.

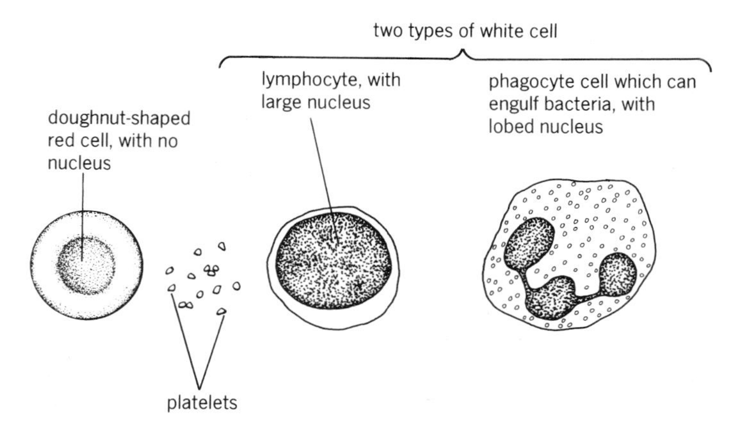

6.12 Blood cells

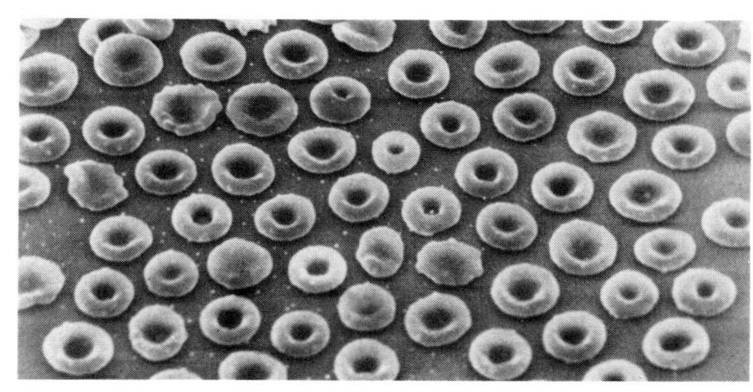

6.13 Scanning electronmicrograph of normal red blood cells

Table 6.2 Some of the main components of blood plasma

Substance	*Source*	*Destination*	*Notes*
Water	Absorbed in colon.	All cells.	Excess is removed by kidneys.
Plasma proteins eg. fibrinogen, antibodies	Fibrinogen made in the liver. Antibodies made by lymphocytes.	Remain in the blood.	Fibrinogen helps in blood clotting. Antibodies kill bacteria.
Lipids, including cholesterol and fatty acids	Absorbed in the ileum. Also derived from fat reserves in the body.	To the liver, for breakdown. To adipose tissue, for storage.	Breakdown of fats yields energy. High cholesterol levels in the blood may increase the chances of heart disease.
Carbohydrates eg. glucose	Absorbed in the ileum. Also derived from glycogen breakdown in the liver.	To all cells, for energy release by respiration.	Excess glucose is converted to glycogen and stored in liver and muscles.
Excretory substances eg. urea	Derived from amino acid deamination in the liver.	To kidneys for excretion.	
Mineral ions eg. Na^+, Cl^-	Absorbed in the ileum and colon.	To all cells.	Excess ions are excreted by the kidneys.
Hormones	Secreted into the blood by endocrine glands.	To all parts of the body.	Hormones only affect their own target organs. Hormones are broken down in the liver, and they are also excreted by the kidneys.
Dissolved gases eg. carbon dioxide	Carbon dioxide is released from all cells as a waste product of respiration.	To the lungs for excretion.	Most carbon dioxide is carried as hydrogen carbonate ions (HCO_3^-) in the plasma.

Old red blood cells are broken down in the liver, spleen and bone marrow. Some of the iron from the haemoglobin is stored, and used for making new haemoglobin. Some of it is turned into bile pigment and excreted.

6.16 White blood cells fight infection.

White cells are made in the bone marrow and in the **lymph nodes** (see section 6.25). White cells do have a nucleus, which is often quite large and lobed (see Fig 6.12). They can move around, like an *Amoeba*, and can squeeze out through the walls of blood capillaries into all parts of the body. Their function is to fight infection, and to clear up any dead body cells.

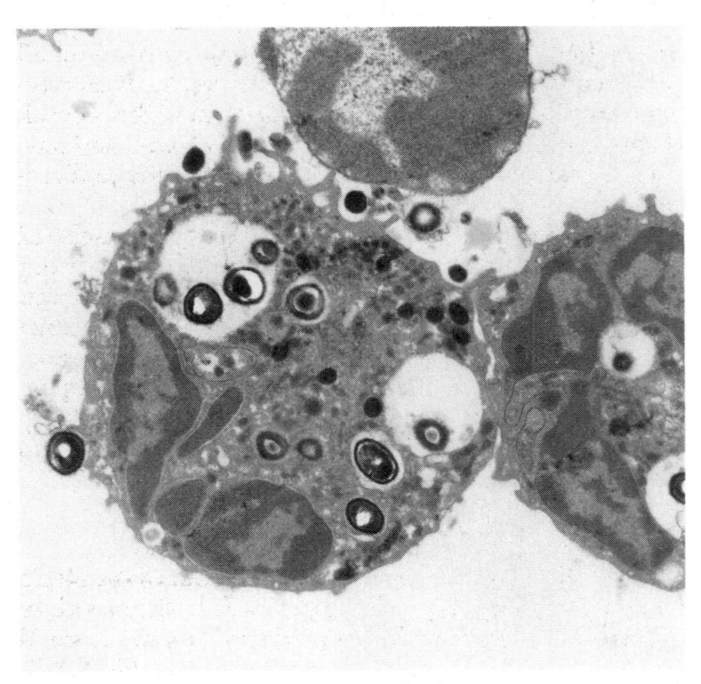

6.14 White cells taking in and digesting bacteria

6.17 Platelets help blood to clot.

Platelets are small fragments of cells, with no nucleus. They are made in the red bone marrow, and they are involved in blood clotting (see Fig 14.23).

Questions

1 List five components of plasma.
2 Where are red blood cells made?
3 What is unusual about red blood cells?
4 What is haemoglobin?
5 Where are white blood cells made?
6 What are platelets?

6.18 Blood has many functions.

Blood has three main functions. These are transport, defence against disease, and regulation of body temperature.

6.19 Many substances are transported by blood.

Transport of oxygen In the lungs, oxygen diffuses from the alveoli into the blood (see section 5.19). The oxygen diffuses into the red blood cells, where it combines with the haemoglobin (Hb) to form oxyhaemoglobin (oxyHb).

The blood is then taken to the heart in the pulmonary veins and pumped out of the heart in the aorta. Arteries branch from the aorta to supply all parts of the body with oxygenated blood. When it reaches a tissue which needs oxygen, the oxyHb gives up its oxygen, to become Hb again.

Because capillaries are so narrow, the oxyHb in the red blood cells is taken very close to the tissues which need the oxygen. The oxygen only has a very short distance to diffuse.

OxyHb is bright red, whereas Hb is purplish-red. The blood in arteries is therefore a brighter red colour than the blood in veins.

Table 6.3 Components of blood

Component	*Structure*	*Functions*
Plasma	Water, containing many substances in solution	1 Liquid medium in which cells and platelets can float 2 Transports CO_2 in solution 3 Transports food materials in solution 4 Transports urea in solution 5 Transports hormones in solution 6 Transports heat 7 Transports substances needed for blood clotting 8 Transports antibodies
Red cells	Biconcave discs, with no nucleus, containing haemoglobin	1 Transport oxygen 2 Transport small amount of CO_2
White cells	Variable shape, with nucleus	1 Engulf and destroy bacteria (phagocytosis) 2 Make antibodies
Platelets	Small particles, with no nucleus	1 Help in blood clotting

Transport of carbon dioxide Carbon dioxide is made by all the cells in the body as they respire. The carbon dioxide diffuses through the walls of the capillaries, into the blood.

Most of the carbon dioxide is carried by the blood plasma in the form of hydrogen carbonate ions, HCO_3^-. A small amount is carried by Hb in the red cells.

Blood containing carbon dioxide is returned to the heart in the veins, and then to the lungs in the pulmonary arteries. The carbon dioxide diffuses out of the blood and is passed out of the body on expiration.

Transport of food materials Digested food is absorbed in the ileum (see section 2.37). It includes amino acids, fatty acids and glycerol, monosaccharides (such as glucose), water, vitamins and minerals. These all dissolve in the plasma in the blood capillaries in the villi.

These capillaries join up to form the hepatic portal vein. This takes the dissolved food to the liver. The liver processes the food (see Table 10.2) and returns some of it to the blood.

The food is then carried, dissolved in the blood plasma, to all parts of the body.

Transport of urea Urea, a waste substance (see section 10.4), is made in the liver. It dissolves in the blood plasma, and is carried to the kidneys. The kidneys excrete it in the urine.

Transport of hormones Hormones are made in endocrine glands (see section 12.24). The hormones dissolve in the blood plasma, and are transported all over the body.

Transport of heat Some parts of the body, such as the liver, muscles and brown fat, make a lot of heat. The blood transports the heat to all parts of the body. This prevents the liver, muscles and brown fat becoming too hot, and helps to keep the rest of the body warm.

6.20 Blood defends the body.

The way in which the blood defends the body are described in Chapter 14. They include blood clotting, phagocytosis, and the production of antibodies.

6.21 Blood helps to regulate temperature.

The capillaries in the skin help to keep your body temperature constant at about 36.8°C. This is described in Chapter 9.

Questions

1 Why is the blood in arteries a brighter red than the blood in veins?
2 Which vessel transports digested food to the liver?
3 How is urea transported?
4 Name two functions of blood other than transport.

Lymph and tissue fluid

6.22 Tissue fluid is leaked plasma.

Capillaries leak! The cells in their walls do not fit together exactly, so there are small gaps between them. Plasma can therefore leak out from the blood.

White blood cells can also get through these gaps. They are able to move, like an *Amoeba,* and can squeeze through, out of the capillaries. Red blood cells cannot get out. They are too large and cannot change their shape very much.

So plasma and white cells are continually leaking out of the blood capillaries. The fluid formed in this way is called **tissue fluid.** It surrounds all the cells in the body (see Fig 6.15).

6.23 The functions of tissue fluid.

Tissue fluid is very important. It supplies cells with all their requirements. These requirements, such as oxygen and food materials, diffuse from the blood, through the tissue fluid, to the cells. Waste products, such as carbon dioxide, diffuse in the opposite direction.

The tissue fluid is the immediate environment of every cell in your body. It is easier for a cell to carry out its functions properly if its environment stays constant. For example this means it should stay at the same temperature, and at the same osmotic concentration.

Several organs in the body work to keep the composition and temperature of the blood constant, and therefore the tissue fluid as well. This process is called **homeostasis**, and is described in section 10.19.

6.24 Lymph is drained tissue fluid.

The plasma and white cells which leak out of the blood capillaries must eventually be returned to the blood. In the tissues, as well as blood capillaries, are other small vessels. They are **lymphatic capillaries** (see Fig 6.15). The tissue fluid slowly drains into them. The fluid is now called **lymph.**

The lymphatic capillaries gradually join up to form larger **lymphatic vessels.** These carry the lymph to the subclavian veins which bring blood back from the arms (see Fig 6.17). Here the lymph enters the blood again.

The lymphatic system has no pump to make the lymph flow. Lymph vessels do have valves in them, however, to make sure that movement is only in one direction. Lymph flows much more slowly than blood.

Fact! A human being contains about 70ml of blood per kilogram of body weight. For an adult, this is about 4 or 5 litres.

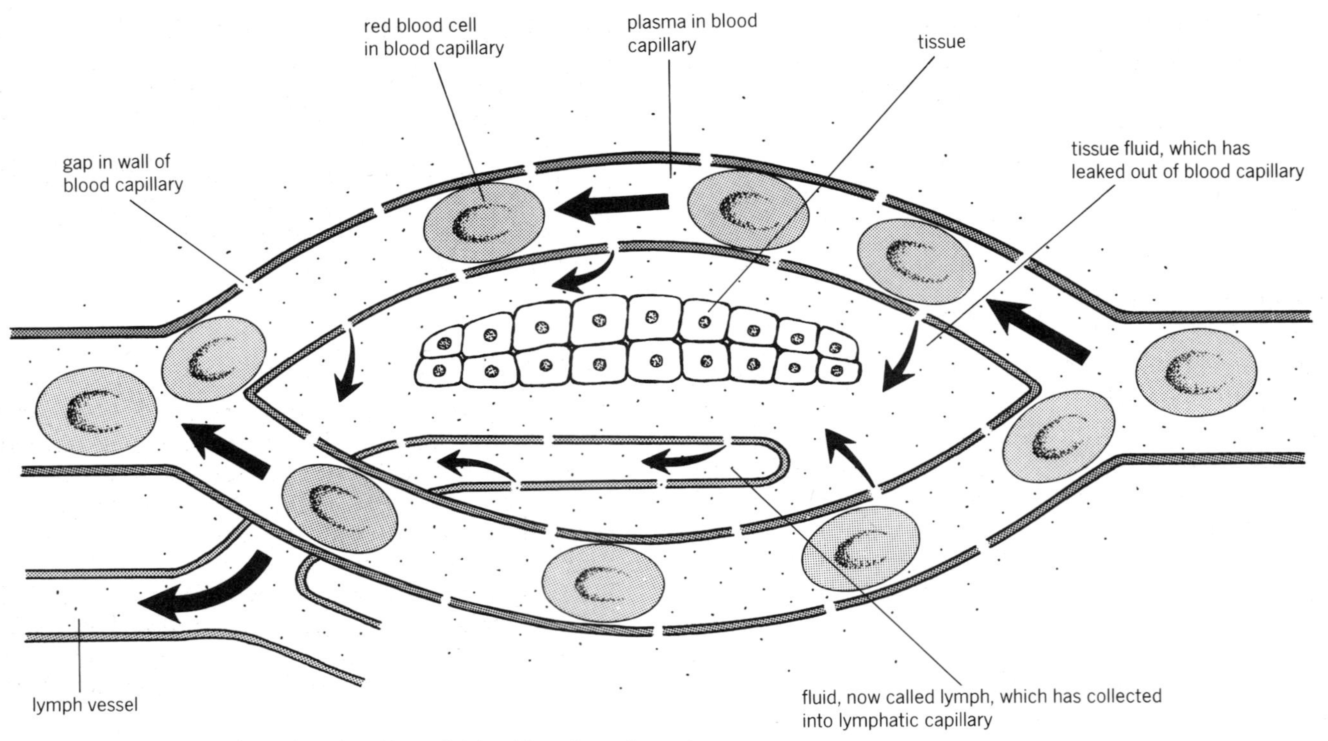

6.15 Part of a capillary network, to show how tissue fluid and lymph are formed

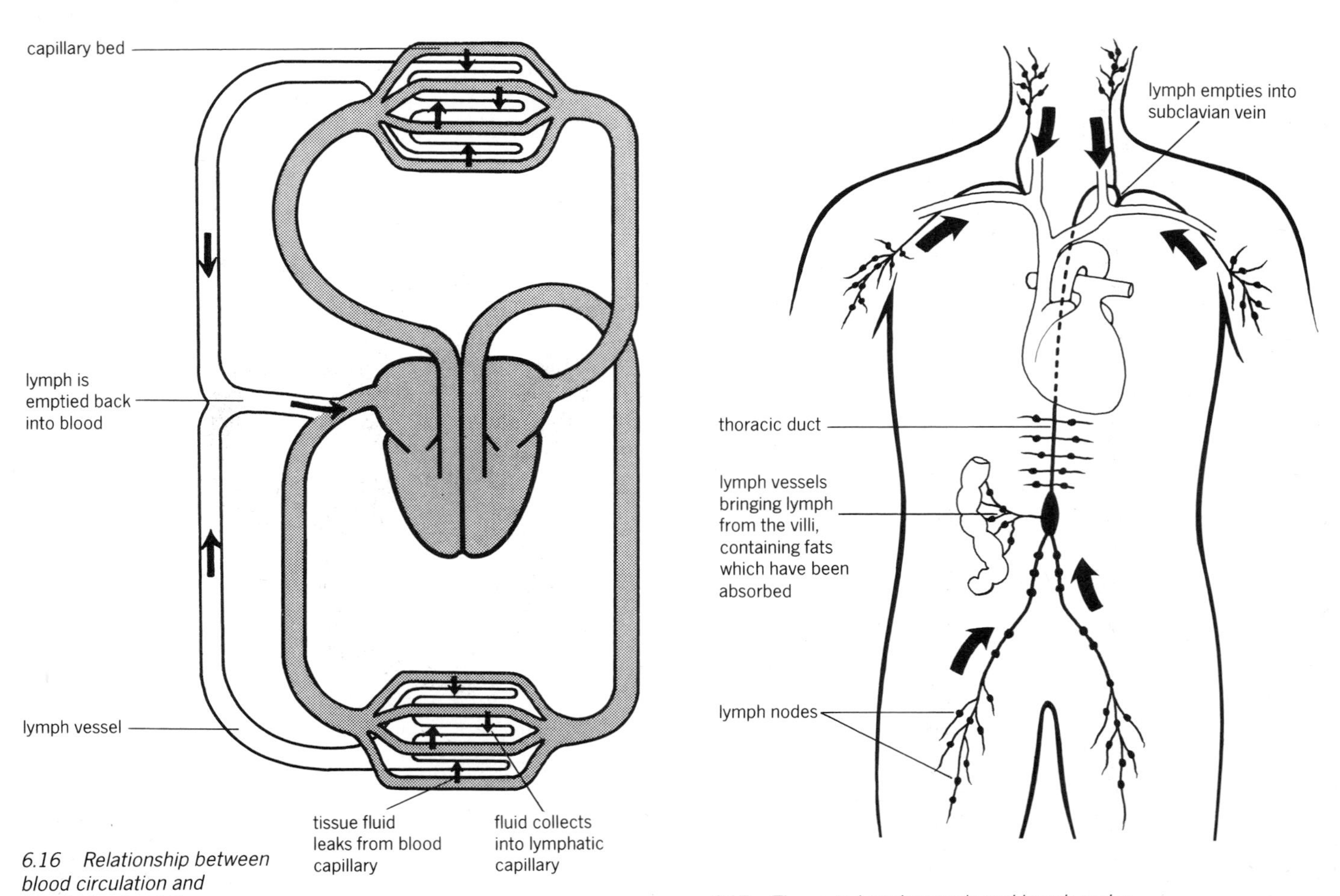

6.16 Relationship between blood circulation and lymphatic circulation

6.17 The main lymph vessels and lymph nodes

6.25 Lymph nodes contain white blood cells.

On its way from the tissues to the subclavian vein, lymph flows through several **lymph nodes.** Some of these are shown in Fig 6.17.

Lymph nodes contain large numbers of white cells. Most bacteria or toxins (see section 14.27) in the lymph can be destroyed by these cells.

Questions

1. What is tissue fluid?
2. Give two functions of tissue fluid.
3. What is lymph?
4. Why do lymphatic capillaries have valves in them?
5. Name two places where lymph nodes are found.
6. What happens inside lymph nodes?

Transport in a flowering plant

6.26 Plants have two transport systems – phloem and xylem.

Transport systems in plants are less elaborate than in mammals. Plants are less active than mammals, and so their cells do not need to be supplied with materials so quickly. Also, the branching shape of a plant means that all the cells can get their oxygen for respiration, and carbon dioxide for photosynthesis, directly from the air, by diffusion.

Plants have two transport systems. The **xylem vessels** carry water and minerals, while the **phloem tubes** carry food materials which the plant has made.

6.27 Xylem helps to support plants.

A xylem vessel is like a long drainpipe (see Figs 6.18 and 6.19). It is made of many hollow cells, joined end to end. The end walls of the cells have disappeared, so a long, open tube is formed. Xylem vessels run from the roots of the plant, right up through the stem. They branch out into every leaf.

Xylem vessels contain no cytoplasm or nuclei. Their walls are made of cellulose and **lignin.** Lignin is very strong, so xylem vessels help to keep plants upright. Wood is made almost entirely of lignified xylem vessels.

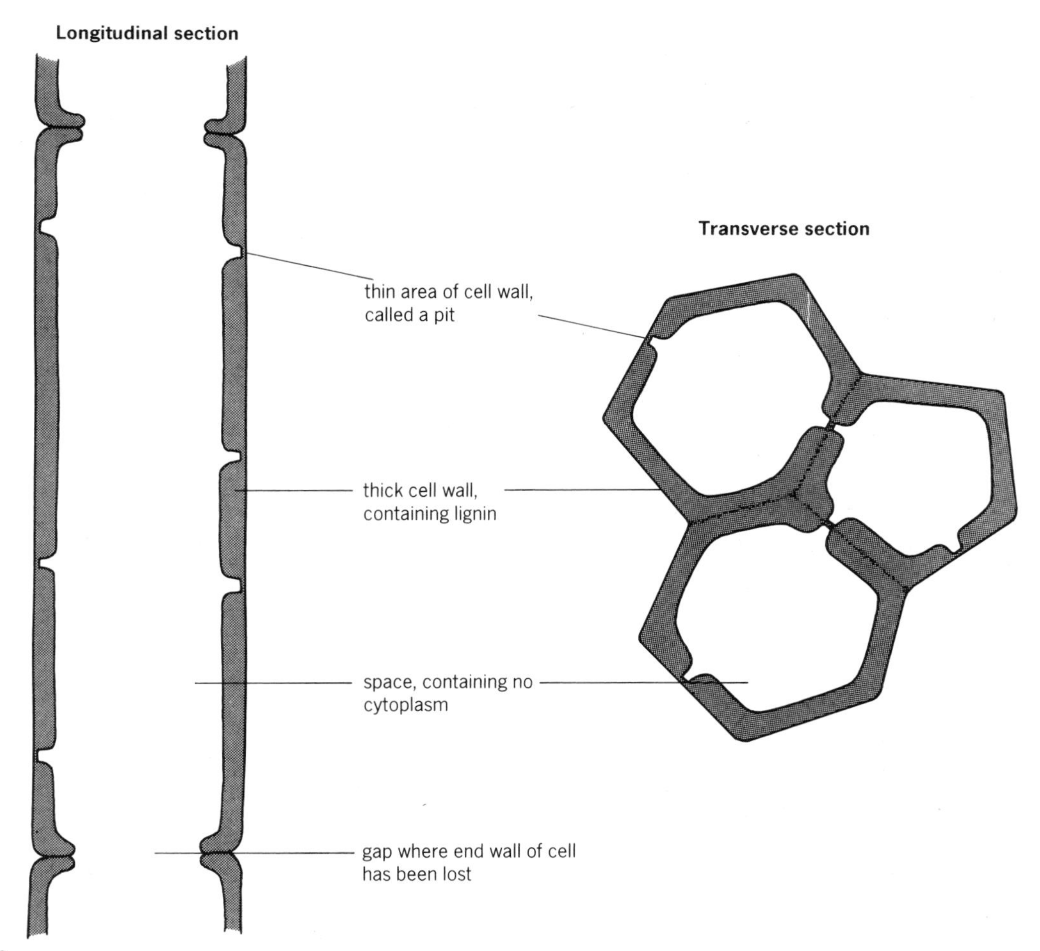

6.18 Xylem vessels

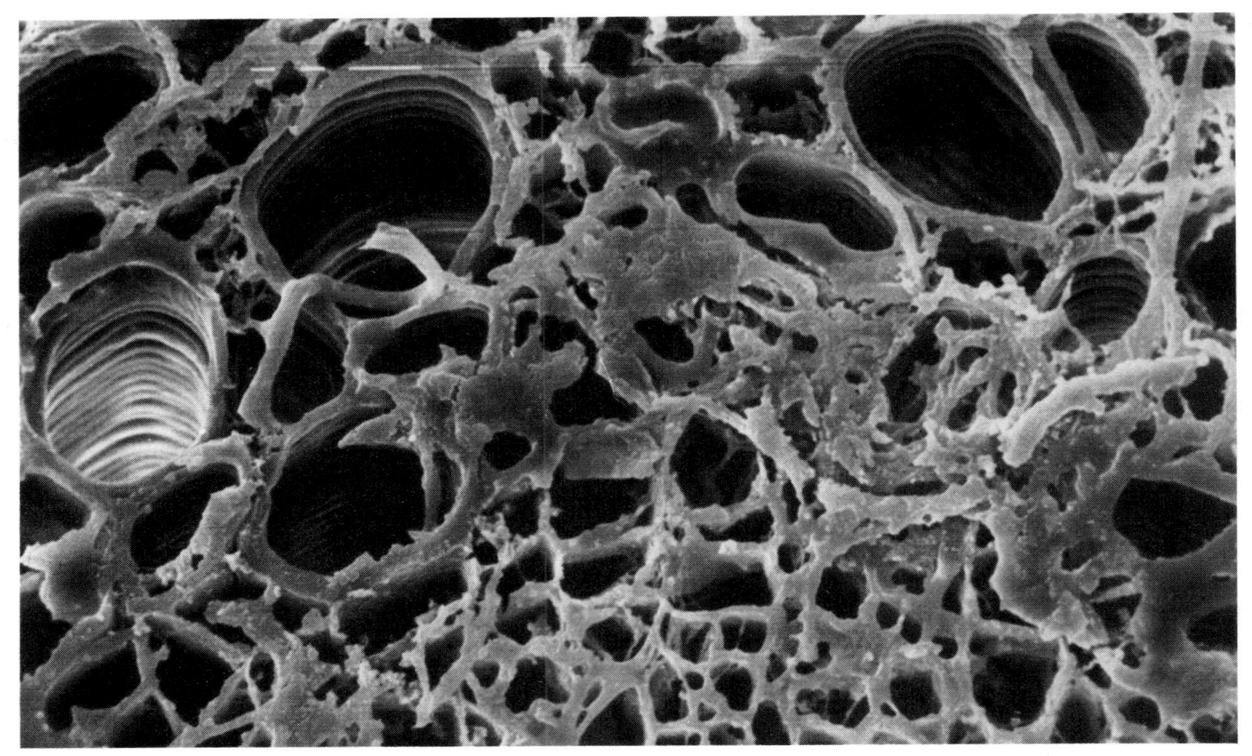

6.19 Scanning electronmicrograph of the tube-like xylem vessels in a plant stem; the large vessels at the top show the strengthening rings of lignin in the cell walls

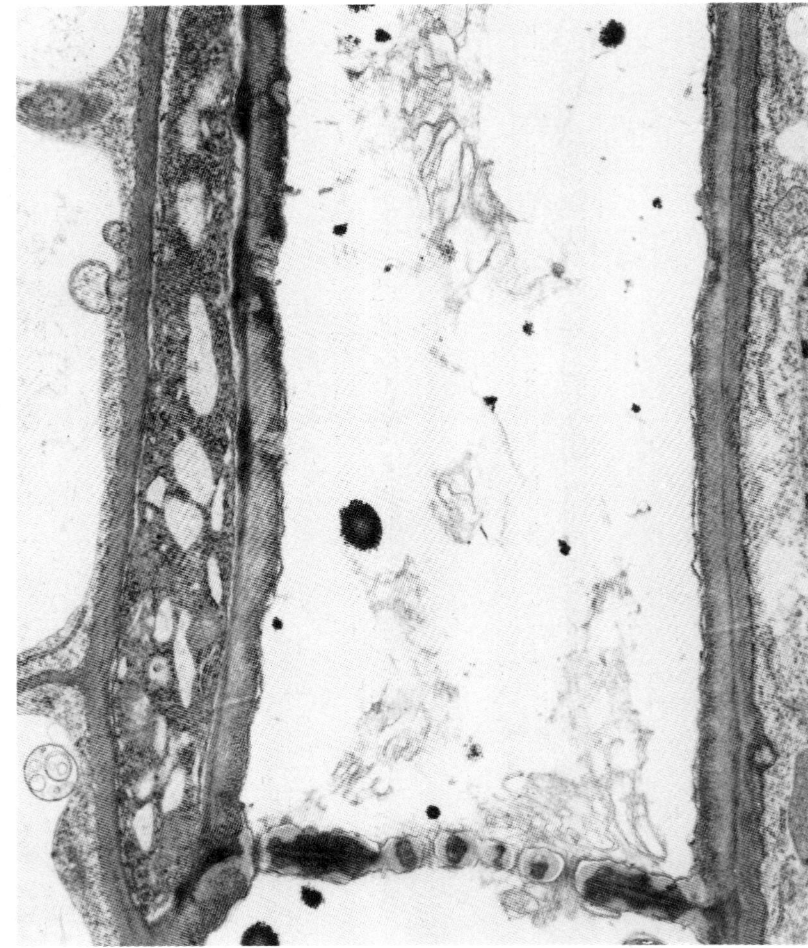

6.20 Electron micrograph of a phloem tube, which carries food such as sugars; a companion cell lies alongside the tube on the left, and a sieve plate can be seen at the bottom. The holes in the plate allow food to pass along the tube.

6.28 Phloem contains sieve-tube elements.

Like xylem vessels, phloem tubes are made of many cells joined end to end. However, their end walls have not completely broken down. Instead, they form **sieve plates** (see Figs 6.20 and 6.21), which have small holes in them. The cells are called **sieve tube elements.** Sieve tube elements contain cytoplasm, but no nucleus. They do not have lignin in their cell walls.

Each sieve tube element has a **companion cell** next to it. The companion cell does have a nucleus, and also contains many other organelles. Companion cells probably supply sieve tube elements with some of their requirements.

6.29 Vascular bundles contain xylem and phloem.

Xylem vessels and phloem tubes are usually found close together. A group of xylem vessels and phloem tubes is called a **vascular bundle.**

The positions of vascular bundles in roots and shoots is shown in Figs 6.22 and 6.23. In a root, vascular tissue is found at the centre, whereas in a shoot vascular bundles are arranged in a ring near the outside edge. They help to support the plant (see section 11.18).

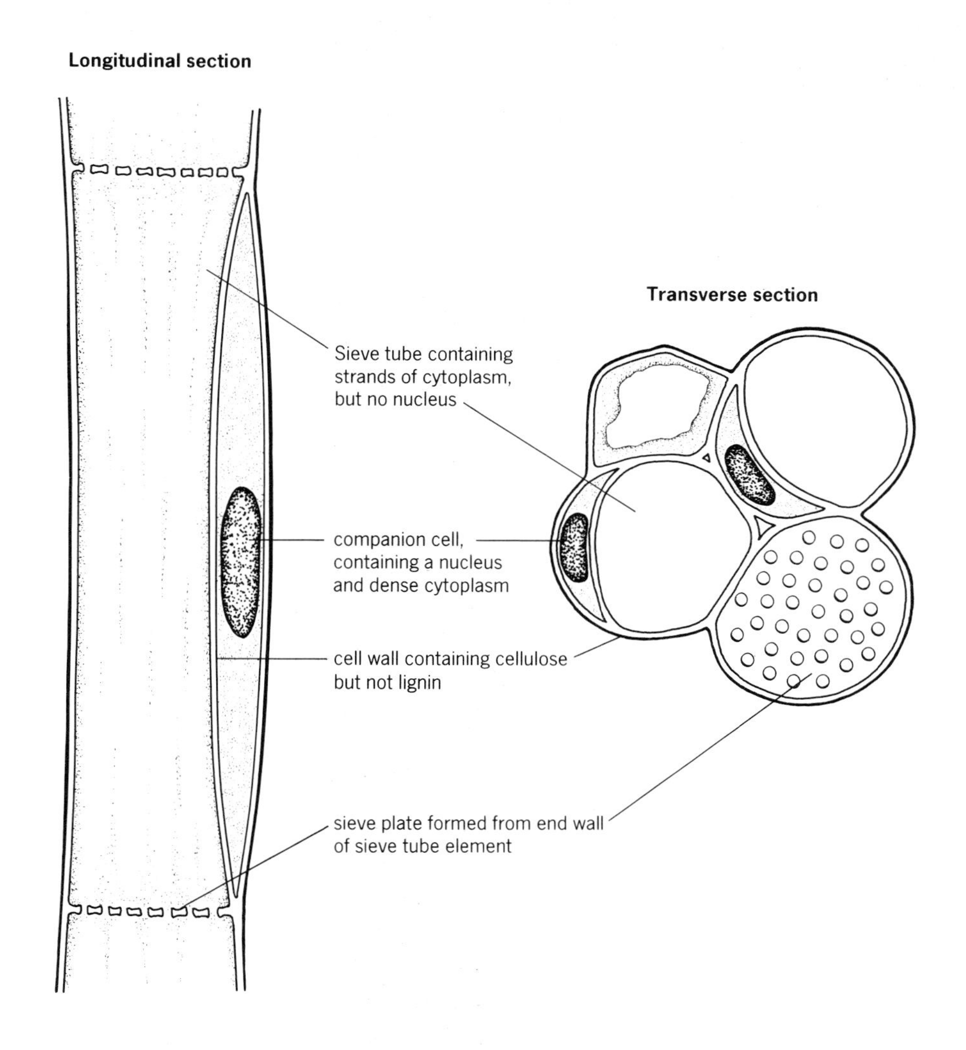

6.21 Phloem tubes

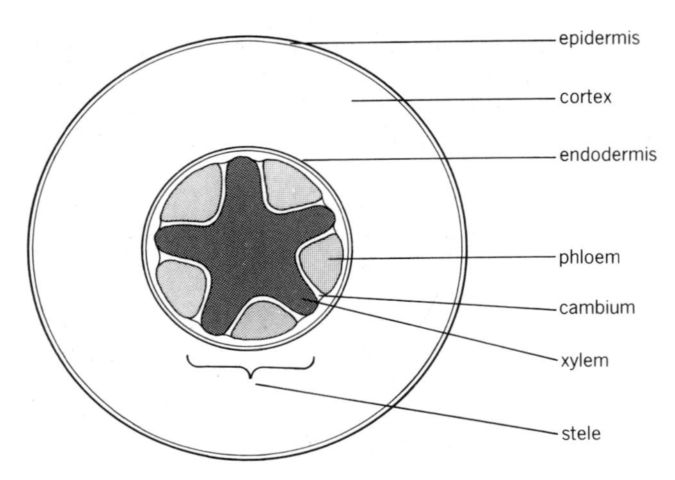

6.23 Transverse section of a root

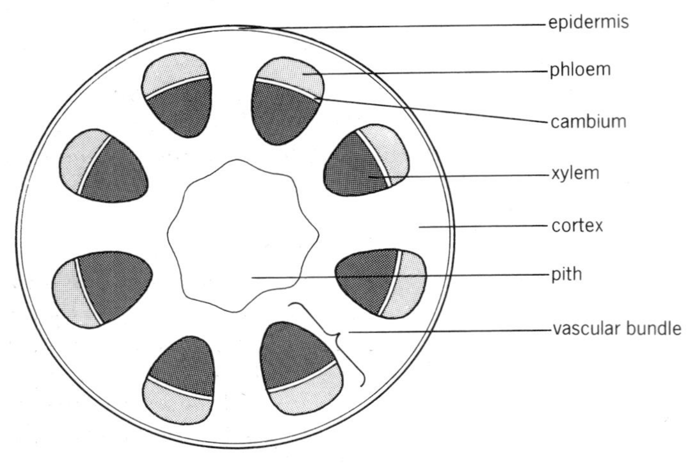

6.22 Transverse section of a stem

Questions

1. Why do plants not need such elaborate transport systems as mammals?
2. What do xylem vessels carry?
3. What do phloem tubes carry?
4. What substance makes up the cell walls of xylem vessels?
5. Give three ways in which phloem tubes differ from xylem vessels.
6. What is a vascular bundle?

Fact!

The longest roots which have ever been measured belonged to a fig tree growing in East Transvaal, South Africa. They went down to a depth of 120 m.

The heaviest carrot ever grown in Britain weighed 3.501 kg.

The transport of water

6.30 Most plants have fibrous or tap roots.

Plants take in water from the soil, through their **root hairs.** The water is carried in the xylem vessels to all parts of the plant.

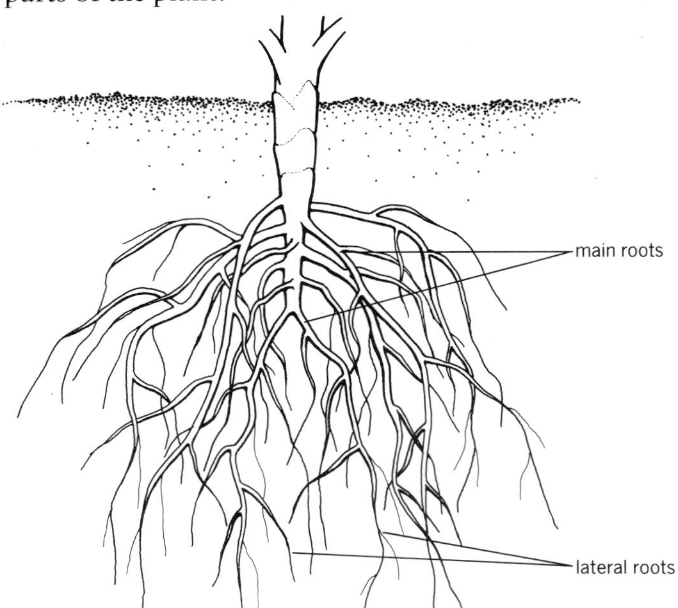

6.24 Fibrous root system of a groundsel plant

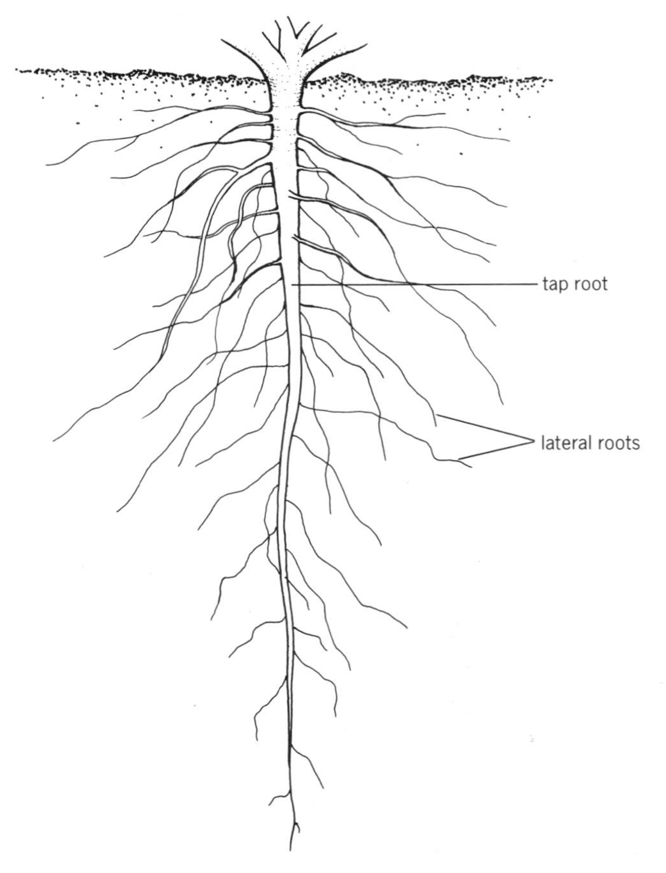

6.25 Tap root of a dandelion plant

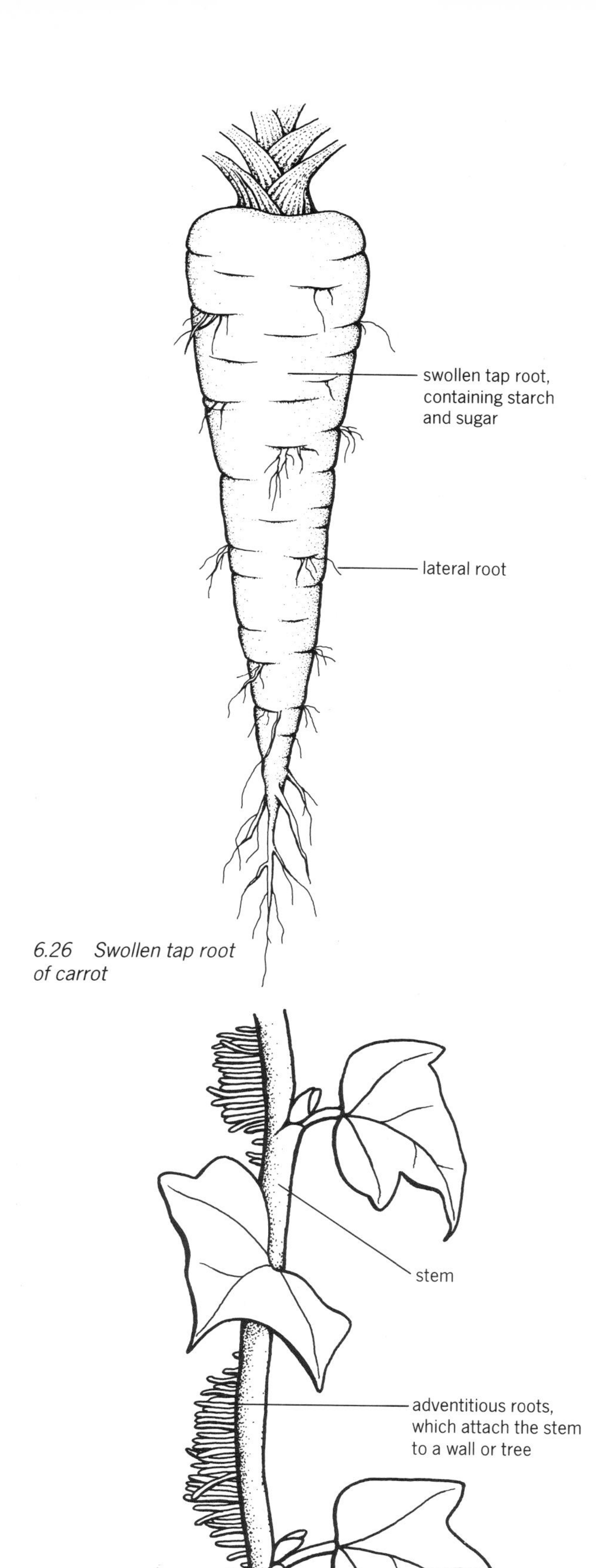

6.26 Swollen tap root of carrot

6.27 Adventitious roots of ivy

There are several different types of root system. Plants such as groundsel (see Fig 6.24) and grasses have **fibrous root systems.** There are several main roots, which have **lateral** or side roots branching from them.

Dandelions and carrots have **tap roots.** A tap root is a single main root, again with smaller lateral roots growing from it. Tap roots are often swollen with stored food.

If roots grow straight out of a stem, they are called **adventitious roots.** Ivy has adventitious roots, which attach it to walls or trees.

6.31 The structure of a root.

Fig 6.28 shows the structure of a root. At the very tip is the **root cap.** This is a layer of cells which protects the root as it grows through the soil. The rest of the root is covered by a layer of cells called the **epidermis.**

The **root hairs** are a little way up from the root tip. Each root hair is a long epidermal cell (see Fig 6.29). Root hairs do not live for very long. As the root grows, they are replaced by new ones.

6.32 Root hairs absorb water by osmosis.

The function of a root hair is to absorb water and minerals from the soil. Water gets into a root hair by osmosis. The cytoplasm and cell sap inside it are quite strong solutions. The water in the soil is normally a weaker solution. Water therefore diffuses into the root hair, down its concentration gradient, through the selectively permeable cell membrane.

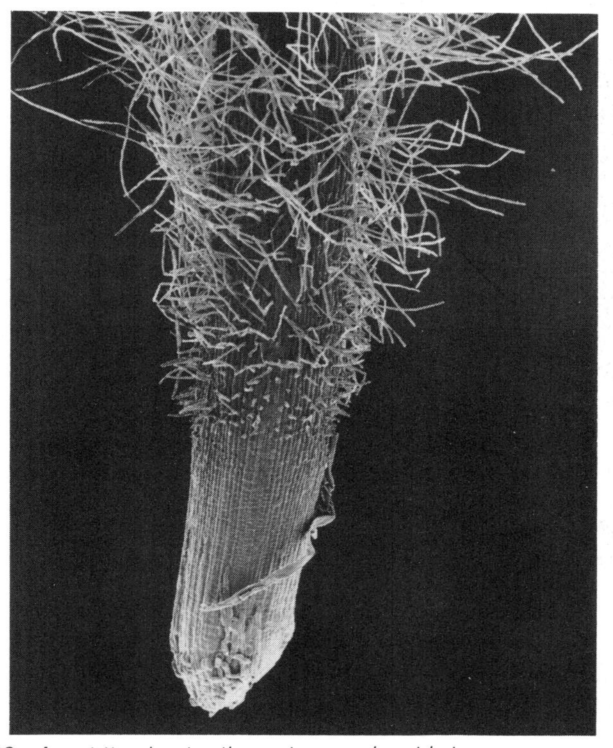

6.28 A root tip, showing the root cap and root hairs

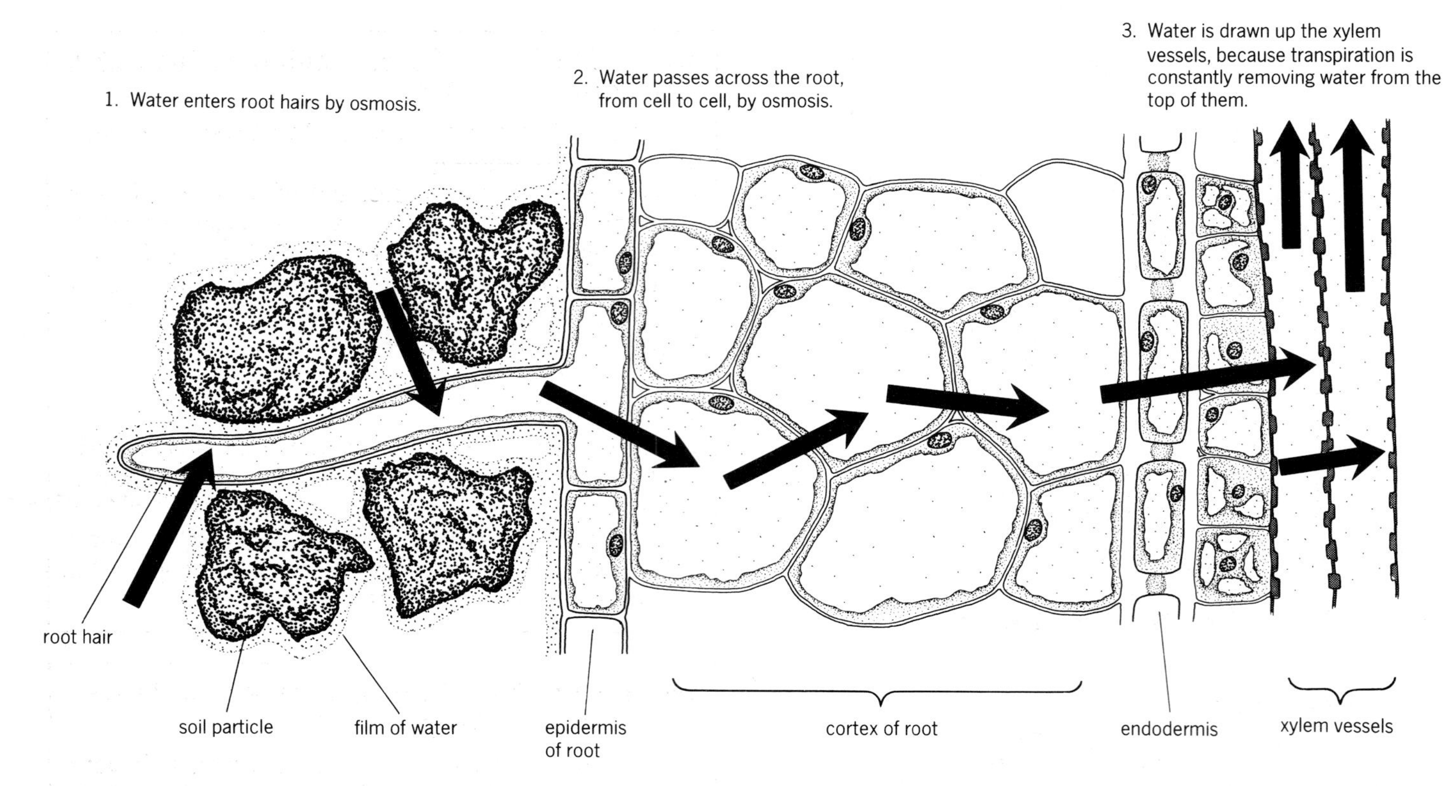

6.29 How water is absorbed by a plant

6.33 Absorbed water enters the xylem.

The root hairs are on the edge of the root. The xylem vessels are in the centre. Before the water can be taken to the rest of the plant, it must travel to these xylem vessels.

The path it takes is shown in Fig 6.29. It travels by osmosis through the cortex, from cell to cell. Some of it may also travel through the spaces between the cells.

6.34 Water is sucked up the xylem.

Water moves up xylem vessels in the same way that a drink moves up a straw when you suck it. When you suck a straw, you are reducing the pressure at the top of the straw. The liquid at the bottom of the straw is at a higher pressure, so it flows up the straw into your mouth.

The same thing happens with the water in xylem vessels. The pressure at the top of the vessels is lowered, while the pressure at the bottom stays high. Water therefore flows up the xylem vessels.

How is the pressure at the top of the xylem vessels reduced? It happens because of transpiration.

6.35 Transpiration is evaporation from leaves.

Transpiration is the evaporation of water from a plant. Most of this evaporation takes place from the leaves.

If you look back at Fig 3.4, you will see that there are openings on the underside of the leaf called **stomata.** The cells inside the leaf are each covered with a thin film of moisture. This is necessary, so that gas exchange can take place.

Some of this film of moisture evaporates from the cells, and diffuses out of the leaf through the stomata. Water from the xylem vessels in the leaf will travel to the cells by osmosis to replace it.

Water is constantly being taken from the top of the xylem vessels, to supply the cells in the leaves. This reduces the effective pressure at the top of the xylem vessels, so that water flows up them. This process is known as the **transpiration stream** (see Fig 6.30).

6.36 A potometer compares transpiration rates.

It is not easy to measure how much water is lost from the leaves of a plant. It is much easier to measure how fast the plant takes up water. The rate at which a plant takes up water depends on the rate of transpiration – the faster a plant transpires, the faster it takes up water.

Fig 6.32 illustrates apparatus which can be used to compare the rate of transpiration in different conditions. It is called a **potometer.** By recording how fast the air/water meniscus moves along the capillary tube, you can compare how fast the plant takes up water in different conditions.

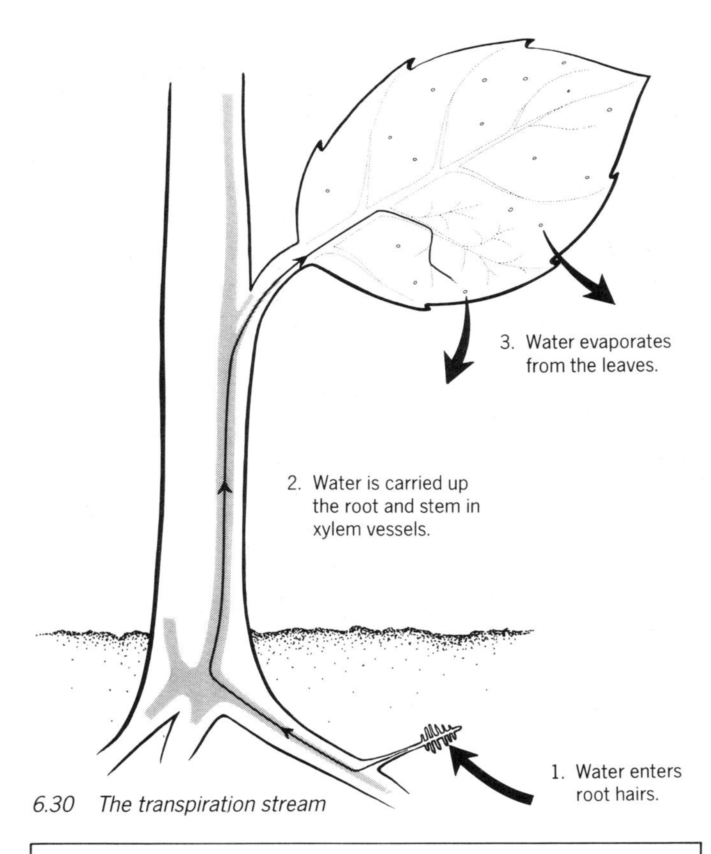

6.30 The transpiration stream

Investigation 6.2 To see which part of a stem transports water and solutes

1 Take a plant, such as groundsel, with a root system intact. Wash the roots thoroughly.
2 Put the roots of the plant into eosin solution. Leave overnight.
3 Set up a microscope.
4 Remove the plant from the eosin solution, and wash the roots thoroughly.
5 Use a razor blade to cut across the stem of the plant about half way up. Take great care when using a razor blade and do not touch its edges.
6 Now cut very thin sections across the stem. Try to get them so thin that you can see through them.
7 Choose your best section, and mount it in a drop of water on a microscope slide. Cover with a coverslip.
8 Observe the section under a microscope. Compare what you can see with Fig 6.22. Make a labelled drawing of your section.

Questions

1 Which part of the stem contained the dye? What does this tell you about the transport of water and solutes (substances dissolved in water) up a stem?
2 Why was it important to wash the roots of the plant:
(a) before putting it into the eosin solution, and
(b) before cutting sections?
3 Try to design an experiment to find out how quickly the dye is transported up the stem.

Investigation 6.3 To see which surface of a leaf loses most water

Cobalt choride paper is blue when dry and pink when wet. Use forceps to handle it.

1 Use a healthy, well watered potted plant, with leaves which are not too hairy. Fix a small square of cobalt chloride paper onto each surface of one leaf, using clear sticky tape. Make sure there are no air spaces around the paper.
2 Leave the paper on the leaf for a few minutes.

Questions

1 Which piece of cobalt chloride paper turned pink first? What does this tell you about the loss of water from a leaf?
2 Why does this surface lose water faster than the other?
3 Why is it important to use forceps, not fingers, for handling cobalt chloride paper?

Investigation 6.4 To measure the rate of transpiration of a potted plant

1 Use two similar well watered potted plants. Enclose one plant entirely in a polythene bag, including its pot. This is the control.
2 Enclose only the pot of the second plant in a polythene bag. Fix the bag firmly around the stem of the plant, and seal with Vaseline (see Fig 6.31).
3 Place both plants on balances, and weigh them.
4 Weigh each plant every day, at the same time, for at least a week.
6 Draw a graph of your results.

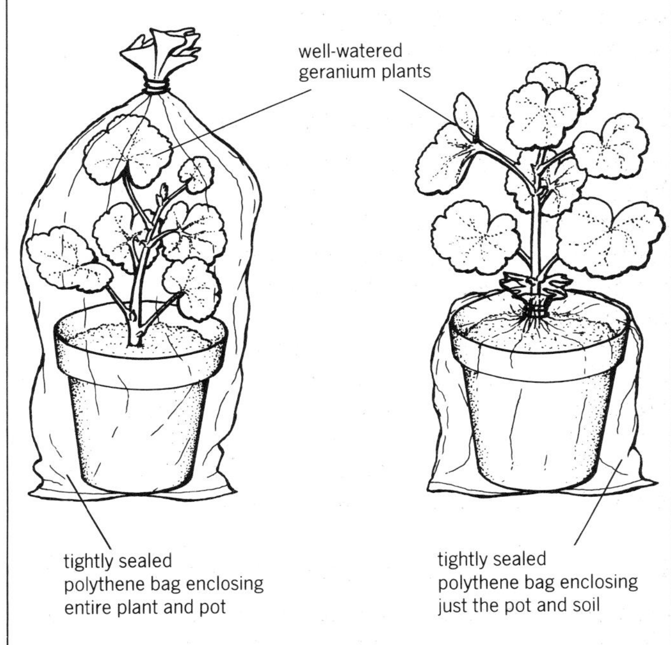

6.31 To measure the rate of transpiration of a potted plant

continued

Questions

1 Which plant lost weight? Why?
2 Do you think this is an accurate method of measuring transpiration rate? How could it be improved?

Fact! An oak tree 16 m high and with a trunk about 1 m in diameter would transpire about 150 litres of water on a warm, sunny day.

Investigation 6.5 Using a potometer to compare rates of transpiration under different conditions

1 Set up the potometer as in Fig 6.32. The stem of the plant must fit exactly into the rubber tubing, with no air gaps. Vaseline will help to make an air tight seal.
2 Fill the apparatus with water, by opening the screw clip.
3 Close the clip again, and leave the apparatus in a light, airy place. As the plant transpires, the water it loses is replaced by water taken up the stem. Air will be drawn in at the end of the capillary tube.
4 When the air/water meniscus reaches the scale, begin to record the position of the meniscus every two minutes.
5 When the meniscus reaches the end of the scale, refill the apparatus with water from the reservoir as before.
6 Now repeat the experiment, but with the apparatus in a different situation. You could try each of these.
(a) blowing it with a fan
(b) putting it in a cupboard
(c) putting it in a refrigerator
7 Draw a graph of your results.

Questions

1 Under which conditions did the plant transpire (a) most quickly, and (b) most slowly?
2 You have been using the potometer to compare the rate of uptake of water under different conditions. Does this really give you a good measurement of the rate of transpiration? Explain your answer.

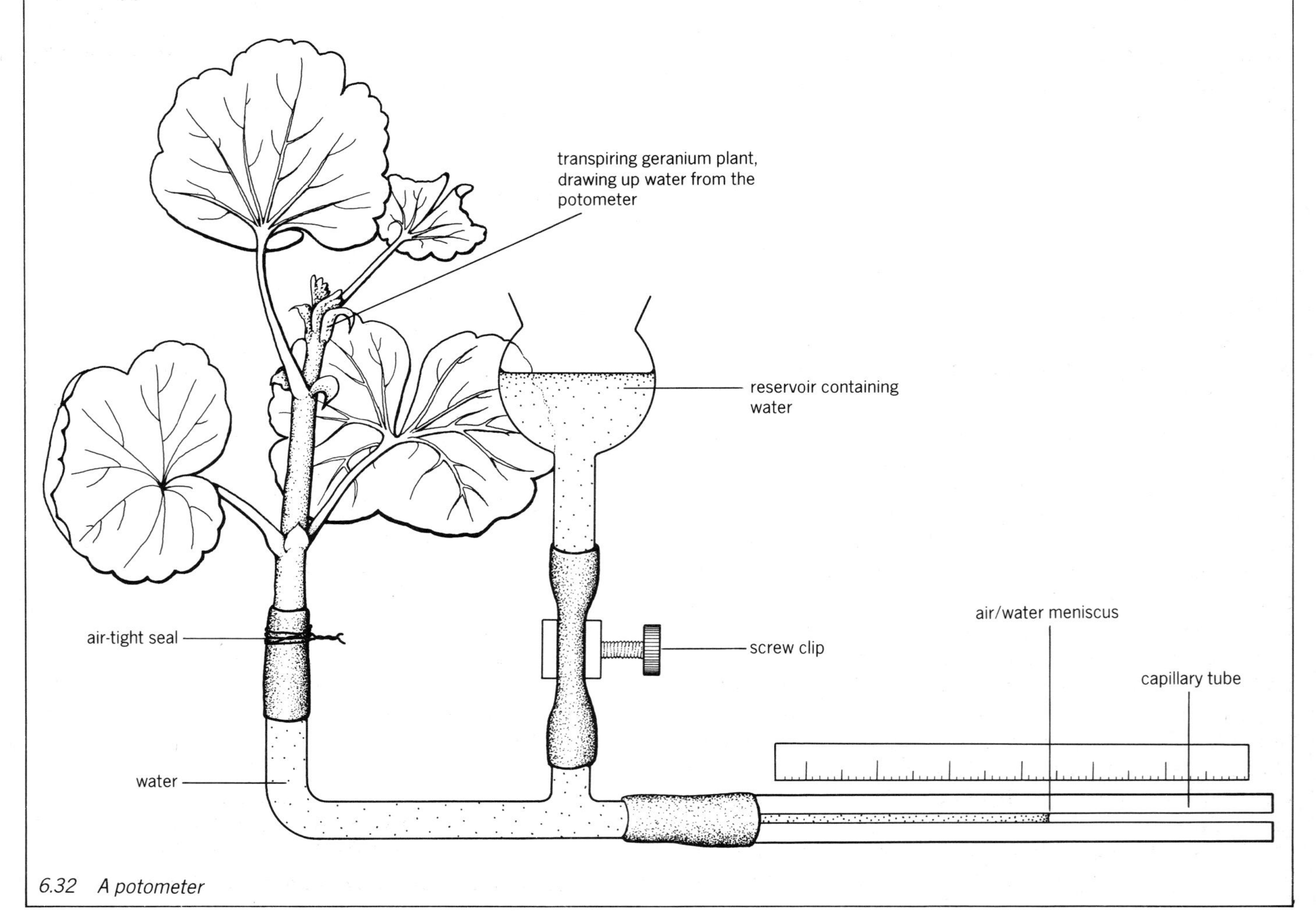

6.32 A potometer

6.37 Conditions which affect transpiration rate.

Temperature On a hot day, water will evaporate quickly from the leaves of a plant. Transpiration increases as temperature increases.

Humidity Humidity means the moisture content of the air. The higher the humidity, the less water will evaporate from the leaves. Transpiration decreases as humidity increases.

Wind speed On a windy day, water evaporates more quickly than on a still day. Transpiration increases as wind speed increases.

Light intensity In bright sunlight, a plant may open its stomata to supply plenty of carbon dioxide for photosynthesis. More water can therefore evaporate from the leaves.

Water supply If water is in short supply, then the plant will close its stomata. This will cut down the rate of transpiration. Transpiration decreases when water supply decreases below a certain level.

Transpiration is useful to plants, because it keeps water moving up the xylem vessels. But if the leaves lose too much water, the roots may not be able to take up enough to replace it. If this happens, the plant **wilts.** Many plants have other ways of cutting down their rate of transpiration, and these are described in section 10.18.

Uptake of mineral salts

6.38 Root hairs absorb minerals by active transport.

As well as absorbing water by osmosis, root hairs absorb mineral salts. These are in the form of ions (see Table 2.3) dissolved in the water in the soil. They travel to the xylem vessels along with the water which is absorbed, and are transported to all parts of the plant.

These minerals are usually present in the soil in quite low concentrations. The concentration inside the root hairs is higher. In this situation the mineral ions would normally diffuse out of the root hair into the soil. Root hairs can, however, take up mineral salts against their concentration gradient. It is the cell membrane which does this. Special carrier molecules in the cell membrane of the root hair carry the mineral ions across the cell membrane into the cell, against their concentration gradient.

This is called **active transport.** It uses a lot of energy. The energy is supplied by mitochondria in the root hairs.

Transport of manufactured food

6.39 Phloem translocates organic foods.

Leaves make carbohydrates by photosynthesis. They also use some of these carbohydrates to make amino acids, proteins, oils and other organic substances.

Some of the organic food material, especially sugar, that the plant makes is transported in the phloem tubes. It is carried from the leaves to whichever part of the plant needs it. This is called **translocation.** The sap inside phloem tubes therefore contains a lot of sugar, particularly sucrose.

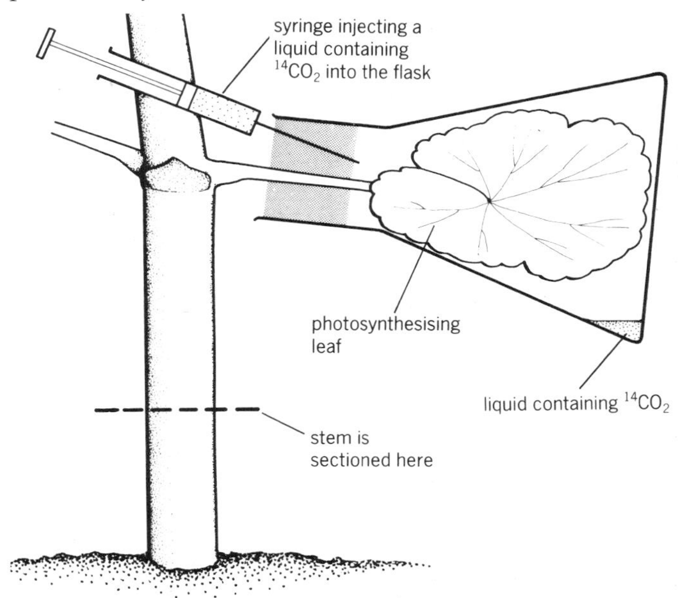

The leaf is supplied with radioactive carbon in the form of $^{14}CO_2$.

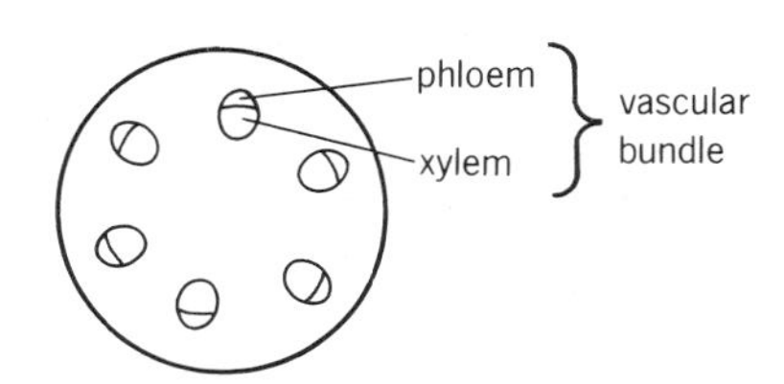

When the leaf has had a chance to photosynthesise, a thin section is cut from the stem.

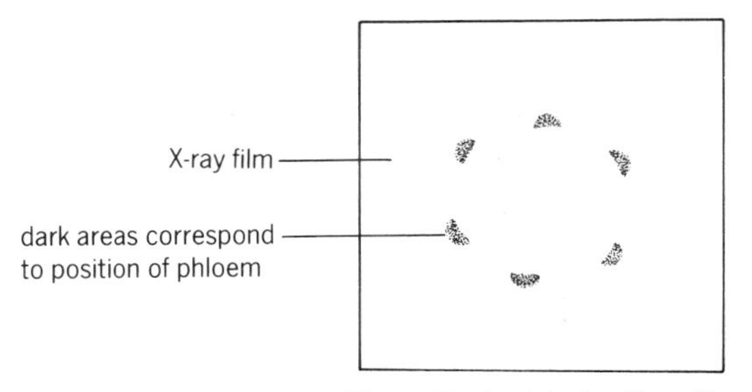

The section is placed on X-ray film. The radioactivity darkens the film.

6.33 Using radioisotopes to demonstrate that phloem transports organic substances

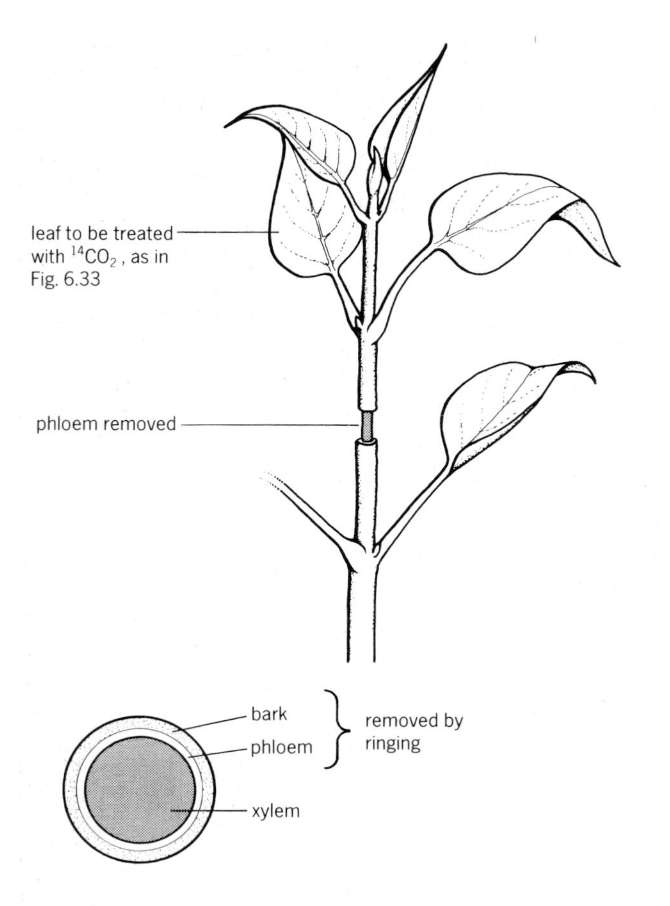

Phloem is removed from a short length of woody stem by ringing. A leaf above the ring is fed with $^{14}CO_2$. The stem is then sectioned both above and below the ring. Radioactivity can be detected above, but not below the ring. This shows that the removal of phloem stops the movement of the products of photosynthesis.

6.34 A ringing experiment

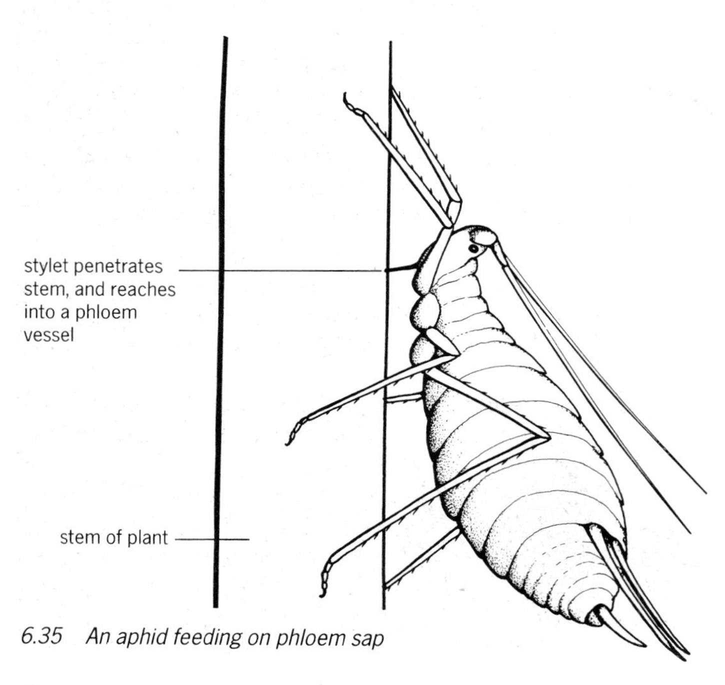

6.35 An aphid feeding on phloem sap

6.40 Evidence that phloem translocates organic food.

Using radio-isotopes The carbon isotope ^{14}C is radioactive (see section 3.12). If a leaf is supplied with $^{14}CO_2$, it will make sugars containing ^{14}C. After a short while, sections can be cut of the stem. If the sections are placed on X-ray photographic film (see Fig 6.33), this will show that all the radioactivity is in the phloem tubes. This means that the food that the leaves have made is being transported in the phloem.

Ringing experiments The phloem tubes in a woody stem are just underneath the bark. If a ring of bark is taken off, the phloem tubes are removed too. Fig 6.34 illustrates the results of one of these experiments.

Using aphids Aphids, such as greenfly, feed on plant juices. They have special mouthparts called **stylets.** They push their stylets into the phloem tubes of a plant, and suck up the sap from them (see Fig 6.35).

A feeding aphid can be anaesthetised, and its mouth parts cut off. Phloem sap keeps flowing out of the phloem tubes through the stylets. The sap can then be analysed. It is found to contain sugars, and many other organic materials.

Questions

1 What are adventitious roots?
2 What is the function of a root cap?
3 Explain how water goes into root hairs.
4 What is transpiration?
5 What are stomata?
6 What is a potometer used for?
7 Explain how (a) temperature, and (b) light affect transpiration.
8 What is active transport?

Chapter revision questions

1 Using Fig 6.10 to help you, list in order, the blood vessels and parts of the heart which:
(a) a glucose molecule would travel through on its way from your digestive system to a muscle in your leg
(b) a carbon dioxide molecule would travel through on its way from the leg muscle to your lungs.

2 Explain the difference between each of the following pairs:
(a) blood, lymph;
(b) diastole, systole;
(c) artery, vein;
(d) deoxygenated blood, oxygenated blood;
(e) atrium, ventricle;
(f) hepatic vein, hepatic portal vein;
(g) red blood cell, white blood cell;
(h) xylem, phloem;
(i) tap root, adventitious root;
(j) diffusion, active transport.

3 Arteries, veins, capillaries, xylem vessels and phloem tubes are all tubes used for transporting substances in mammals and flowering plants. Describe how each of these tubes is adapted for its particular function.

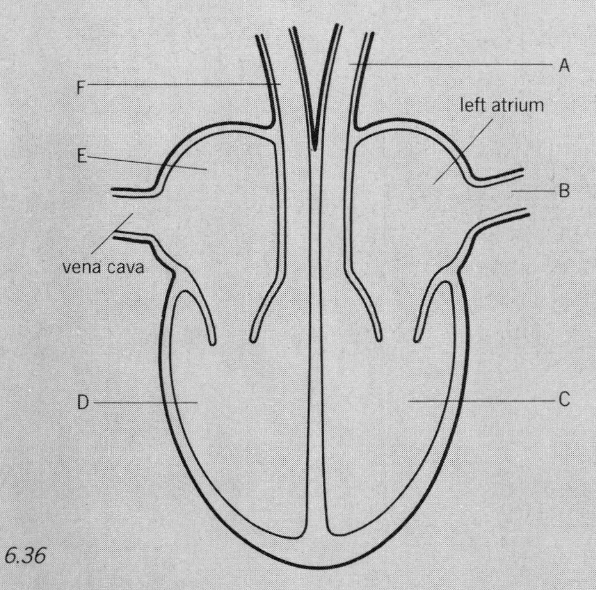

6.36

4 (a) What is meant by a double circulatory system?
(b) Copy the diagram in Fig 6.36 of a section through a heart. Fill in the labels A to F, and draw in arrows to show the direction of blood flow through it.
(c) In an unborn child, the lungs do not work. The baby gets its oxygen from the mother, to which it is connected by the umbilical cord. This cord contains a vein, which carries the oxygenated blood to the baby's vena cava.
(i) Which chamber of the heart does oxygenated blood enter in an adult person?
(ii) Which chamber of the heart does oxygenated blood enter in an unborn baby?
(iii) In an unborn child, there is a hole in the septum between the left and right atria. What purpose do you think this has?
(iv) As soon as the baby takes its first breath, this hole closes up. Why is this important?

5 An experiment was performed where a solution of human haemoglobin was exposed to samples of air containing different amounts of oxygen (measured in kilopascals, kPa). At each oxygen concentration, the haemoglobin sample was tested to see how much oxygen it had absorbed. If it had absorbed as much oxygen as it could possibly carry, it was said to be 100% saturated. If it only carried half this amount, it was 50% saturated, and so on. The graph shows the results obtained.

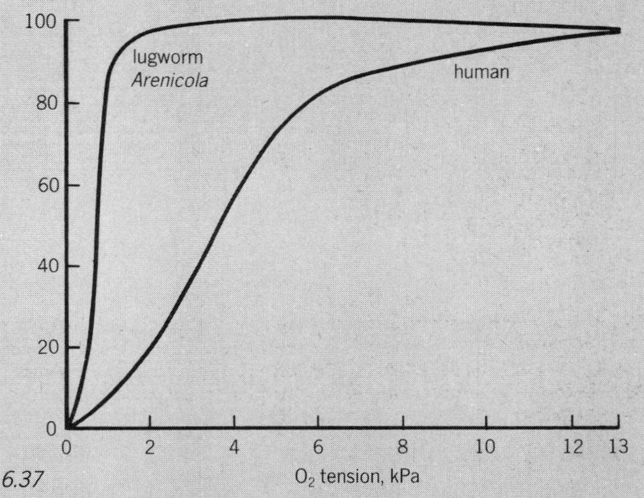

6.37

(a) What is the approximate percentage saturation of human haemoglobin with oxygen, at an oxygen tension of (i) 2 kPa and (ii) 6 kPa?
(b) If haemoglobin which had been exposed to an oxygen tension of 6 kPa was then exposed to an oxygen tension of 2 kPa, would it absorb or give up oxygen?
(c) In the lungs, the oxygen tension is usually about 13 kPa. In the muscles, it is around 4 kPa. Explain why.
(d) Using the information in part (c), and your answer to part (b), explain how haemoglobin transports oxygen from the lungs to a muscle.
(e) What would be the disadvantage of having haemoglobin which absorbed a lot of oxygen at low oxygen tensions?
(f) The lugworm, *Arenicola,* lives in burrows on muddy beaches. There is often only a very little oxygen available. *Arenicola* is not a very active animal. The blood of *Arenicola* contains haemoglobin, which behaves rather differently from human haemoglobin (see graph). Can you explain why this type of haemoglobin suits *Arenicola's* way of life better than human haemoglobin would?

7 Growth

7.1 Growth does not just mean getting bigger.

All living organisms are said to grow; but what exactly is growth?

Growth does not simply mean getting bigger. Puffer fish, for example, swallow water when a predator attacks them. They swell up, and get much larger. This often frightens the predator away, and the fish shrinks back to its normal size again. This is not growth, because the fish only got bigger for a very short time.

An organism can be said to be growing if it gets bigger, and stays bigger. Growth is permanent increase in size.

7.2 Growth usually involves cell division.

When an organism grows, the cells it is made of get bigger. However, a cell can only grow to a certain size. If it gets too big, then its surface area to volume ratio (see section 5.14) becomes so small that gas exchange cannot take place fast enough.

Because of this growth also involves the division of cells to make new ones. This in itself does not make the organism any bigger, because the two new cells are the same size as the one big cell which divided to form them. However, the new cells can then grow bigger.

The way in which cells divide to provide new cells for growth is called **mitosis.**

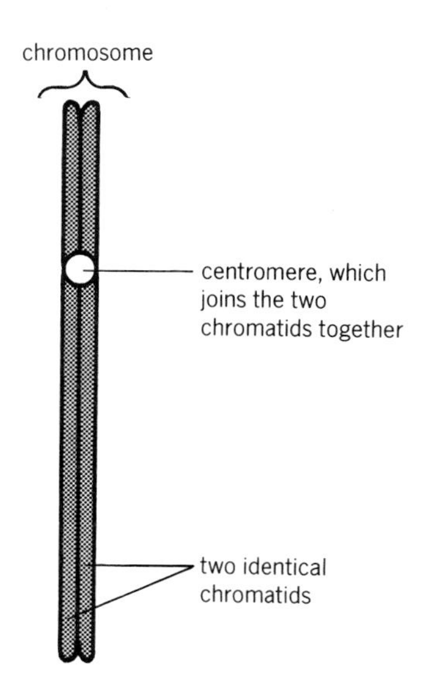

7.1 A chromosome just before cell division

7.3 In growth, cells divide by mitosis.

Mitosis is the way in which any cell, plant or animal, divides when an organism is growing, or repairing a damaged part of its body.

The nucleus of a cell contains **chromosomes.** These chromosomes have **genes** on them, which give instructions to the cell about what sort of proteins to make. When a cell divides, it is very important that the two new cells each get a complete set of chromosomes from their parent cell.

Mitosis is a method of cell division which makes two new cells with exactly the same number and kinds of chromosomes as the original cell. The way in which mitosis happens is shown in Fig 7.2.

Questions

1. What is growth?
2. Why can cells not grow beyond a certain size?
3. What is mitosis?
   Examine Fig 7.2 overleaf.
4. Why can you not see chromosomes in a cell when it is not dividing?
5. What do the centrioles do during mitosis?
6. What happens to the two chromatids of each chromosome during anaphase of mitosis?
7. What is the end result of mitosis?

Fact!

Human giantism can be caused by an imbalance of hormones. The tallest man was Robert Wadlow, who died in 1940 at the age of 22. He was 2.72 metres (8′ 11″) tall when he died.

The fastest growth rate in the animal kingdom is shown by the blue whale calf. It puts on 26 tonnes of weight in 22¾ months – 10¾ months in its mother's body, and the first 12 months after birth.

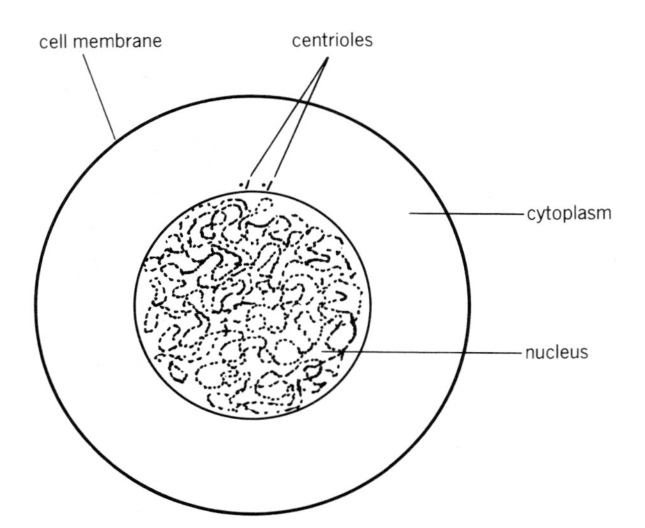

Interphase When a cell is not dividing, no chromosomes can be seen clearly in the nucleus. They are there, but are so long and thin that they are invisible.

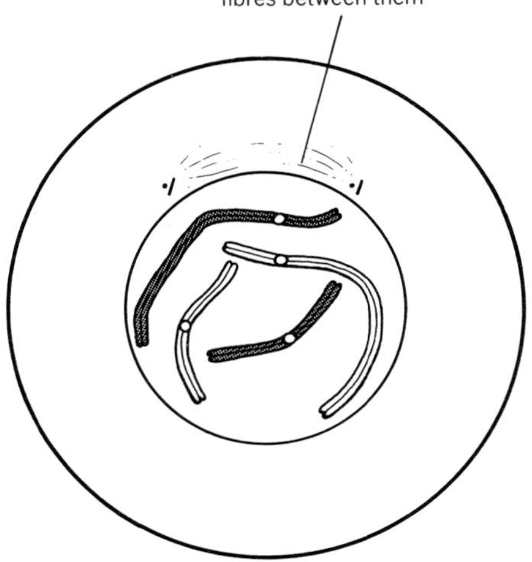

Prophase The chromosomes get short and fat, so they can now be seen with a light microscope. Each chromosome contains two chromatids (see Fig 7.1).

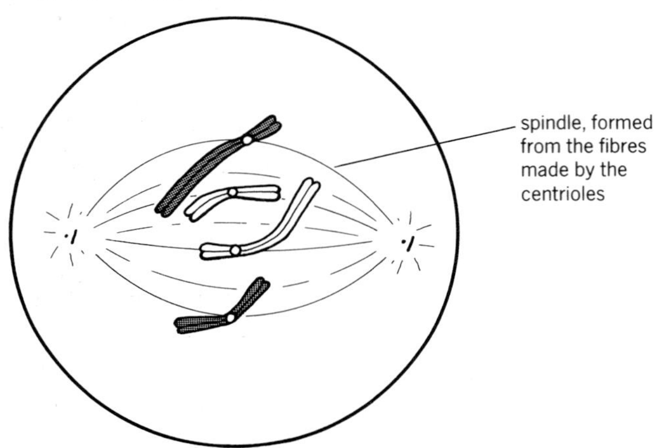

Metaphase The nuclear membrane vanishes. The chromosomes line up on the equator of the spindle.

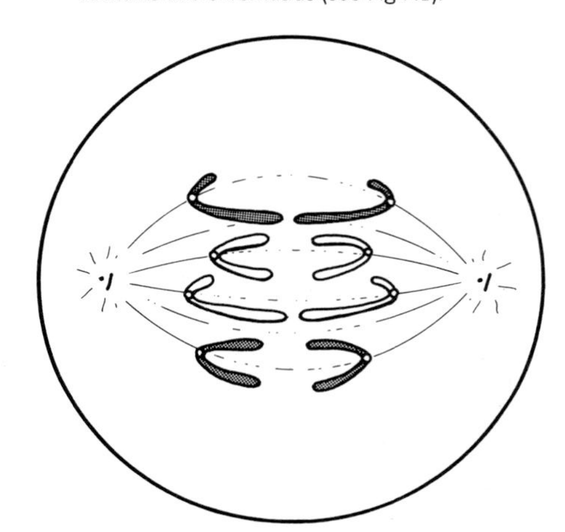

Anaphase The centromere of each chromosome splits, so the two chromatids separate. The chromatids move away from each other, along the spindle fibres.

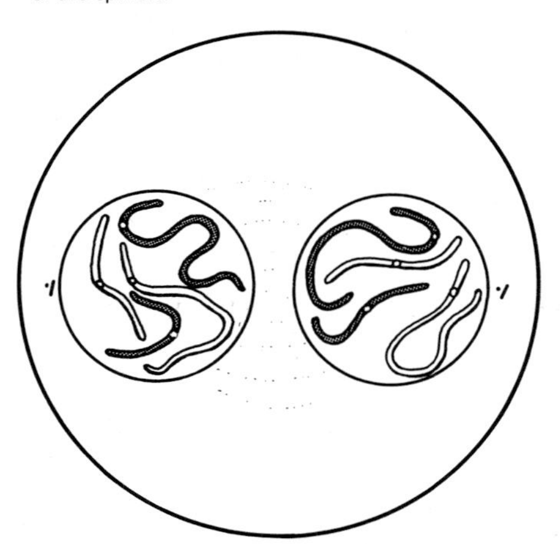

Telophase The chromatids arrive at opposite ends of the cell, and form into groups. A nuclear membrane appears round each group. The spindle fibres fade away.

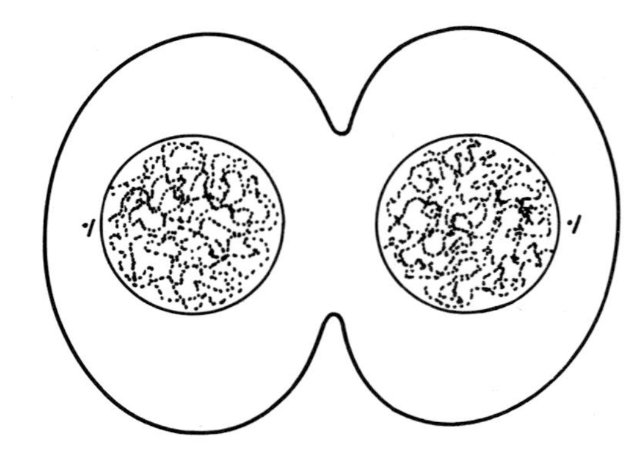

Late telophase The chromosomes become long and thin again, so that they are invisible. The cytoplasm divides, forming two daughter cells. Each cell now goes into interphase again.

7.2 Mitosis in an animal cell with four chromosomes

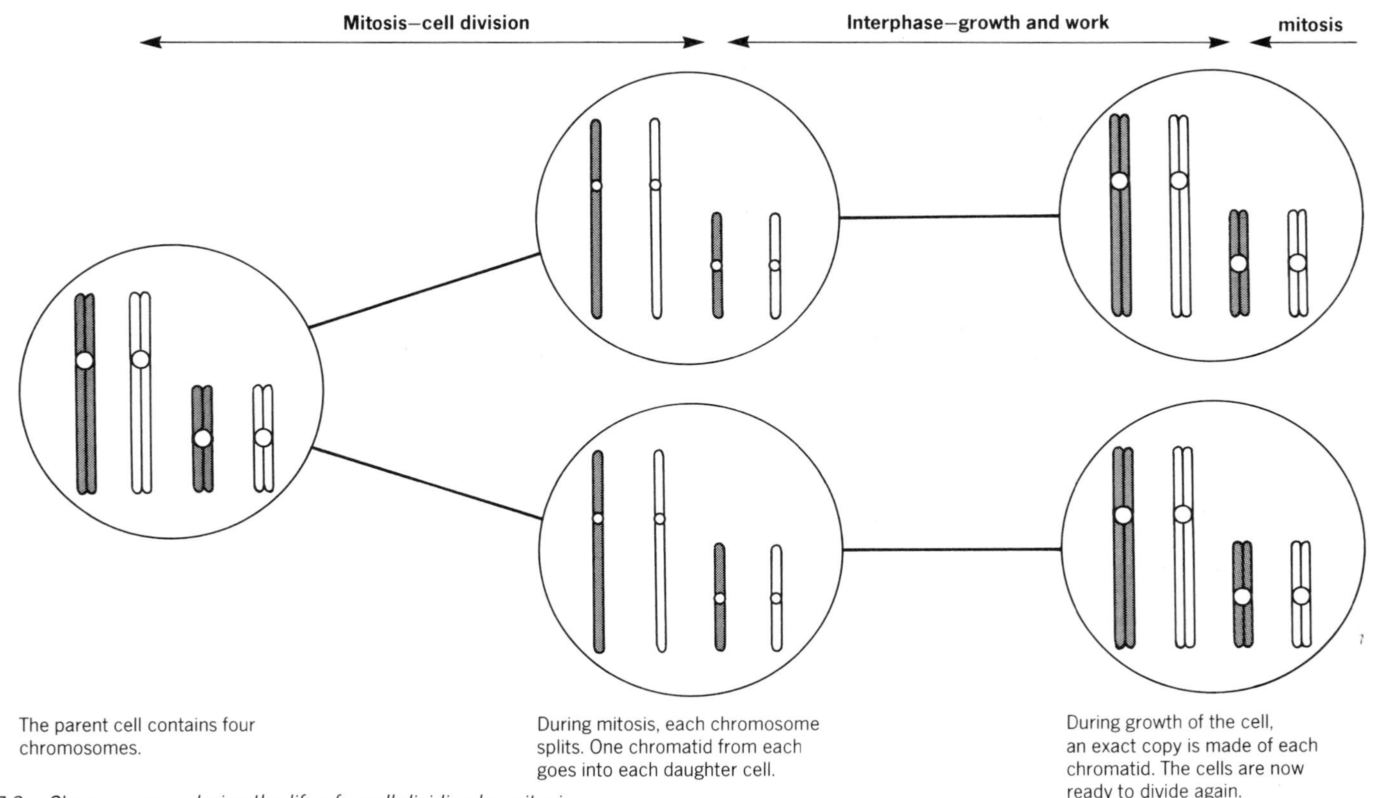

7.3 Chromosomes during the life of a cell dividing by mitosis

The growth of a flowering plant

7.4 Only some parts of plants can grow.

In animals, most cells can divide by mitosis. All the parts of an animal's body can grow. In a plant, though, not all cells can divide by mitosis. Only the cells in certain places can do this. These places are called **meristems.**

Fig 7.4 shows where the meristems are in a typical plant. Most of them are at the tips of shoots, and just behind the tips of roots. Because of this, most plants grow in a branching shape. Animals, however, because all the parts of their bodies can grow, usually grow into a more compact shape.

7.5 A flowering plant grows from a seed.

Many plants begin their life as a seed. The way in which a seed is formed is explained in section 8.40.

A seed contains an embryo plant. The embryo consists of a **radicle,** which will grow into a root, and a **plumule,** which will grow into a shoot (see Fig 7.5).

There is also food for the embryo. In a French bean seed, the food is stored in two cream coloured **cotyledons.** These contain starch and protein. The cotyledons also contain enzymes.

Surrounding the cotyledons is a tough, protective covering called the **testa.** The testa stops the embryo from being damaged and it prevents bacteria and fungi from entering the seed.

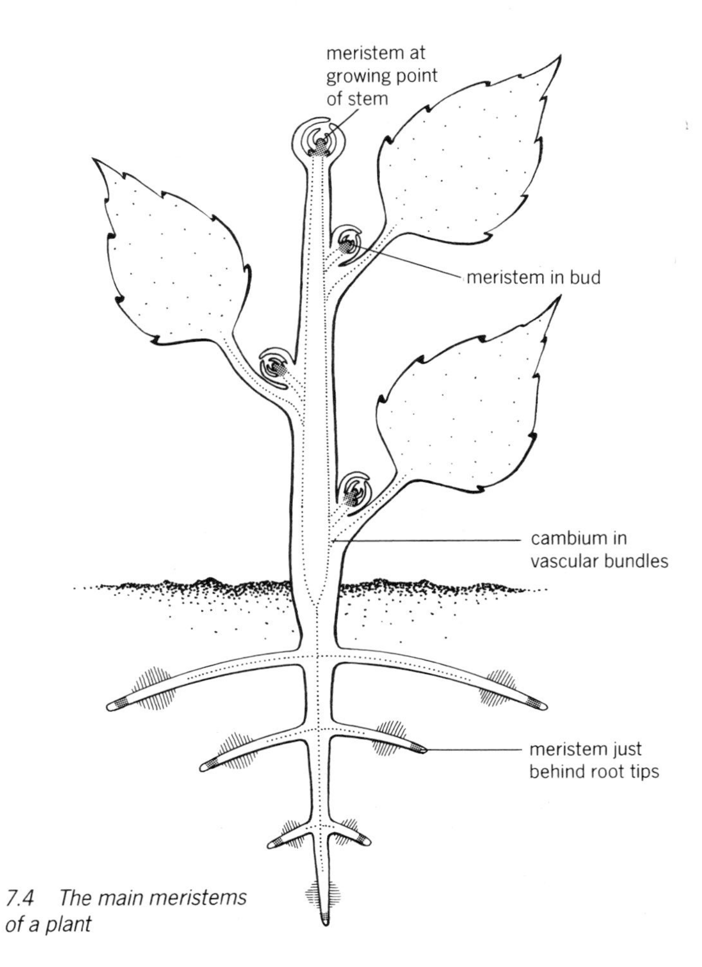

7.4 The main meristems of a plant

The testa has a tiny hole in it call the **micropyle.** Near the micropyle is a scar, the **hilum,** where the seed was joined onto the pod.

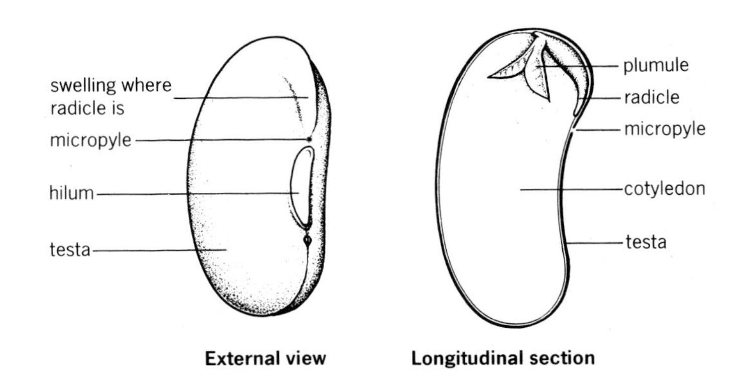

7.5 Structure of a French bean seed

7.6 Uptake of water begins seed germination.

A seed contains hardly any water. When it was formed on the plant, the water in it was drawn out, so that it became dehydrated. Without water, almost no metabolic reactions can go on inside it. The seed is inactive or **dormant.** This is very useful, because it means that the seed can survive harsh conditions, such as cold or drought, which would kill a growing plant.

A seed must be in certain conditions before it will begin to germinate. You can find out what they are if you do Investigation 7.1.

Investigation 7.1 To find the conditions necessary for the germination of mustard seeds

1 Set up five tubes as shown in Fig 7.6.
2 Put tubes A, D and E in a warm place in the laboratory, in the light.
3 Put tube B in a refrigerator.
4 Put tube C in a warm, dark cupboard.
5 Fill in the results table to show what conditions the seeds in each tube have. The first line has been done for you.
6 Leave all the tubes for several days, then examine them to see if the seeds have germinated or not.

Results table

	Tube A	B	C	D	E
Water	✓	✓	✓	✓	×
Warmth					
Oxygen					
Light					
Did seeds germinate?					

Questions

1 What three conditions do mustard seeds need for germination?
2 Read paragraphs 7.6 and 7.7, and then explain why each of these conditions is needed for successful germination.

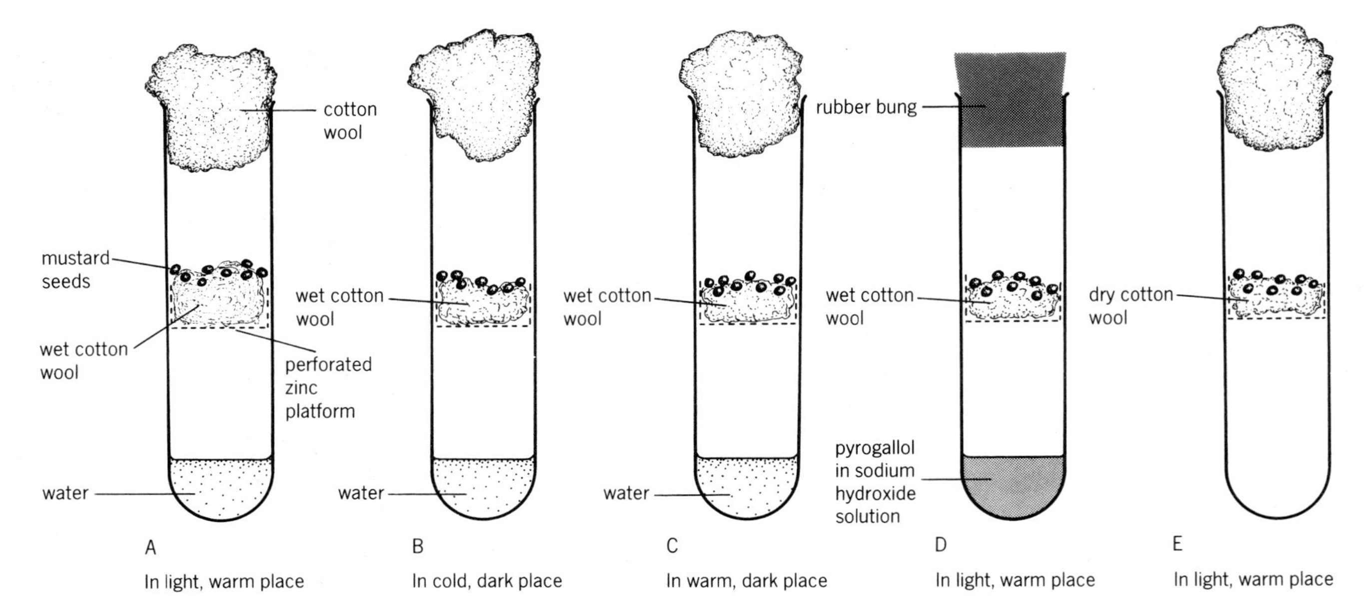

7.6 Experiment to find the conditions necessary for the germination of mustard seeds

When a seed germinates, it first takes up water through the micropyle. As the water goes into the cotyledons, they swell. Eventually, they burst the testa (see Fig 7.7).

Once there is sufficient water, the enzymes in the cotyledons become active. Amylase begins to break down the stored starch molecules to maltose. Proteases break down the protein molecules to amino acids.

Maltose and amino acids are soluble, so they dissolve in the water. They diffuse to the embryo plant, which uses these foods for growth. The way in which the embryo plant grows is shown in Fig 7.7.

7.7 During germination, enzymes digest food stores.

When a seed first begins to germinate, it increases in weight. This is because it absorbs water from the soil.

As soon as it begins to grow, it starts to use its food stores. The stored protein is broken down to amino acids, which are used to make new protein molecules for cell membranes and cytoplasm. The stored starch is broken down to maltose and then to glucose. Some of the glucose will be made into cellulose, to make cell walls for the new cells.

All this requires energy. The seed, like all living organisms, get its energy by breaking down glucose, in respiration. Quite a lot of the glucose from the stored starch will be used up in respiration, so the seed loses weight.

After a few days, the plumule of the seed grows above the surface of the ground. The first leaves open out and begin to photosynthesise. The plant can now make its own food faster than it is using it up. It begins to increase in weight.

Fig 7.8 summarises the changes in weight of an annual plant, such as a French bean, from germination until death. An annual plant is one which lives for less than one year.

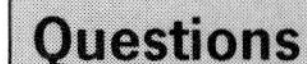

Questions

1 What is a meristem?
2 List three places where meristems may be found on a growing plant.
3 What do the cotyledons of a French bean seed contain?
4 What does dormant mean?
5 What is the advantage of dormancy?
6 What activates the enzymes in the cotyledons?
7 What do the enzymes do?

7.8 Plants grow lengthways by cell division and elongation.

During germination, the radicle and plumule grow lengthways, quickly becoming the root and shoot of the young plant. Roots and shoots continue to grow throughout the life of the plant.

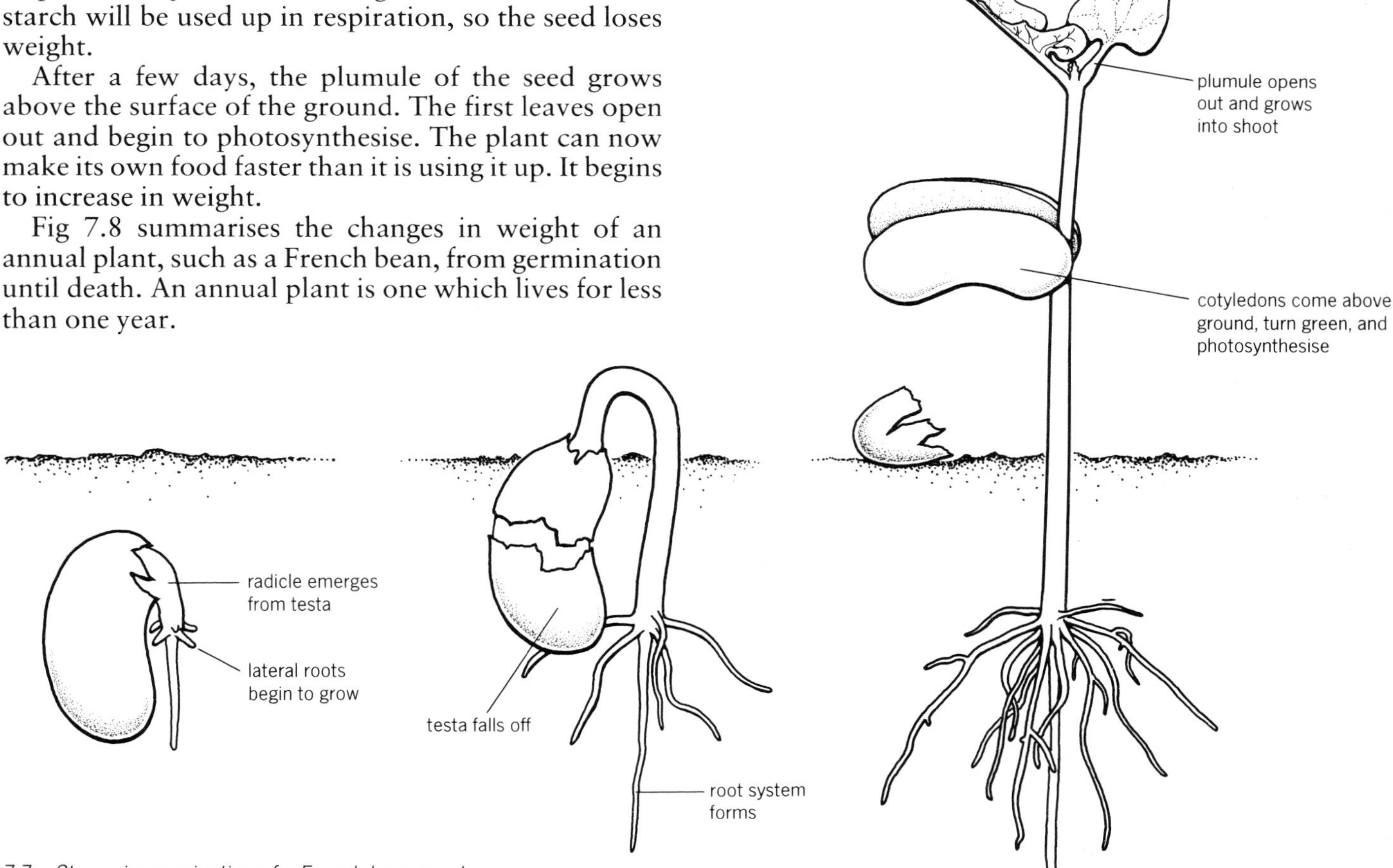

7.7 Stages in germination of a French bean seed

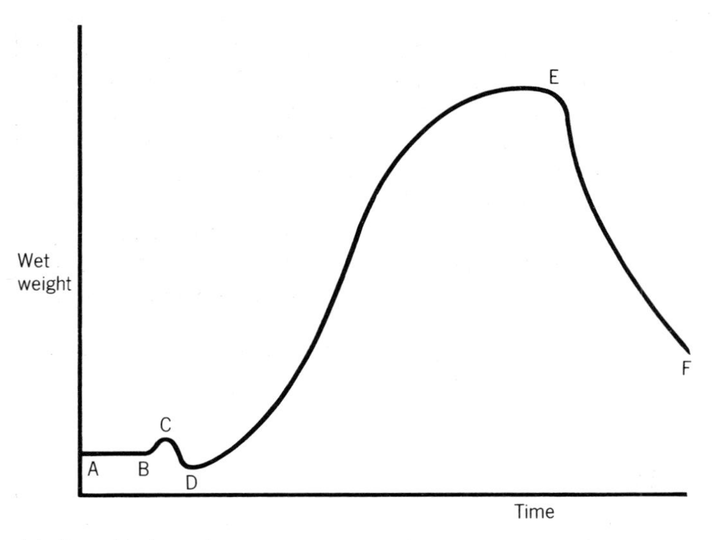

A to B: seed is dormant
B to C: germination begins, and seed increases in weight as it absorbs water
C to D: seed loses weight, as it uses up food stores to provide energy for growth
D to E: plant is photosynthesising, building up new cells
E to F: plant loses weight quickly, as it flowers and produces seeds and fruits, which are dispersed; plant gradually dies

7.8 Growth curve for an annual plant

Fig 7.9 shows how the cells are arranged near the tip of a root. The very tip of the root is covered by a **root cap.** This protects the root as it grows through the soil.

Just behind the tip is an area where the cells are very small. They are small because they are young; they have been made by other cells dividing. These small cells go on dividing, making many new cells. This part of the root is called the region of cell division or root meristem. It always remains just behind the root tip, no matter how the root grows.

A little way up from these dividing cells, the cells are larger. The further up the root you look, the larger the cells are. These larger cells have grown from the small cells made in the region of cell division.

This part of the root is called the **region of cell elongation.**

Shoots grow in a very similar way. There is a region of cell division at the tip (there is no need for a protective layer at the tip), and a region of cell elongation behind it.

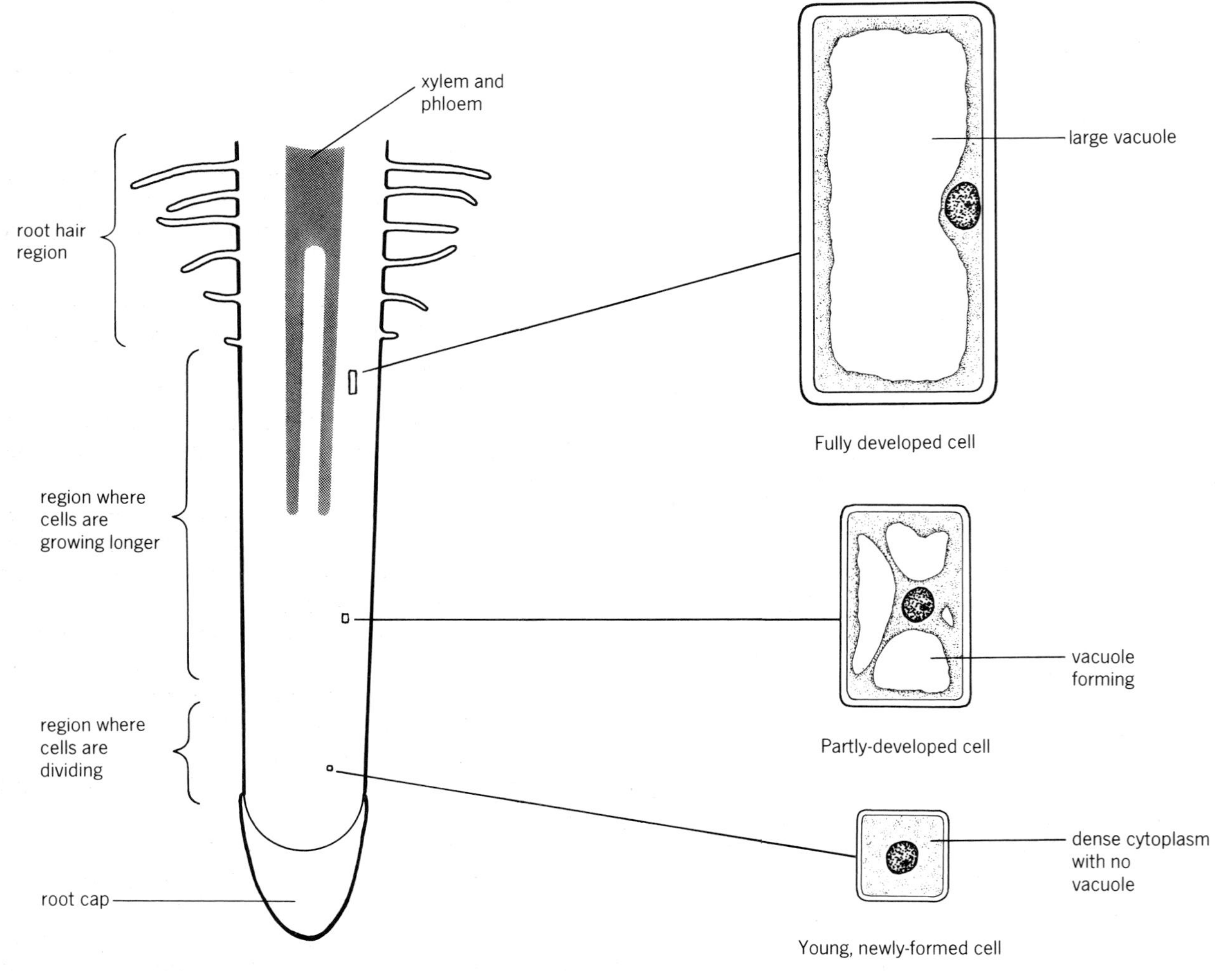

7.9 Longitudinal section through a root tip

Investigation 7.2 To find which part of a root is the growing region

1 Germinate some broad bean seeds, by soaking them in water and leaving them in a gas jar supported by wet blotting paper. Leave them until their radicles (young roots) are about 4 cm long.
2 Copy out the results table.
3 Choose a bean seedling with a really straight, healthy-looking radicle. Carefully remove it from the gas jar, without damaging the radicle. Dry the radicle gently, by patting it with blotting paper.

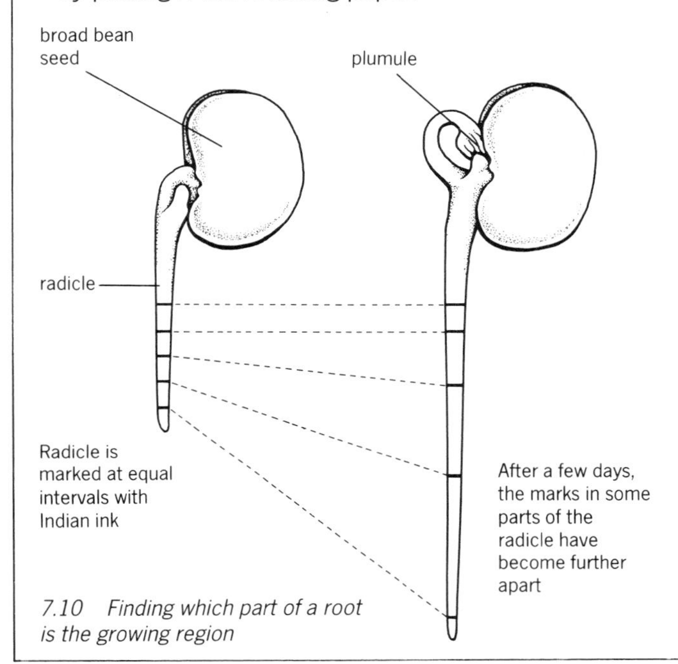

7.10 Finding which part of a root is the growing region

4 Using Indian ink, mark the radicle all the way along, at exactly equal intervals (see Fig 7.10). The interval you choose will depend on the length of your radicle. Record this distance in your results table.
5 Carefully replace the bean in the gas jar, as before. Leave it for several days.
6 Remove the bean again, and measure the distance between your markings. Fill in these results in the table.
7 Calculate the increase in distance between each pair of marks.
8 Draw a graph of your results. Put the number of the mark on the bottom axis, and increase in distance on the vertical axis.

Results table

Interval (numbered from bottom upwards)	1	2	3	4	5
Initial distance apart in mm					
Final distance apart in mm					
Increase in distance apart					

Questions

1 Which part of the bean radicle grew most quickly?
2 What was happening to the cells in this region?
3 Were there any parts of the bean radicle which did not grow? Explain why.

Questions

1 Why does a seed increase in weight at the beginning of germination?
2 Why does a seed decrease in weight for a time, once germination is under way?
3 What is a root cap, and what is its function?
4 In which part of a root are new cells formed?
5 Apart from size, what difference is there between a newly formed cell in a root, and a mature cell?

7.9 Roots and stems can grow wider.

When a young plant first begins to grow, its shoots and roots mainly grow longer. This is called primary growth. Some plants, such as trees, have stems and roots which also grow much wider. This is called **secondary growth.**

Fig 7.11(a) shows a section through a young stem of a plant. Between the xylem and phloem tissue of the vascular bundles, there is a very narrow band of **cambium.** The cells in the cambium can divide; they are meristematic cells.

As they divide, they make two sorts of new cells. The ones towards the middle of the stem become xylem cells. The ones towards the outside of the stem become phloem cells.

As more and more xylem cells are made, the cambium and phloem get pushed outwards. The circumference of the stem gets larger and larger. The epidermis can no longer cover it.

Underneath the epidermis another layer of meristematic cells is formed. They are called the **cork cambium.** They divide to make cork cells.

Cork cells have a waterproof substance called **suberin** in their cell walls. No water can get to the inside of the cell through the suberin, and so the cell contents die. Cork cells are therefore hollow.

A thick layer of cork cells builds up, to replace the old epidermis. It makes up the bark of the tree. The bark protects the cells underneath it from damage, and stops bacteria and fungi from getting in. It also prevents too much water being lost from the tree trunk by evaporation, and it insulates the tree. The hollow cork

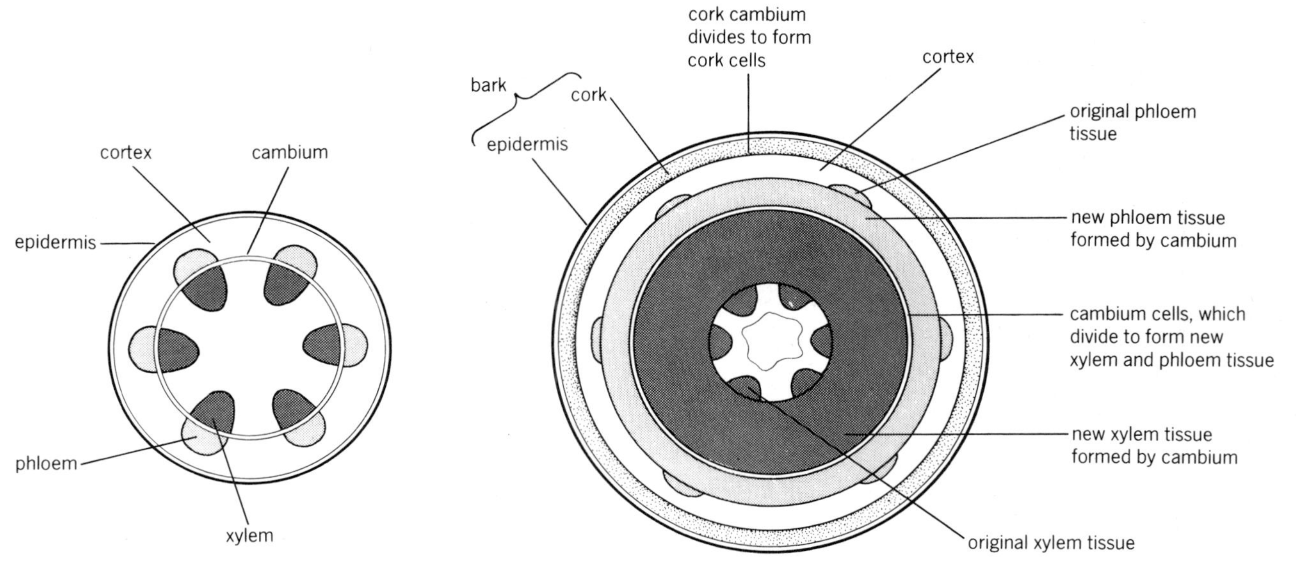

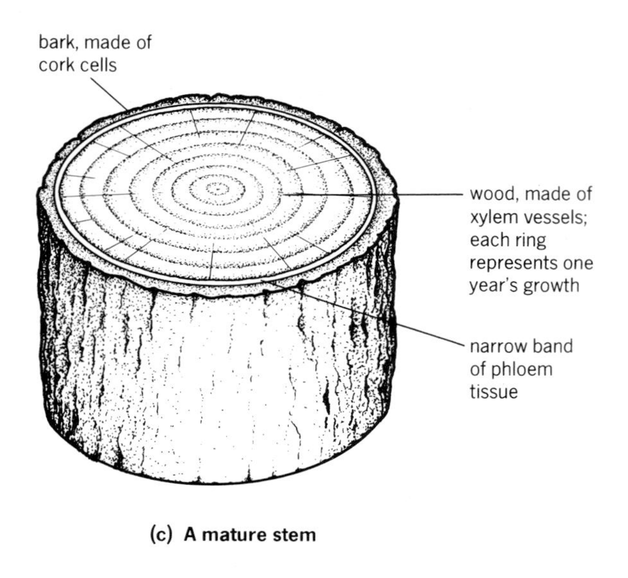

7.11 Secondary growth

cells are rather like expanded polystyrene, and make an excellent insulating material.

Fig 7.11(c) shows the final structure of the trunk of a tree. Almost all of it is xylem tissue, called **wood.** The phloem and cambium are found in a thin, soft layer between the cork and the wood.

7.10 Yearly growth shows as rings in wood.

In temperate countries the secondary growth of a tree is faster at some times of year than at others. In the spring, a lot of new large xylem vessels are made. They are needed to supply water to the expanding buds. In summer, not so many new xylem vessels form, and they are smaller. In autumn and winter, there is almost no growth.

You can see these different types of xylem vessel on a cross section of a tree trunk. Each year's growth forms a ring. Nearest the inside of the ring is the spring wood, with the summer wood towards the outside. The number of rings tells you the age of the tree.

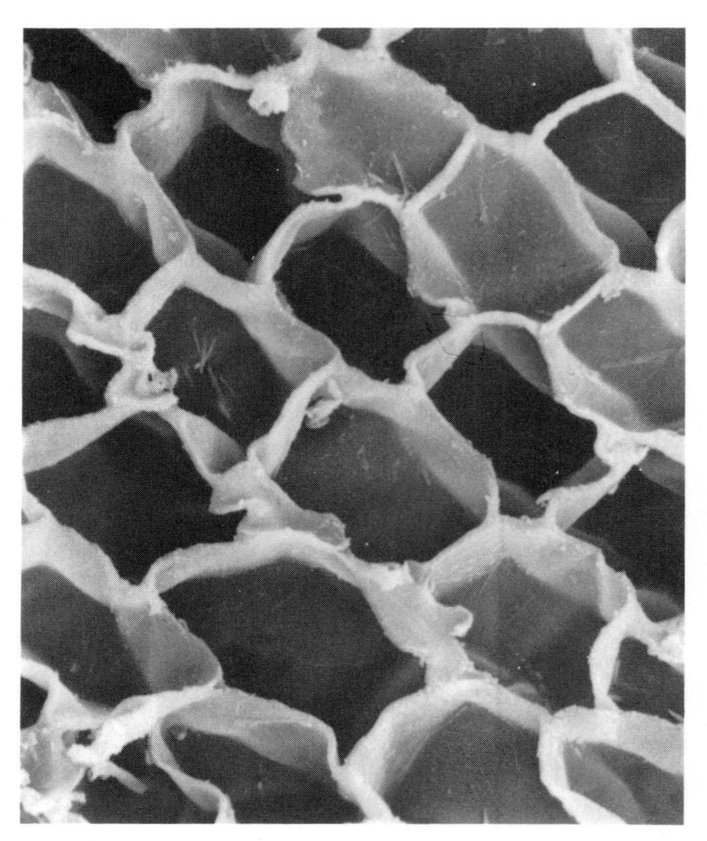

7.12 Scanning electron micrograph of empty, box-like cork cells

7.13 *Section of a tree trunk; the dark outer layer is the bark, with a thin layer of phloem beneath; the rest is wood, or xylem*

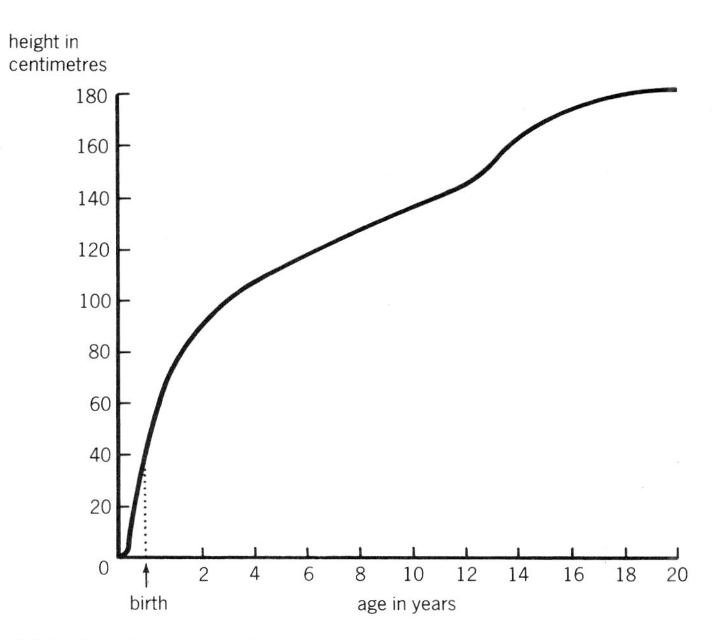

7.14 *Growth curve of a human*

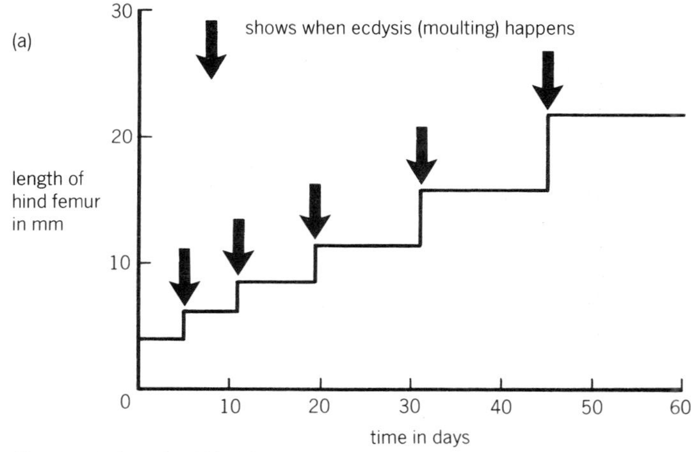

Changes in length of hind femur

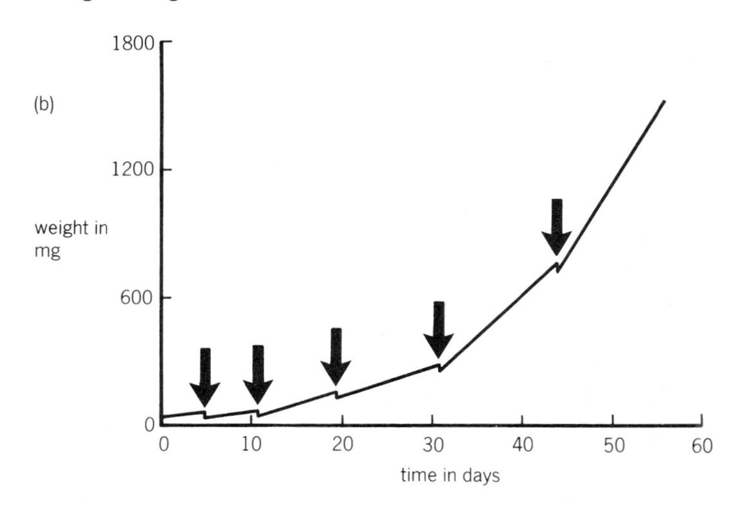

Weight changes

7.15 *Growth curves of a locust*

Questions

1. Explain the difference between primary and secondary growth.
2. Which tissue is responsible for secondary growth in a stem or root?
3. What does cork cambium do?
4. What is suberin?
5. List four functions of bark.
6. What is wood made of?

The growth of animals

7.11 Mammals grow rapidly when young.

Fig 7.14 shows a growth curve for a human. The curve is quite smooth, showing that growth takes place fairly steadily until a person is about twenty years old. There is a growth spurt at adolescence.

7.12 Insects periodically shed their skeletons.

Fig 7.15 shows growth curves for a locust. It is very different from the smooth curve in Fig 7.14. Locusts, like all insects, have a hard covering on the outside of their bodies. It is called an **exoskeleton.** The exoskeleton cannot grow. As the insect grows bigger, it has to shed its exoskeleton every now and then, and produce a new one. This is called **ecdysis,** or moulting.

7.13 New exoskeletons can stretch.

As the young locust feeds, it increases in weight inside its exoskeleton. When it has filled up all the space inside the exoskeleton, it stops feeding. It climbs onto a branch, and hangs upside down.

Underneath the exoskeleton, a liquid called **moulting fluid** is made. It dissolves the inner layer of the exoskeleton, so that it is loosened from the insect's body. In the space that the moulting fluid has made, a new, soft exoskeleton forms.

The locust now pumps blood into its thorax, so that it expands. The new exoskeleton stretches, and the old one splits. The locust wriggles out of its old exoskeleton. Underneath is the new, soft one. This expands, so the locust is suddenly a lot bigger. It takes a few hours for the new exoskeleton to harden. The young locust begins to feed, and will need to moult again in a few weeks' time.

7.14 Insects grow in stages.

Fig 7.15(a) shows the length changes during the growth of a locust. Between moults, it stays the same size, because the exoskeleton cannot grow. At ecdysis, the new exoskeleton expands quickly, so there is a sudden increase in length.

Fig 7.15(b) shows the weight changes. The locust gradually increases in weight underneath its exoskeleton. When it moults, its old exoskeleton is shed, so it loses a little weight. As soon as it begins feeding again, its weight goes up.

7.15 Metamorphosis is a change from young to adult.

When a locust hatches from an egg, it is not a miniature version of an adult locust. It is called a **nymph**. Quite a

7.16 A locust, newly emerged from its old skin, expands its wings

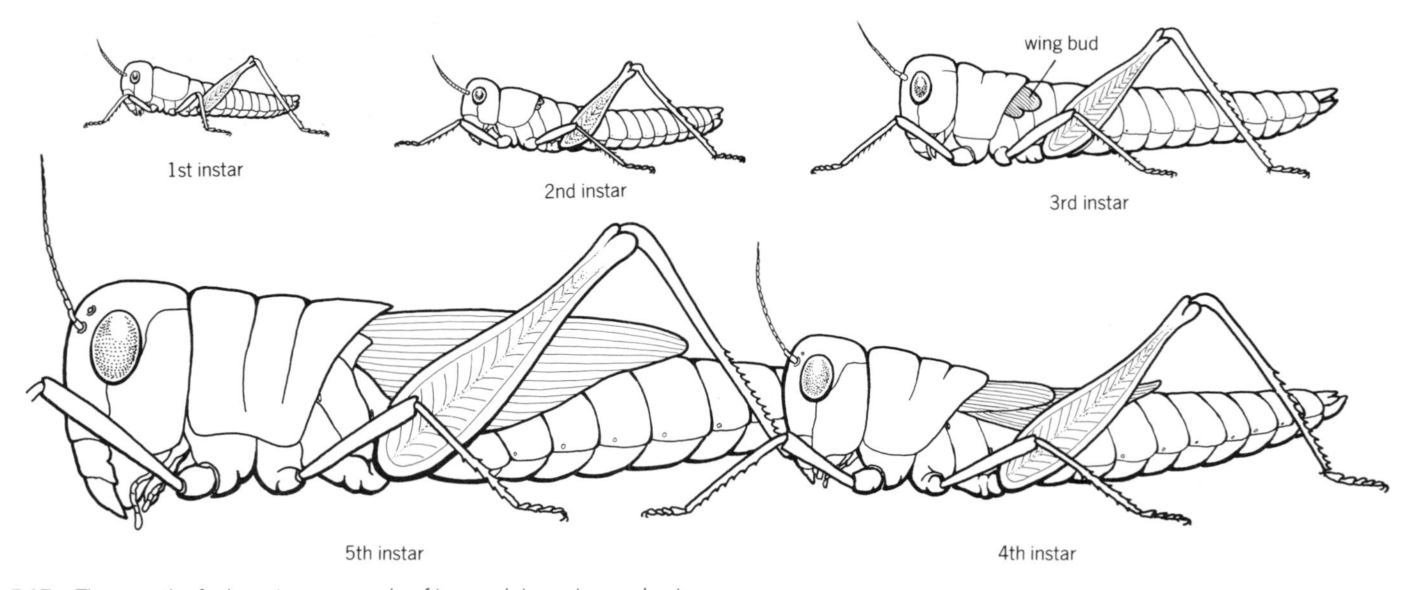

7.17 The growth of a locust–an example of incomplete metamorphosis

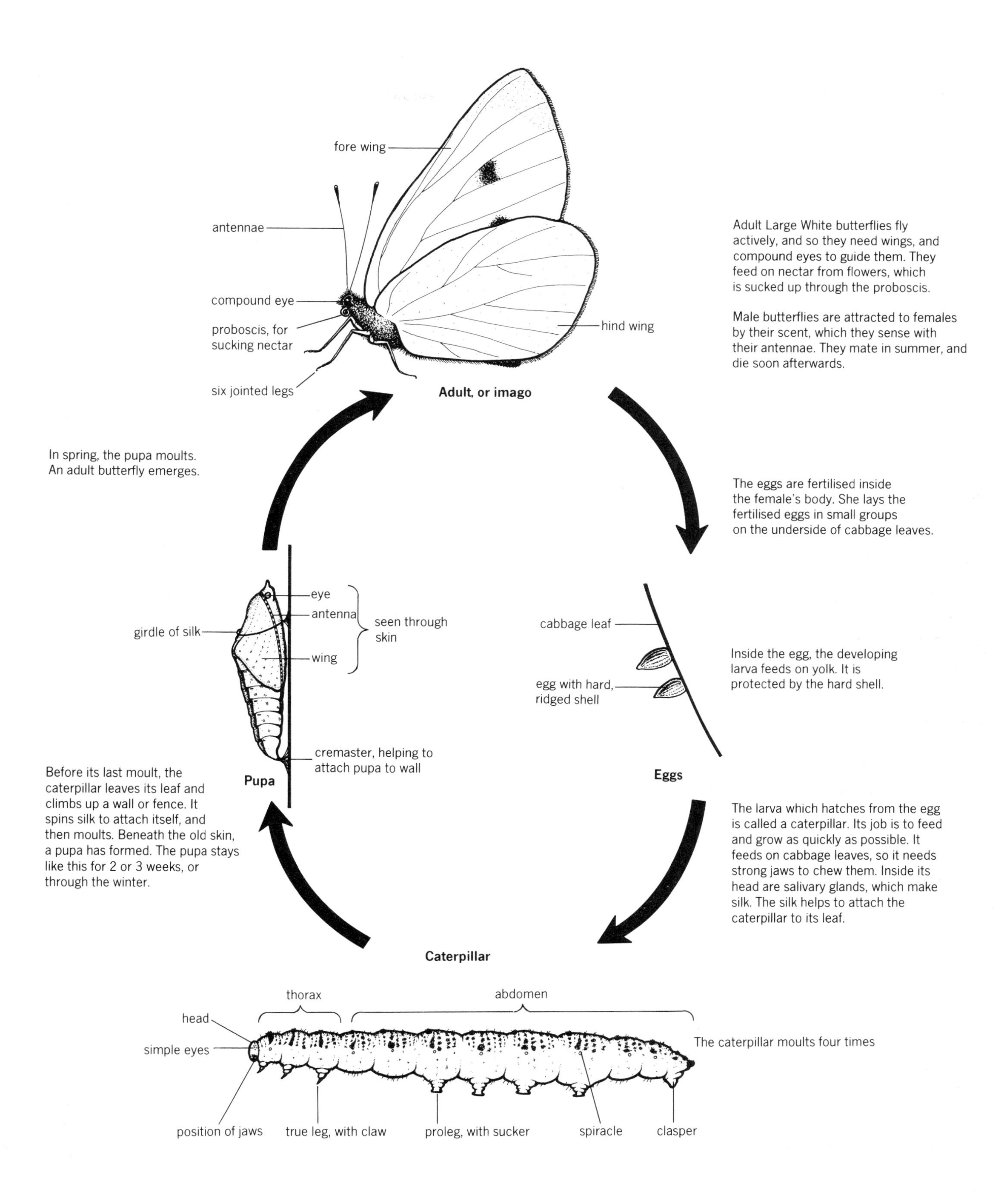

7.18 Life cycle of the Large White butterfly–an example of complete metamorphosis

lot of changes will have to take place before it becomes an adult locust. The changes are called **metamorphosis.**

7.16 Locusts show incomplete metamorphosis.

A locust moults five times before it is adult. Each stage in between moults is called an **instar.**

The five instars of a locust are shown in Fig 7.17. At each moult, the instars become more like the adult. This gradual change from a nymph to an adult insect is called **incomplete metamorphosis.**

7.17 Butterflies show complete metamorphosis.

Many insects, such as butterflies, undergo more dramatic changes when they moult. The larva or caterpillar which hatches from a butterfly's egg is very different from the adult. When fully grown the caterpillar changes into a pupa. After some time the adult butterfly emerges from inside the pupa's skin. These changes from a larva to an adult are called **complete metamorphosis** (see Fig 7.18).

Amphibians, such as frogs, also undergo metamorphosis. This is described in section 8.28.

7.18 Metamorphosis reduces competition.

Why do insects and amphibians undergo metamorphosis? Would it not be simpler if they hatched as miniature adults, and then just grew? One reason is that if the young and the adults are different from one another, then they do not compete with each other. The adult Large White butterfly, for example, feeds on nectar, whereas the caterpillar feeds on cabbage leaves. There is more food available for each of them, because they need different things.

Another reason is that the young and the adult have different functions. The caterpillar has to grow, and it is adapted for eating and growing as fast as possible. The adult has to reproduce. It needs wings to find a mate, and to find cabbage leaves on which to lay its eggs. The dormant pupa is adapted to survive the cold weather of the winter, when neither caterpillar nor adult food is available.

The same is true for tadpoles and frogs. Tadpoles are adapted to feed and grow, while the adult frog's function is to reproduce.

Questions

1. Why is the growth curve for an insect not a smooth one?
2. What is a nymph?
3. Explain the difference between incomplete and complete metamorphosis.
4. Give two reasons why some animals undergo metamorphosis.

7.19 Growth is controlled by hormones.

Living organisms do not grow randomly. Growth is usually carefully controlled. For example, different parts of an animal's body grow at different speeds. The shoots of a plant grow towards the light. A tadpole does not simply grow bigger, but develops into a frog.

The growth of animals and plants is usually controlled by **hormones.** These are chemicals which are made in one part of the organism, and affect the way in which other parts of it behave. Hormones are described in Chapter 12.

Measuring growth

7.20 There are many ways of measuring growth.

Whatever way you choose to measure the growth of an animal or plant, you must take your measurements regularly. This might be every hour, or at the same time every day, or at the same time every week. The time interval you choose will depend on how fast the organism grows.

There are several different sorts of measurements which you could make. Each has its advantages.

7.21 Height is a quick measurement of growth.

Height and length are quick and easy ways to measure growth in many organisms, such as humans and plants. They also have the advantage that they do not damage the animal or plant. However, they do not take into account any sideways growth, nor do they include the roots of a plant unless you uproot it.

Sometimes, the length of a certain part of an animal's body can be measured, instead of its entire length. It is quick and easy, for example, to measure the length of a mouse's tail, and this does give a reasonable idea of its overall size.

7.22 Wet weight measures all the body contents.

Increase in weight gives a better estimate of the overall increase in size of an organism than height or length do. It does, however, often take a little longer to measure. If you measure the weight of a complete plant, you must uproot it, which may kill it, and will certainly slow down its growth.

The easiest way of weighing an organism is to measure its **wet weight**. The wet weight of an organism is its weight including all the water in its body. You are measuring your wet weight when you stand on the bathroom scales.

7.23 Dry weight is the most accurate method.

However, the amount of water in an organism's body is variable. Try weighing yourself before and after you have a bath, or after you have been sweating a lot.

Table 7.1 **Some methods of measuring growth**

Method	*Advantages*	*Disadvantages*	*Suitable organisms*
Height or length	Quick, easy; doesn't damage organism; can be used in the field	Only measures growth in one dimension	Small plants, most animals
Length of one part of body	As above	Only measures growth of one part of body	Many animals, such as small mammals and birds
Wet weight	Gives more accurate measure of overall size than height or length; fairly quick for small animals	Plants must be uprooted, and roots cleared of soil; difficult for large animals without special equipment; fluctuations in water content may affect results	Most animals; small plants
Dry weight	Gives the best measurement of the amount of living material in an organism	Time consuming; kills individual, so large numbers are needed.	Plants; small animals such as insects

These weight changes, which can be several kilos, will be quickly reversed so that your weight is normal again.

Similarly, a well watered plant will weigh more than one which has been short of water, even though they may be the same size.

To get rid of any fluctuations in weight which are simply due to changes in water content, **dry weight** can be measured. All of the water is removed from the plant. It is dried gently in a cool oven, until all the water has evaporated from its tissues. To check that the water has all been lost, the dried remains are weighed, put back in the oven for a little longer, and then weighed again. If the two weights are the same, then all the water has gone. This is called **drying to constant weight.**

7.24 Using the dry weight method.

There is one very obvious disadvantage to the dry weight method of measuring growth. It kills the organism you are measuring! How can you measure growth if you kill the organism whenever you measure it?

The answer is to use a large number of organisms. If you wanted to measure the growth of a plant, for example, using the dry weight method, you may need to grow several hundred plants. They would need to be as identical as possible – all from the same pure-breeding stock (see section 15.17), all planted at exactly the same time, and all grown under identical conditions. Each day, you could take a sample of these (the larger the sample the better) and measure their dry weight. If the average dry weight is then worked out for each day, you will get a very good idea of how any one plant increases its dry weight.

Fact! The slowest growth rate in the animal kingdom is shown by the deep-sea clam, *Tindaria callistiformis.* It can take 100 years to grow 8mm.

Questions

1 Give one advantage of using height or length as a method for measuring growth.
2 Why is wet weight not a very accurate method for measuring growth?
3 What is meant by drying to constant weight?

Chapter revision questions

1 Explain the difference between each of the following pairs of terms:
(a) incomplete metamorphosis, complete metamorphosis;
(b) larva, nymph;
(c) wet weight, dry weight;
(d) primary growth, secondary growth;
(e) radicle, plumule.

2 For a *named* seed
(a) List three conditions which need to be fulfilled before the seed will germinate.
(b) What are the main food reserves in this seed?
(c) In which part of the seed are these food reserves found?
(d) How are these food reserves mobilised when germination begins?
(e) With the aid of diagrams, describe how the radicle of this seed grows in length.

3 (a) Give an example of an insect which undergoes complete metamorphosis.
(b) Draw a large, labelled diagram of the adult of this insect.
(c) List three features shown on your diagram which are characteristic of all insects.
(d) What are the advantages to this insect of having a life cycle which includes complete metamophosis?

continued

4 Use the table below to plot growth curves for (a) a human male, and (b) a human female. Plot both curves on the same axes.

Age in years	*Height in cm*	
	Male	*Female*
0	53	53
1	61	61
2	71	71
3	91.5	86.5
4	99	91.5
5	104.5	96
6	108.5	101
7	114	111
8	122	119
9	124.5	124
10	124.5	127.5
11	127	130
12	129.5	132
13	131.5	134
14	137	137
15	142	142
16	147	147
17	155	152.5
18	162.5	157
19	170	160
20	172.5	161.5
21	175	162
22	175	162
23	175	162
24	175	162

(a) At which age does growth appear to stop?
(b) At which age is the difference in height between male and female (i) least, and (ii) greatest?
(c) How much does the female grow between the ages of 9 and 17?
(d) What is the average rate of growth per year for the female between the ages of 9 and 17?
(e) Do you think that height is a good way of measuring human growth? Give reasons for your answer.

5 (a) Use the following table to plot a growth curve for a pea plant.

Time in weeks from planting of seed	*Dry weight in g*
0	1.4
0.5	0.8
1	1.6
1.5	2.5
2	5.2
3	21.5
4	31.6
5	41.3
6	49.0
7	53.2
8	61.1
9	63.4
10	66.5
11	65.2
12	67.4
13	66.5
14	66.4
15	67.2
16	66.0
17	53.2

(b) Explain exactly how these results would have been obtained.
(c) Explain the reasons for the shape of the curve.

8 Reproduction

8.1 Reproduction may be sexual or asexual.

Living organisms may be killed by other organisms, or die of old age. New organisms have to be produced to replace those that die. This is reproduction.

Each species reproduces in a different way. However, there are only two basic types – **asexual reproduction,** and **sexual reproduction.**

Asexual reproduction

8.2 Mitosis takes place in asexual reproduction.

Asexual reproduction involves only one parent. The parent organism simply produces new cells by mitosis, which grow into a new organism. Because mitosis produces new cells exactly like the parent cell (see section 7.3), the new organism is exactly like its parent.

8.3 *Amoeba* reproduces asexually.

Amoeba reproduces by **binary fission** (see Fig 8.1). This happens when the cell gets so large that its surface area to volume ratio is too small to allow gas exchange to occur easily. In good conditions, division will take place every two or three days.

8.4 Some plants reproduce by vegetative propagation.

Many flowering plants can produce offspring by a form of asexual reproduction called **vegetative propagation.** New plants are formed from an outgrowth of the old one. Some examples of the outgrowths are **runners** (see Fig 8.2) and **bulbs** (see Fig 8.3).

Gardeners sometimes use artificial methods of vegetative propagation to increase their stock of a plant. The advantage of this method is that the new plants will be just like their parent. **Cuttings** (see Fig 8.4) are the commonest method, because many young plants can be quickly produced from just one parent plant. **Grafting** is used if you want to grow more of a plant with good flowers or fruit, but do not want it to grow on its own roots. This might be because its own root system is too weak, not allowing it to grow, or it may be too strong, tending to make an enormous tree.

A piece of the plant is grafted onto a root stock from another sort of tree (see Fig 8.5).After a while, the two

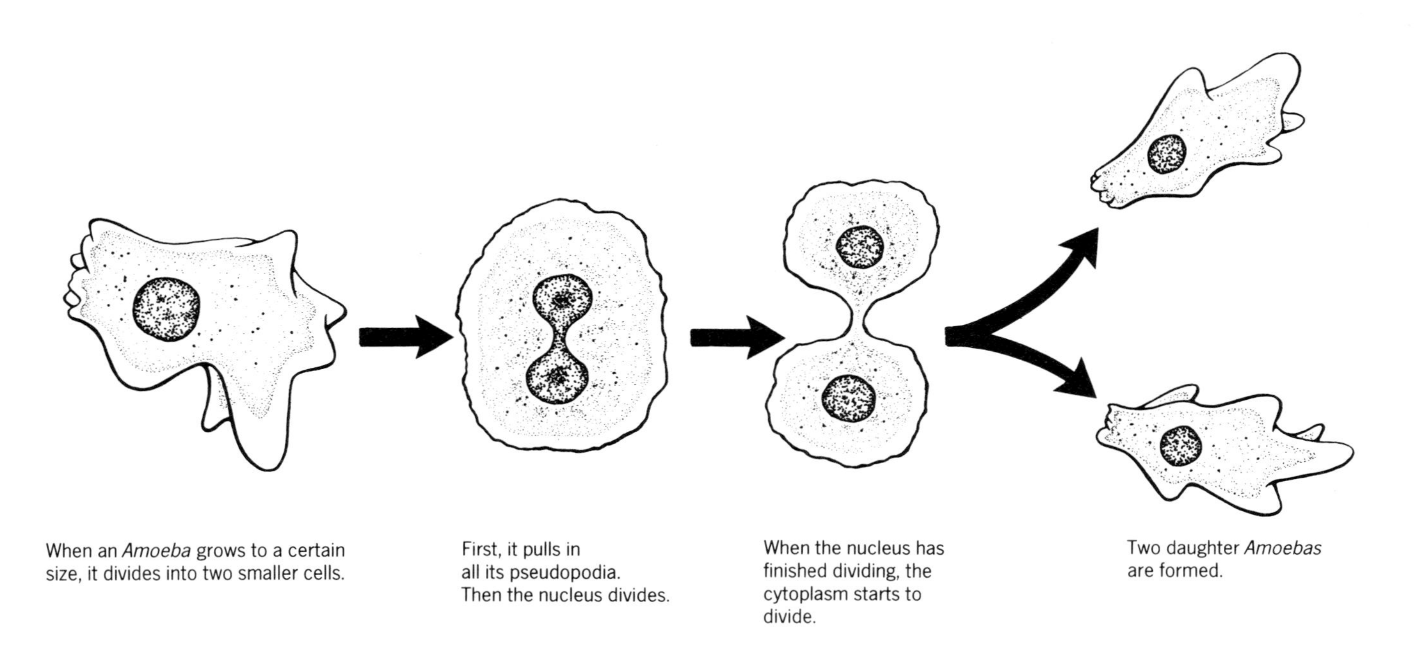

8.1 Amoeba reproducing by binary fission

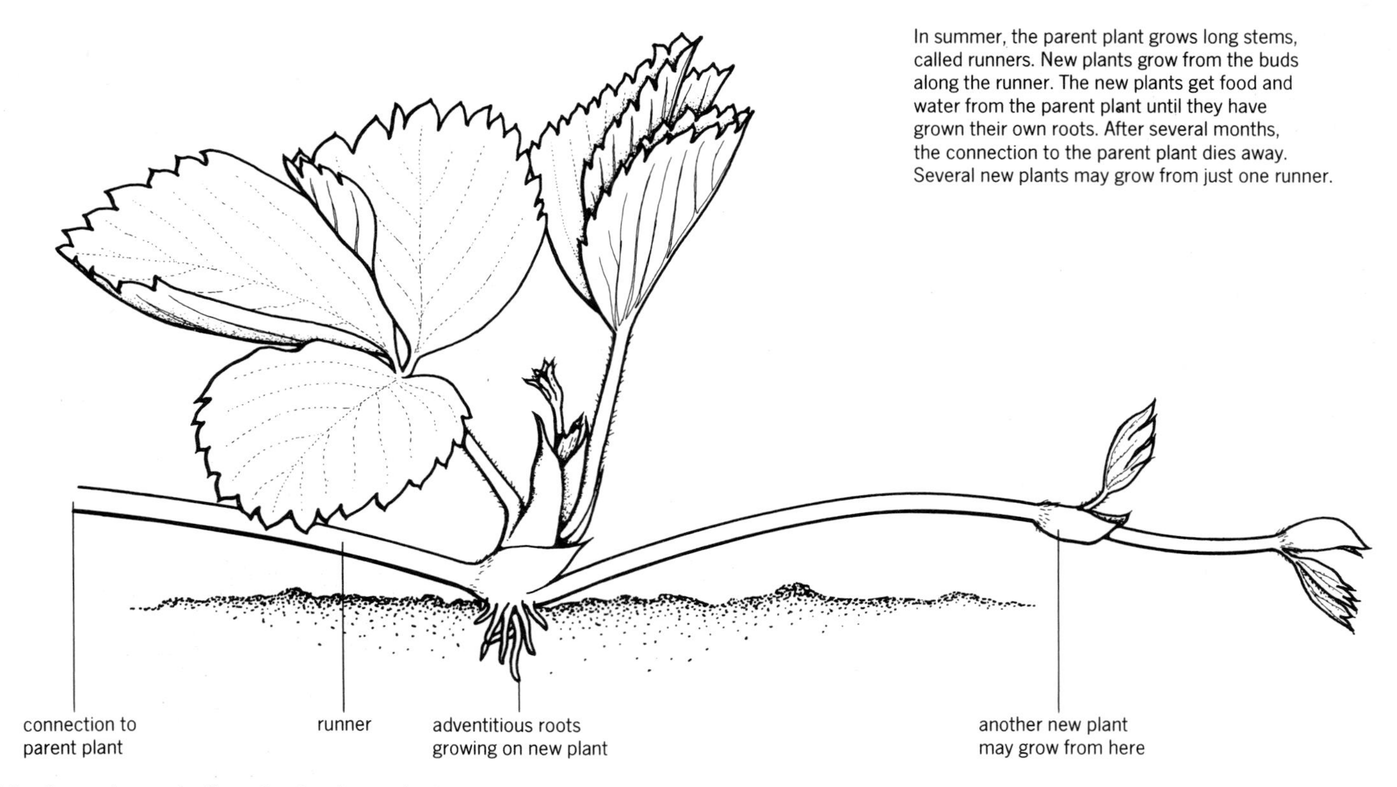

8.2 Asexual reproduction of a strawberry plant

A bulb is an underground bud. The stem is very short, and never grows above the ground. The leaves are thick and fleshy because they contain stored food. They are white, because being underground they cannot photosynthesise, and so do not need chlorophyll.

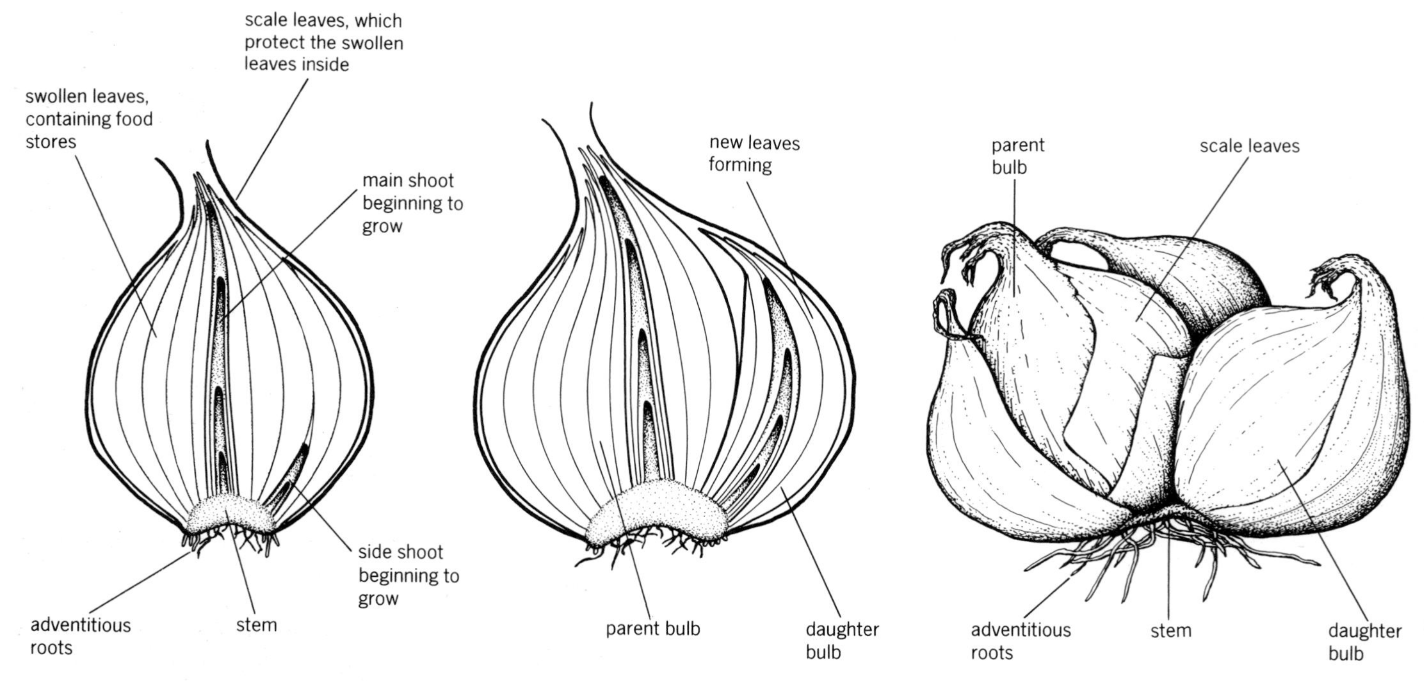

Section through a shallot bulb in early spring. The main shoot and side shoots will grow above the ground, turn green, and photosynthesise.

Later, food from the photosynthesising shoots is used to produce a daughter bulb at the base of the side shoot.

A group of shallot bulbs.

8.3 Asexual reproduction of a shallot bulb

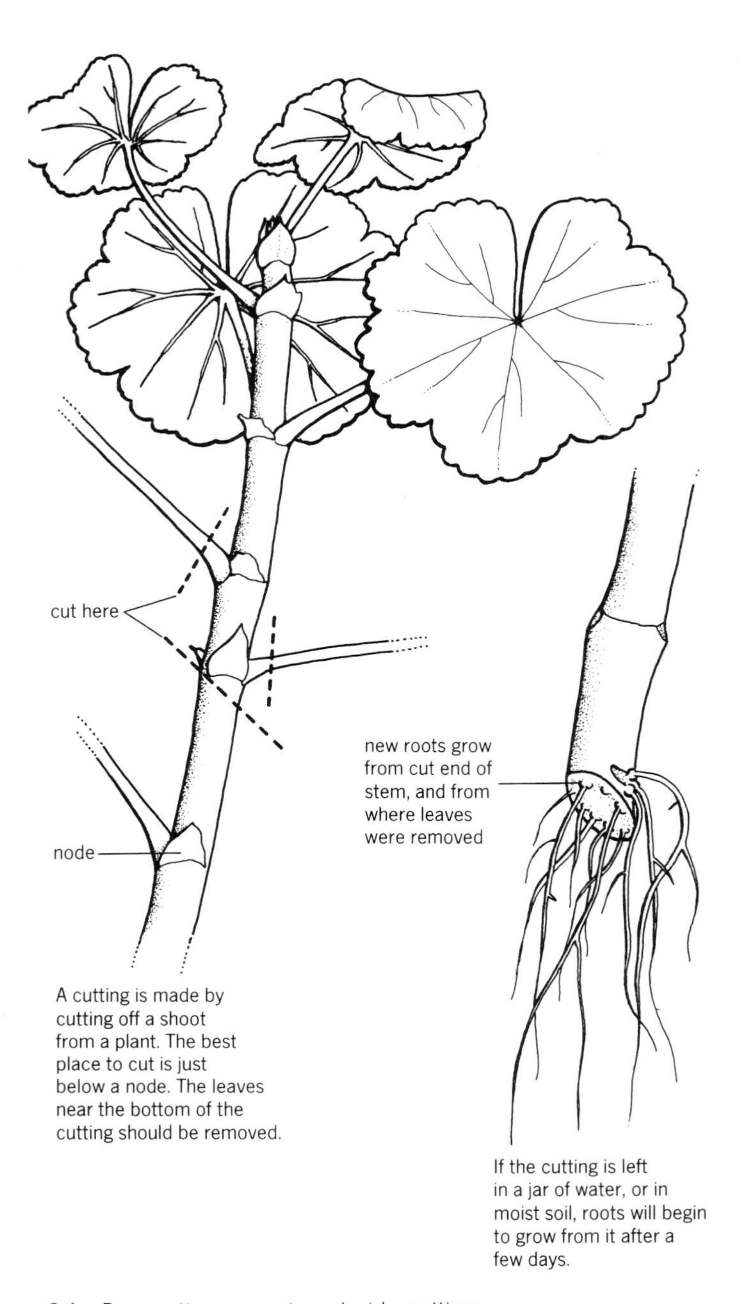

8.4 Propagating a geranium plant by cuttings

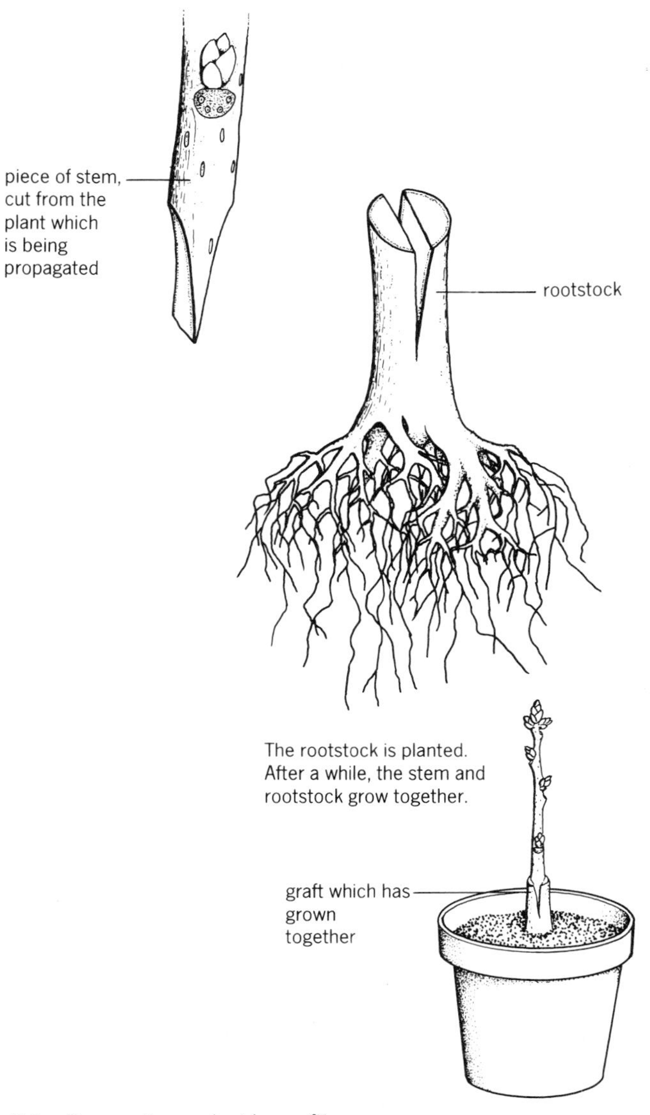

8.5 Propagating a plant by grafting

will grow together. The new plant will have all the features of its parent, but its size will be governed by the root stock it has been grafted onto.

Most apple trees are grafted onto root stocks which stop the trees from growing too tall, so that the apples are easy to pick.

8.5 Aphids may reproduce by parthenogenesis.

Sometimes, an egg will develop into a new organism without being fertilised. This is called **parthenogenesis.**

Aphids (greenfly) can reproduce in this way. When they have plenty of food, the female aphids produce large numbers of eggs. The eggs are not fertilised. They stay inside the female's body, where they grow into young aphids. The young are then born, fully developed (see Fig 8.6 overleaf).

Questions

1. What kind of cell division is involved in asexual reproduction?
2. What is vegetative reproduction?
3. Give two reasons why gardeners often use cuttings to produce new plants.
4. What is parthenogenesis?

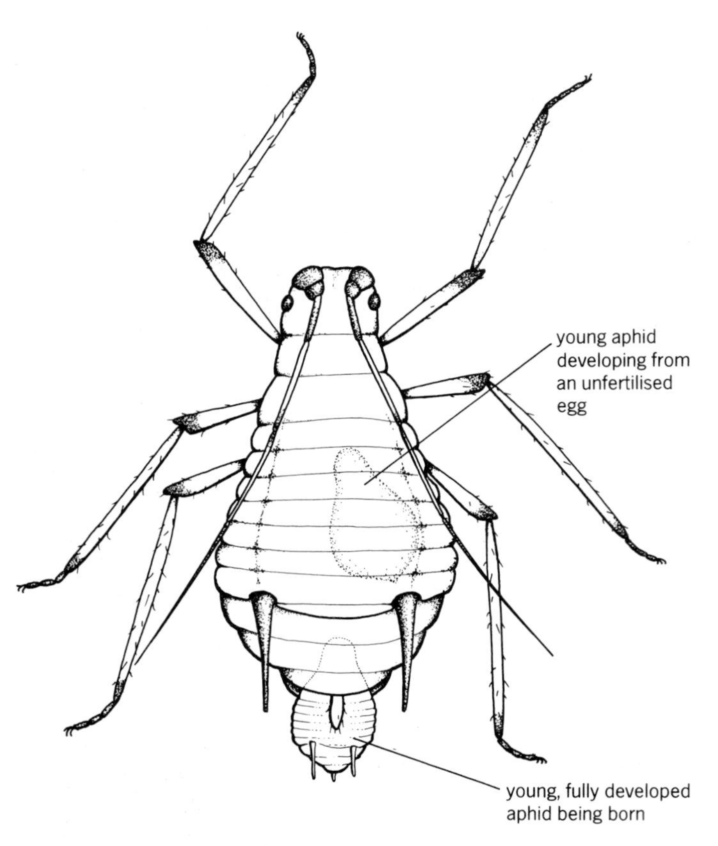

8.6 An aphid (greenfly) reproducing by parthenogenesis

The cells in a human body each contain 46 chromosomes.

body cell

46

body cell

46

In sexual reproduction, cells in testes and ovaries divide by meiosis producing gametes, with half the number of chromosomes.

meiosis

meiosis

sperm

23

23

egg

When the male and female gametes join together, a zygote is formed which has the full number of chromosomes.

fertilisation

46

zygote

8.7 Sexual reproduction

Sexual reproduction

8.6 Sexual reproduction involves fertilisation.

In sexual reproduction, the parent organism produces sex cells or **gametes.** Eggs and sperm are examples of gametes. Two of these gametes then join together. This is called **fertilisation.** The new cell which is formed by fertilisation is called a **zygote.** The zygote divides again and again, and eventually grows into a new organism.

8.7 Gametes have half the normal number of chromosomes.

Gametes are different from ordinary cells, because they contain only half as many chromosomes as usual. This is so that when two of them fuse together, the zygote they form will have the correct number of chromosomes.

Humans have 46 chromosomes, for example, in each of their body cells. But human egg and sperm cells only have 23 chromosomes each. When an egg and sperm fuse together at fertilisation, the zygote which is formed will therefore have 46 chromosomes, the normal number (see Fig 8.7).

The 46 chromosomes in an ordinary human cell are of 23 different kinds. There are two of each kind. The two chromosomes of one kind are called **homologous chromosomes.** A cell which has the full number of chromosomes, with two of each kind, is called a **diploid cell.**

An egg or sperm, though, only has 23 chromosomes, one of each kind. It is called a **haploid cell.** Gametes are always haploid. When two gametes fuse together, they form a **diploid zygote.**

8.8 Gametes are made by meiosis.

Gametes are made by ordinary body cells dividing. For example, human sperm are made when cells in a testis divide.

Because gametes need to have only half as many chromosomes as their parent cell, division by mitosis will not do. When gametes are being made, cells divide in a different way, called **meiosis.** This process is shown in Fig 8.8.

In flowering plants and animals, meiosis only happens when gametes are being made. Meiosis produces new cells with only half as many chromosomes as the parent cell.

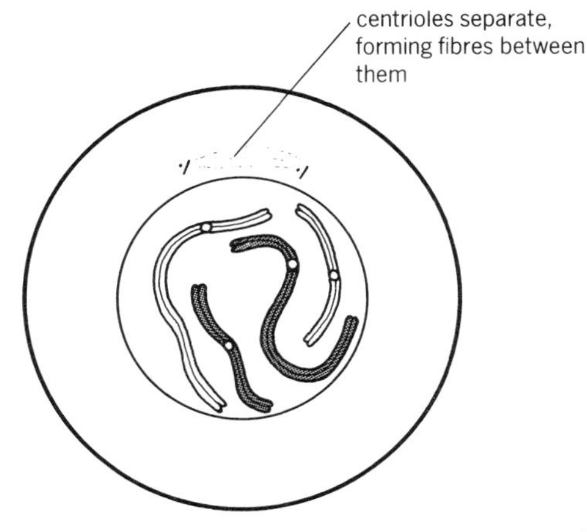

Early Prophase I The chromosomes get short and fat, so they can be seen with a light microscope. Each chromosome contains two chromatids, just as in mitosis.

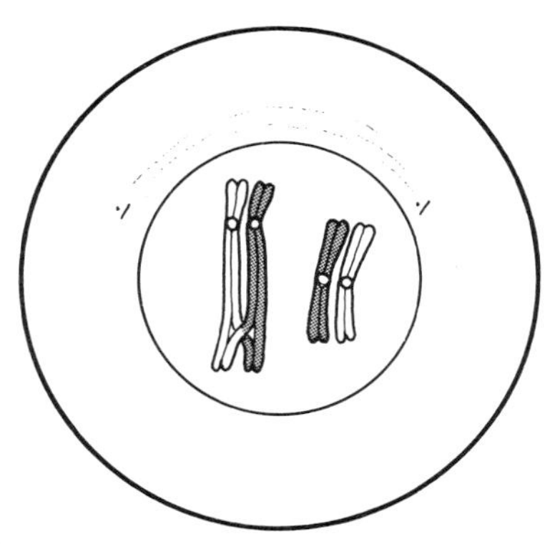

Late Prophase I Homologous chromosomes come together, forming two bivalents. Chromatids of homologous chromosomes may break and rejoin with each other, forming crossover points.

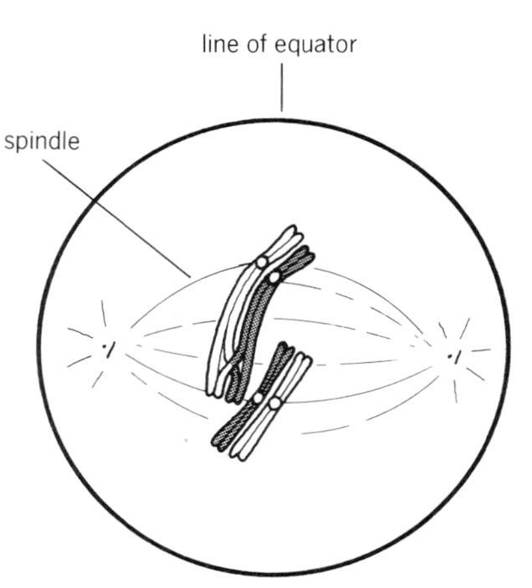

Metaphase I The bivalents line up on the equator of the spindle.

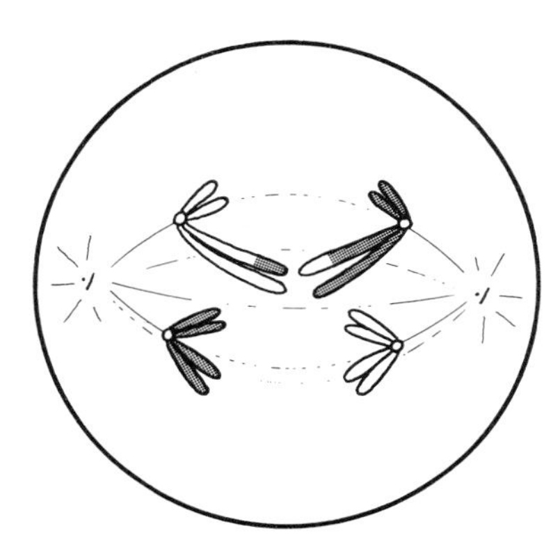

Anaphase I The bivalents separate, and the homologous chromosomes move away from one another along the spindle fibres. Notice that the centromeres do not split, so the two chromatids of each chromosome are still joined together.

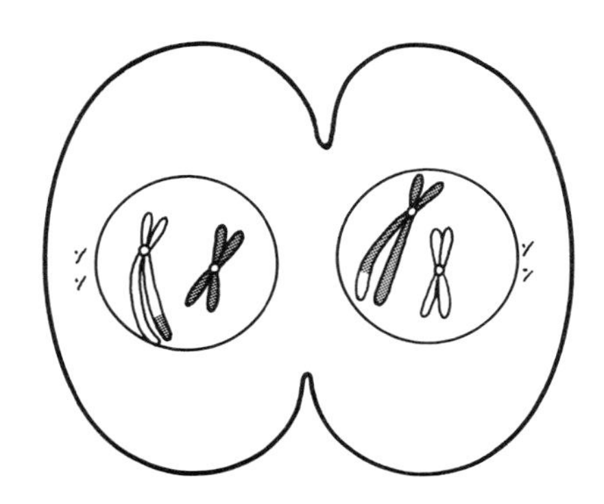

Telophase I The chromosomes arrive at opposite ends of the cell. A nuclear membrane forms round each group. The spindle fibres fade away. The centrioles divide, and so does the cytoplasm.

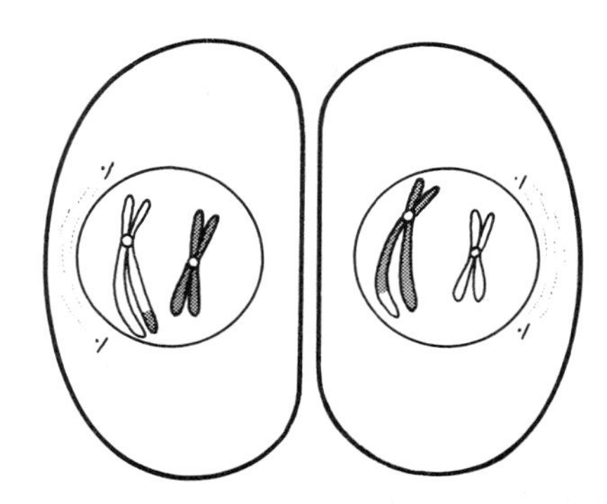

Prophase II The centrioles begin to form new spindles at right angles to the first one.

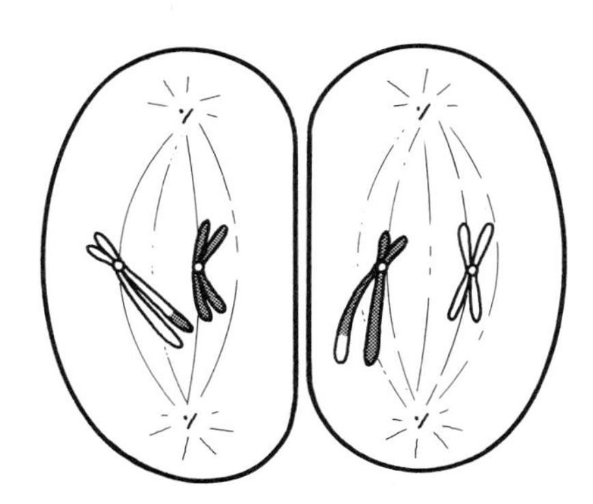

Metaphase II The chromosomes line up on the equators of the spindles.

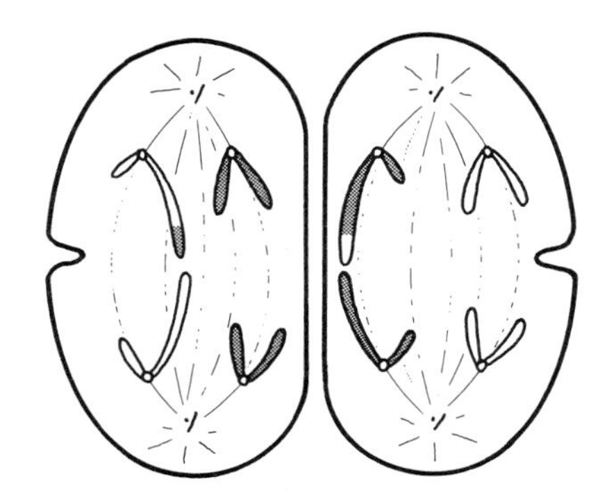

Anaphase II The centromeres of each chromosome split, so the two chromatids separate. The chromatids move away from each other, along the spindle fibres.

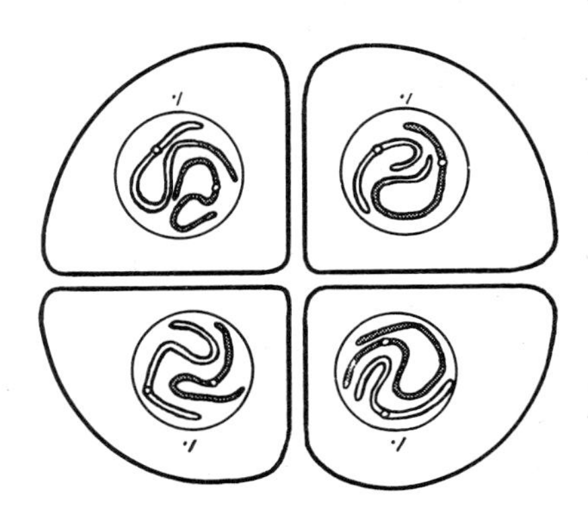

Telophase II The chromatids arrive at opposite ends of the cell, and nuclear membranes form around them. The cytoplasm divides.

Four daughter cells have been formed, each with half the number of chromosomes of the parent cell.

8.8 Meiosis in an animal cell with four chromosomes. In meiosis, the cell divides twice. In the first division, homologous chromosomes separate from one another. In the second division, chromatids separate, as in mitosis.

8.9 Male gametes move – female ones stay still.

In many organisms, there are two different kinds of gamete. One kind is quite large, and does not move much. This is called the female gamete. In humans, the female gamete is the egg.

The other sort of gamete is smaller, and moves actively in search of the female gamete. This is called the male gamete. In humans, the male gamete is the sperm.

Often, one organism can only produce one kind of gamete. Its sex is either male or female, depending on what kind of gamete it makes. All mammals, for example, are either male or female.

Sometimes, though, an organism can produce both sorts of gamete. Earthworms, for example, can produce both eggs and sperms. An organism which produces both male and female gametes is a **hermaphrodite.** Many flowering plants are also hermaphrodite.

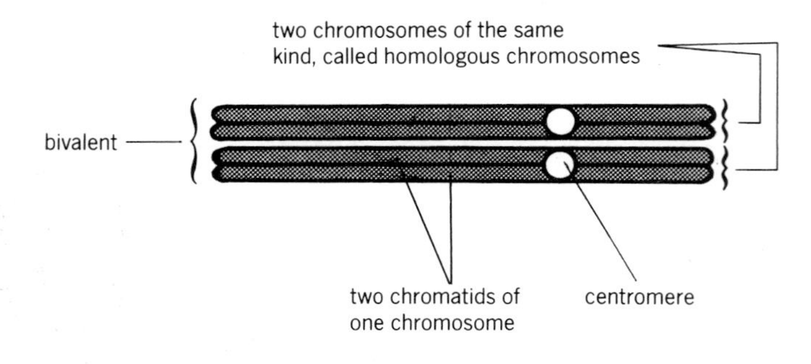

8.9 A bivalent. During meiosis, homologous chromosomes come together in pairs, forming bivalents.

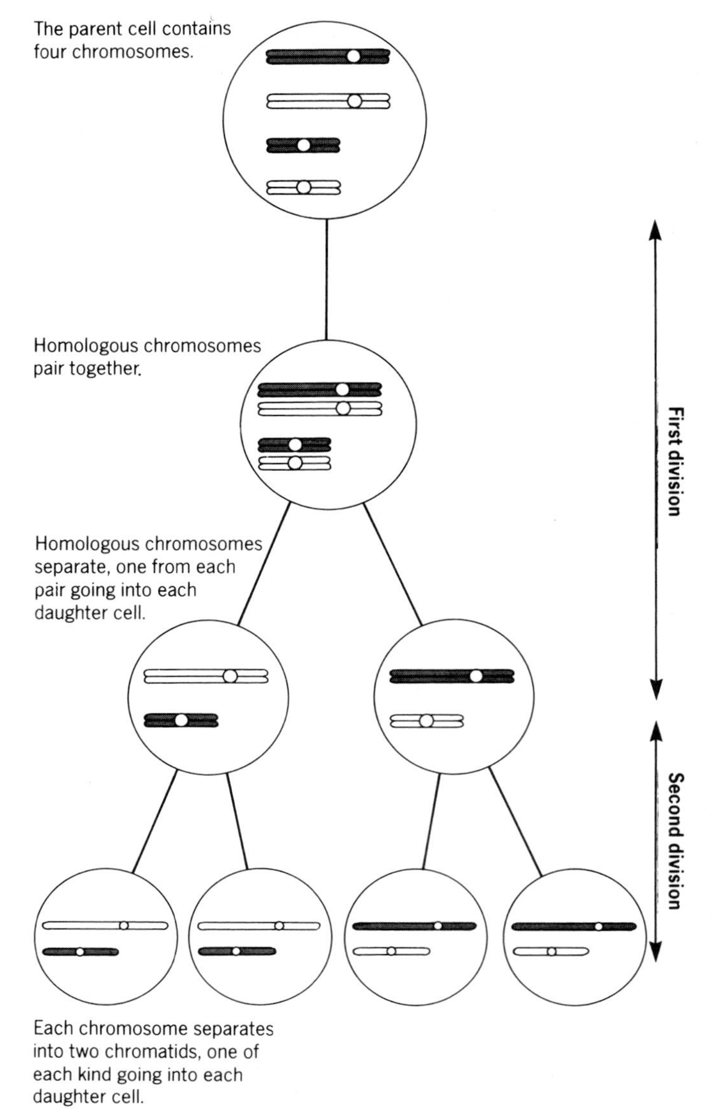

8.10 Summary of chromosome behaviour during meiosis

Table 8.1 A comparison of chromosome behaviour during mitosis and meiosis

Mitosis	*Meiosis*
Prophase Chromosomes appear. They do not associate with one another.	*Prophase I* Chromosomes appear. Homologous chromosomes pair up and form bivalents.
Metaphase Chromosomes line up individually on equator of spindle.	*Metaphase I* Bivalents line up on equator of spindle.
Anaphase Centromeres split. *Chromatids* separate and travel to opposite ends of the cell.	*Anaphase I* Centromeres do not split. Homologous *chromosomes* separate and travel to opposite ends of the cell.
Telophase Two groups of chromatids come together at opposite ends of the cell, and begin to uncoil.	*Telophase I* Two groups of chromosomes come together at opposite ends of the cell, but do not uncoil.
There is no second division of the nucleus. Division has now been completed.	*Second division* The cell goes into another division, at right angles to the first one. The chromatids separate as in mitosis.
Interphase Chromatids are completely uncoiled, and are not visible.	*Interphase* Chromatids are completely uncoiled, and are not visible.
Result Two new cells are formed with exactly the same number and kind of chromosomes as their parent cell.	*Result* Four new cells are formed, each with only half the number of chromosomes of their parent cell.

Questions

1 What is a gamete?
2 What is a zygote?
3 Why do gametes contain only half the normal number of chromosomes?
4 What is meant by a diploid cell?
5 Name one part of your body where you have diploid cells.
6 What is meant by a haploid cell?
7 Give one example of a haploid cell.
8 When do cells divide by meiosis?
9 What is the purpose of meiosis?
10 What does hermaphrodite mean?
11 Give one example of a hermaphrodite organism.

Fact! Length for length, a sperm swimming up the uterus travels as fast as a nuclear submarine.

Sexual reproduction in a mammal

8.10 The female reproductive organs.

Fig 8.11 shows the reproductive organs of a woman. The female gametes, called eggs or **ova,** are made in the two **ovaries.** Leading away from the ovaries are the **oviducts,** sometimes called Fallopian tubes. They do not connect directly to the ovaries, but have a funnel shaped opening just a short distance away.

The two oviducts lead to the womb or **uterus.** This has very thick walls, made of muscle. It is quite small–only about the size of a clenched fist–but it can stretch a great deal when a woman is pregnant.

At the base of the uterus is a narrow opening, guarded by muscles. This is the neck of the womb, or **cervix**. It leads to the **vagina,** which opens to the outside.

The opening from the bladder, called the **urethra,** runs in front of the vagina, while the **rectum** is just behind it. The three tubes open quite separately to the outside.

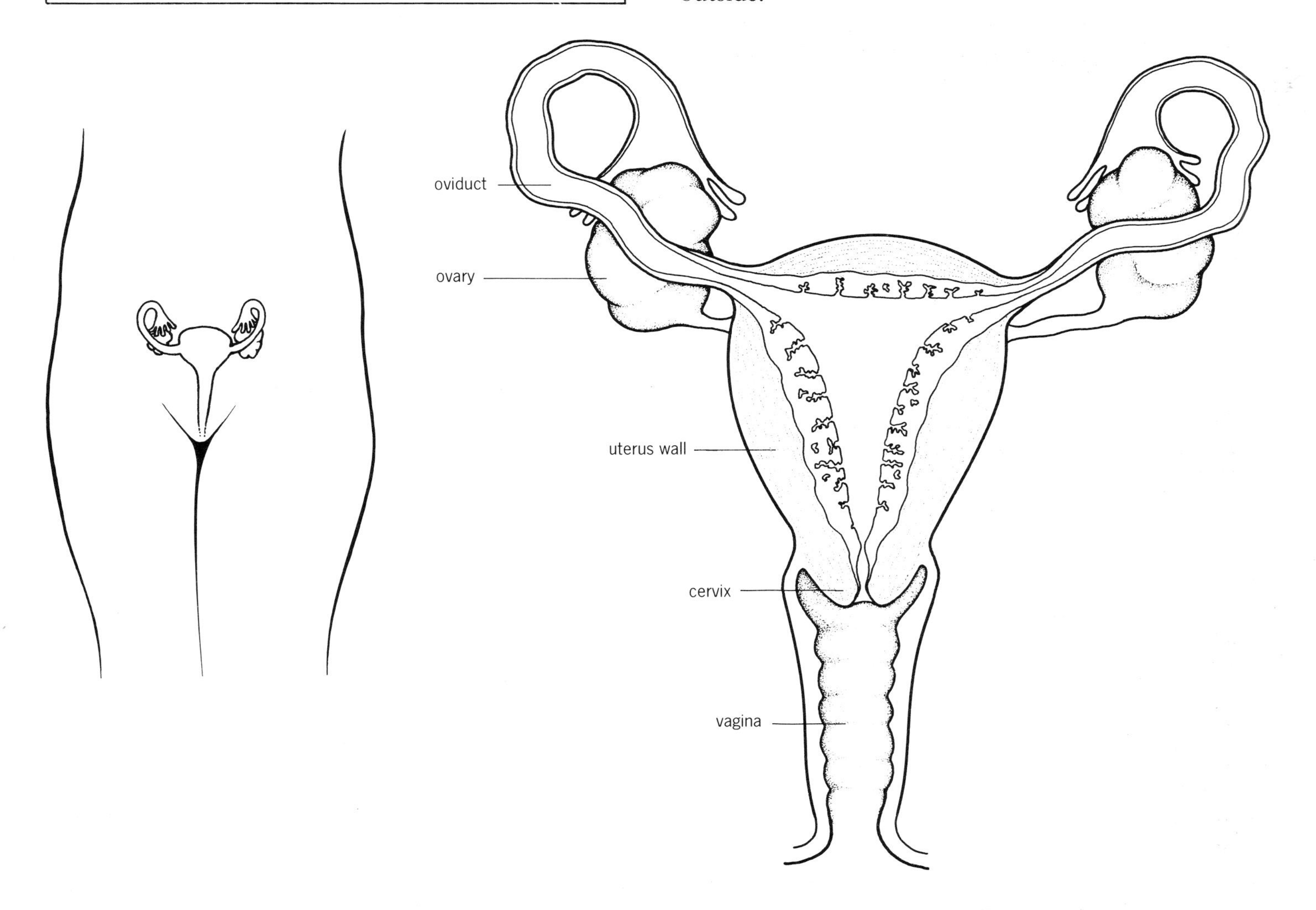

8.11 The female reproductive organs

8.11 The male reproductive organs.

Fig 8.12 shows the reproductive organs of a man. The male gametes, called **spermatozoa** or sperm, are made in two **testes.** These are outside the body, in two sacs of skin called the **scrotum.**

The sperms are carried away from each testis in a tube called the **vas deferens.** The vasa deferentia from the testes join up with the **urethra** just below the bladder. The urethra continues downwards, and opens at the tip of the **penis.** The urethra can carry both urine and sperms at different times.

Where the vasa deferentia joins the urethra, there is a gland called the **prostate gland.** This makes a fluid which the sperms swim in. Just behind the prostrate gland are the **seminal vesicles,** which also secrete fluid.

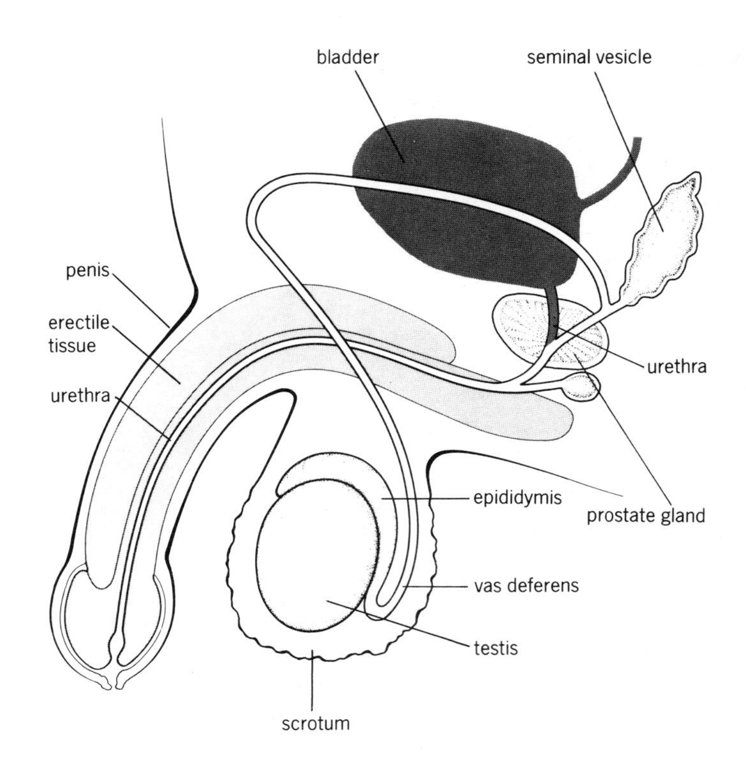

8.12 The male reproductive organs

8.12 Ovaries make eggs.

Fig 8.13 shows a section through a human ovary. The eggs are made from cells in the outside layer, or **epithelium,** of the ovary. Some of these cells move towards the centre of the ovary. A small space, filled with liquid, forms around each one. The space and the cell inside it is called a **follicle.**

This has happened inside a girl's ovaries before she is born. At birth, she will already have many thousands of follicles inside her ovaries.

When she reaches puberty (see section 8.23), some of these follicles will begin to develop. Usually, only one develops at a time. The cell inside the follicle grows bigger, and so does the fluid filled space around it. The follicle moves to the edge of the ovary.

It is now called a **Graafian follicle.** It is little more than 1 cm across, and bulges from the outside of the ovary.

The cell inside it undergoes meiosis. Only one of the cells which are made becomes an egg. The follicle bursts, and the egg shoots out of the ovary. This is called **ovulation.** In humans, it happens about once a month.

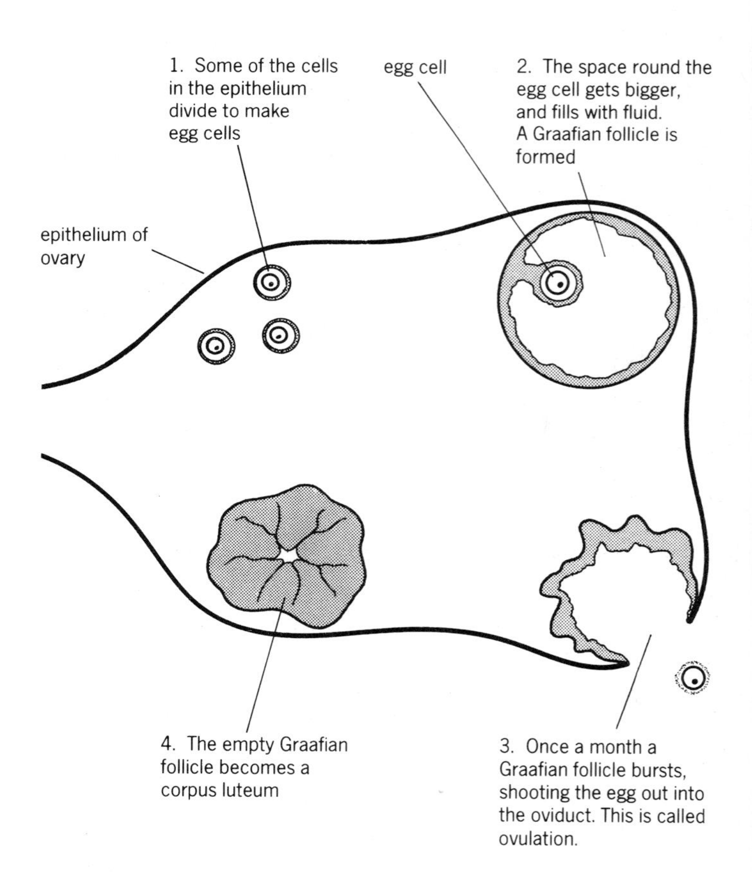

8.13 How eggs are made

8.13 Testes make sperms.

Fig 8.15 shows a section through a testis. It contains thousands of very narrow, coiled tubes or **tubules.** These are where the sperms are made.They develop from cells in the walls of the tubules, which divide by meiosis. Sperms are made continually from puberty onwards.

Sperm production is very sensitive to heat. If they get too hot, the cells in the tubules will not develop into sperms. This is why the testes are outside the body, where they are cooler than they would be inside.

8.14 Mating introduces sperms into the vagina.

After ovulation, the egg is caught in the funnel of the

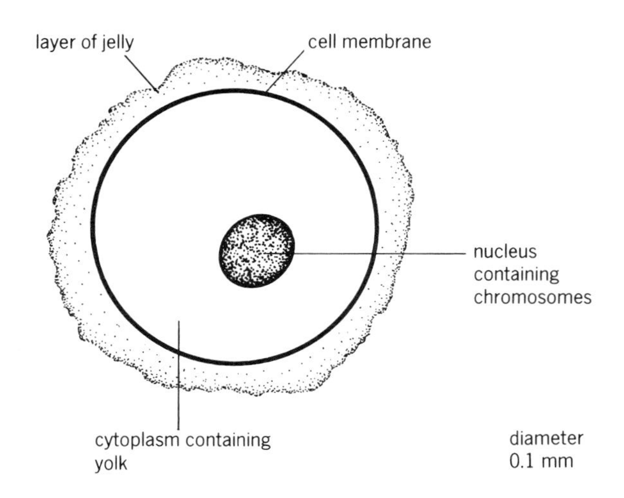

8.14 *An egg or ovum*

oviduct. The funnel is lined with cilia (see section 11.10) which beat rhythmically, wafting the egg into the entrance of the oviduct.

Very slowly, the egg travels towards the uterus. Cilia lining the oviduct help to sweep it along. Muscles in the wall of the oviduct also help to move it, by peristalsis (see section 2.31).

If the egg is not fertilised by a sperm within 8–24 hours after ovulation, it will die. By this time, it has only travelled a short way along the oviduct. So a sperm must reach an egg while it is quite near the top of the oviduct if fertilisation is to be successful.

When the man is sexually excited, blood is pumped into spaces inside the penis, so that it becomes erect. To bring the sperms as close as possible to the egg, the man's penis is placed inside the vagina of the woman.

Sperms are pushed out of the penis into the vagina. This happens when muscles in the walls of the tubes containing the sperms contract rhythmically. The wave of contraction begins in the testes, travels along the vasa deferentia, and into the penis. The sperms are squeezed along, and out of the man's urethra into the woman's vagina. This is called **ejaculation.**

The fluid containing the sperms is called **semen.** Ejaculation deposits the semen at the top of the vagina, near the cervix.

8.15 Fertilisation happens in the oviduct.

The sperms are still quite a long way from the egg. They swim, using their tails, up through the cervix, through the uterus, and into the oviduct (see Fig 8.18).

Sperms can only swim at a rate of about 4 mm per minute, so it takes quite a while for them to get as far as

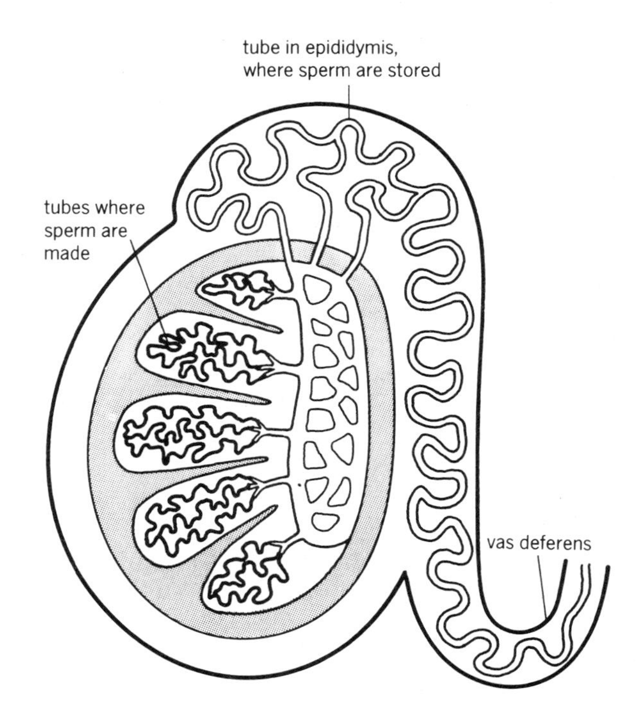

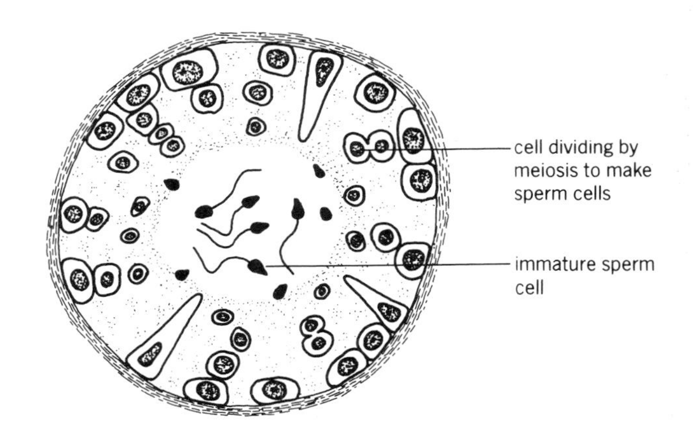

8.15 *How sperm are made*

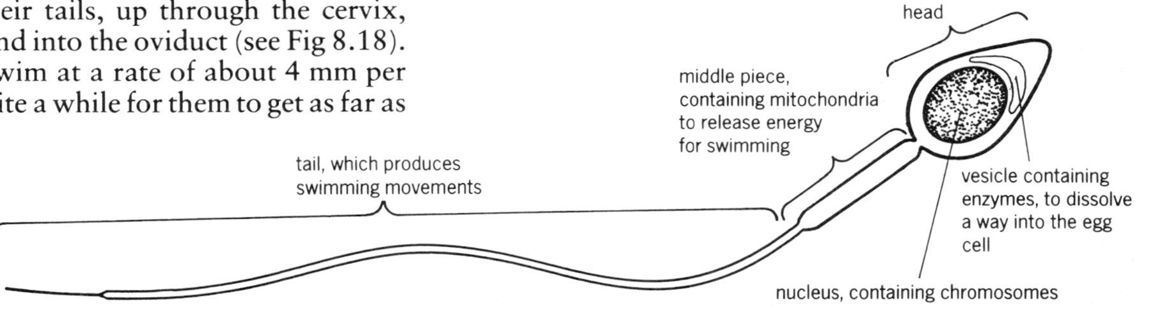

8.16 *A sperm cell*

8.17 *Sperm cells swimming over the ciliated cells of the oviduct*

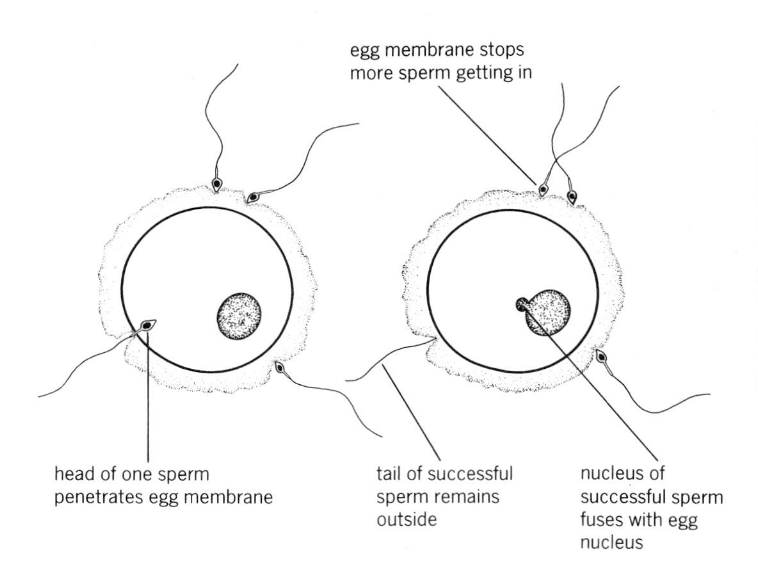

8.19 *Fertilisation*

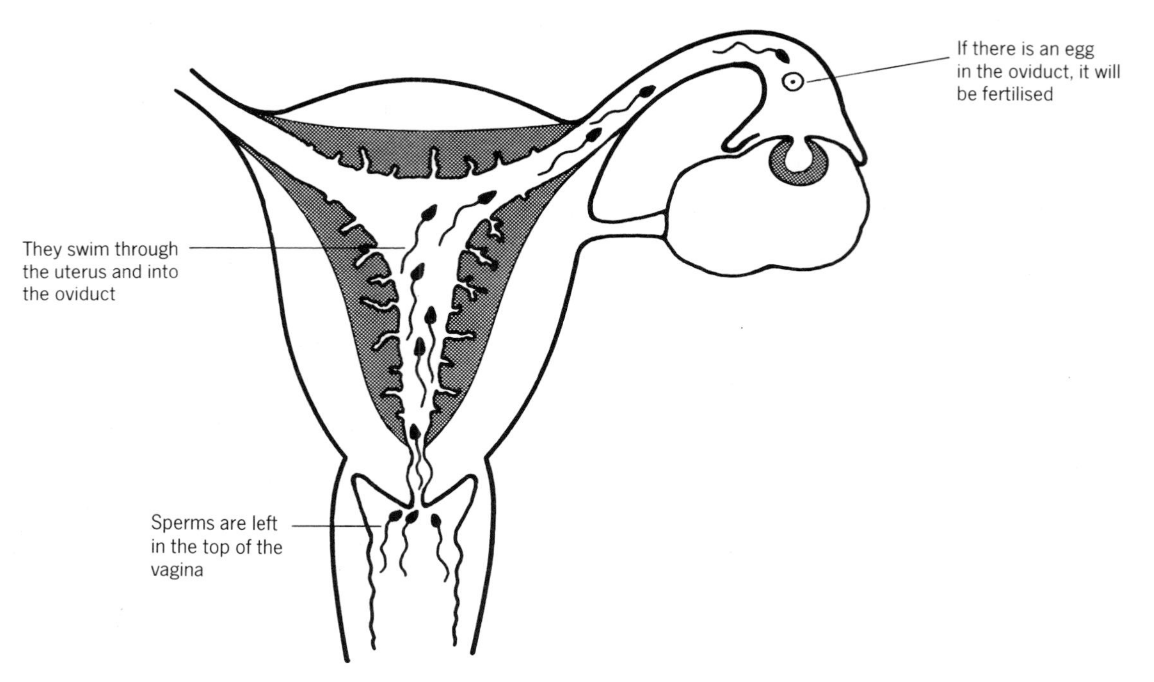

8.18 *How sperms get to the egg (sperms and egg drawn to different scales)*

the oviducts. Many will never get there at all. But one ejaculation deposits about a million sperms in the vagina, so there is a good chance that some of them will reach the egg.

One sperm enters the egg. Only the head of the sperm goes in; the tail is left outside. The nucleus of the sperm fuses with the nucleus of the egg. This is **fertilisation** (Fig 8.19).

As soon as the successful sperm enters the egg, the egg membrane becomes impenetrable, so that no other sperm can get in. The unsuccessful sperms will all die.

Questions

1 What is the name for the narrow opening between the uterus and the vagina?
2 Where is the prostate gland, and what is its function?
3 What is a Graafian follicle?
4 Explain how ovulation happens.
5 Where are sperms made?
6 How does an egg travel along the oviduct?
7 What is semen?
8 Where does fertilisation take place?

8.16 The zygote implants in the uterus wall.

When the sperm nucleus and the egg nucleus have fused together, they form a zygote. The zygote continues to move slowly down the oviduct. As it goes, it divides by mitosis. After several hours, it has formed a ball of cells. This is called an **embryo.** The embryo obtains food from the yolk of the egg.

It takes several hours for the embryo to reach the uterus, and by this time it is a ball of 16 or 32 cells. The uterus has a thick, spongy lining, and the embryo sinks into it. This is called **implantation** (see Fig 8.20).

8.17 The embryo's life-support system is its placenta.

The cells in the embryo, now buried in the soft wall of the uterus, continue to divide. As the embryo grows, a **placenta** also grows, which connects it to the wall of the uterus (see Figs 8.21 and 8.22). The placenta is soft and dark red, and has finger-like projections called **villi.** The villi fit closely into the uterus wall.

The placenta is joined to the embryo by the **umbilical cord.** Inside the cord is an artery and a vein. The artery takes blood from the embryo into the placenta, and the vein returns the blood to the embryo.

In the placenta are capillaries filled with the embryo's blood. In the wall of the uterus are large spaces filled with the mother's blood. The embryo's and mother's blood do not mix. They are separated by the wall of the placenta. But they are brought very close together, because the wall of the placenta is very thin.

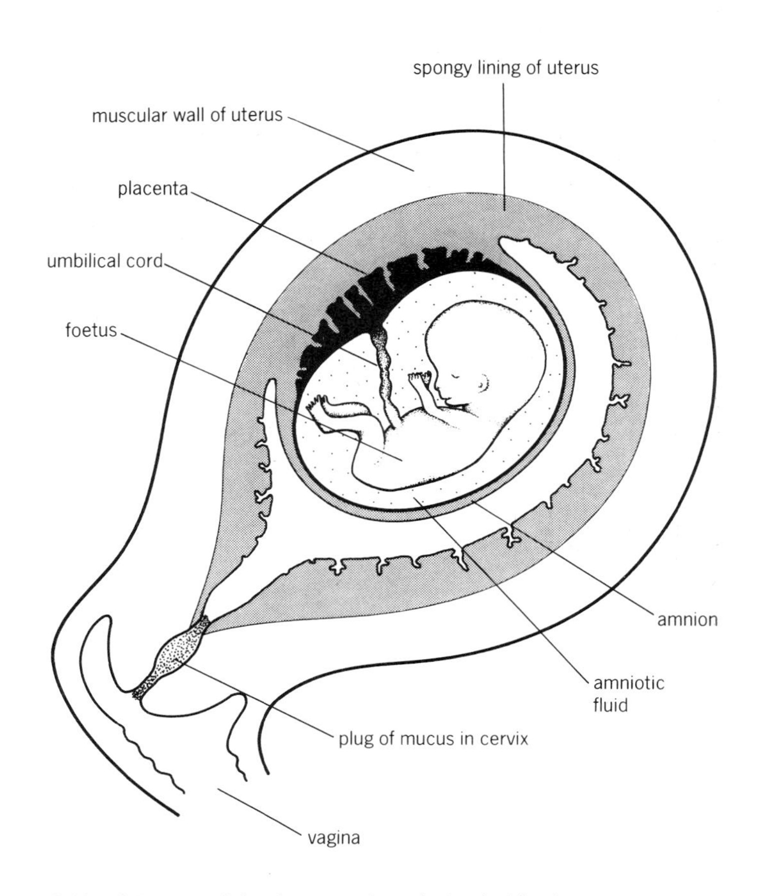

8.21 Side view of developing embryo foetus inside uterus

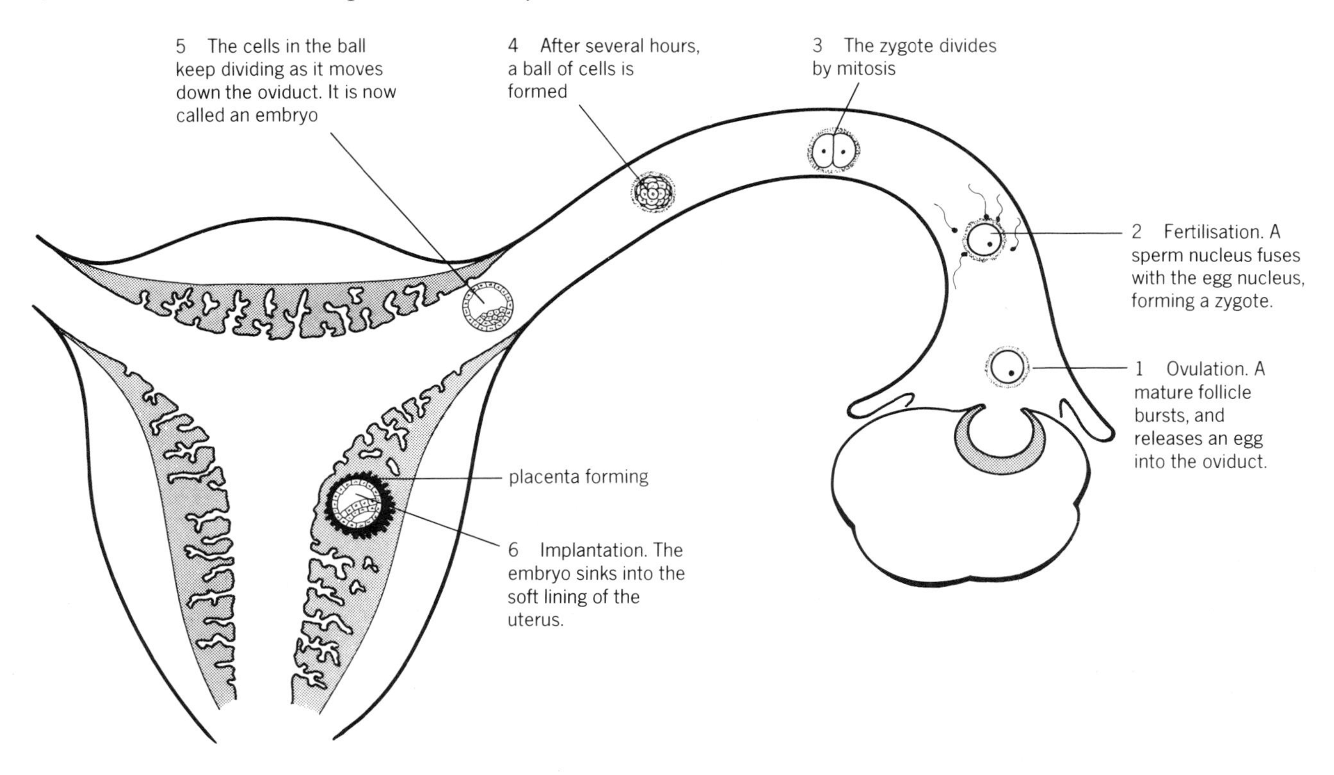

8.20 Stages leading to implantation

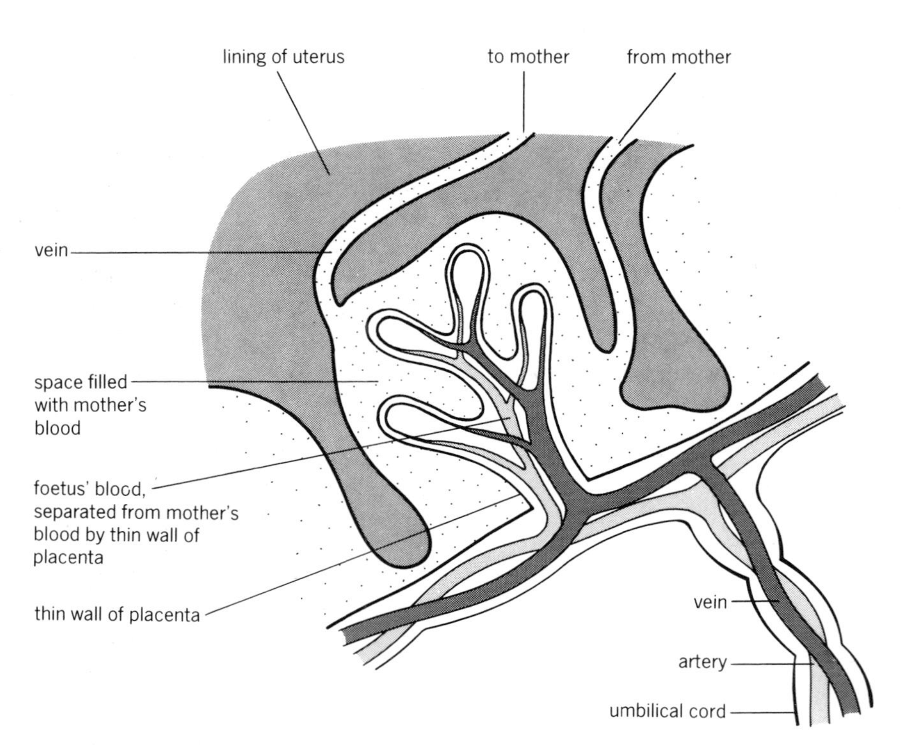

8.22 Part of the placenta

Oxygen and food materials in the mother's blood diffuse across the placenta into the embryo's blood, and are then carried along the umbilical cord to the embryo. Carbon dioxide and waste materials diffuse the other way, and are carried away in the mother's blood.

As the embryo grows, the placenta grows too. By the time the embryo is born, the placenta will be a flat disc, about 12 cm in diameter, and 3 cm thick.

8.18 An amnion protects the embryo.

The embryo is surrounded by a strong membrane, called the **amnion.** Inside the amnion is a liquid called **amniotic fluid.** This fluid helps to support the embryo, and to protect it.

8.19 A baby develops during gestation.

No-one fully understands how the cells in the ball which embedded itself in the wall of the uterus become arranged to form a baby. The cells gradually divide and grow. By eleven weeks after fertilisation they have become organised into all the different organs. By this stage the embryo is called a foetus.

After this, the foetus just grows. It takes nine months before it is ready to be born. This length of time between fertilisation and birth is called the **gestation period.**

Fact! The mammal with the longest gestation period is the Asian elephant – it is 609 days on average, but may last as long as 760 days.

8.20 Muscular contractions cause birth.

A few weeks before birth, the foetus usually turns over in the uterus, so that it is lying head downwards. Its head lies just over the opening of the cervix.

Birth begins when the strong muscles in the wall of the uterus start to contract. This is called **labour.** To begin with, the contractions are quite gentle, and only happen about once an hour. Gradually, they become stronger and more frequent. The contractions of the muscles slowly stretch the opening of the cervix (see Fig 8.23).

After several hours, the cervix is wide enough for the head of the baby to pass through. Now, the muscles start to push the baby down through the cervix and the vagina. This part of the birth happens quite quickly.

The baby is still attached to the uterus by the umbilical cord and the placenta. Now that it is in the open air, it can breathe for itself, so the placenta is no longer needed. The placenta falls away from the wall of the uterus, and passes out through the vagina. It is called the **afterbirth.**

The umbilical cord is cut, and clamped just above the point where it joins the baby. This is completely painless, because there are no nerves in the cord. The stump of the cord forms the baby's navel.

The contractions of the muscles of the uterus are sometimes painful. They feel rather like cramp. However, there is now no need for any mother to suffer really bad pain. She can help herself a lot by preparing her body with exercises before labour begins, by breathing in a special way during labour, and she can also be given pain killing drugs if she needs them.

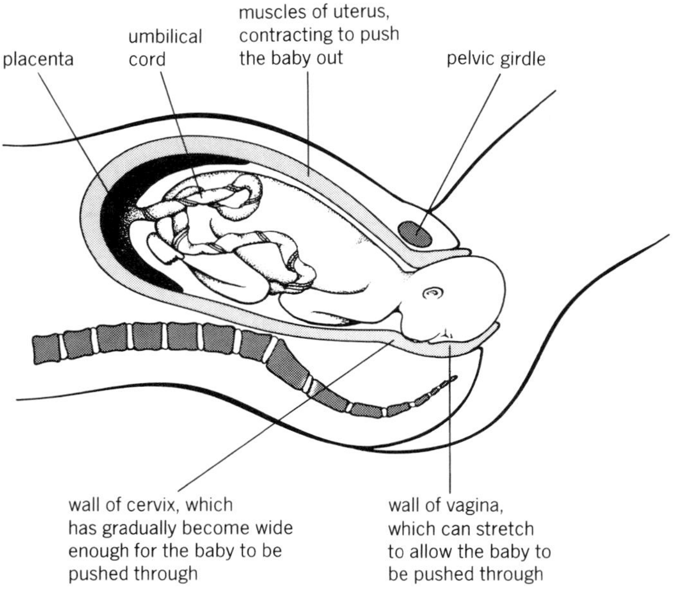

8.23 Birth

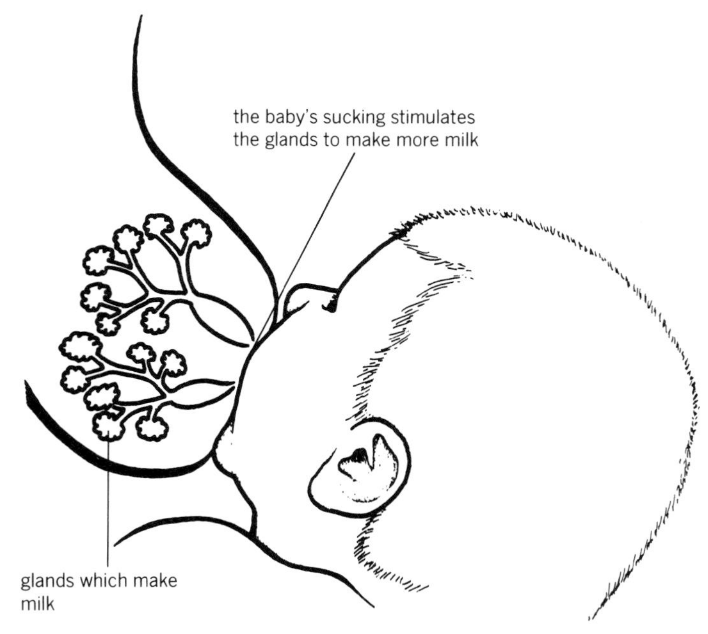

8.24 Lactation

8.21 Mammals care for their young.

Although it has been developing for nine months, a human baby is very helpless when it is born. Usually, both parents help to care for it.

During pregnancy, the glands in the mother's breasts will have become larger. Soon after the birth of the baby, they begin to make milk. This is called **lactation.** Lactation happens in all mammals, but not in other animals (see Fig 8.24).

Milk contains all the nutrients that the baby needs. It also contains **antibodies** (see section 14.28) which will help the baby to resist infection.

As well as being fed, the baby needs to be kept warm. Because it is so small, a baby has a large surface area in relation to its volume, so it loses heat very quickly.

It is extremely important that a young baby is cared for emotionally, as well as physically. Babies need a lot of close contact with their parents.

Most mammals care for their young by feeding them and keeping them warm. In humans, parental care also involves teaching the baby and young child how to look after itself, and how to live in society. This continues into its 'teens – a much longer time than for any other animal.

Questions

1. What is formed when an egg and sperm fuse together?
2. What kind of cell division takes place in the growth of an embryo?
3. From where does the very young embryo obtain its food?
4. What is implantation?
5. What is a foetus?
6. How is the foetus connected to the placenta?
7. Describe two ways in which the structure of the placenta helps diffusion between the mother's and the foetus's blood to take place quickly.
8. List two substances which pass from the mother's blood into the foetus's blood.
9. What is the function of the amnion?
10. How long is the gestation period in humans?
11. Describe what happens to each of the following during the birth of a baby.
    (a) muscles in the uterus wall
    (b) the cervix
    (c) the placenta
12. Why must babies be kept warm?

Fact!

The most children one woman has ever had is 69. A Russian woman who lived between 1707 and 1782 had sixteen pairs of twins, seven sets of triplets and four sets of quadruplets, all born between 1725 and 1765.

The most children surviving at a single birth is six. This has happened three times – once in South Africa in 1974, once in 1980, in Italy, and once in England in 1983.

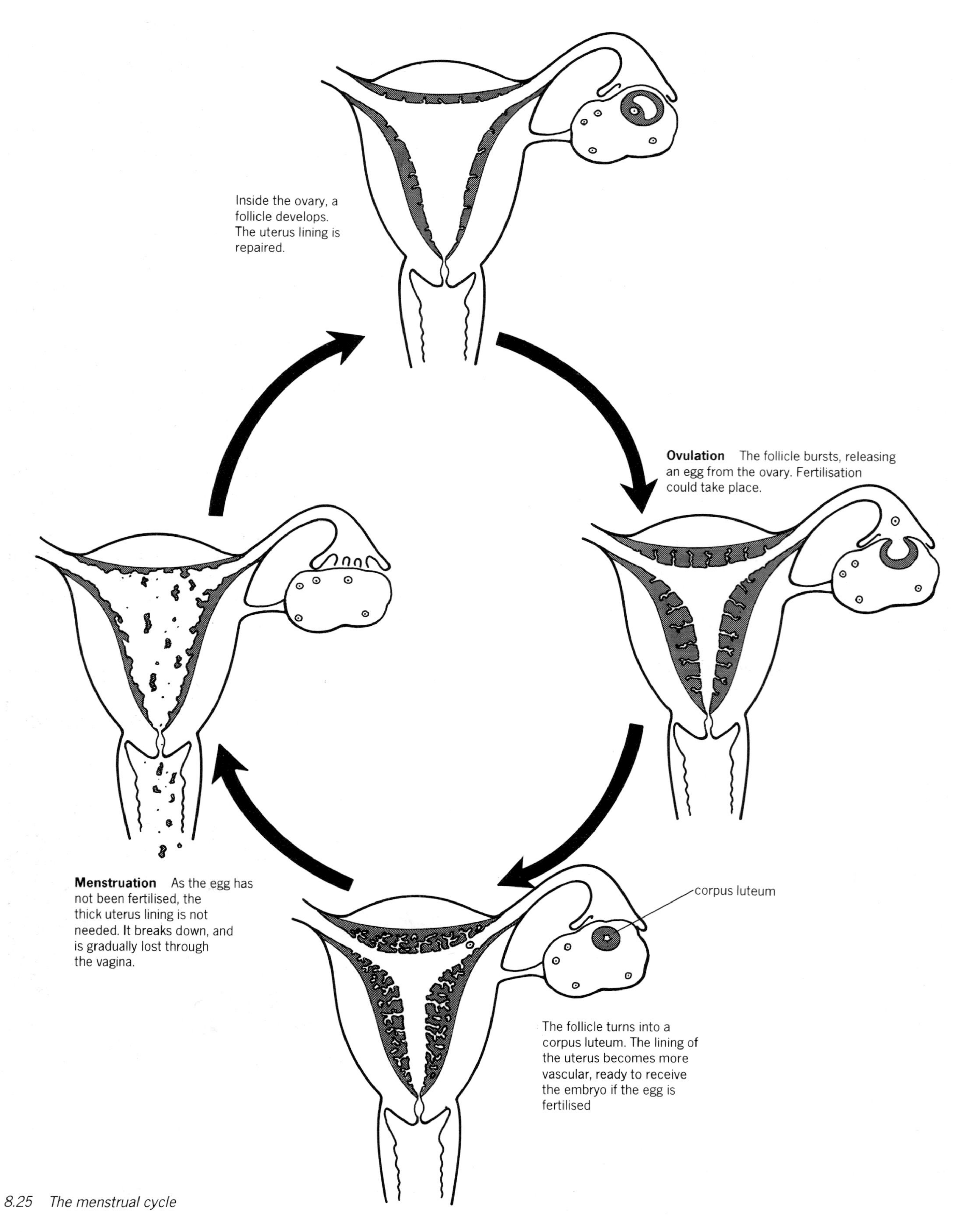

8.25 The menstrual cycle

8.22 The menstrual cycle.

Usually, one egg is released into the oviduct every month in an adult woman. Before the egg is released, the lining of the uterus becomes thick and spongy, to prepare itself for a fertilised egg. It is full of tiny blood vessels, ready to supply the embryo with food and oxygen if it should arrive.

If the egg is not fertilised, it is dead by the time it reaches the uterus. It does not sink into the spongy wall, but continues onwards, down through the vagina. As the spongy lining is not needed now, it gradually disintegrates. It, too, is slowly lost through the vagina. This is called **menstruation,** or a period. It usually lasts for about five days.

After menstruation, the lining of the uterus builds up again, so that it will be ready to receive the next egg, if it is fertilised.

The menstrual cycle is controlled by hormones. This is described in section 12.29.

8.23 Sexual maturity is reached at puberty.

The time when a person approaches sexual maturity is called **adolescence.** Sperm production begins in a boy, and ovulation in a girl.

During adolescence, the secondary sexual characteristics develop. In boys, these include growth of facial and pubic hair, breaking of the voice, and muscular development. In girls, pubic hair begins to grow, the breasts develop, and the pelvic girdle becomes broader.

These changes are brought about by hormones. The male hormones are called **androgens.** The female hormones are called **oestrogens.**

The point at which sexual maturity is reached is called **puberty.** This is often several years earlier for girls than for boys. At puberty, a person is still not completely adult, because emotional development is not complete.

Questions

1 Why does the uterus wall become thick and spongy before ovulation?
2 What happens if the egg is not fertilised?
3 What is meant by (a) adolescence, and (b) puberty?
4 What are androgens?
5 List two effects of androgens.

Sexual reproduction in a fish

8.24 Fish use external fertilisation.

The herring is a marine fish, which lives in the northern parts of the Atlantic Ocean. Between spring and summer, large numbers of adult herring collect at their spawning grounds. The females have enormous numbers of eggs in their ovaries, while the males have sperm in their testes.

The female lays her eggs in water, and the male releases sperm onto them. This is called **spawning.** The sperm swim to the eggs, and fertilise them. This is called **external fertilisation,** because it happens outside the female's body.

8.25 Few eggs survive to grow into fish.

The eggs sink to the bottom of the sea. There are so many of them that they attract other fish, such as haddock, which eat them. The parent fish do not look after the eggs.

The eggs contain **yolk,** which supplies the developing embryo with food. They are surrounded by a layer of jelly-like **albumen,** which helps to protect the developing fish. Albumen is a protein. The embryo obtains its oxygen by diffusion, from the water.

After a few days, the young fish hatches from the egg. It is now called a **larva.** The remains of the yolk are still attached to it, and it uses this for food for the next day or so.

The young larvae feed on microscopic plants floating in the water, called **phytoplankton.** Many of them are eaten by predators. Of the thousands of eggs laid by the female fish, only a very few will survive to become adult herrings.

Fact! The fish which lays the largest number of eggs is the Ocean Sunfish – it may lay up to 300 000 000 at one time.

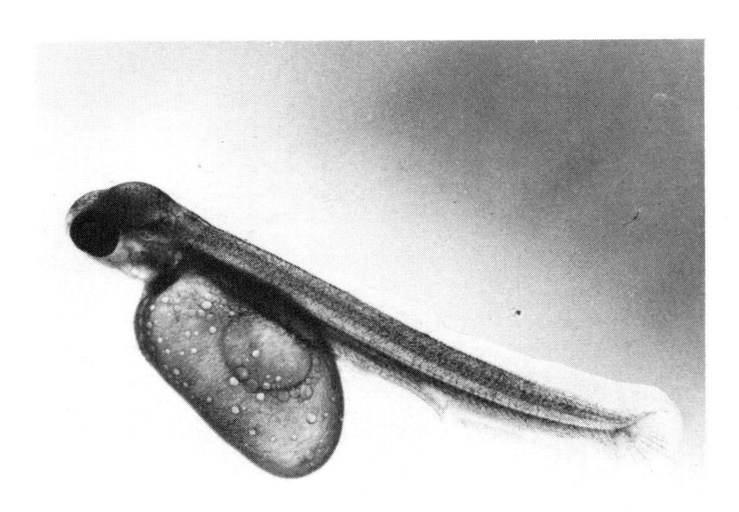

8.26 A small fish larva with its yolk sac containing food

Sexual reproduction in an amphibian

8.26 Frogs return to water to fertilise eggs.

The common frog is an amphibian which lives near ponds and slow moving streams. Like fish, frogs use external fertilisation. In spring, male and female frogs collect together in ponds. The male frogs, which are usually slightly smaller than the females, climb onto the females' backs. The male clings firmly to the female using his front legs. He has a horny pad on each thumb, which helps him to grip.

The female lays her eggs in the water. The male releases sperm onto them immediately. He must do this very quickly, because the eggs swell as soon as they get into the water. Once they are swollen, the sperm cannot get in.

8.27 Tadpoles are adapted to life in water.

Like a fish's egg, a frog's egg contains an embryo, yolk and albumen (see Fig 8.28). The albumen helps to stick the eggs together in a large group.

The embryo is ready to hatch after about ten days. It is now called a larva or tadpole.

Fig 8.28 shows how the tadpole develops. The tadpoles are well adapted for life in water. They respire using gills, first external ones and later internal ones. Their mouths and digestive systems are designed to eat first water plants, and then small aquatic (water living) animals. They have muscular tails to help them to swim.

8.27a Mating frogs

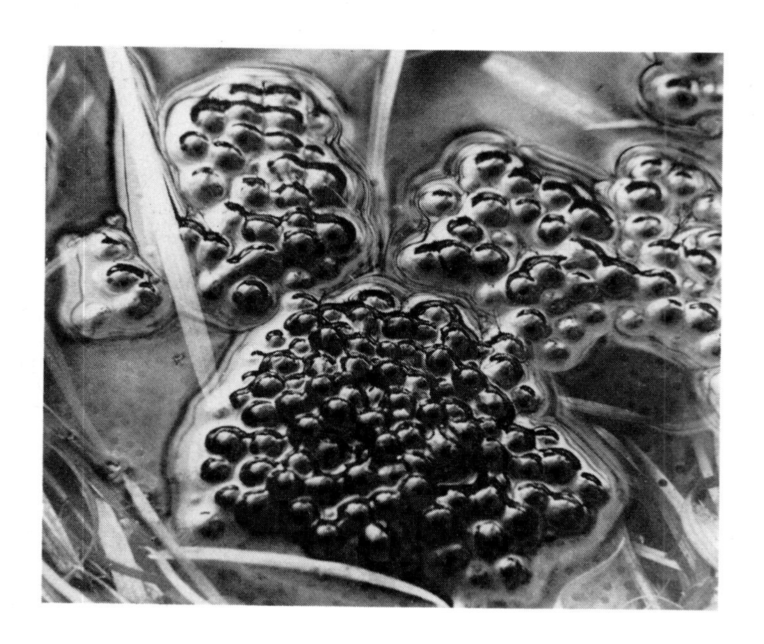

8.27b Mass of frog spawn one day after laying

8.27c Ropes of toad spawn laid by the Common Toad

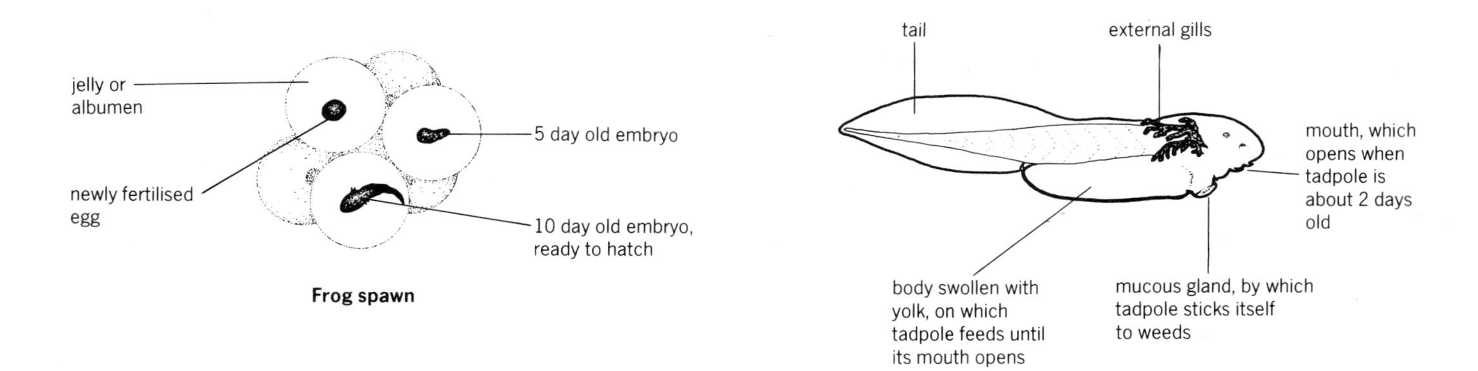

Frog spawn

A two day old tadpole

muscular tail

spiracle—water which has passed over the gills comes out through here

operculum—a flap of skin covering the gills

nostril, used for smell

hind limb bud

anus

coiled intestine, seen through skin; it is long, to deal with the tadpole's herbivorous diet

mouth with horny lips, for scraping algae from the surface of water and weed.

A four week old tadpole

front limb developing underneath operculum

mouth used for eating meat

fully developed hind limbs

intestine shorter, because diet is now carnivorous

An eight week old tadpole

tail begins to be absorbed into body

pelvic girdle showing through skin

eye moves to top of head

mouth widens

fully developed fore limb

A twelve week old frog

strong pelvic girdle, to transmit forces from hind limbs when leaping

eye near top of head, so that frog can lie in water and see above it

ear drum

brownish yellow markings on skin, for camouflage

nostril, used for smell

wide, gaping mouth, containing sticky tongue for catching insects

short fore limbs, to steer when swimming, and to absorb shock when landing from leap

long, muscular hind limbs for swimming and leaping

webbed hind feet for swimming

thin, moist skin, used for gas exchange

An adult frog

8.28 Development of a frog

8.28 Frogs are adapted to spend some time on land.

In late spring and early summer, the tadpoles change into frogs. This change is called **metamorphosis.** Metamorphosis changes the aquatic tadpole into a terrestrial (land living) frog.

The tadpole's tail is digested by special cells which secrete enzymes into it, and it is absorbed into the tadpole's body. It would be a hindrance when the frog is living on land.

The gills are also reabsorbed, and are replaced by lungs, adapted for breathing in the air. The adult frog also breathes through its thin, moist skin.

The mouth becomes wider, and a sticky tongue develops, fixed to the front of the mouth. The tongue can be flicked out to catch insects. The eyes move to the top of the head, so that the frog can see above the surface of the water.

The hind limbs develop strong muscles, which enable the frog to leap. Webs grow between the toes, to provide a large surface to push against the water when swimming. The front limbs are smaller. They are used for balance on land, and for steering when the frog is swimming.

After metamorphosis, the young frog is still very small, and does not become sexually mature for about four years.

Questions

1 What type of fertilisation do herrings use?
2 Explain how an embryo herring is (a) supplied with food, and (b) protected.
3 Why do female herrings lay so many eggs?
4 What type of fertilisation do frogs use?
5 How do tadpoles breathe when they are (a) two days old, and (b) four weeks old?
6 Why do young tadpoles have a mucous gland?
7 What do four week old tadpoles feed on?
8 What do eight week old tadpoles feed on?
9 Describe two differences in the appearance of four week old and eight week old tadpoles, which are associated with the change in their diet.
10 What is metamorphosis?
11 List four changes that occur at metamorphosis, which are associated with the change from an aquatic to a terrestrial life.

Sexual reproduction in birds

8.29 Robins establish breeding territories.

Robins (see Fig 8.29) live in woods and gardens throughout Great Britain and much of Europe. They breed in April and May. The male robin begins to claim a territory in August of the previous year. He does this by singing. His song warns off rival males, and advertises his presence to females. He sings all through the autumn, winter and spring.

During the winter, a female bird will pair up with a male. The two birds then share the territory, although they usually take very little notice of one another at this time. Only males which have a territory will obtain mates. This ensures that the parents will be able to find enough food to support their young in the spring.

In April, the two birds build a nest. It is usually built in a hollow, and made of dead leaves and moss.

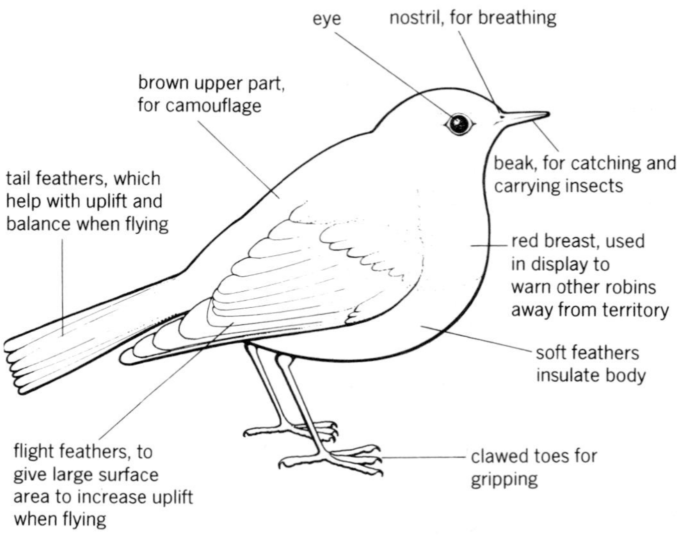

8.29 A robin

8.30 Birds use internal fertilisation.

During the nest building period, eggs are developing in the female robin's ovary. Each egg has a very large amount of yolk.

Robins, like all birds, use **internal fertilisation.** The male robin balances on the female's back, and passes sperm into her oviduct. The sperm swim up the oviduct, where fertilisation takes place.

The fertilised egg then travels down the oviduct. As it goes, the walls of the oviduct secrete a layer of albumen, which surrounds the yolk. Just before it is laid, the oviduct walls secrete a calcium rich substance over it, which hardens into the shell (see Fig 8.30).

Fact! The largest bird's egg is that of the ostrich, which is about 15–20 cm long. It weighs around 1.7 kg, and takes 40 minutes to boil.

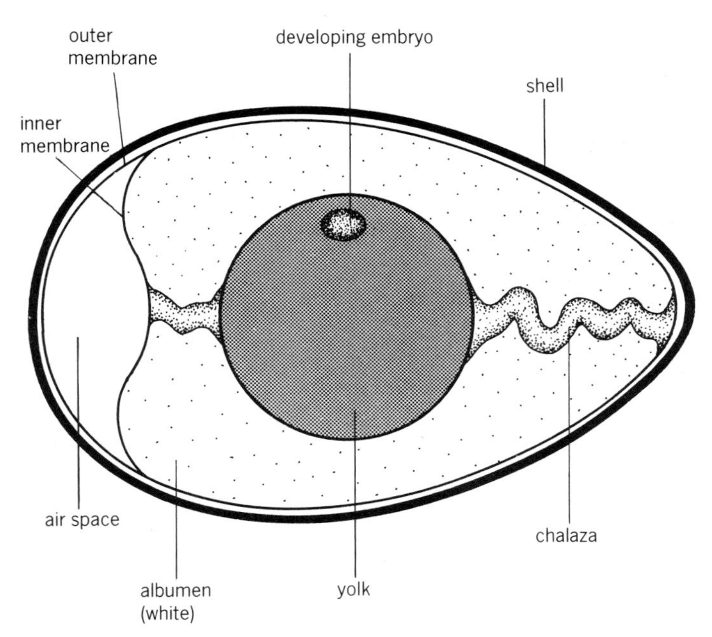

8.30 A section through a fertilised bird's egg

8.31 Bird embryos develop inside shelled eggs.

The female robin lays about five eggs. The shells are white, with reddish brown markings. This helps to camouflage the eggs from predators.

As soon as she has laid all her eggs, the female robin begins to incubate them. Her body heat keeps them warm, and the embryo in each egg begins to develop.

The embryo grows on top of the yolk (see Fig 8.30). It grows blood vessels over the surface of the yolk, which absorb food from it.

The albumen helps to protect the embryo from damage. It also insulates it from extreme temperatures, and stops it from drying out.

The chalaza supports the embryo in the centre of the egg. When the egg is turned, the yolk always swings round so that the embryo lies on top. The robin turns her eggs frequently.

The embryo obtains its oxygen by diffusion through the shell and membranes.

The embryo's nitrogenous waste (see section 10.4) is excreted in the form of **uric acid.** This is a white paste, which collects in one part of the egg, where it will not interfere with the embryo's growth.

8.32 Birds care for their young.

Robins like most birds, take great care of their eggs and young. The young robins hatch after thirteen or fourteen days. They break the shell of the egg using a hard growth on the top of their beak, called an **egg tooth.** The parent birds remove the egg shells from the nest, as these might attract predators.

The female continues to sit on the young birds just as she did on the eggs, until they are about a week old. By this time, they have brown feathers, which camouflage them.

Both the male and female robins bring food to their young. The young robins leave the nest after about a fortnight. Within five weeks of hatching they are completely independent of their parents.

Investigation 8.1 Examining the structure of a hen's egg

You will need a hard-boiled egg, and a fresh egg.

1. Firstly, examine the hard-boiled egg. Very carefully remove the shell. Look at a piece of the shell under a binocular microscope, and notice the pores in the shell. What do you think goes in and out of the egg through these pores? What other functions does the shell have?
2. What is lying between the shell and the white of the egg? What is its function?
3. The white of the egg is made of protein. What is the name of this protein?
4. Put the shelled egg onto a tile. Cut it through lengthways. Make a labelled drawing of the surface of one half. Label membranes, white and yolk.
5. Now examine a fresh, unfertilised egg. Put it into a petri dish, on a paper towel to hold it steady. Using the blunt end of your forceps, make a small crack in the shell. Carefully pick off pieces of shell with forceps. Can you see anything lying on top of the yolk?
6. Gently tip the contents of the shell into a dish containing a little water. Examine the white, and look for the twisted parts called the chalaza. What is their function?
7. Look at a fertilised egg, in which a window has been made. Look for a small spot or circle near the top of the yolk. What is it? Why could you not see this on the yolk of the other egg?

Questions

1. How do male robins claim a territory?
2. What is the advantage of claiming a territory?
3. What type of fertilisation do robins display?
4. How is an embryo robin (a) supplied with food, (b) protected, and (c) supplied with oxygen?
5. List three ways in which the parent birds care for their young after they have hatched.

Fact! The smallest bird's egg belongs to the Vervain humming bird. It is less than 10 mm long and weighs 0.37 g.

Sexual reproduction in flowering plants

8.33 Flowers are for sexual reproduction.

Many flowering plants can reproduce in more than one way. Often, they can reproduce asexually, by vegetative propagation (see section 8.4) and also sexually, by means of flowers.

The function of a flower is to make gametes, and to ensure that fertilisation will take place. Fig 8.32 illustrates the structure of a wallflower. On the outside of the flower are four **sepals.** The sepals protect the flower while it is a bud. In wallflowers, the colour of the sepals depends on the colour of the petals. In other flowers, the sepals are often green.

Just inside the sepals are four **petals.** These are brightly coloured, and have lines on them running from top to bottom. The petals attract insects to the flower. The lines are called **guide-lines**, because they guide the insect to the base of the petal. Here, there is a gland called a **nectary.** The nectary makes a sugary liquid called nectar, which insects feed on.

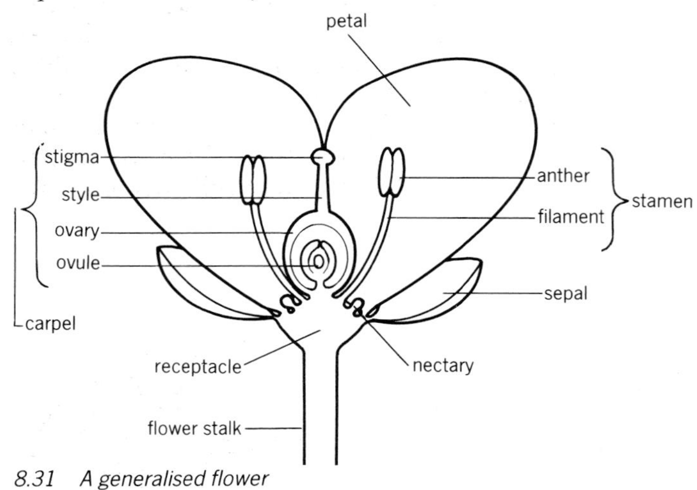

8.31 A generalised flower

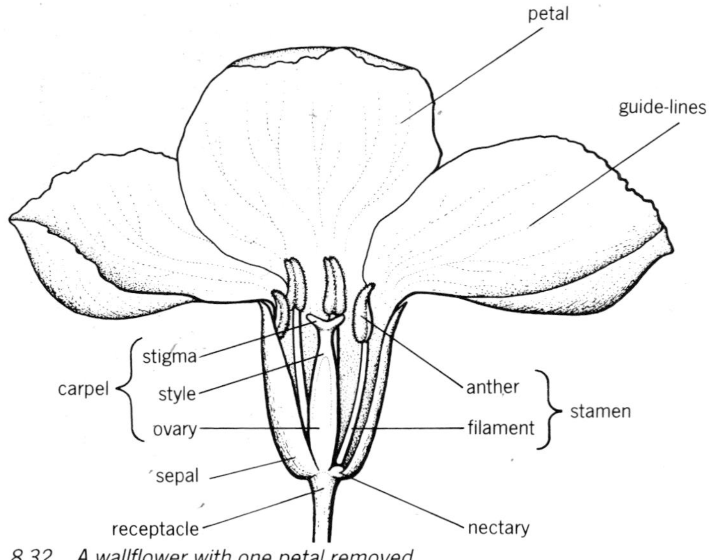

8.32 A wallflower with one petal removed

Inside the petals are six **stamens.** These are the male parts of the flower. Each stamen is made up of a long **filament**, with an **anther** at the top. The anthers contain **pollen grains,** which contain the male gametes.

The female part of the flower is in the centre. It is called a **carpel.** Most of it consists of an **ovary.** Inside the ovary are many **ovules**, which contain the female gametes. At the top of the ovary is a short **style**, with a forked **stigma** at the tip. The function of the stigma is to catch pollen grains.

The wallflower makes both male and female gametes, so it is a hermaphrodite flower. Most, but not all, flowers are hermaphrodite.

Investigation 8.2 Investigating the structure of a wallflower

1 Take an open, fresh looking flower. Can you suggest two ways in which the flower advertises itself to insects?
2 Gently remove the four sepals from the outside of the flower. Look at the sepals on a flower bud, near the top of the stem. What is the function of the sepals?
3 Now remove the four petals from your flower. Make a labelled drawing of one of them, to show the markings. What is the function of these markings?
4 Find the six stamens. If you have a young flower there will be pollen on the anthers at the top of the stamens. Dust some onto a microscope slide, and look at it under a microscope. Draw a few pollen grains.
5 Now remove the six stamens. What do you think is the function of the filaments?
6 Using a hand lens, try to find the nectaries at the bottom of the flower. What is their function?
7 The carpel is now all that is left of the flower. Find the ovary, style and stigma. Look at the stigma under a binocular microscope. What is its function, and how is it adapted to perform it?
8 Using a sharp razor blade, make a clean cut lengthways through the ovary, style and stigma. You have made a longitudinal section. Find the ovules inside the ovary. How big are they? What colour are they? About how many are there?

8.34 Pollen grains contain male gametes.

The male gametes are inside the pollen grains, which are made in the anthers.

Fig 8.34 illustrates a young anther, as it looks before the flower bud opens. The anther has four spaces or **pollen sacs** inside it. Some of the cells around the edge of the pollen sacs divide by meiosis to make pollen grains. When the flower bud opens, the anthers split open (see Fig 8.34(c)). Now the pollen is on the outside of the anther.

The pollen looks like a fine, yellow powder. Under the microscope, you can see the shape of individual

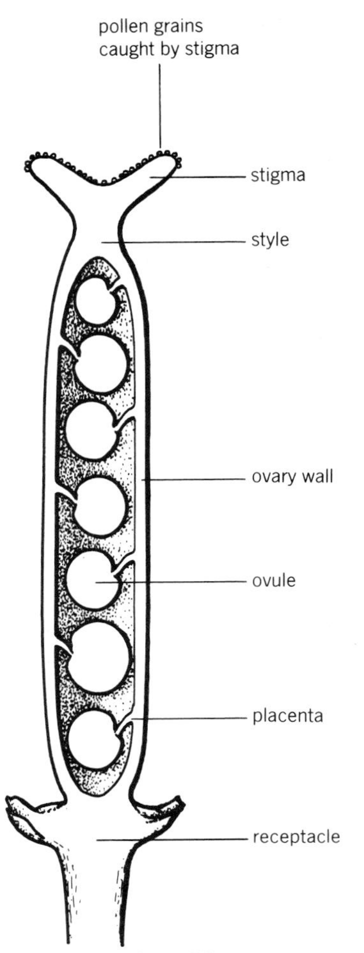

8.33 Section through carpel of a wallflower

grains. Pollen grains from other kinds of flowers have different shapes.

Each grain is surrounded by a hard coat, so that it can survive in difficult conditions if necessary. Wallflower pollen has a smooth, sticky coat, so that it will stick to insects' bodies.

8.35 Each ovule contains a female gamete.

The female gametes are inside the ovules, in the ovary. They have been made by meiosis. Each ovule contains just one gamete.

8.36 Pollen must be carried from anther to stigma.

For fertilisation to take place, the male gametes must travel to the female gametes. The first stage of this journey is for pollen to be taken from the anther where it was made, to a stigma. This is called **pollination.**

In wallflowers, pollination is carried out by insects. Small insects, such as beetles and honey bees, come to the flowers, attracted by their colour and strong, sweet scent. The bee follows the guide lines to the nectaries, brushing past the anthers as it goes. Some of the pollen will stick to its body.

The bee will probably then go to another wallflower, looking for more nectar. Some of the pollen it picked up at the first flower will stick onto the

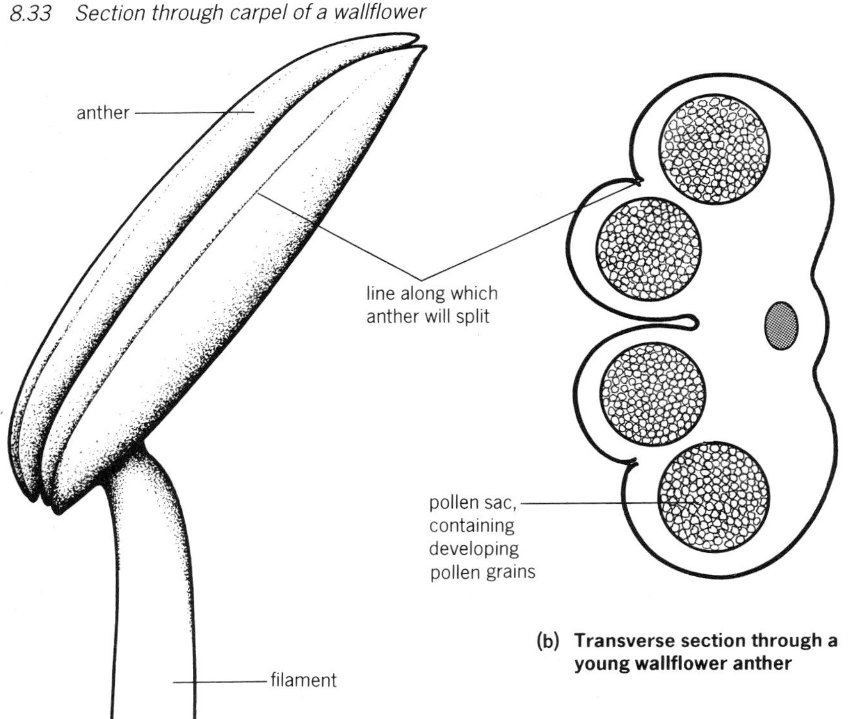

(a) A young wallflower anther

(b) Transverse section through a young wallflower anther

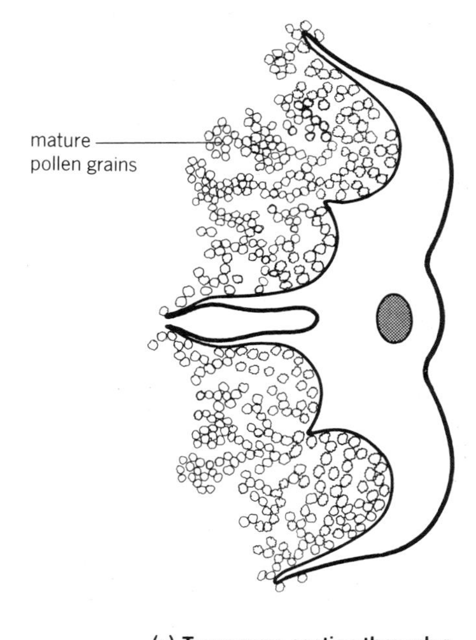

(c) Transverse section through a mature wallflower anther

8.34 How pollen is made

stigma of the second flower when the bee brushes past it. The stigma is sticky, and many pollen grains get stuck on it (see Fig 8.36).

8.37 Flowers can be self- or cross-pollinated.

Sometimes, pollen is carried to the stigma of the same flower, or to another flower on the same plant. This is called **self-pollination.**

If pollen is taken to a flower on a different plant of the same species, this is called **cross-pollination.** If pollen lands on the stigma of a different species of plant, it usually dies.

8.38 Some flowers are wind pollinated.

In the wallflower, pollen is carried from an anther to a stigma by insects. In some flowers, it is the wind which does this.

Fig 8.37 illustrates an oat flower which is an example of a wind-pollinated flower. Table 8.2 compares insect pollinated and wind pollinated flowers.

8.35 Scanning electronmicrograph of wallflower pollen on a petal

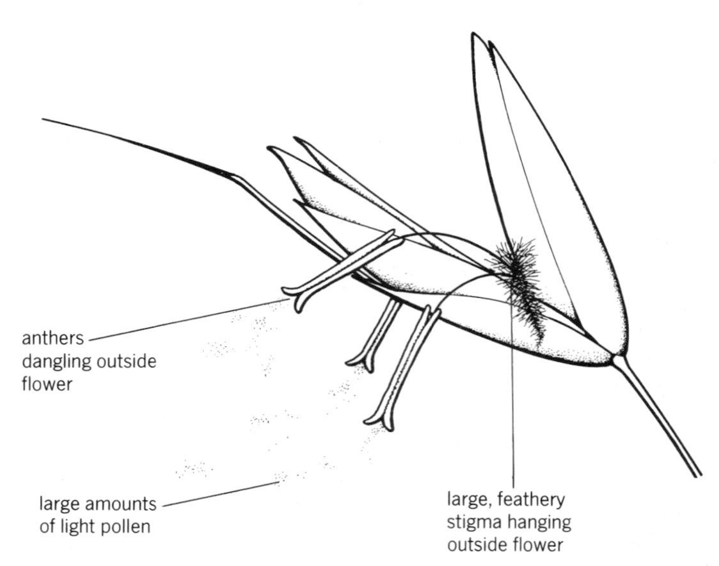

8.37 An example of a wind-pollinated flower—an oat flower

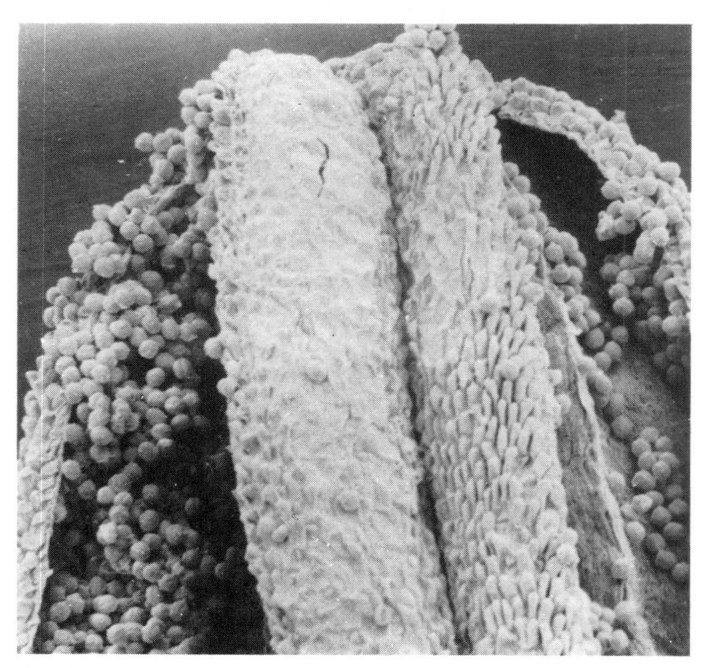

8.36a Tip of wallflower anther split to release pollen grains

8.36b Pollen grains on the sticky surface of a wallflower stigma

Table 8.2 A comparison between wind-pollinated and insect-pollinated flowers

Insect-pollinated, e.g. wallflower	*Wind-pollinated, e.g. grass*
Large, conspicuous petals, often with guide-lines	Small, inconspicuous petals, or no petals at all
Often strongly scented	No scent
Often have nectaries at base of flower	No nectaries
Anthers inside flower, where insect has to brush past them to get to nectar	Anthers dangling outside flower, where they catch the wind
Stigma inside flower, where insect has to brush past it to get to nectar	Stigma large and feathery, dangling outside flower, where pollen in the air may land on it
Sticky or spiky pollen grains, which will stick to insects	Smooth, light pollen, which can be blown in the wind
Quite large quantities of pollen made, because some will be eaten, or carried to the wrong sort of flower	Very large quantities of pollen made, because most will be blown away and lost
Flowers usually appear at warm times of year, when there are plenty of active insects	Flowers sometimes appear at colder times of year

8.39 Pollen tubes take male gametes to ovules.

After pollination, the male gamete inside the pollen grain on the stigma still has not reached the female gamete. The female gamete is inside the ovule, and the ovule is inside the ovary.

If it has landed on the right kind of stigma, the pollen grain begins to grow a tube. You can try growing some pollen tubes, in Investigation 8.3. The pollen tube grows down through the style and the ovary, towards the ovule (see Fig 8.38). It secretes enzymes to digest a pathway through the style.

The ovule is surrounded by a double layer of cells called the **integuments.** At one end, there is a small hole in the integuments, called the **micropyle.** The pollen tube grows through the micropyle, into the ovule.

The male gamete travels along the pollen tube, and into the ovule. It fuses with the female gamete. Fertilisation has now taken place.

One pollen grain can only fertilise one ovule. If there are many ovules in the ovary, then many pollen grains will be needed to fertilise them all.

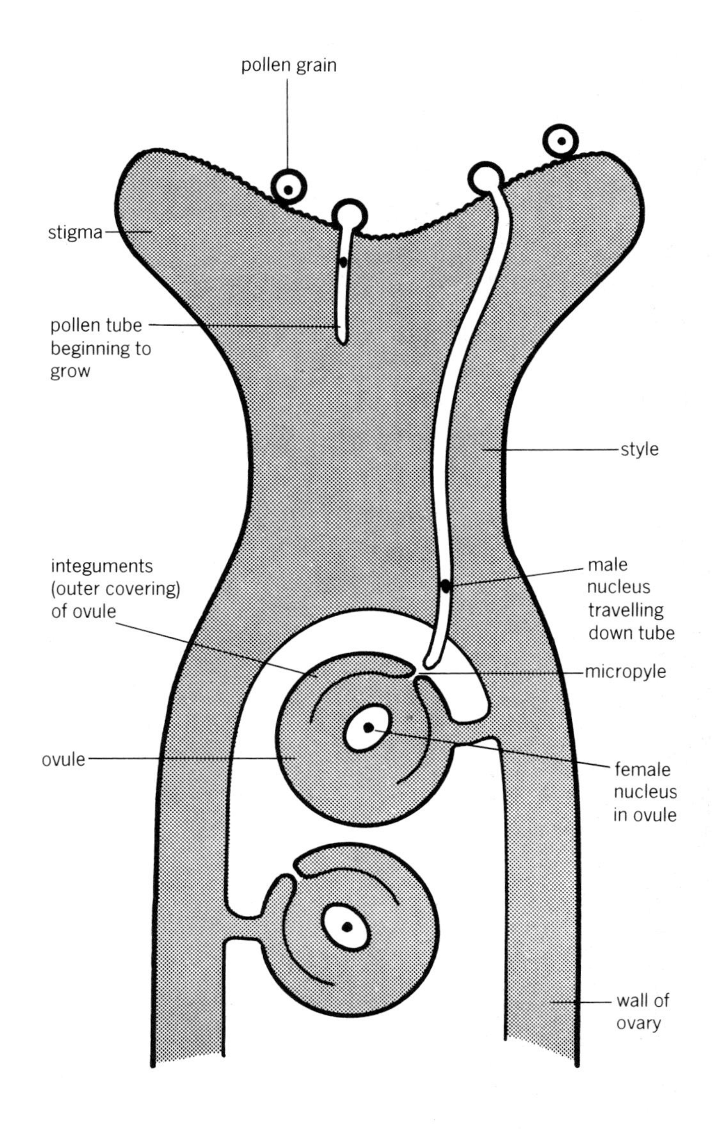

8.38 Fertilisation in a wallflower

8.40 Fertilised ovules become seeds.

Once the ovules have been fertilised, many of the parts of the flower are not needed any more. The sepals, petals and stamens have all done their job. They wither, and fall off.

Inside the ovary, the ovules start to grow. Each ovule now contains a zygote, which was formed at fertilisation. The zygote divides by mitosis to form an embryo plant. The structure of the embryo is shown in Fig 7.5.

The ovule is now called a **seed**. The integuments of the ovule become hard and dry, to form the testa of the seed. Water is withdrawn from the seed, so that it becomes dormant (see section 7.6).

The ovary also grows. It is now called a **fruit.** The wall of the fruit is called the **pericarp.**

Investigation 8.3 Growing pollen tubes

When a stigma is ripe, it secretes a fluid which stimulates pollen grains on it to grow tubes. The fluid contains sugar. In this experiment, you can try germinating different kinds of pollen grains in different strengths of sugar solution.

It is best if the class is divided into groups. Each group should use sugar solution of just one strength.

1 Collect four cavity slides. Using your finger, make a neat ring of Vaseline around the outer edge of each cavity.
2 Stick a label on each slide. Write your initials on it, and the strength of sugar solution your group is using.
3 Fill the cavity in each slide with sugar solution.
4 Choose one flower of each kind which has pollen on its anthers. Dust pollen from one flower onto the solution on one of your slides. Gently lower a coverslip over it, without squashing the Vaseline ring. Write the name of the flower on the label.
5 Repeat step 4 with the other three flowers.
6 Place each slide in a warm incubator, and leave for at least an hour.
7 Set up a microscope. Examine each of your slides under the microscope. Look carefully for pollen tubes. Record your results in the table, and collect results from groups using other strengths of sugar solution.

Results table

Strength of sugar solution	distilled water	0·1M solution	0·5M solution	1M solution	etc.
Flower A					
Flower B					
Flower C					
Flower D					

Questions

1 Why was a ring of Vaseline put around the cavity in each slide?
2 In which solution did each of the four types of pollen germinate best?
3 Can you suggest why pollen dies if it lands on an unripe stigma, or a stigma of the wrong sort of flower?
4 Why do pollen grains grow tubes?

Questions

1 What is the function of a flower?
2 In which part of a flower are male gametes made?
3 In which part of a flower are female gametes made?
4 What is pollination?
5 Why do wind pollinated flowers usually produce more pollen than insect pollinated ones?
6 After pollination, how does the male gamete reach the ovule?
7 What is a micropyle?
8 What happens to each of the following once a flower's female gametes have been fertilised?
   (a) petals
   (b) stamens
   (c) zygote
   (d) ovule
   (e) integuments of the ovules
   (f) ovary

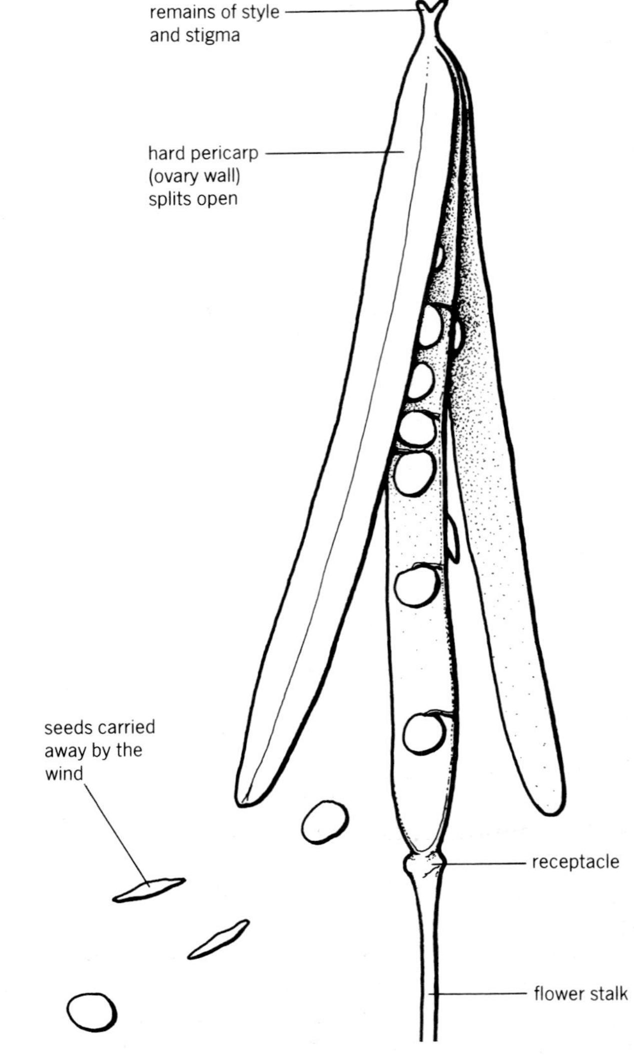

8.39 A wallflower fruit

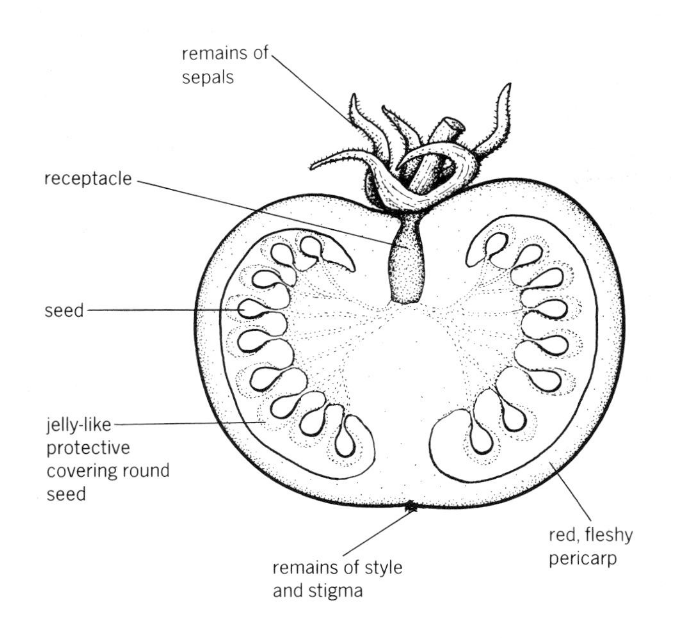

Animal dispersal–tomato

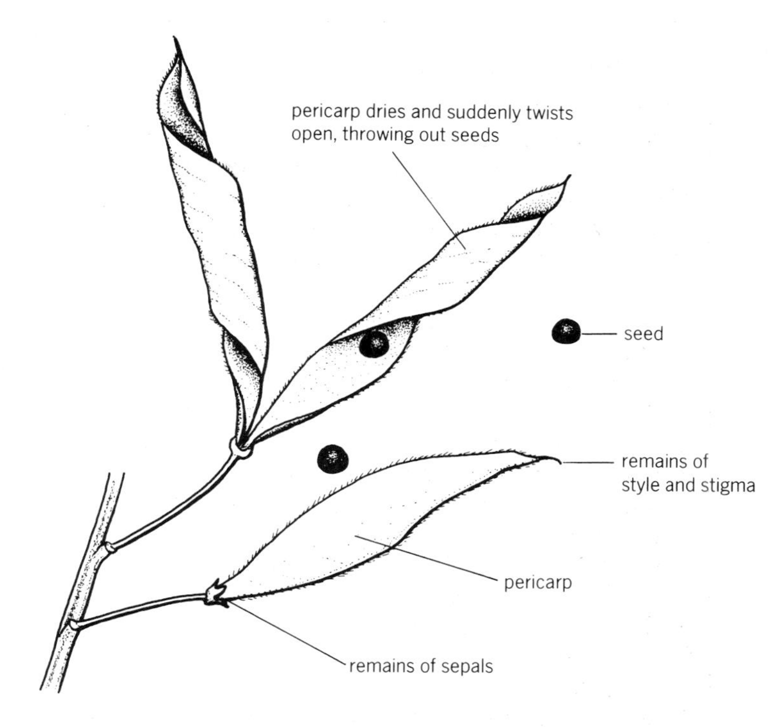

Self dispersal–broom

Animal dispersal–goosegrass

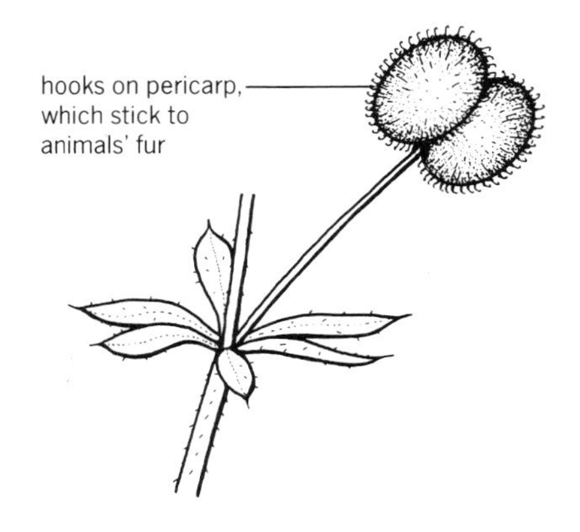

8.40a Fruits dispersed by animals

Wind dispersal–sycamore

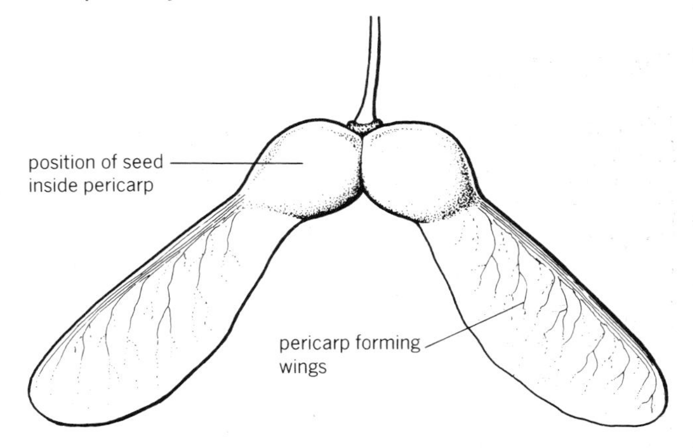

8.40b Fruits dispersed through the air

8.41 Fruits protect and disperse seeds.

The function of the fruit is to protect the seeds inside it until they are ripe, and then to help disperse the seeds. In the wallflower, the fruit is hard and dry. When the seeds are ripe, the fruit splits open. The seeds are very small and light, and can be carried a short distance on the wind (see Fig 8.39).

Dispersal of seeds is important, because it prevents too many plants growing close together. If this happens, they compete for light, water and nutrients, so that none of them can grow properly. Dispersal also allows the plant to colonise new areas.

8.42 Fruits are ovaries after fertilisation.

Plants have an enormous variety of fruits, all adapted to disperse their seeds as effectively as possible.

It is important to remember that, in biology, the word 'fruit' has a very particular meaning. Most people use the word to mean any sweet part of a plant which you can eat. Some of these are not real fruits at all. Rhubarb, for example, is really a leaf stalk, or petiole.

The biological definition of a fruit is an ovary after fertilisation, containing seeds. Blackberries, plums, and oranges are true fruits, but so also are tomatoes, cucumbers and pea pods! You can tell a fruit because (a) it contains one or more seeds, and (b) it has two scars – one where it was attached to the plant, and one where the style and stigma were attached to it.

Sometimes, it is not easy to tell a fruit from a seed. A seed, though, only has one scar, called the **hilum,** where it was joined on to the fruit.

Questions

1 Give two functions of a fruit.
2 List three different ways in which seeds may be dispersed, giving one example for each.
3 Give two differences between fruits and seeds.
4 Which of the following are fruits, and which are not?
(a) blackberry (b) tomato (c) potato (d) cabbage
(e) pear (f) bean pod (g) rhubarb (h) cucumber

A comparison of various methods of sexual reproduction

This chapter has described sexual reproduction in a fish, an amphibian, a bird, a mammal and a flowering plant. Basically, their methods of reproduction are the same. First, gametes are made by meiosis, then fertilisation takes place. A zygote is formed, which grows into a new organism.

There are, however, some important differences in the details of reproduction in these five organisms. What are they, and what are the reasons for them?

8.43 Terrestrial organisms use internal fertilisation.

Male gametes need water to swim through to get to the female gametes. If an organism lives in water, then it can use **external fertilisation,** releasing male gametes into the water near the female. Most fish and amphibians do this.

A terrestrial organism, though, cannot use external fertilisation. The male gametes must be provided with a liquid to swim in, and not allowed to dry out. This is why birds and mammals use **internal fertilisation.** Liquid is provided by the male, with the gametes, and also by the female, inside her body.

Flowering plants have a different solution to this problem. They cannot move around, so the animal type of internal fertilisation is not possible. Instead, the male gamete is enclosed in a tough, waterproof case, forming a pollen grain. Insects or the wind carry the pollen grain to a stigma, where the male gamete can travel along a pollen tube which grows towards the ovule.

8.44 Many gametes are lost in external fertilisation.

All organisms produce more male gametes than female ones. Female gametes are larger and more 'expensive' to make, so fewer are made. Many more male gametes are needed, because many will be lost on their way to the female gametes.

Animals with external fertilisation waste more gametes than animals with internal fertilisation. Fish and frogs therefore produce more gametes than birds or mammals.

Even after fertilisation, not all the embryos will survive. Organisms which do not look after their offspring, such as herrings and frogs, must produce a large number of eggs to make up for ones which are eaten by predators. Birds and mammals care for their eggs and young, so fewer eggs need be produced. The embryo of a flowering plant is protected inside the seed and fruit. Extra seeds, though, are needed, to allow for ones which will land in places unsuitable for growth, or which are eaten by animals.

8.45 Birds' eggs are large to store food.

Female gametes are larger than male ones, because they contain food stores for the embryo. In animal eggs, this is the yolk. Fish and frog eggs have a fairly large yolk, to supply the embryo for several days. Birds' eggs have a very large yolk, to supply the embryo for several weeks before it hatches. Human eggs only have a tiny amount. This is used to feed the embryo only until it arrives at the uterus wall and implants.

The ovules of a flowering plant contain very little stored food, because the ovule is still attached to the parent plant and can get its food from it. Once the ovule is fertilised, it is supplied with food from the parent, to build up food stores.

Fact!

The plant with the largest seed is the Coco de Mer, which grows on the Seychelle islands in the Indian Ocean. The seeds (coconuts) weigh up to 18 kg.

The plants with the smallest seeds are epiphytic orchids. Some kinds have 35 000 000 seeds to the ounce.

Table 8.3 A comparison of sexual reproduction in fish, amphibians, birds, mammals and flowering plants

	Fish, (e.g. herring)	*Amphibian, (e.g. frog)*	*Bird, (e.g. robin)*	*Mammal, (e.g. human)*	*Flowering plant, (e.g. wallflower)*
Number of eggs	Large	Quite large	Small	Small	Small
Size of eggs	Quite large	Quite large	Large	Small	Quite small
Fertilisation	External	External	Internal	Internal	Internal
How embryo feeds	Yolk in egg	Yolk in egg	Yolk in egg	Yolk in egg, then by diffusion from mother's blood through placenta	From parent plant, then from cotyledons of seed
How embryo obtains water	By osmosis from sea water	By osmosis from pond water	From albumen	By osmosis from mother's blood through placenta	By osmosis from parent until seed is fully developed, but then dries. At germination, water is absorbed from soil by osmosis.
How embryo obtains oxygen	By diffusion from sea water	By diffusion from pond water	By diffusion from air	By diffusion from mother's blood through placenta	By diffusion from air
Protection of embryo	By albumen	By albumen	By albumen, shell, nest and parents	By amniotic fluid and mother's body wall	By testa of seed and pericarp of fruit
Protection of young organism	None	None	By nest and parents	By parents	None

Sexual and asexual reproduction

8.46 Sexual reproduction produces variation.

This chapter has described several methods of asexual reproduction, and of sexual reproduction. There are some very important differences between them.

In asexual reproduction, some of the parent's cells divide by mitosis. This makes new cells with the same number and kind of chromosomes as the parent's cells. The new organisms are just like their parents. **Asexual reproduction does not produce variation.**

But in sexual reproduction some of the parent's cells divide by meiosis. The cells which are made are called gametes, and they have only half as many chromosomes as the parent cell. When two sets of chromosomes in two gametes combine during fertilisation, a new combination of genes is produced. The new organism will be different from either of its parents. **Sexual reproduction produces variation.**

8.47 Sexual and asexual reproduction each have advantages.

Is it useful or not to have variation amongst offspring? Sometimes, it is a good thing not to have variation. If a plant, for example, is growing well in a particular place, and if there is plenty of room for it, then it is advantageous if it produces a lot more plants just like itself. The plant is well adapted to living in these conditions, and if its offspring are identical to it, then they will be well adapted, too. Also, asexual reproduction is likely to be a quicker method than sexual reproduction, because there is no need to find a mating partner.

However, if the plant is having difficulty in surviving, or if space is very limited, then sexual reproduction might be more advantageous. The seeds produced could be scattered over a wide area. The new plants which grow from them will all be slightly different from one another, and there is a good chance that some of them will be well adapted to the new conditions they find themselves in.

In general, asexual reproduction is beneficial in an unchanging environment, or when spreading out in a new area where the parent organism is well adapted to survive. Sexual reproduction is most useful in an unstable environment, where variation in the offspring might produce organisms able to survive in a variety of conditions. This helps organisms to begin to colonise new areas.

You will find more about variation, and its importance in evolution, in Chapter 16.

Chapter revision questions

1 Match each of these words with its definition.
Zygote, mitosis, meiosis, gamete, albumen, pollination, fertilisation, pericarp, fruit, seed
(a) a sex cell, containing only half the normal number of chromosomes
(b) an ovary after fertilisation
(c) a diploid cell, formed by the fusion of two gametes
(d) a jelly-like protein which protects the embryos of fish, amphibians and birds
(e) a type of cell division which produces daughter cells just like the parent cell
(f) a type of cell division which produces daughter cells with only half the number of chromosomes as the parent cell
(g) an ovary wall after fertilisation
(h) the transfer of pollen from an anther to a stigma
(i) an ovule after fertilisation
(j) the fusion of two gametes

2 (a) With the aid of diagrams, explain how the embryos of each of the following animals (i) obtains food, (ii) obtains oxygen, (iii) gets rid of waste products, and (iv) is protected.

a named amphibian
a named bird
a named mammal

(b) Why do amphibians lay so many more eggs than birds?

3 (a) Which type of cell division is involved in (i) the production of a new organism by asexual reproduction, (ii) the production of gametes, and (iii) the growth of a zygote?
(b) With the aid of diagrams, describe one way in which a named plant naturally reproduces asexually.
(c) What advantages are there to the plant in reproducing in this way?
(d) Many plants also reproduce sexually. What are the advantages to a plant in reproducing in this way?

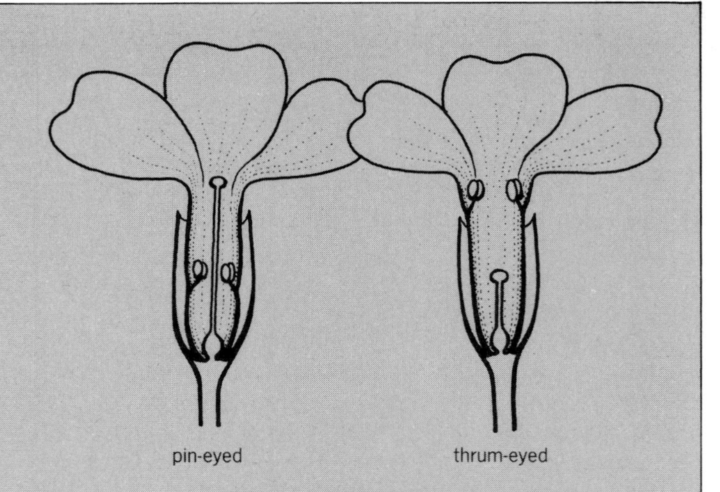

4 The diagram shows two types of primrose flower. These types of flower are often found growing close together. Any one primrose plant, however, only has one type of flower.

(a) Describe the difference in the arrangement of the anthers and stigmas in the pin-eyed and thrum-eyed primrose.
(b) Primroses are pollinated by insects, which reach into the bottom of the flower to get nectar. Which part of the insect's body would pick up pollen in (i) a pin-eyed primrose, and (ii) a thrum-eyed primrose?
(c) Which part of the insect's body would touch the stigma in (i) a pin-eyed primrose, and (ii) a thrum-eyed primrose?
(d) Explain how this will help to ensure that cross-pollination takes place.
(e) Self pollination does sometimes occur in primroses. Would you expect it to occur more often in pin-eyed or thrum-eyed flowers? Explain your answer.
(f) Why is cross pollination usually preferable to self pollination?

9 Living organisms and temperature

9.1 Reactions go faster at higher temperatures.

Most chemical reactions happen faster when the temperature is higher. At higher temperatures molecules move around faster, which makes it easier for them to react together. Usually, a rise of 10°C will double the rate of a chemical reaction (see Fig 9.1(a)).

(a) No enzyme involved

The rate of reaction doubles with every 10 ° C rise in temperature. This is because the molecules which are reacting move faster and have more energy at higher temperatures.

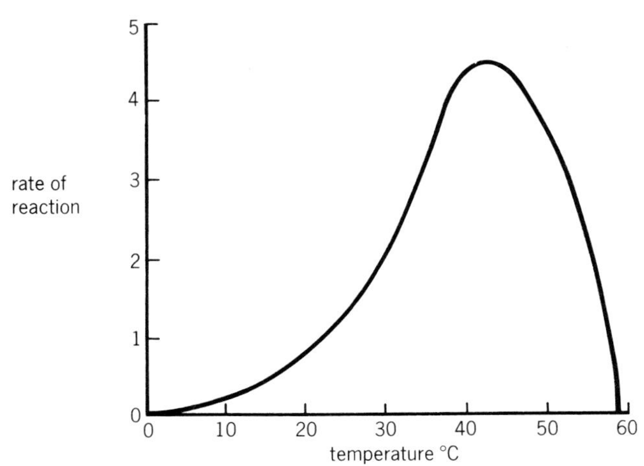

(b) Enzyme-catalysed reaction

Between 0–40 °C, the rate of reaction rises in just the same way as in graph (a) for just the same reasons. But at 40 °C, the enzyme begins to be damaged, so the reaction slows down. By 60 °C, the enzyme is completely destroyed.

40 °C is the **optimum** temperature for this enzyme–the temperature at which the rate of reaction is greatest.

9.1 How temperature affects chemical reactions

All living organisms are affected by the temperature of their environment. This is because there are chemical reactions taking place inside their bodies, called metabolic reactions.

9.2 Enzymes are sensitive to heat.

Most of the chemical reactions happening inside a living organism are controlled or catalysed by enzymes (see sections 2.23–2.26). Enzymes are very sensitive to heat. Once the temperature gets to about 40 °C, they begin to be damaged. When this happens to an enzyme, it cannot catalyse its reaction so well, so the reaction slows down. At higher temperatures, it will stop completely because the enzymes are destroyed (see Fig 9.1(b)).

Investigation 9.1 Investigating the effect of temperature on enzyme activity

The enzyme in saliva is called **amylase**, and it breaks down starch to maltose.

1 Copy the results table.
2 Put 2 cm^3 of starch solution into each of four test tubes. Label them A, B, C and D.
3 Put tube A into refrigerator. Record the temperature of the refrigerator.
4 Put tube B into a test tube rack on your bench. Record the room temperature.
5 Put tube C into a water bath at about 35°C. Record the temperature of the water bath.
6 Put tube D into a water bath at about 80 °C. Record the temperature of the water bath.

Results table

Test tube	A	B	C	D
Temperature in °C				
Time when saliva added				
Results of testing				
after 2 minutes				
after 4 minutes				
etc.				

continued

7 Collect about 5 cm^3 of saliva in a clean boiling tube. Dilute it with an approximately equal amount of distilled water, and mix thoroughly.
8 Put a drop of iodine solution into each of the cavities on a spotting tile.
9 Add 2 cm^3 of saliva solution to each of the four tubes. Record the time at which saliva was added to each tube.
10 Stir each one with its own glass rod.
11 At two minute intervals, check each tube to see if it still contains starch. Do this by dipping the glass rod into the tube, and then into a spot of iodine solution on the spotting tile. If it turns black, there is still starch there. If it stays brown, the starch has gone.
12 Stop the experiment after 30 minutes.

Questions

1 In which tubes did the starch disappear by the end of the experiment?
2 What had happened to the starch in these tubes?
3 How could you check what the starch had been turned into?
4 Why must each tube have the same amounts of starch, and the same amounts of saliva?
5 Why did each tube have its own glass rod?
6 At which temperature did this enzyme seem to work best? Can you explain why?
7 At which temperatures did this enzyme seem to work most slowly? Can you explain why?

Questions

1 Why do chemical reactions happen faster at high temperatures?
2 What is meant by optimum temperature?
3 Why is the optimum temperature for many enzyme controlled reactions about 40 °C?

Animals and temperature

9.3 Poikilotherms do not have constant temperatures.

Invertebrate animals, and also fish, amphibians and reptiles, have very little control over their body temperature. If their environment is cold, their bodies are cold, and so their metabolism slows down. Because the chemical reactions in the cells have slowed down, all their activities slow down. The animals become very sluggish.

If the environment is warm, the animals' cells become warm, and their metabolism speeds up. The animals become more active. If it gets too hot, though, their enzymes may be damaged, which could kill the animals.

Animals like this, whose body temperatures are the same as the temperature of their environment, are called **poikilothermic animals.**

Outside temperature 0 °C

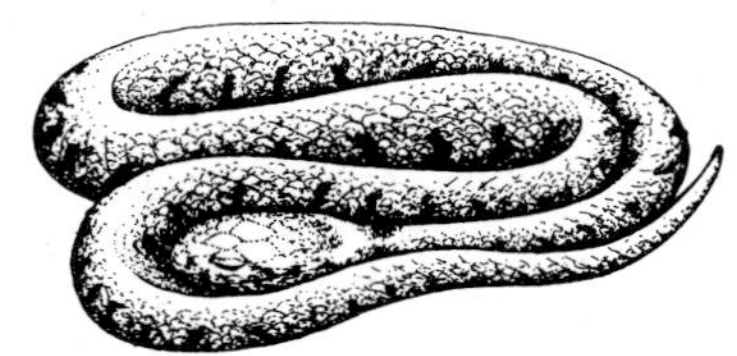

At 0 °C, a poikilothermic animal's metabolic rate slows down, because its body temperature is also 0 °C. The animal is inactive.

At 0 °C, a homeothermic animal remains active. Its cells produce heat by breaking down food through respiration. Its body temperature stays high enough to keep its metabolism going.

Outside temperature 20 °C

At 20 °C, a poikilothermic animal's body temperature is also 20 °C. Its metabolic rate speeds up, and it becomes active.

At 20 °C, a homeothermic animal is no more active than at 0 °C, because its body temperature does not change. It may even be less active, to avoid overheating.

9.2 Poikilothermic and homeothermic animals

9.3 *A tortoiseshell butterfly warms itself in sunshine*

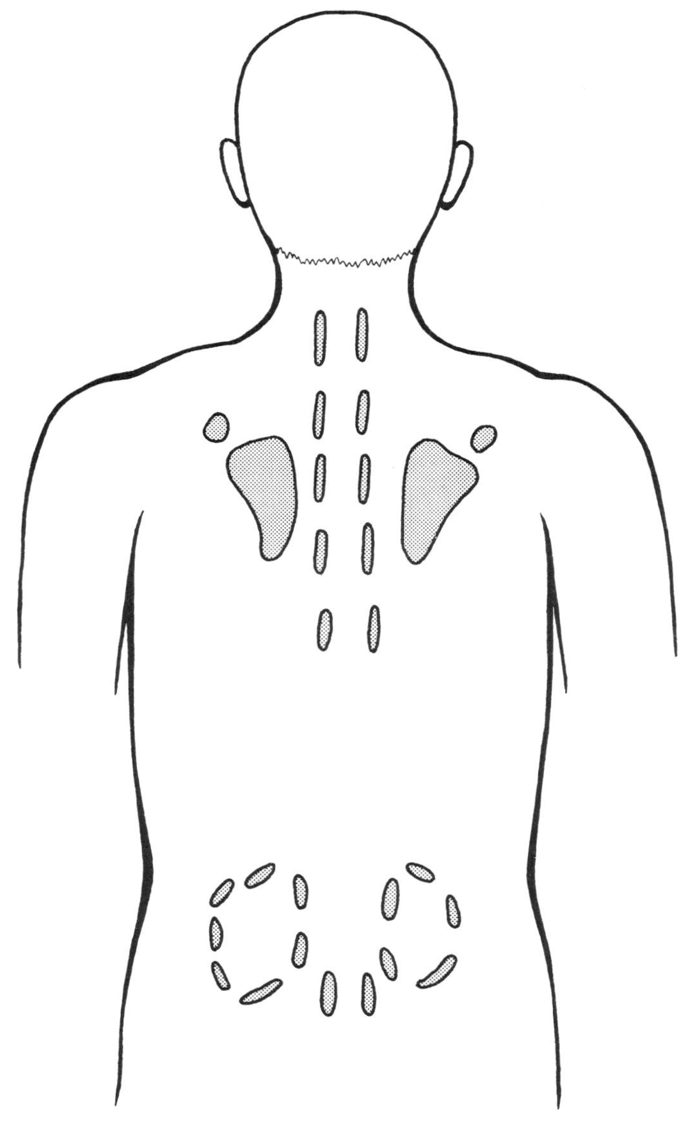

9.4 *Position of the main areas of brown fat*

9.4 Poikilotherms control temperature by behaviour.

Many poikilothermic animals can do something to adjust their body temperature. For example, on a cool but sunny day in early summer, you might see Small Tortoiseshell butterflies resting in the sun with wings spread (see Fig 9.3). The large surface area of their wings is warmed by the sun, so their whole body warms up. They need to do this before they can fly.

Animals can also move from one place to another. If it is very hot in the sun, for example, an animal can move into the shade.

9.5 Homeotherms keep body temperature constant.

Some animals are very good at controlling their body temperature. They can keep their temperature almost constant, even though the temperature of their environment changes. Animals which can do this are called **homeothermic animals.** Mammals and birds are homeothermic.

Being homeothermic has great advantages. If the body temperature can be kept at around 37 °C, then enzymes can always work very efficiently, no matter what the outside temperature is. Metabolism can keep going, even when it is cold outside. In cold weather, a homeothermic animal can be active when a poikilothermic animal is too cold to move.

But there is a price to pay. The energy to keep warm has to come from somewhere. Homeothermic animals get their heat energy from food, by respiration. In mammals, this happens mostly in their brown fat (see Fig 9.4). Because of this, homeothermic animals have to eat far more food than poikilothermic ones.

Questions

1. What is meant by poikilothermic? Give one example of a poikilothermic animal.
2. What is meant by homeothermic? Give one example of a homeothermic animal.
3. What is the advantage of being homeothermic?
4. Why does a homeothermic animal need to eat more food than a poikilothermic animal of the same body weight?

Fact!

The domestic goat has the highest normal body temperature of any mammal, at about 39.9 °C.

The highest body temperature a person has ever had and survived is 46.5°C.

How mammals control their body temperature

9.6 Skin has two layers.

One of the most important organs involved in temperature regulation in mammals is the skin. Fig 9.5 shows a section through human skin.

Human skin is made up of two layers. The top layer is called the **epidermis,** and the lower layer the **dermis.**

9.7 The epidermis protects the deeper layers.

All the cells in the epidermis have been made in the layer of cells at the base of it, called the **Malpighian layer.** These cells are always dividing by mitosis. The new cells which are made gradually move towards the surface of the skin. As they go, they die, and fill up with a protein called **keratin.** The top layer of the skin is made up of these dead cells. It is called the **cornified layer.**

The cornified layer protects the softer, living cells underneath, because it is hard and waterproof. It is always being worn away, and replaced by cells from beneath. On the parts of the body which get most wear, for example the soles of the feet, it grows thicker.

Some of the cells in the epidermis contain a dark brown pigment, called **melanin.** Melanin absorbs the harmful ultra-violet rays in sunlight, which would damage the living cells in the deeper layers of the skin.

Here and there, the epidermis is folded inwards,

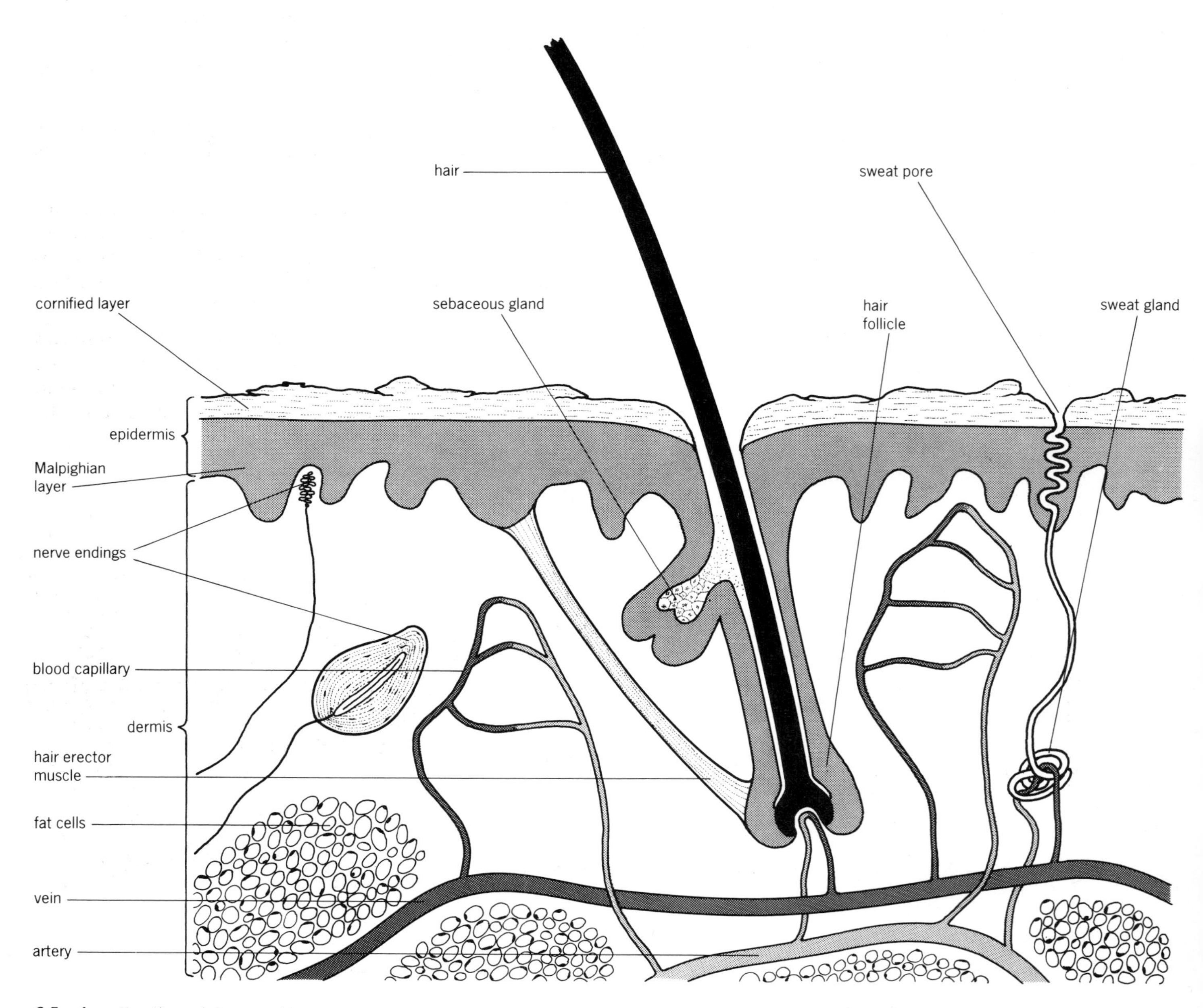

9.5 A section through human skin

forming a **hair follicle.** A hair grows from each one. Hairs are made of keratin.

Each hair follicle has a **sebaceous gland** opening from the side of it. These glands make an oily liquid called **sebum.** Sebum keeps the hair and skin soft and supple.

9.8 The dermis has many functions.

Most of the dermis is made of **connective tissue.** This tissue contains elastic fibres. As a person gets older, the fibres lose their elasticity, so the skin becomes loose and wrinkled.

The dermis also contains **sweat glands.** These secrete a liquid called **sweat.** Sweat is mostly water, with small amounts of salts and urea dissolved in it. It travels up the sweat ducts, and out onto the surface of the skin through the sweat pores. Sweat helps in temperature regulation (see section 9.11).

The dermis contains **blood vessels,** and **nerve endings.** The nerve endings are sensitive to touch, pain, pressure and temperature, so they help to keep you aware of changes in your environment (see section 12.2).

Underneath the dermis is a layer of **fat.** This is made up of cells which contain large drops of oil. This layer helps to insulate your body, and also acts as a food reserve (see section 2.13).

9.9 The hypothalamus coordinates temperature control.

To keep your body temperature at 37 °C, the skin and several other parts of your body must work together. They are controlled and coordinated by a part of the brain called the **hypothalamus.**

The hypothalamus can sense the temperature of the blood running through it. If it is above or below 37 °C, then the hypothalamus sends messages, along nerves, to the parts of the body which have the job of regulating your body temperature.

9.10 When cold the body produces and saves heat.

If your body temperature drops below 37 °C, messages from the hypothalamus cause the following things to happen.

Fat respiration increases Respiration speeds up in the brown fat and liver cells. They break down more glucose, and release more heat. The heat warms your blood, and the blood transfers it all over the body. Brown fat produces more heat after a meal, so eating food warms the body.

Muscles work Muscles in some parts of the body contract and relax very quickly. This also produces heat. It is called shivering.

Hair stands up The **erector muscles** in the skin contract, pulling the hairs up on end. In humans this does not do anything very useful–just produces 'goose pimples'. In a hairy animal, though, like a cat, it traps a thicker layer of warm air next to the skin.

Blood system conserves heat The blood capillaries near to the surface of the skin become narrower, or **constricted.** Only a very little blood can flow in them, so your skin looks blue or white instead of red. The

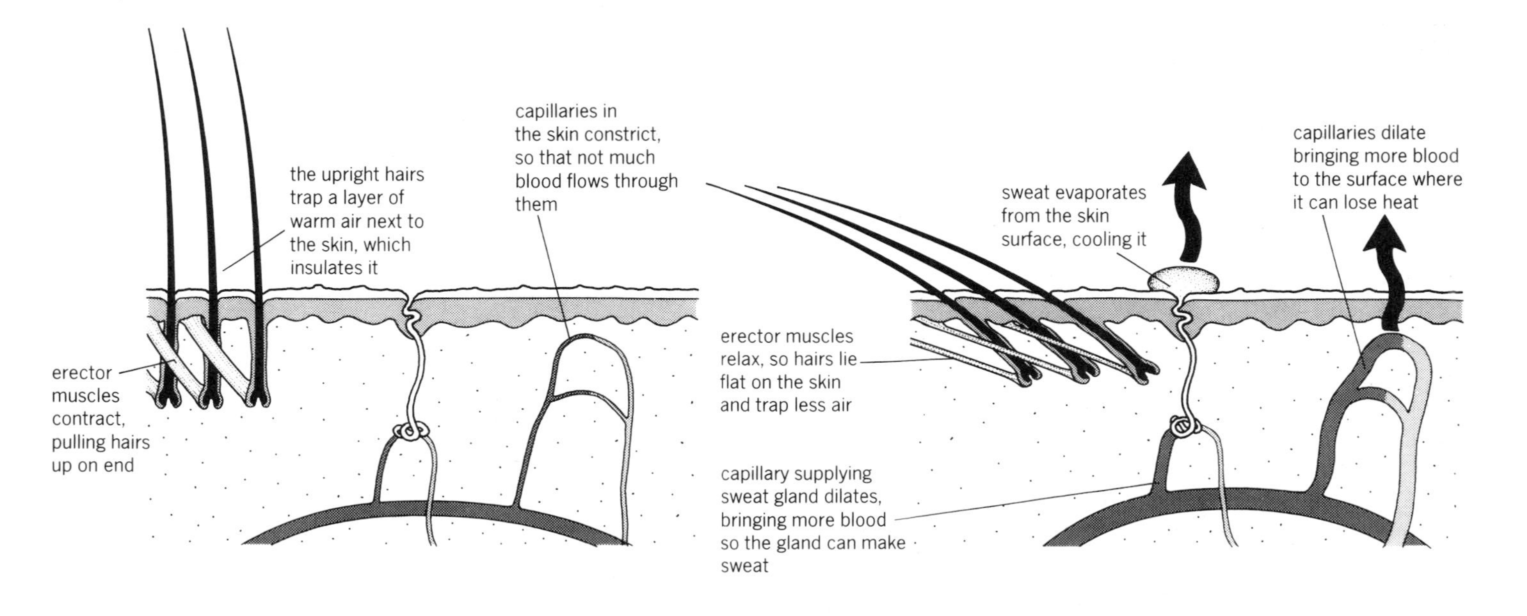

9.6 How the skin helps in temperature regulation

blood flows through the deep-lying capillaries instead. Because these are deep under the skin, the blood does not lose so much heat to the air.

9.11 When hot the body loses more heat.

Respiration slows Respiration in the brown fat and liver cells slows down, so less heat is released.

Hair lies flat The erector muscles in the skin relax, so that the hairs lie flat on the skin.

Blood system loses heat The capillaries near the surface of the skin get wider, or become **dilated.** More blood therefore flows through them. Because a lot of blood is so near the surface of the skin, your skin looks red. Heat is lost from the blood into the air.

Sweat The sweat glands secrete sweat. The sweat lies on the surface of the hot skin, and water in it evaporates. As it does so, it takes heat from the skin with it, cooling the body.

9.12 Low temperatures make life difficult in winter.

In Britain, the most difficult time for any animal to survive is the winter, when temperatures fall, and food is in short supply.

Poikilothermic animals are so cold during the winter months that their metabolic rate is far too low for them to be active. Reptiles, amphibians and most invertebrates spend the winter in an inactive or dormant state. The Large White butterfly, for example, spends the winter as a pupa (see Fig 7.18). It emerges in spring when the weather is warmer.

Homeothermic animals can often remain active in the winter, providing they can get enough food to keep their bodies warm. Many birds, such as the robin, do this, and so do some mammals, such as rabbits.

Other mammals, though, eat the sort of food which is not available in winter. Hedgehogs, for example, feed on worms, slugs and beetles. They would not be able to find enough of this food during the winter to keep their body temperature high enough to be active. Because of this hedgehogs **hibernate**. Towards the end of the summer, when food is plentiful, they eat as much as they can, and build up large fat stores in their bodies. In autumn, they find a sheltered, well insulated place, such as a pile of dead vegetation, and make a nest. They curl into a small ball, to keep in as much body heat as possible, and go to sleep (see Fig 9.7).

It is not an ordinary sleep, however. The hedgehogs' metabolic rate slows right down, and their body temperature drops well below normal. The hedgehogs do not usually wake at all until spring comes.

9.7 A hibernating hedgehog amongst dry leaves under a log

Questions

1. In which part of the skin are new cells produced?
2. What is keratin?
3. What is melanin?
4. What do sebaceous glands make?
5. Which part of the brain controls your body temperature?
6. Why do you get 'goose pimples' when you are cold?
7. Why does your skin go red when you are hot?
8. How do most poikilothermic animals spend the winter in Britain?
9. What is hibernation?

Fact!

During hibernation, a hedgehog's body temperature drops to 5.5 °C. Its heart beats at a rate of between 2–6 beats per minute, and it takes only 1–3 breaths per minute.

Hamsters have the lowest body temperature of any hibernating mammal – it sometimes drops as low as 3.5 °C.

Plants and temperature

9.13 Plants tolerate temperature changes.

Plants have very little control over their temperatures. They are all poikilothermic – their temperature is the same as the temperatrure of their environment. They cannot even move to another place to get away from extremes of temperature, as most animals can. Plants must simply tolerate changes in temperature.

9.14 Plants lose too much water when hot.

Most plants cannot survive temperatures much above 40 °C. In Britain, temperatures never get this high. But even our summer temperatures can be harmful to plants.

The main danger is that the plant will lose too much water by transpiration. If water is lost from the leaves faster than it can be replaced by the roots, the plant will wilt and eventually die. Most plants which are likely to experience hot, dry conditions have some way of cutting down their transpiration rate. This is described in section 10.18.

Plants living in hot, wet places, thrive and grow luxuriantly, producing tropical jungle. The high humidity reduces the danger of excessive water loss by transpiration (see section 6.37).

9.15 Plants survive the winter in different ways.

In Britain, the most difficult time of the year for most plants is winter. Temperatures drop, the ground may freeze, and snow may fall.

Many plants survive the winter only as seeds. These are called **annual plants** (see Fig 9.8). Annual plants, such as corn poppies, grow during spring, and flower in summer. In autumn, their seeds are dispersed, and the plant dies. The seeds are dormant and survive in the ground until the spring. They are not harmed by frost. When the weather becomes warmer, they germinate.

Other plants live through the winter. They are called **perennial plants** (see Fig 9.9). Perennial plants usually have a variety of adaptations to help them to survive through the difficult conditions of the winter months (see sections 9.17–20).

9.16 Plants are dormant at low temperatures.

Low temperatures slow down metabolic reactions. This means that, during the winter, there is very little activity in the cells of a plant. Active transport (see section 6.38) virtually stops, so minerals are not brought into the root hairs. Respiration slows down, and so does photosynthesis. Photosynthesis would not be able to take place efficiently anyway, because there is less light available in winter.

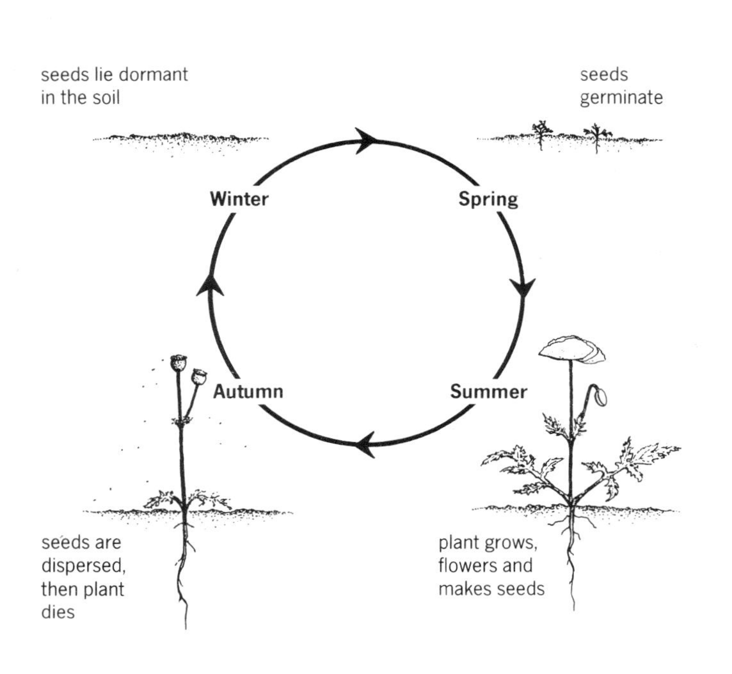

9.8 Life cycle of a corn poppy–an annual plant

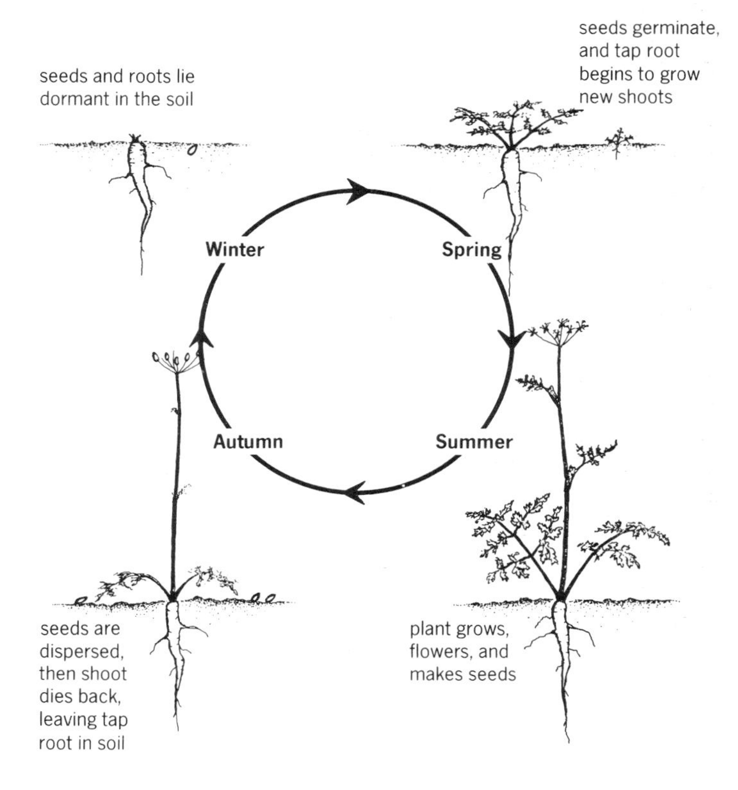

9.9 Life cycle of a cow parsley–a perennial plant

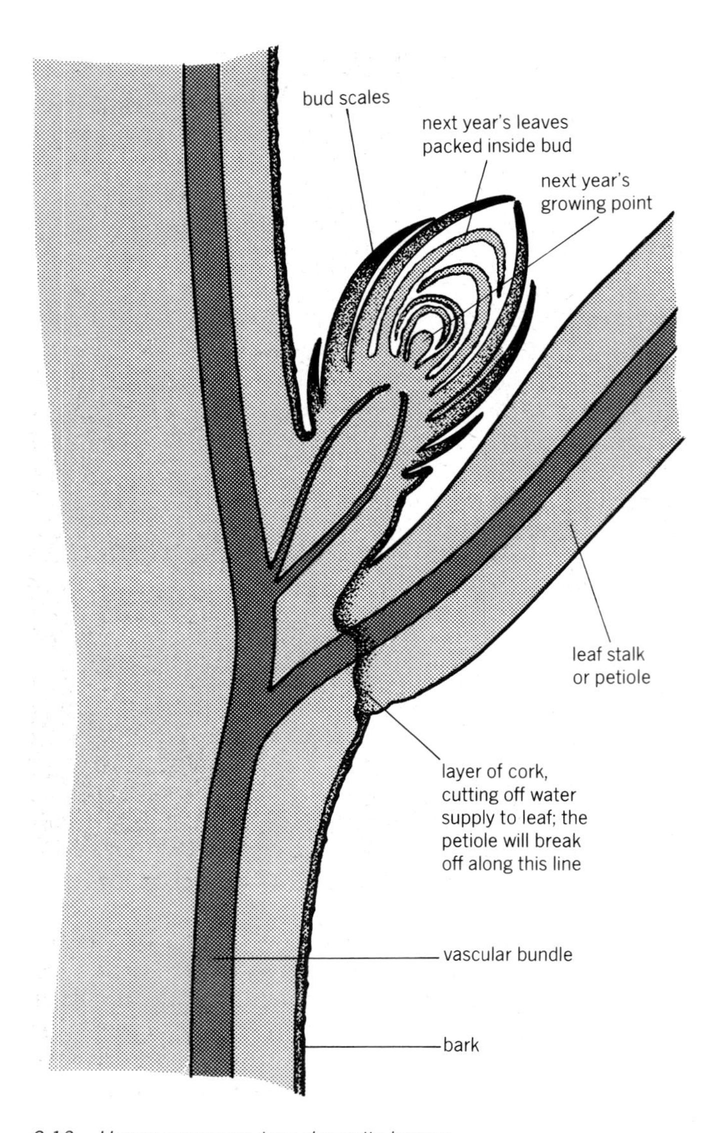

9.10 How a sycamore tree drops its leaves

So most perennial plants spend the winter in a more or less dormant state, with very little metabolic activity going on inside them.

9.17 Some plants can avoid frost damage.

The biggest danger for most plants in winter is freezing. If ice crystals form inside their cells, the cells will be damaged beyond repair. Perennial plants are adapted to prevent this from happening.

Dying back Many perennial plants die back during the winter, so that the only parts which stay alive are under the ground. The parts which survive might be roots, or bulbs (see Fig 8.3) for example. In Britain, the ground rarely freezes deeper than about 10 cm, so these parts are usually safe from frost.

Dropping leaves Many trees drop their leaves in winter, so that they will not freeze. They are called **deciduous trees.**

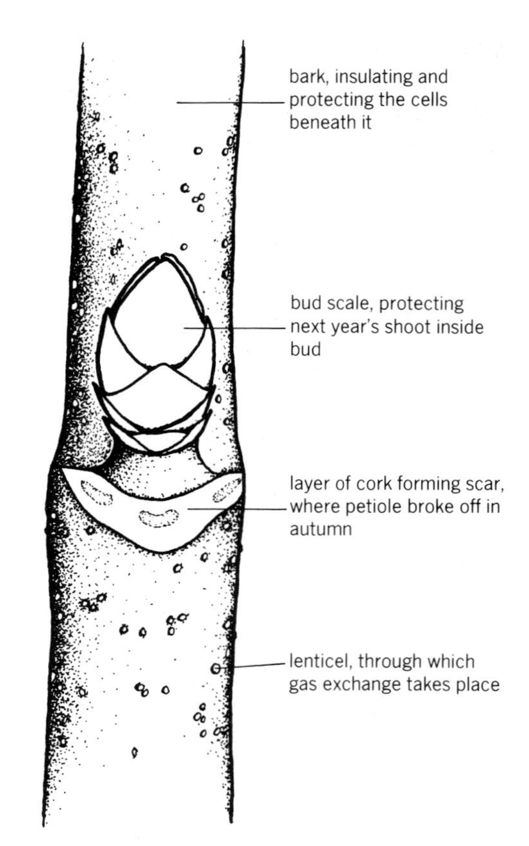

9.11 Part of a sycamore twig in winter

In autumn, a layer of cork grows across the end of the petiole (see Fig 9.10), blocking off the vascular bundles and stopping water getting through to the leaf. The cells in the leaf gradually die. Eventually, the petiole snaps off where the cork layer has grown. The cork layer remains on the tree as a scar, stopping bacteria and fungi getting into the tree.

Many trees take advantage of leaf fall to get rid of some of their excretory products (see section 10.3). The waste materials build up in the leaves, and are got rid of when the leaves drop. However, some useful chemicals are reabsorbed into the plant before leaf fall.

Forming buds Any living parts of the plant which remain above ground in winter must be well protected. The parts which will grow into next year's flowers and leaves spend the winter as buds.

A bud is a shoot, with all the parts packed tightly together. It is protected with a layer of **scale leaves,** which help to insulate the young leaves inside, and also protect them from attack by bacteria or fungi. If you open a Horse Chestnut bud, you will find that the young leaves are surrounded by a white, fluffy packing material which also helps to insulate them.

Forming buds helps a plant to start growth again quickly when the weather warms up. All the parts for the new leaves and flowers are there – they just need to expand when spring comes. The leaves can open out and begin to photosynthesise very quickly.

Bark Woody plants have bark to insulate their stems during the winter. Bark contains cork cells, which are mainly filled with air. They stop the cells inside a tree trunk from getting frozen, unless temperatures drop very low.

9.18 Heavy snow can damage plants.

In some winters in Britain, there are quite heavy falls of snow. If snow builds up on the branches of a tree or even a smaller plant, the weight of the snow may break branches. This is less of a problem for deciduous trees than for ones which keep their leaves, because the loss of leaves decreases the surface on which snow may build up.

Trees which keep their leaves during winter, such as conifers, are usually of a shape which allows snow to slip off (see Fig 9.12).

9.19 Soil water may be frozen in winter.

In some places, though not in most of Britain, water is in short supply in winter. This is because it is all frozen, so plant roots cannot absorb it. Plants which are native to areas where this happens – such as conifers, which live in northern Europe, north Asia and Canada – are adapted to cut down water loss as much as possible. For this reason conifers are also found in hot arid areas such as parts of Greece, Turkey and the Middle East. Some of these adaptations for reducing water loss are described in section 10.18.

9.20 Plants build up food stores for winter.

Plants make their own food by photosynthesis. In winter, this is not possible, because temperature and light intensity are too low.

9.12 Trees in winter

9.13 Food stored in bulbs helps snowdrops to start growth early

During the winter months, this does not matter much, because the plant is dormant and does not need food. In spring, though, it is important that the plant starts growing as quickly as possible. It will help if it has some food reserves to start it off, before longer days and higher temperatures allow it to start photosynthesising more efficiently.

Most perennial plants store food for the winter. Onions and daffodils store food in bulbs (see Fig 8.3). Carrots store food in tap roots (see Fig 6.26). Trees store food in their roots.

Questions

1 What is the main problem for plants living in very hot places?
2 How do annual plants survive the winter?
3 Why does the uptake of minerals by plant roots slow down in winter?
4 Why is frost dangerous to many plants?
5 Explain how deciduous trees drop their leaves.
6 What is a bud?
7 How does bark help a tree to survive the winter?
8 Why do many perennial plants store food during the winter?

Chapter revision questions

1 Explain the differences between each of the following pairs of terms:
(a) homeothermic, poikilothermic;
(b) dilate, constrict;
(c) epidermis, dermis;
(d) sebaceous gland, sweat gland;
(e) annual, perennial.

2 Explain the following observations.
(a) Homeothermic animals eat more food in proportion to their body weight than poikilothermic animals.
(b) Your skin becomes red when you are hot.
(c) Many trees drop their leaves in winter.
(d) Hedgehogs hibernate.
(e) Most insects only become active when it is quite warm.

3 An experiment was performed, to find how the rate of a chemical reaction was affected by temperature. The results are shown on the graph.
(a) Is this reaction catalysed by an enzyme? Give reasons for your answer.
(b) What is the optimum temperature for this reaction?
(c) Suggest one reaction which might give results like this.

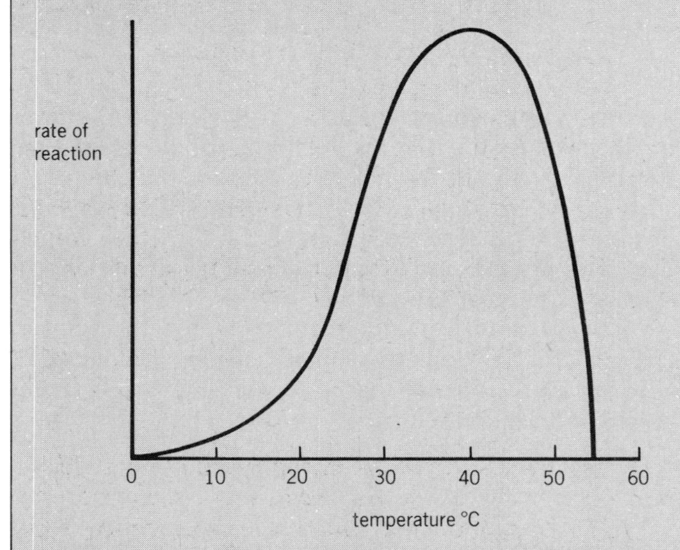

4 (a) List two ways in which high temperatures are dangerous to plants.
(b) List two ways in which low temperatures are dangerous to plants.
(c) Briefly describe how a named annual plant survives the winter.
(d) Draw a large, labelled diagram of part of a named perennial plant in winter.
(e) Explain how any two of the structures you have labelled in your diagram help this plant to survive the winter.

5 In an experiment, a number of newly hatched sparrows were kept at a temperature of 20 °C. Each day, their body temperature and the amount of oxygen they used were measured. The average temperature and oxygen consumption for each day were plotted on a graph.

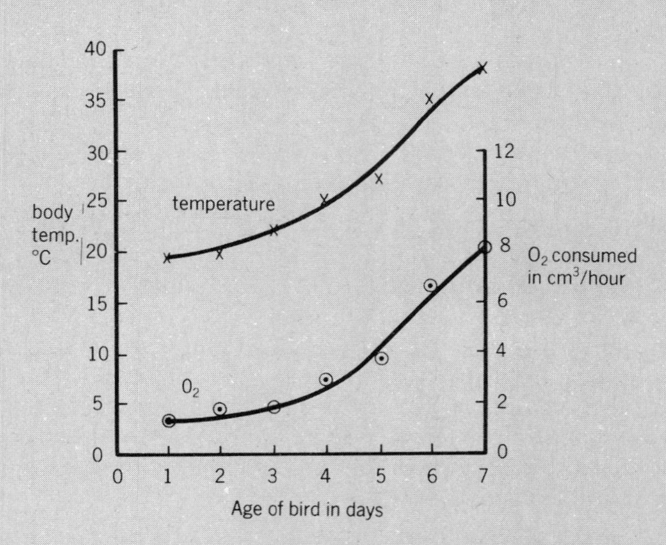

(a) What is (i) the body temperature, and (ii) the oxygen consumption of a three day old sparrow, when the air temperature is 20 °C?
(b) Adult sparrows are homeothermic, keeping their body temperature at around 38 °C. At what age do young sparrows become able to maintain this body temperature?
(c) Using the information provided by the graph, and your own knowledge of how homeothermic animals maintain their body temperature, explain why one day old sparrows consume less oxygen than seven day old sparrows.
(d) What would you expect to happen to (i) the body temperature, and (ii) the oxygen consumption of a one day old sparrow, if the air temperature was raised to 25 °C? Explain your answer.

Fact! The lowest body temperature which has been survived is 16.0 °C. This has happened at least twice. On 1st February, 1951, a 32-year old woman was found in an alley in Chicago with this body temperature. Her pulse rate was only twelve beats per minute – but she survived. On the 21st January, 1956, a 2-year old girl was found in a house in Iowa, again with a body temperature of 16.0 °C – again, she survived.

10 Excretion and osmoregulation

10.1 Waste products of metabolism are excreted.

All living cells have a great many metabolic reactions going on inside them. The reactions of respiration (see section 5.4) for example, provide energy for the cell. The reactions of photosynthesis (see section 3.4) provide plant cells with sugars.

However, these reactions often produce other substances as well, which the cells do not need. Respiration in animal cells, for example, not only makes energy, but also water and carbon dioxide.

$$C_6H_{12}O_6 + 6\,O_2 \rightarrow 6\,CO_2 + 6\,H_2O + \text{energy}$$

glucose oxygen carbon dioxide water

The animal cells need the energy, and may be able to make use of the water. They do not, however, need the carbon dioxide. The carbon dioxide is a waste product. Plant cells may be able to use the carbon dioxide in photosynthesis.

A waste product like carbon dioxide, which is made in a cell as a result of a metabolic reaction, is called an **excretory product.** The removal of excretory products is called **excretion.**

10.2 Egestion is not excretion.

Many animals have another kind of waste material to get rid of. Almost always, some of the food which an animal eats cannot be digested. Humans for example, cannot digest cellulose. The cellulose in our food goes straight through the alimentary canal, and out of the anus as part of the faeces.

This cellulose is not an excretory product. It has never been involved in any metabolic reaction in your cells. It has not even been inside a cell – it has simply passed, unchanged, through your digestive system. So getting rid of undigested cellulose in faeces is not excretion. It is called **egestion.**

Excretion in plants

10.3 Excretory products of plants.

Carbon dioxide Like all cells, plant cells are always respiring. Respiration produces carbon dioxide as a waste product.

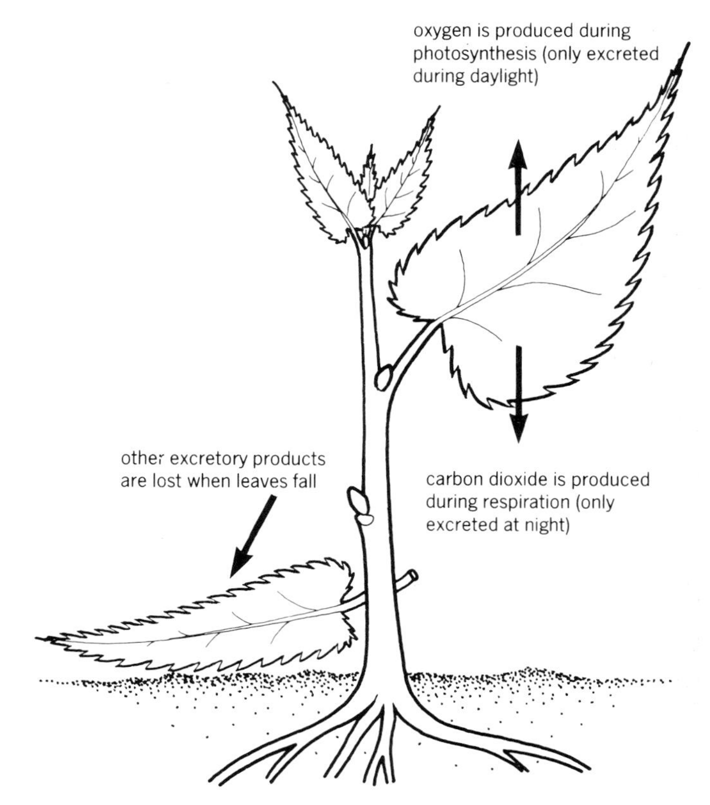

10.1 Excretory products of plants

During the daytime, the carbon dioxide is used by the chloroplasts for photosynthesis (see section 5.33). At night, though, the plant has no use for it, and so it is excreted through the stomata, by diffusion.

Oxygen In the daylight, plants produce oxygen as a by-product of photosynthesis.

$$6\,CO_2 + 6\,H_2O \xrightarrow[\text{chlorophyll}]{\text{sunlight}} C_6H_{12}O_6 + 6\,O_2$$

Quite a bit of this oxygen is used by the plant for respiration. But there is often plenty left over, and this is excreted through the stomata, by diffusion.

Other substances Plants can also excrete other substances. In deciduous plants, for example, chemicals may be left in the leaves in autumn. When the leaves fall, these substances are lost from the plant.

Excretion in animals

10.4 Excretory products of animals.

Carbon dioxide All animal cells are always respiring. This process makes carbon dioxide. The carbon dioxide is excreted across the animal's respiratory surface. In *Amoeba,* it diffuses across its cell membrane. In humans, it dissolves in the blood plasma, and is taken to the lungs. It is excreted when you breathe out.

Bile pigments In mammals, old red blood cells are broken down in the liver and spleen. The iron in the haemoglobin is kept, to be reused. The rest of the haemoglobin is turned into greenish yellow pigments. These pigments go into the bile. They are excreted when the bile pours into the duodenum, and then finally out of the body in the faeces.

Nitrogenous waste materials All animals need protein in their diets. But if an animal eats more protein than it needs, it cannot store the extra quantity. It must get rid of it. The protein is usually turned into a different substance before it is excreted. In mammals, this substance is **urea.**

Urea contains nitrogen, and so it is called a **nitrogenous excretory product.**

Plants do not have the problem of getting rid of excess proteins. This is because they make their own proteins, and so they make only as much as is needed.

10.5 Excess proteins are converted to urea.

When you eat proteins, digestive enzymes in your stomach, duodenum and ileum break them down into amino acids. The amino acids are absorbed into the blood capillaries in the villi in your ileum (see section 2.38). The blood capillaries all join up to the hepatic portal vein, which takes the absorbed food to the liver.

The liver allows some of the amino acids to carry on, in the blood, to other parts of your body. But if you have eaten more than you need, then some of them must be got rid of.

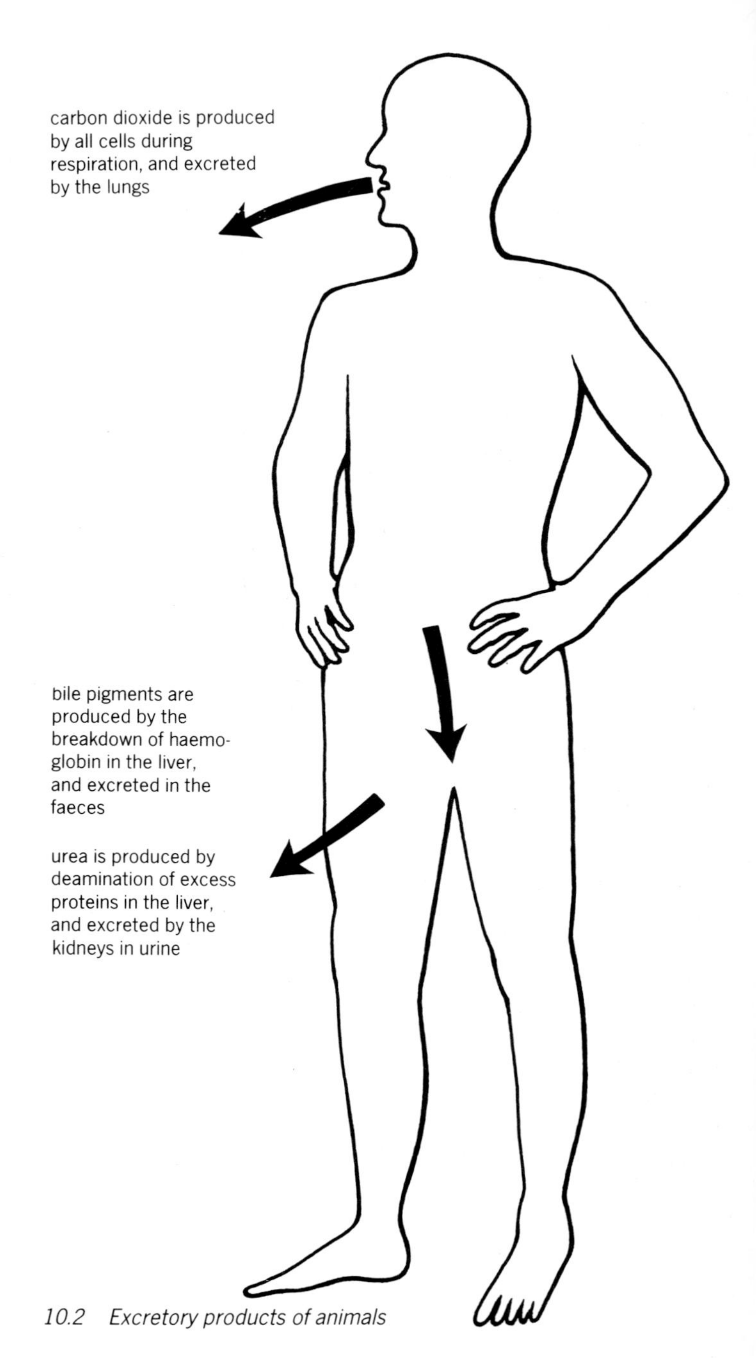

10.2 Excretory products of animals

Table 10.1 Excretory products

Excretory product	*Where it is made*	*How it is made*	*Where it is excreted*	
			In animals	*In plants*
Carbon dioxide	in all living cells	by respiration	from lungs or other respiratory surface	from stomata, but only at night
Oxygen	in green plant cells	by photosynthesis		from stomata, but only in the daytime
Nitrogenous waste products, e.g. urea	in liver of mammals	by deamination	from kidneys, in urine	
Bile pigments	in liver and spleen of mammals	from haemoglobin	in bile, which flows into duodenum and out of body in faeces	

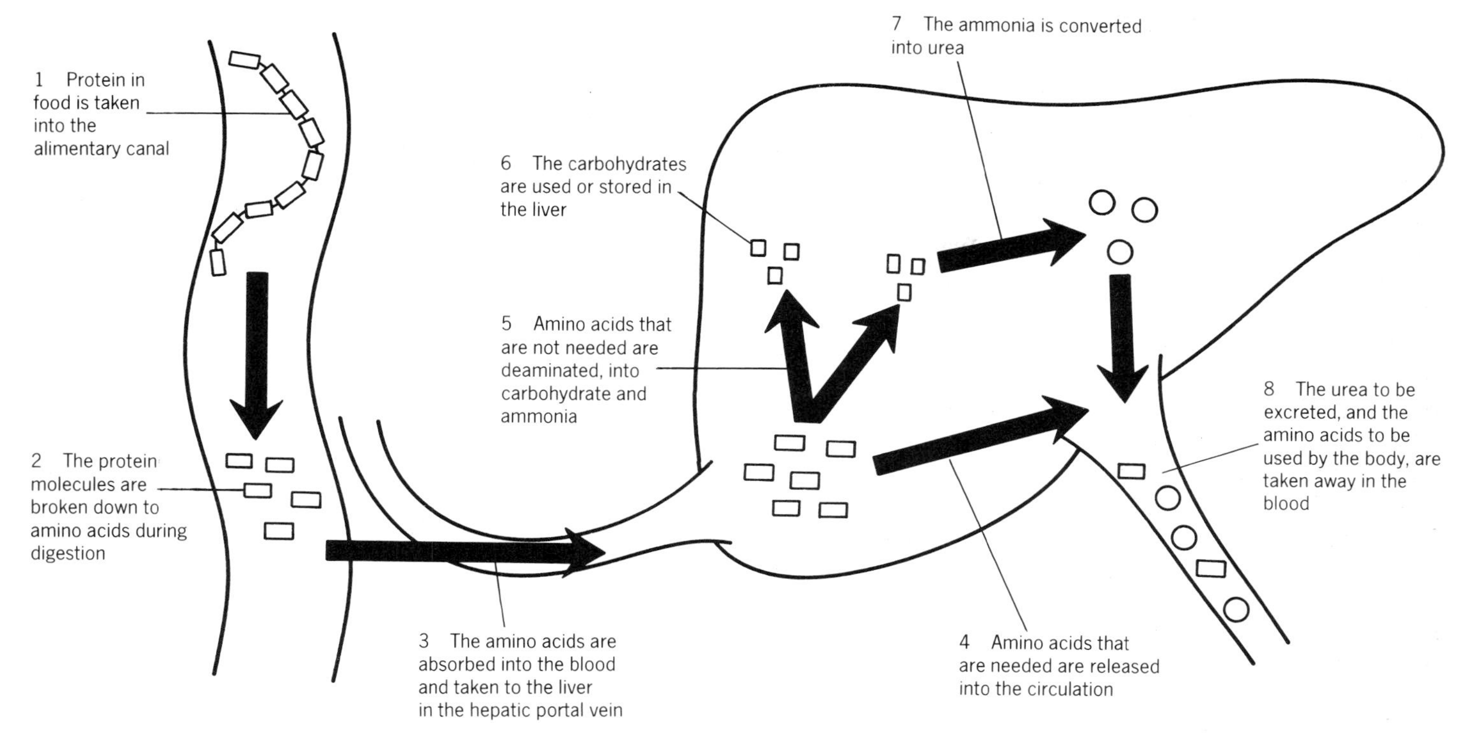

10.3 How urea is made

It would be very wasteful to excrete the extra amino acids just as they are. They contain energy which, if it is not needed straight away, might be needed later.

So enzymes in the liver split up each amino acid molecule (see Figs 10.3 and 10.4). The part containing the energy is kept, turned into carbohydrate or fat, and stored. The rest, which is the part which contains nitrogen, is turned into urea. This process is called **deamination.**

The urea dissolves in the blood plasma, and is taken to the kidneys to be excreted. A small amount is also excreted in sweat.

The liver has many other functions, as well as deamination. Some of them are listed in Table 10.2.

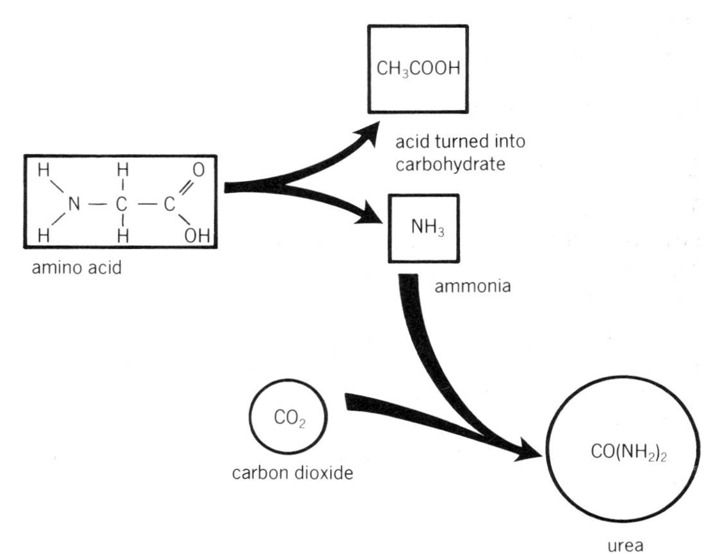

10.4 Deamination and urea formation

Table 10.2 Some functions of the liver

1 Converts excess amino acids into urea and carbohydrate, in a process called deamination
2 Controls the amount of glucose in the blood, with the aid of the hormone insulin
3 Stores carbohydrate as the polysaccharide glycogen
4 Makes bile
5 Breaks down old red blood cells, storing the iron, and excreting the remains of the pigments in bile
6 Breaks down harmful substances such as alcohol
7 Stores vitamins D and A
8 Makes cholesterol, which is needed to make and repair cell membranes
9 Produces heat as a result of the many metabolic reactions which take place in the liver cells

Questions

1 What is meant by an excretory product?
2 Explain the difference between excretion and egestion.
3 What are the two main excretory products of plants?
4 Through which part of the plant are these substances excreted?
5 Why is urea called a nitrogenous excretory product?
6 Why do plants not have nitrogenous excretory products?
7 What happens to excess amino acids in the liver?

The human excretory system

10.6 The kidneys are part of the excretory system.

Fig 10.5 illustrates the position of the two kidneys in the human body. They are near the back of the abdomen, behind the intestines.

Fig 10.6 illustrates a longitudinal section through a kidney. It has three main parts – the **cortex, medulla** and **pelvis.** Leading from the pelvis is a tube, called the **ureter.** The ureter carries urine that the kidney has made to the **bladder.**

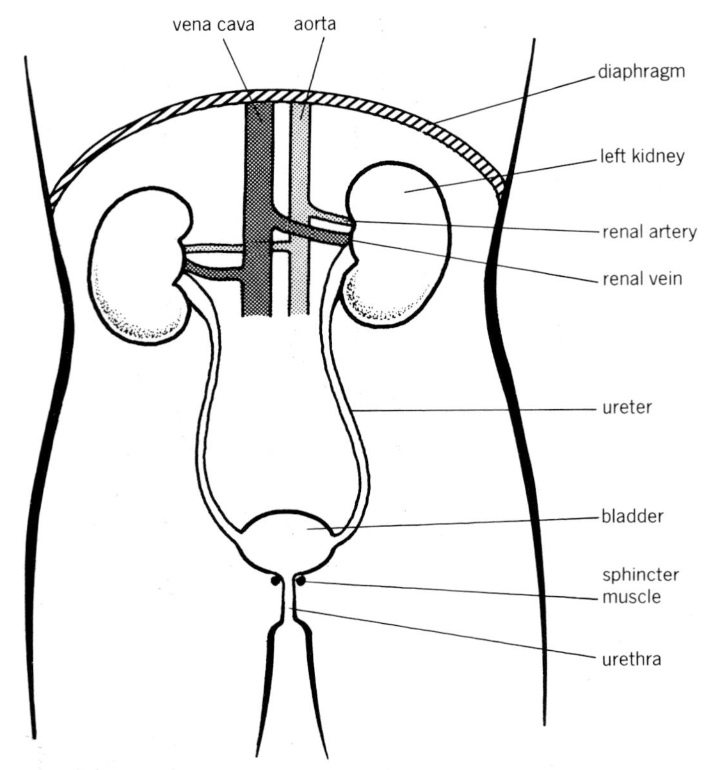

10.5 The human excretory system

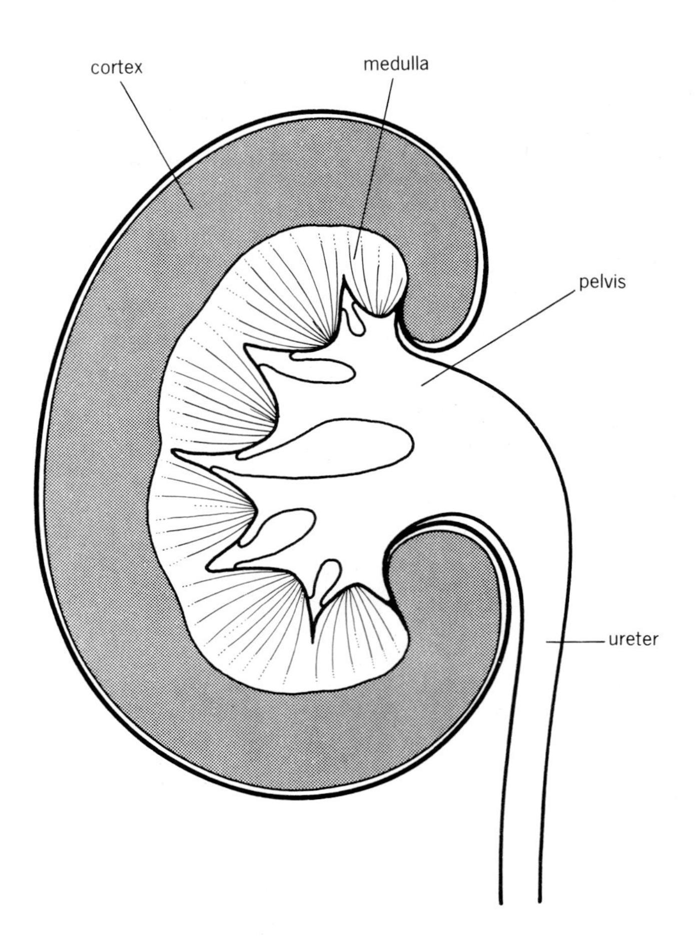

10.6 A longitudinal section through a kidney

10.7 The kidneys are full of tubules.

Although they seem solid, kidneys are actually made up of thousands of tiny tubules, or **nephrons** (see Figs 10.7 and 10.8). Each nephron begins in the cortex, loops down into the medulla, back into the cortex, and then goes down again through the medulla to the pelvis. In the pelvis, the nephrons join up with the ureter.

10.8 Urine is made by filtration and reabsorption.

The job of the kidneys is to take unwanted substances from the blood and to pass them on to the bladder, to be excreted. The way they do this is shown in Fig 10.9.

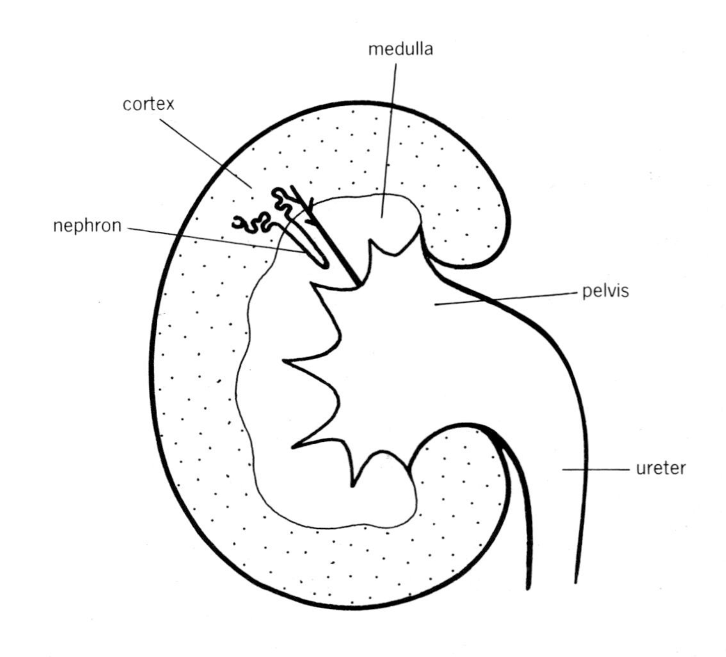

10.7 The position of a nephron in a kidney

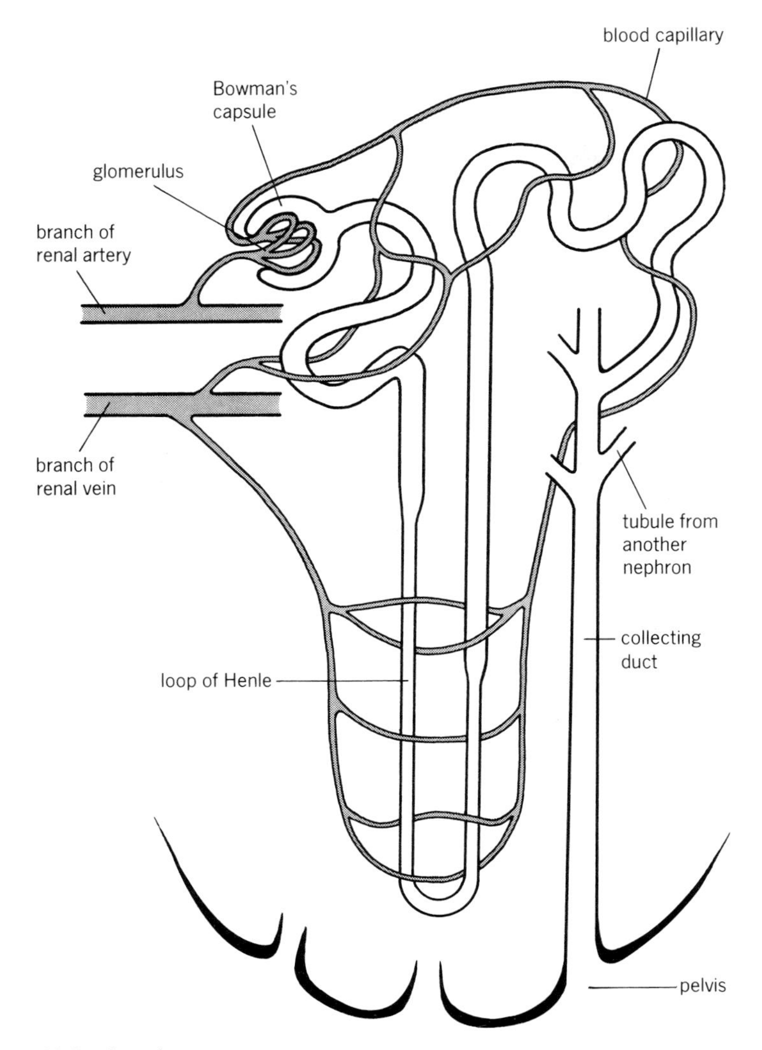

10.8 A nephron

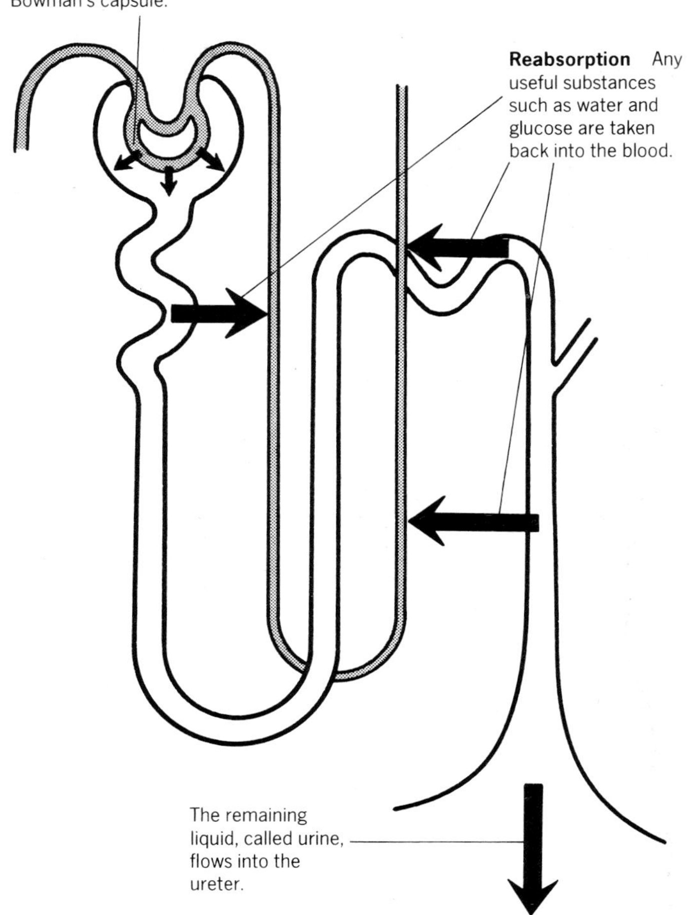

10.9 How urine is made

Blood is brought to the Bowman's capsule in a branch of the renal artery. Small molecules, including water and most of the things dissolved in it, are squeezed out of the blood into the Bowman's capsule. Any useful substances are then taken back into the blood again. The liquid left in the tubule, called **urine,** goes into the ureter and is taken to the bladder.

10.9 Filtration happens in Bowman's capsules.

There are thousands of Bowman's capsules in the cortex of each kidney. Each one is shaped like a cup. It has a tangle of blood capillaries, called a **glomerulus,** in the middle.

The blood vessel bringing blood to each glomerulus is quite wide, but the one taking blood away is narrow. This means that the blood in the glomerulus can not get away easily. Quite a high pressure builds up, squeezing the blood in the glomerulus against the capillary walls.

These walls have small holes in them. So do the walls of the Bowman's capsule. Any molecules small enough to go through these holes will be squeezed through, into the space in the Bowman's capsule.

Only small molecules can go through. These include **water, salts, glucose** and **urea.** Most protein molecules are too big, so they stay in the blood, along with the blood cells.

10.10 Useful substances are reabsorbed.

The fluid in the Bowman's capsule is a solution of glucose, salts and urea, dissolved in water. Some of the substances in this fluid are needed by the body. All of the glucose, some of the water and some of the salts, need to be kept in the blood.

Wrapped around each kidney tubule are blood capillaries. They reabsorb these useful substances back from the fluid in the kidney tubule.

The remaining fluid continues on its way along the tubule. By the time it gets to the collecting duct, it is mostly water, with urea and salts dissolved in it. It is called urine.

10.11 The bladder stores urine.

The urine from all the nephrons in the kidneys flows into the ureters. The ureters take it to the bladder.

The bladder stores urine. It has stretchy walls, so that it can hold quite large quantities.

Leading out of the bladder is a tube called the **urethra.** There is a sphincter muscle at the top of the urethra, which is usually tightly closed. When the bladder is full, the sphincter muscle opens, so that urine flows along the urethra and out of the body.

Adult mammals can consciously control this sphincter muscle. In young mammals, it opens automatically when the bladder gets full.

Questions

1 What is a nephron?
2 Which blood vessels bring blood to the kidneys?
3 What is a glomerulus?
4 How is a high blood pressure built up in a glomerulus?
5 Why is this high blood pressure needed?
6 Name two substances found in the blood which you would not find in the fluid inside a Bowman's capsule.
7 List three substances which are reabsorbed from the nephron into the blood.
8 What is urine?
9 Where are (a) ureters, and (b) the urethra found?

Osmoregulation

10.12 Organisms need the correct content of water.

Living organisms are mostly water. You are about 60% water. A cucumber is 95% water.

Water is very important to living organisms, for many reasons. One of the most important reasons is that water is a solvent. It dissolves salts, sugars, proteins and many other substances. Cytoplasm contains all of these things, dissolved in water.

Unless these substances are dissolved in water, they will not behave in the way they should. Metabolic reactions could not take place, and the cell would die.

Too much water can also be dangerous. Fig 4.4 shows what happens to an animal cell if it has too much water.

10.13 Regulation of water content is osmoregulation.

To keep the amount of water in their bodies just right, living organisms must balance the amount of water coming in with the amount going out. This is called **osmoregulation.**

10.14 How organisms gain water.

By osmosis Plants take in water from the soil, by osmosis, through the cell membrane of their root hairs. Animals which live in fresh water also take in water by osmosis through their cell membranes.

In food Most animals drink water, and also get a lot of water from the food they eat.

By respiration If you look at the respiration equation (see section 10.1) you will see that six molecules of water are made for every molecule of glucose that is broken down in respiration. Because this water is made as a result of a metabolic reaction, it is called **metabolic water.**

Metabolic water is not very important to most organisms. Some desert animals, though, depend on it to keep their cells supplied with water.

10.15 How organisms lose water.

By evaporation Organisms living on land lose water by evaporation from the surface of their bodies. In plants, this is called transpiration.

In animals, most of the water evaporates from their moist respiratory surfaces. If they do not have waterproof coverings to their skins, then water will evapo-

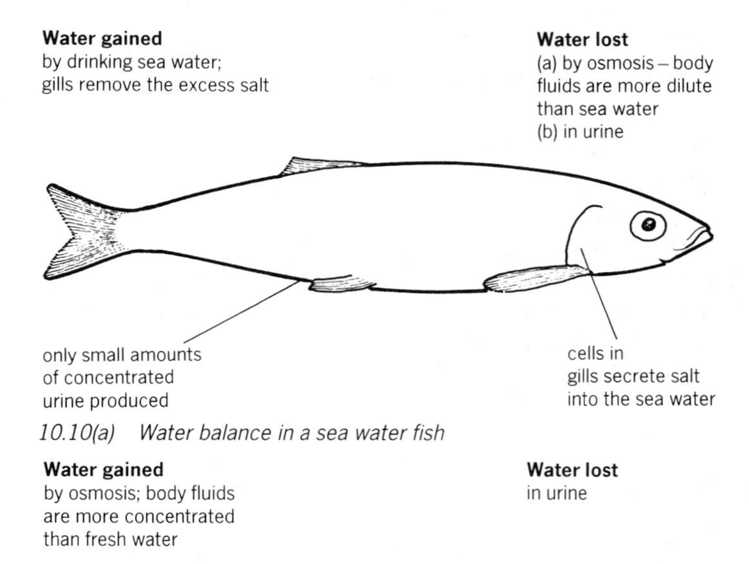

10.10(a) Water balance in a sea water fish

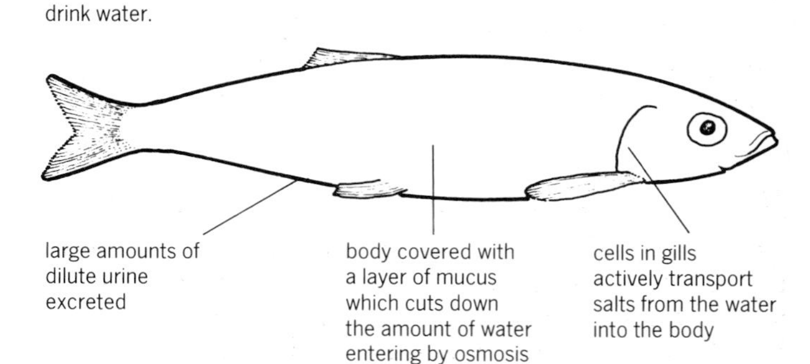

10.10(b) Water balance in a fresh water fish

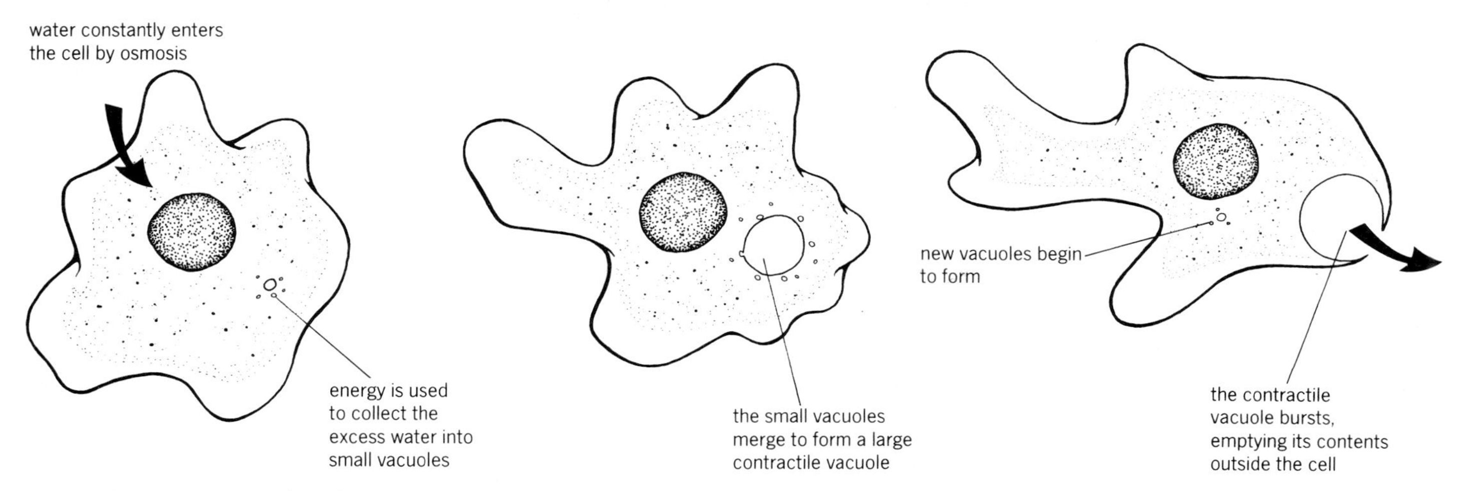

10.11 *Osmoregulation in Amoeba*

rate from there, too. Mammals also lose water when sweat evaporates.

In urine Nitrogenous excretory products, like urea, have to be dissolved in water, so that they can be passed out of the body.

By osmosis Fish like the herring have body fluids which are more dilute than sea water, so they lose water by osmosis (see Fig 10.10).

10.16 *Amoeba* expels excess water.

The cytoplasm of *Amoeba* is a fairly concentrated solution of proteins and other substances in water. The water it lives in is much more dilute. These two solutions are separated by the *Amoeba*'s cell membrane. This membrane is selectively permeable, so water diffuses into the *Amoeba* by osmosis.

The *Amoeba* cannot stop this from happening. All it can do is to bail out the extra water. If it did not, it would burst.

The water is collected into a **contractile vacuole** (see Fig 10.11). When it is full, the contractile vacuole empties the water out.

The contractile vacuole contains almost pure water, so energy is needed to collect the water into it–it is being pushed against the way in which it would naturally go by osmosis. So contractile vacuoles are surrounded by mitochondria to supply the energy.

10.17 Mammalian kidneys regulate water loss.

Amoeba's problem is that it has too much water. Terrestrial animals, such as humans and most insects, have the problem of tending to lose too much water. Fig 10.12 shows how an insect is adapted to conserve water.

In humans, the kidneys have an important part to play in regulating the amount of water in the body. They do this by controlling how much water is lost in the urine. Fig 10.13 overleaf summarises how this happens.

In the hypothalamus in the brain, there are cells which can sense the amount of water in the blood which flows past them. If there is not enough water, then these cells cause the pituitary gland to secrete a hormone called **ADH** (anti-diuretic hormone). The ADH dissolves in the blood and is taken to the kidneys.

When the ADH arrives at the kidneys, it has an effect on the nephrons. It makes the capillaries around them reabsorb more water from the urine. The water goes back into the blood, instead of going down the ureter to the bladder. There will be less urine, and it will be concentrated.

If there is too much water in the blood, the hypothalamus does not secrete ADH. A large amount of urine is made, containing a lot of water.

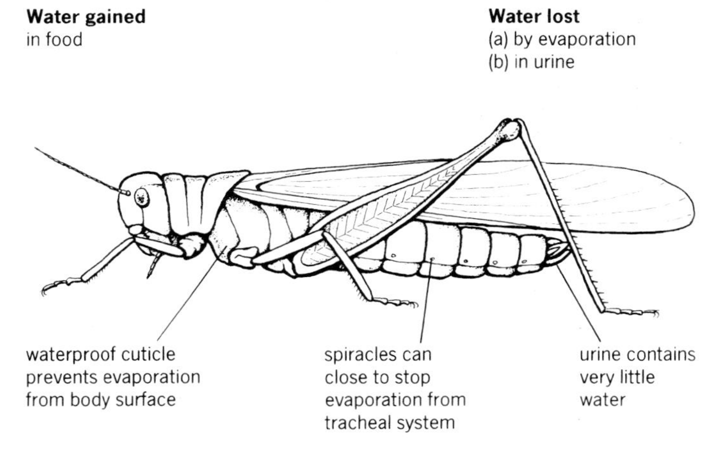

10.12 *Water balance in a locust*

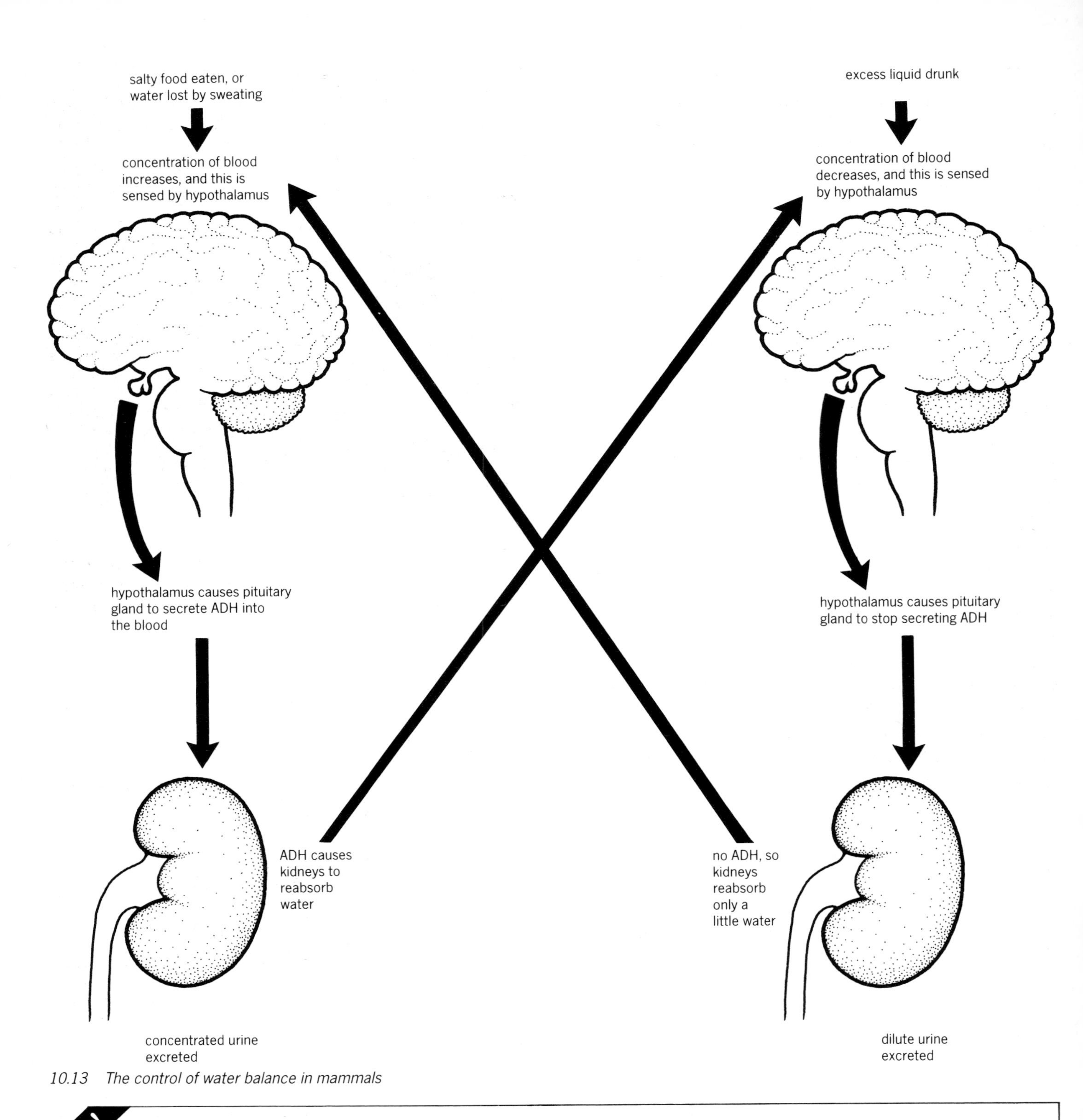

10.13 The control of water balance in mammals

Fact!

Birds excrete their nitrogenous waste as semi-solid uric acid, not as urea. Birds on islands off the west coast of South America built up deposits of such droppings, called guano, over many years. Some of the deposits were as much as 30 m thick. These deposits have been mined as a source of nitrogen-rich fertiliser.

10.18 Many plants can cut down water loss.

No matter where they live, plants do not have a problem of having too much water in their cells. If you look at Fig 4.6, you will see that plant cells simply do not absorb too much water. The cell wall cannot stretch beyond a certain point, so excess water cannot get into the cell.

Terrestrial plants may, however, have the problem of having too little water. They lose water by transpiration almost all the time. They can normally replace this water as fast as they lose it, by absorbing it through their roots. If water is in short supply, they are in danger of becoming dehydrated. They will wilt, and may die.

To stop this happening, many plants have ways of cutting down the rate of transpiration.

Closing stomata Plants lose most water through their stomata. If they close their stomata, then transpiration will slow right down. Fig 10.14 shows how they do this.

However, if its stomata are closed, then the plant cannot photosynthesise, because carbon dioxide cannot diffuse into the leaf. Plants only close their stomata when they really need to, such as when it is very hot and dry, or when they could not photosynthesise anyway, such as at night.

Waxy cuticle Many leaves, such as holly leaves, are covered with a waxy cuticle, made by the cells in the epidermis. The wax waterproofs the leaf.

Hairy leaves Some plants have hairs on their leaves (see Fig 10.15). These hairs trap a layer of moist air next to the leaf.

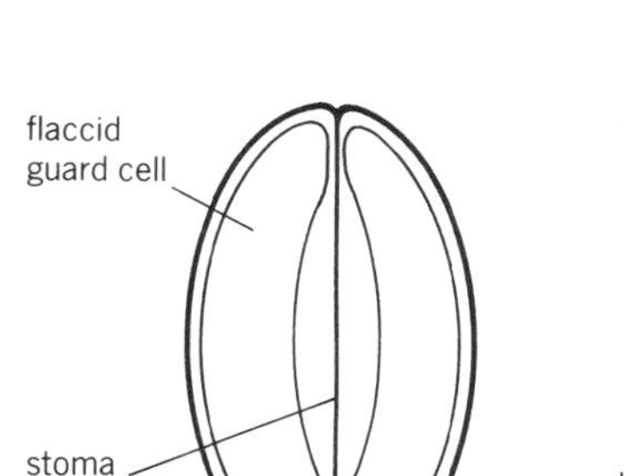

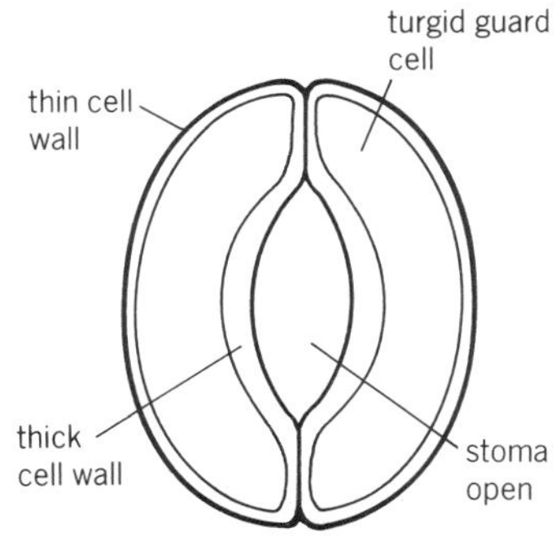

10.14 How stomata open and close

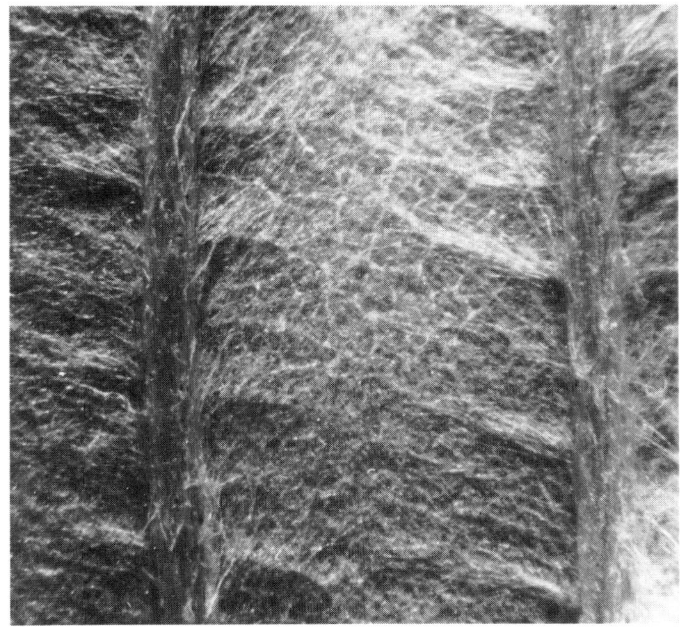

10.15 Hairs on a leaf help to stop too much water being lost

Stomata on underside of leaf In most leaves, there are more stomata on the lower surface than on the upper surface. The lower surface is usually cooler than the upper one, so less water will evaporate.

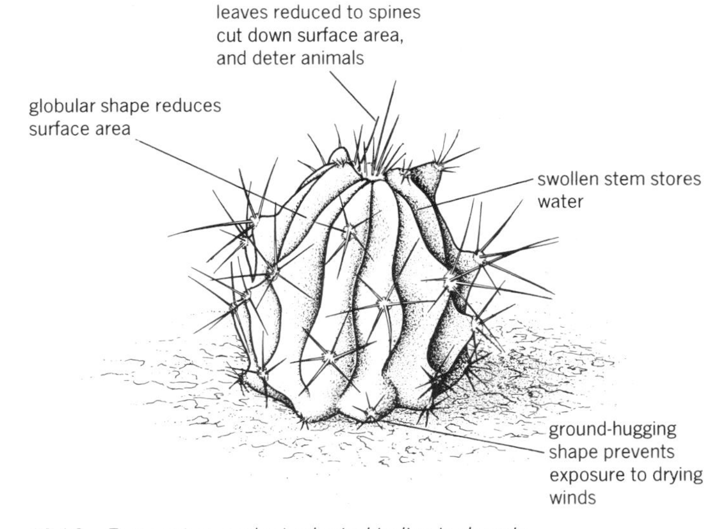

10.16 Ferocactus–a plant adapted to live in deserts

Cutting down the surface area The smaller the surface area of the leaf, the less water will evaporate from it. Plants like cacti (see Fig 10.16) have leaves with a small surface area, to help them to conserve water. However, this slows down photosynthesis, because it means less light and carbon dioxide can be absorbed.

Rolling leaves Marram grass (see Fig 14.32) can roll up its leaves when water is in short supply. The stomata are enclosed inside the leaf, so hardly any water is lost through them.

Homeostasis

10.19 Cells work best in a constant environment.

You are alive because of the metabolic reactions going on inside your cells. These reactions are all controlled by enzymes. To work properly, they must have just the right conditions. Temperature, amount of food and amount of water must all be just right. If they are not, then the metabolic reactions can go wrong, and you will become ill.

Your environment is always changing. You might walk from a warm room into a cold street. You might swim in cool water, and then lie on hot, dry sand to sunbathe.

But the environment of each of your cells must not be allowed to change like this. All of your cells are surrounded by **tissue fluid** (see section 6.22). The temperature and concentration of this fluid must not change, no matter how the outside environment changes. Whether you are swimming in the sea, or sunbathing on the sand, your body keeps the tissue fluid at 37 °C, with just the right amount of water and dissolved food in it.

Keeping the environment of your cells constant is called **homeostasis.** Several organs of your body are involved in homeostasis. The amount of water is controlled by the kidneys, under the control of the hypothalamus. The amount of glucose is controlled by the liver, under the control of the pancreas. The temperature is controlled by the skin, brown fat and muscles, under the control of the hypothalamus.

Questions

1 What is osmoregulation?
2 How does a fresh-water fish gain water?
3 What is metabolic water?
4 Describe three ways in which humans lose water from their bodies.
5 Why does *Amoeba* have excess water to get rid of?
6 Why are contractile vacuoles surrounded by mitochondria?
7 Describe three ways in which insects are adapted to cut down water loss.
8 What sort of urine, and how much, would be produced by a person who had been playing tennis on a hot day? Explain your answer.
9 Why do plant cells not absorb too much water?
10 Explain how the structure of a guard cell enables it to open or close a stoma.
11 What is homeostasis?
12 Why is homeostasis important?
13 List three examples of homeostasis, and explain which organs are responsible for them.

Chapter revision questions

1 Explain the difference between each of the following pairs of terms:
(a) excretion, egestion;
(b) urine, urea;
(c) ureter, urethra;
(d) osmosis, osmoregulation;
(e) ADH, ATP.

2 Explain the following.
(a) There is no glucose in normal urine.
(b) Marine species of *Amoeba* do not have contractile vacuoles.
(c) Plants do not excrete nitrogenous waste material.

3 (a) What is meant by osmoregulation?
(b) Why is osmoregulation necessary for living cells?
(c) Describe how osmoregulation is carried out in (i) a named protistan, and (ii) a named mammal.

4 (a) What are the main ways in which water is (i) gained, and (ii) lost by terrestrial animals?
(b) List three ways in which insects manage to keep their water loss to a minimum.
(c) How do terrestrial plants lose water?
(d) List three ways in which terrestrial plants keep water loss to a minimum.

5 (a) What is the main nitrogenous waste product excreted by human kidneys?
(b) Where is this waste product formed?
(c) Briefly describe how this waste product is formed.
(d) Which blood vessels deliver this waste product to the kidneys?
(e) Name two substances found in the blood plasma which are not found in the urine of a healthy person.
(f) For each of these substances, explain how the structure and function of the kidney ensures that they are not lost in the urine.

11 Support and movement

11.1 Skeletons support organisms.

All living organisms are held in shape, or supported, in some way. Many of them have special structures which do this. These structures are called **skeletons.**

The human skeleton, made of bone and cartilage, is only one kind of skeleton. Because it is inside the body, it is called an **endoskeleton** ('endo-' means inside). Earthworms and plants also have endoskeletons, although they are very different. Insects have a skeleton on the outside of their bodies, called an **exoskeleton** ('exo-' means outside).

Support and movement in mammals

11.2 Bone is made of protein and minerals.

Most of the human skeleton is made of **bone**. Bone is mostly made of mineral substances such as calcium phosphate, with small amounts of magnesium salts. This makes it very hard. Bone also contains **collagen fibres** (see Fig 11.1) which give it elasticity. Collagen is a protein.

11.1 Scanning electron micrograph of collagen fibres

Bone is alive. It contains living cells, which are supplied with food and oxygen by blood vessels. The cells are arranged in rings around the blood vessels (see Fig 11.2).

11.3 The structure of a bone.

Fig 11.3 shows a leg bone, cut in half lengthways. The hardest bone, called **compact bone,** is on the outside. Underneath this is a layer of **spongy bone,** which has spaces in it. This stops the bone from being too heavy.

In the centre is the **bone marrow.** This is very soft, and has a good supply of blood. Red blood cells, white blood cells and platelets are made here. The ends of the bone are covered with a layer of **cartilage.**

11.4 Cartilage contains less minerals than bone.

Cartilage (see Fig 11.4) is much softer than bone. This is because it does not contain very many mineral salts, but like bone, it contains collagen.

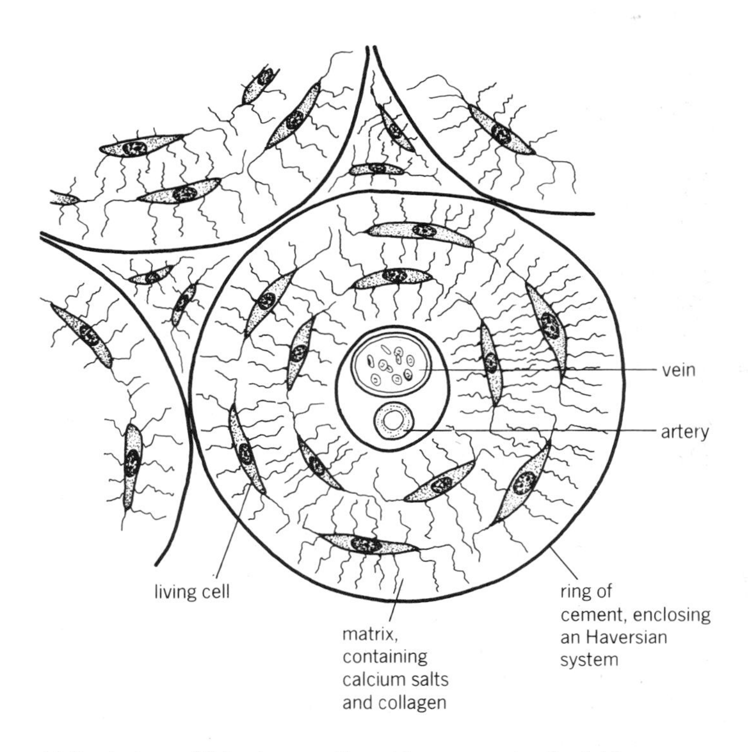

11.2 A piece of living bone as it would appear under the light microscope

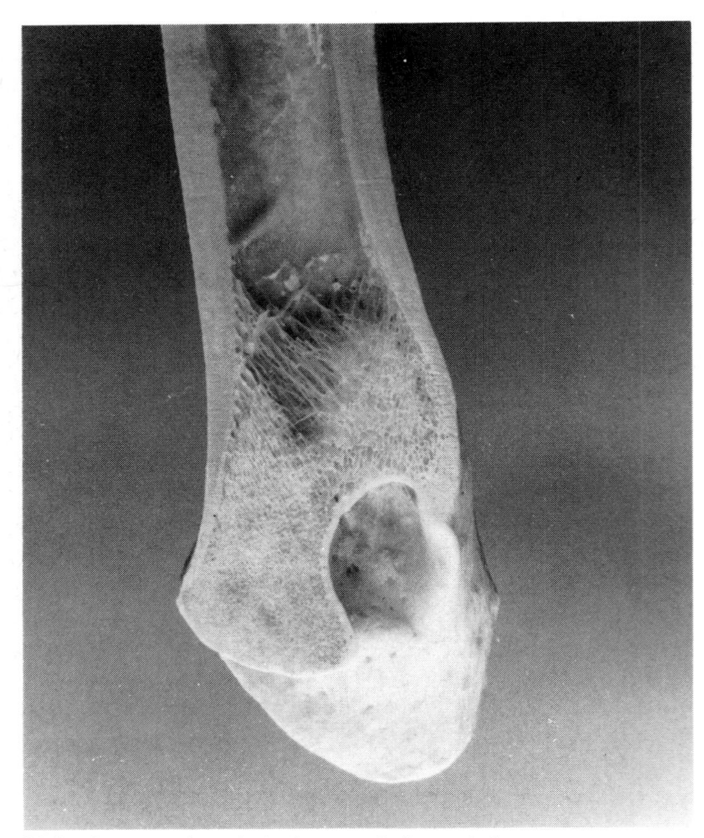

11.3 *A section through the lower end of a cow's femur*

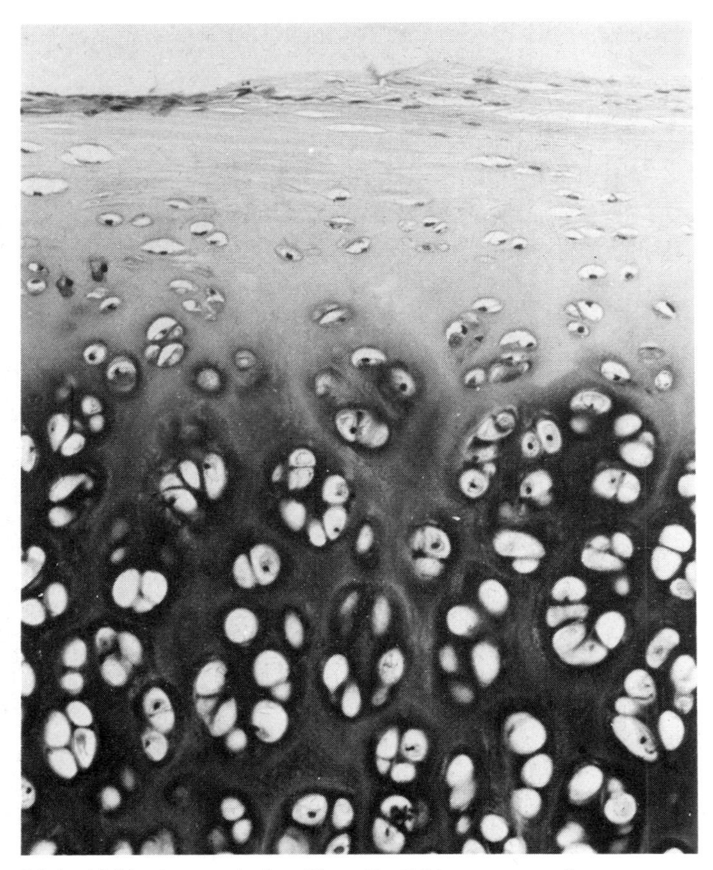

11.4 *Light micrograph of cartilage; the light areas are cells*

Cartilage is found on the ends of bones, where they meet one another at a joint. It allows the bones to move easily over each other because it is smooth. There is also cartilage in the pinnae of your ears, and in the end of your nose.

Fact! There are 206 bones in the human body. The longest is the femur and the smallest is the stapes in the middle ear.

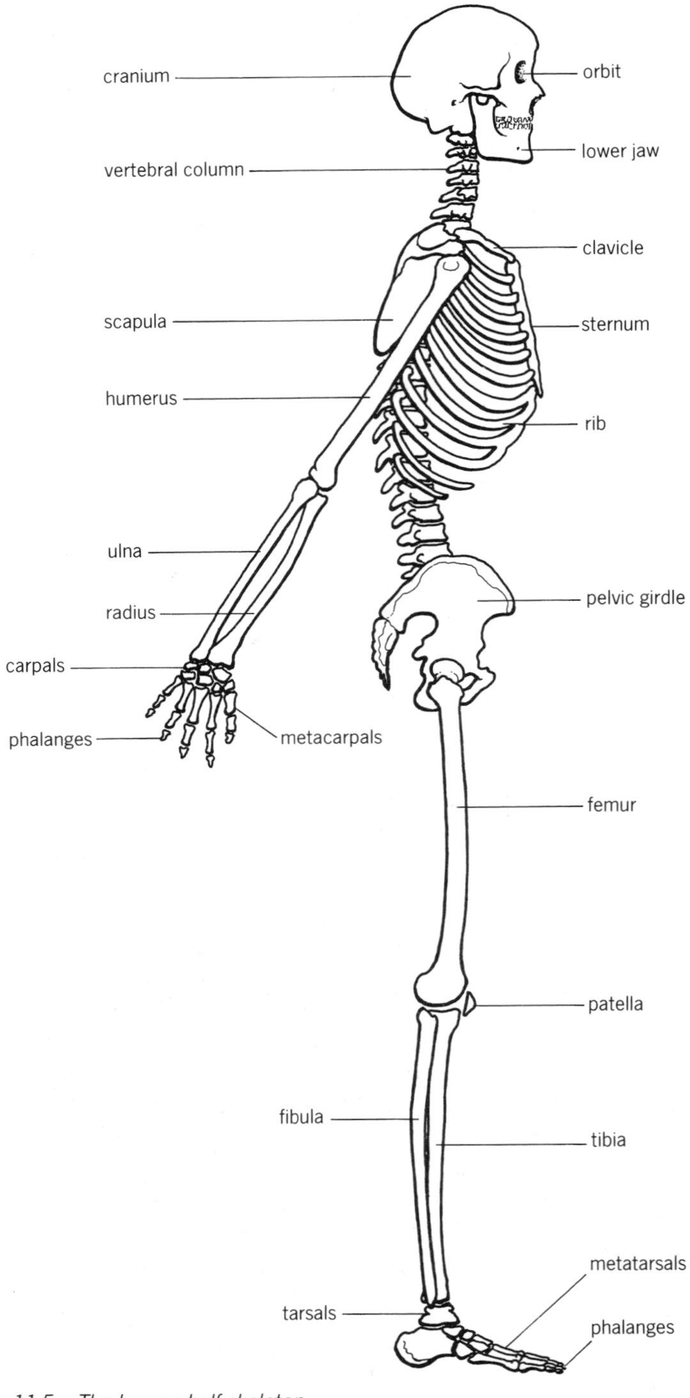

11.5 *The human half-skeleton*

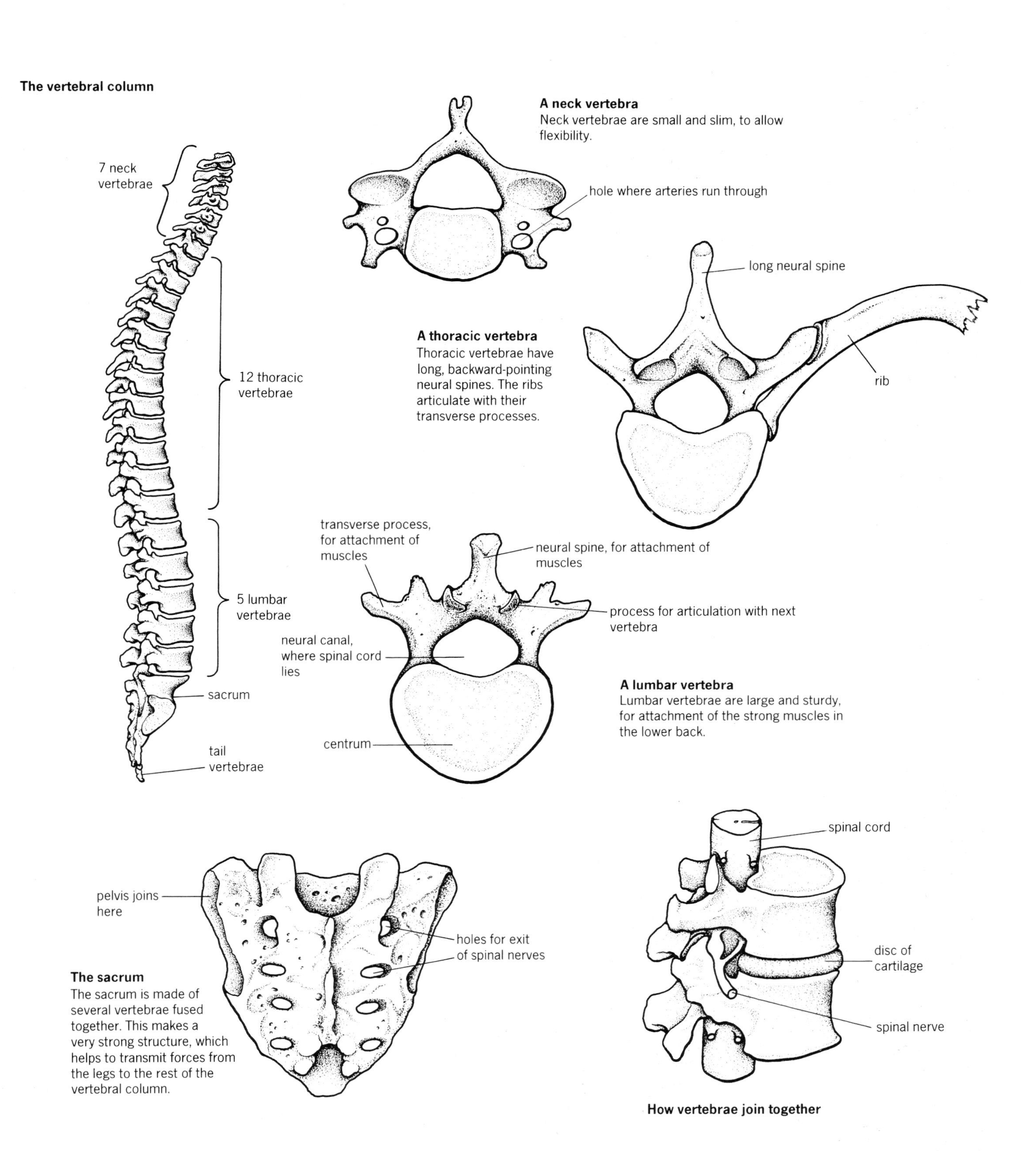

11.6 Vertebrae

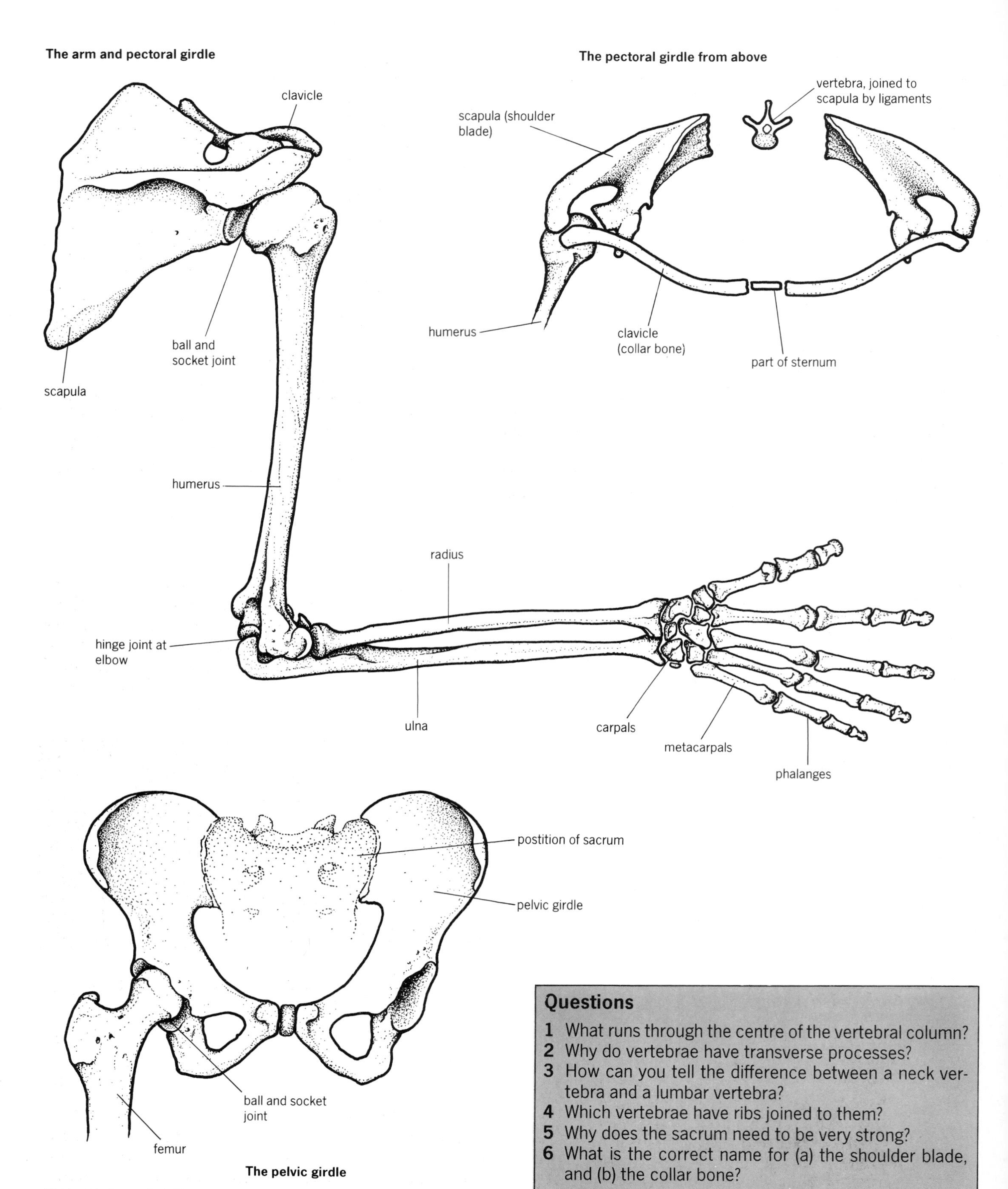

11.7 Limbs and girdles

Questions

1 What runs through the centre of the vertebral column?
2 Why do vertebrae have transverse processes?
3 How can you tell the difference between a neck vertebra and a lumbar vertebra?
4 Which vertebrae have ribs joined to them?
5 Why does the sacrum need to be very strong?
6 What is the correct name for (a) the shoulder blade, and (b) the collar bone?

Table 11.1 Functions of the human skeleton.

Function	*Example*
Support	vertebral column pectoral girdle pelvic girdle leg bones
Movement	leg and arm bones
Protection	skull (protects brain) ribs (protect heart and lungs)
Making red and white blood cells	marrow in leg bones and ribs

Questions

1. What is an endoskeleton? Give an example of an organism which has one.
2. What is an exoskeleton? Give an example of an organism which has one.
3. What are the two main constituents of bone?
4. Give one similarity and one difference between bone and cartilage.

11.5 Bones are joined in different ways.

Wherever two bones meet each other, a **joint** is formed. There are two main kinds of joint.

Fibrous joints Sometimes two bones are joined quite firmly together by fibres. The bones in the cranium of the skull are joined like this. The joins are called **sutures.** The bones are held so tightly together in an adult human that they cannot move at all.

There are also fibrous joints between the vertebrae. The bones are joined by cartilage with fibres in it, called **intervertebral discs.** The cartilage is quite soft in the middle, so the bones can move a little. Although any one joint between two vertebrae only allows a slight movement, the sum total of all these movements makes the backbone quite supple.

Synovial joints **Synovial joints** are found where two bones need to move freely. The elbow joint and shoulder joint are examples of synovial joints.

Fig 11.8 shows the structure of a typical synovial joint. The two bones are held together by **ligaments.** Ligaments are very strong, but can stretch when the bones move.

If the two bones rubbed against one another when they moved, they would quickly be damaged. So the ends of the bones are covered with a layer of cartilage. Between the bones is a small amount of a thick liquid called **synovial fluid.** This lubricates the joint, so that it moves smoothly. The fluid is made and kept in place by the synovial membrane.

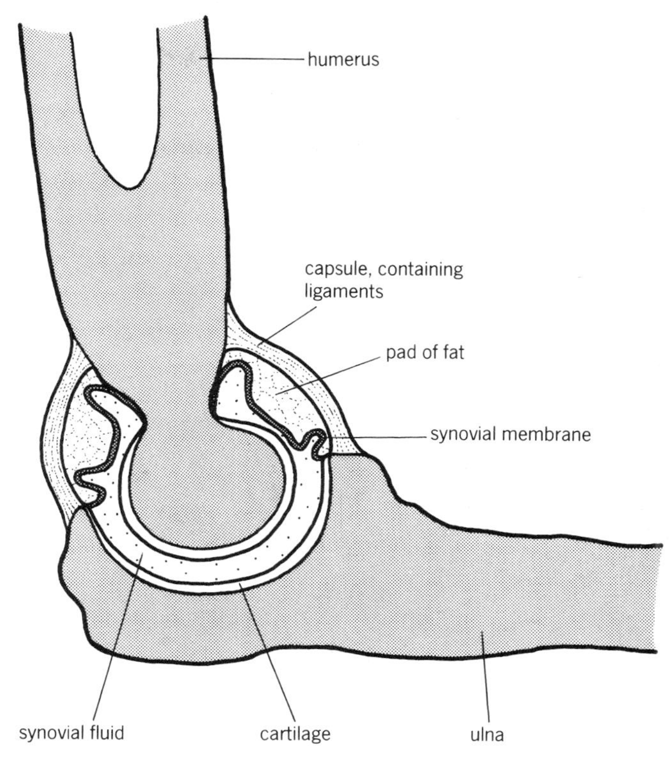

11.8 Section through the elbow joint

Synovial joints are given different names, depending on the kind of movement that takes place. The elbow joint is a **hinge joint,** because the bones can only move in one plane, like a door on hinges. The shoulder and hip joints are **ball and socket joints.** A ball at the end of one bone fits into a socket in the other. This allows a circular movement, or movement in all planes.

Questions

1. What is meant by a fibrous joint?
2. Where can you find sutures in the human skeleton?
3. What is an intervertebral disc?
4. What is meant by a synovial joint?
5. What are ligaments?
6. Describe two features of the elbow joint which help it to move smoothly.

11.6 There are three kinds of muscle in mammals.

The three kinds of muscle in the human body (see Fig 11.9) are listed below.

Cardiac muscle Cardiac muscle is only found in the heart. It makes up the walls of the atria and ventricles.

Smooth muscle Smooth muscle is found in organs such as the walls of the alimentary canal and the bladder. Smooth muscle is also called involuntary muscle, because you do not have conscious control over it.

Striated muscle All of the muscles attached to your bones are striated muscle. They are sometimes called skeletal muscles, or voluntary muscles, as they are normally under conscious control.

'Striated' means striped, and you can see why this kind of muscle has this name if you look at Fig 11.10.

> **Fact!** Muscles make up about 40% of your body weight.

11.7 Each kind of muscle contracts in a different way.

Muscles cause movement by getting shorter, or **contracting.** They need energy to do this. They get their energy from respiration, and so muscles must have a good blood supply, to bring food and oxygen to them. They also need plenty of mitochondria, to use these substances to make ATP.

All of the three types of muscle can contract, but they do it in slightly different ways. Cardiac muscle contracts and relaxes rhythmically all through your life. It never tires. It does not need conscious messages from your brain to make it contract – it will do it anyway. Nervous impulses from the brain can, however, alter its rate of contraction.

Smooth muscle, too, can contract of its own accord. For example, the muscles in the wall of the alimentary canal do this during peristalsis. In other places, however, smooth muscle needs to be stimulated by nerves, in the same way as striated muscle.

The contractions of smooth muscle are much slower than those of cardiac muscle. Smooth muscle contracts and relaxes slowly and rhythmically.

Striated muscle only contracts when messages are sent to it along nerves. Striated muscle can contract quickly, and very strongly. But it gets tired more quickly than smooth or cardiac muscle.

11.8 Movements of the forearm.

Fig 11.11 shows the bones and two of the muscles in your arm. The arm can bend at the elbow, which is a hinge joint.

The **biceps muscle** is attached to the scapula at the top, and the radius at the bottom. When it contracts, it pulls the radius and ulna up towards the scapula, so the arm bends. This is called **flexing** your arm, so the biceps is a **flexor muscle.**

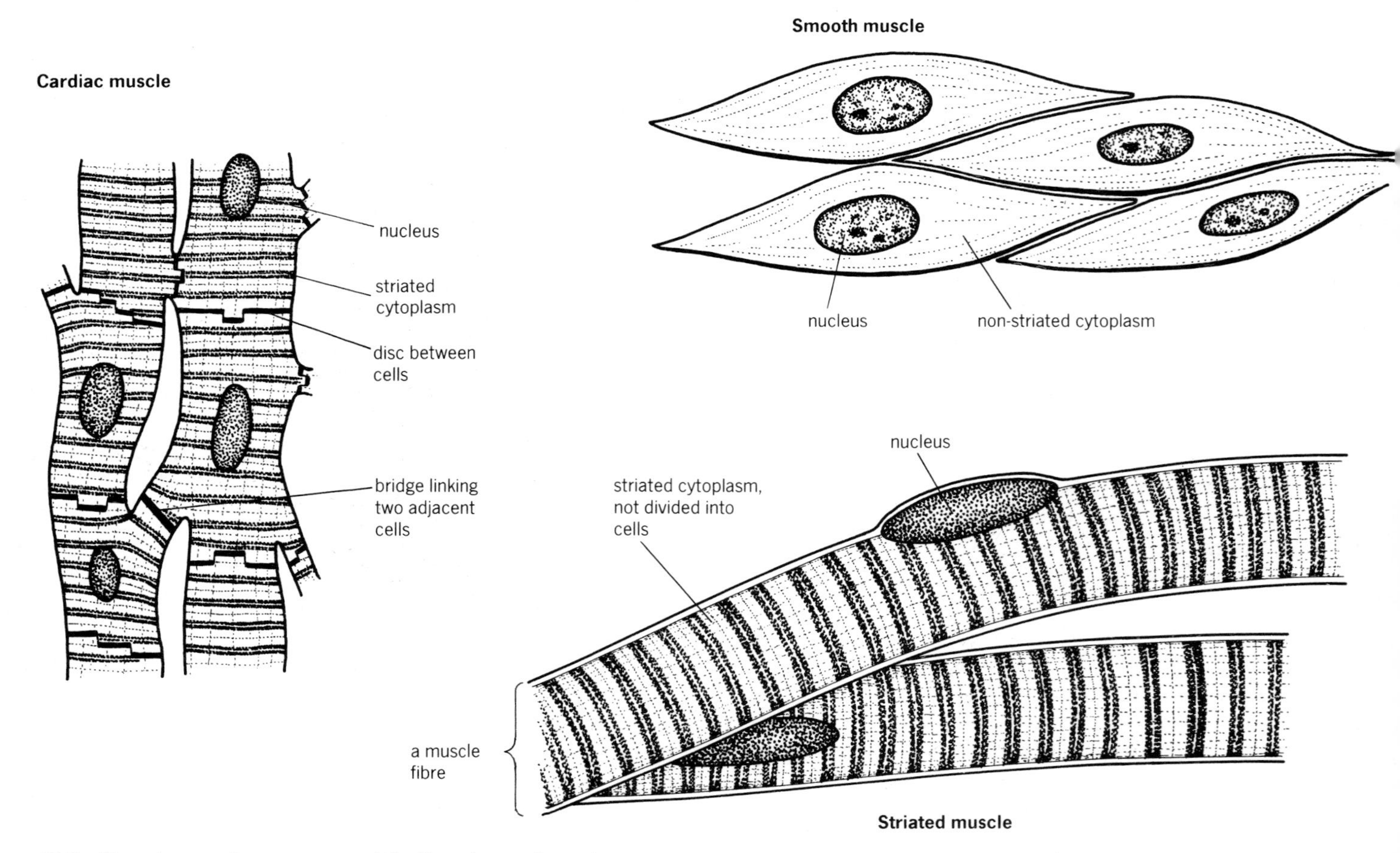

11.9 The microscopic appearance of the three types of muscle

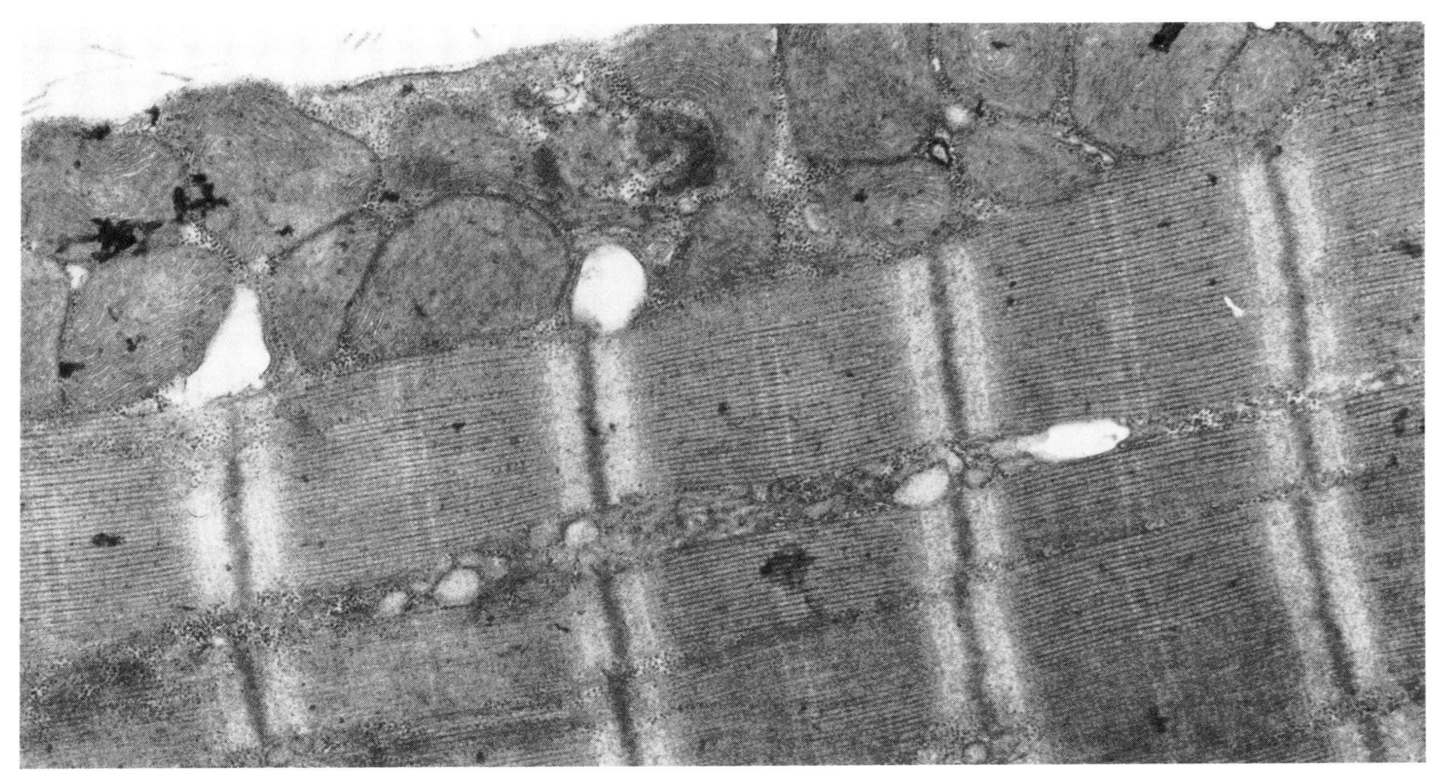

11.10 *Electron micrograph of striated muscle; energy is released by mitochondria (top) so that the fibres (below) can contract*

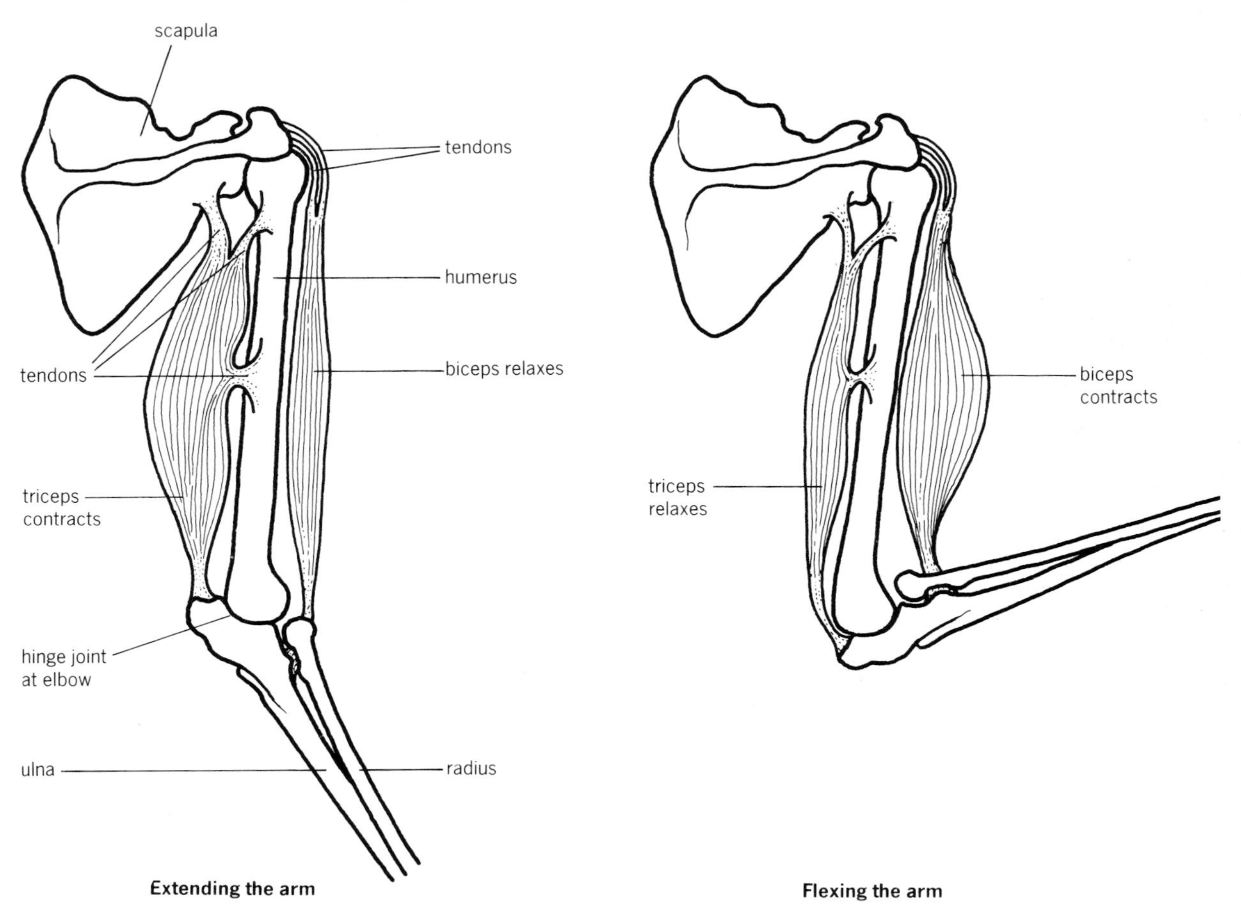

11.11 *Movement of the fore-arm*

Investigation 11.1 Using a model arm to investigate the action of the biceps muscle

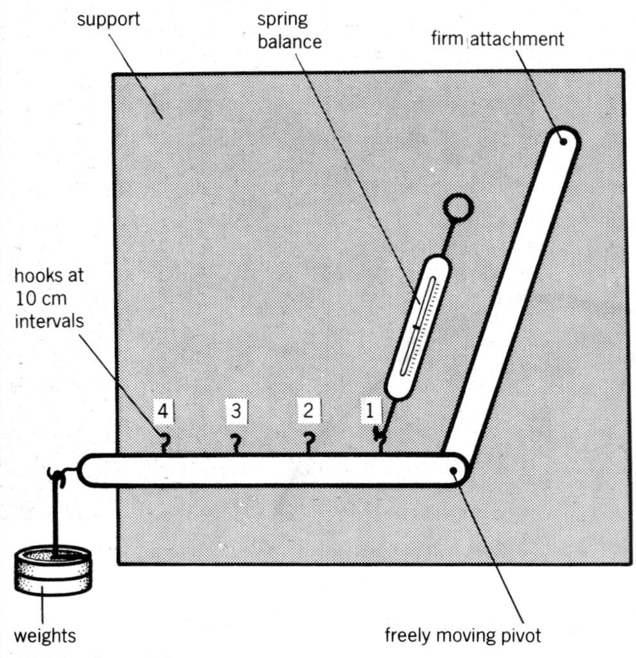

11.12 A model arm

1. Copy the results table.
2. Hang a 100 g weight on the hook at the end of the 'fore arm' on the model.
3. Attach a spring balance to hook 1. Pull upwards with the spring balance, parallel to the 'humerus', until the 'fore arm' is exactly horizontal. Take the reading on the spring balance, and fill it in on the results table.
4. Repeat with the balance pulling on hooks 2, 3 and 4.
5. Replace the 100 g weight with a 500 g weight, and repeat steps 3 and 4.

Results table

Weight (grammes)	Distance from 'elbow' (cm)	Force exerted by spring balance (Newtons)
100	10	
	20	
	30	
	40	
500	10	
	20	
	30	
	40	

continued

Questions

1. What does the spring balance represent?
2. At which position was the force needed to lift the weight greatest?
3. Which position most closely represents the actual position of attachment of the biceps to the radius?
4. Striated muscles are able to exert considerable forces, but they cannot shorten by a very large amount. Can you suggest why the biceps is attached in this position?

   The forces acting on the fore arm when lifting a weight can be shown diagrammatically like this.

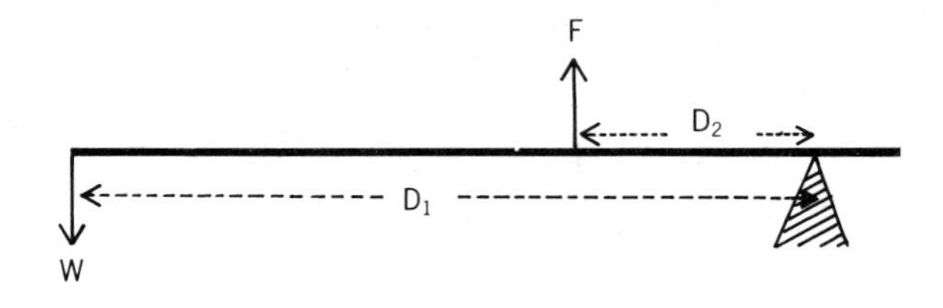

When the arm is horizontal and not moving,
$W \times D_1 = F \times D_2$

5. What does W stand for?
6. What does F stand for?
7. Use the equation to explain your answer to question 2.

But muscles can only pull, not push. The biceps cannot push your arm back down again. Another muscle is needed to pull it down. The **triceps muscle** does this. When it contracts, the triceps straightens or extends your arm. It is called an **extensor muscle.**

The flexor and extensor muscles work together. When the biceps contracts, the triceps relaxes. When the triceps contracts, the biceps relaxes. The muscles are said to be **antagonistic muscles** because, in a way, they work against each other. There are many other examples of antagonistic muscles in your body.

11.9 Tendons are groups of collagen fibres.

Muscles are joined to bones by **tendons.** Tendons are very strong, and do not stretch. They are made of collagen fibres.

Some tendons are quite large. You can feel your Achilles tendon at the back of your ankle. It attaches your calf muscle to your heel bone.

11.10 Cilia can cause movement.

Muscles are not the only things which can move in the human body. Some cells have microscopic threads on them, called **cilia** (see Fig 11.13).

11.13 Scanning electronmicrograph of a dense layer of cilia

Ciliated cells are found in the tubes of the respiratory stystem. There are cilia in your trachea and bronchi, for example. They beat rhythmically, wafting mucus up towards the back of your mouth. The mucus traps bacteria and particles of dirt in the air you breathe in (see section 5.18).

11.11 Some blood cells move like *Amoeba*.

Another type of movement which happens inside the human body is **amoeboid movement.** White blood cells move like this (see Fig 11.14). They can pass out of the blood capillaries into every part of the body, where they destroy bacteria and other unwanted cells.

Questions

1. Why do muscles contain many mitochondria?
2. Which kinds of muscle will contract without a conscious message from the brain?
3. What are the special features of striated muscle?
4. What is meant by (a) a flexor muscle, and (b) an extensor muscle? Give an example of each.
5. What are tendons? How do they differ from ligaments?

Fact! The bird with the fastest wingbeat is the Horned Sun Gem, which lives in South America. Its wings beat at a speed of 90 beats per second.

Condors, which also live in South America, can cruise for up to 96 km without beating their wings.

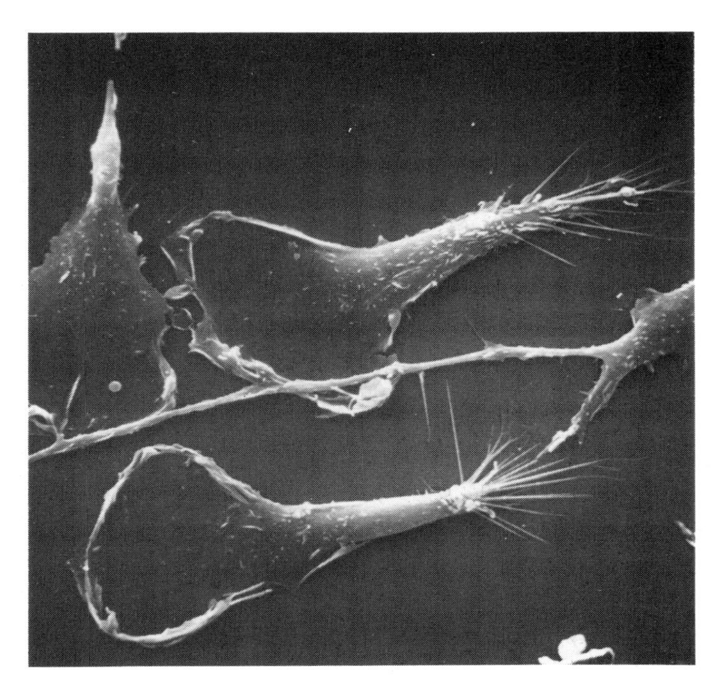

11.14 White cells crawling from right to left on a glass surface

Support and movement in birds

11.12 Birds are adapted for flight.

Fig 11.15 shows part of the skeleton of a bird. It is basically very similar to the human skeleton. But there are some important differences, which enable birds to fly.

The bones contain large air spaces This makes them lighter. Think of the weight of a chicken bone, compared with the bone in a pork chop.

The forelimbs have become wings They have very strong muscles attached to them, which pull the wings up and down. The sternum is much larger than in other vertebrates, forming a **keel.** The keel is needed for the attachment of these strong flight muscles. You eat the flight muscles when you eat chicken breast.

The body is covered with feathers These have two important functions. Firstly, birds, like mammals, are homeothermic (see section 9.5). The feathers help to insulate their bodies, like hair on a mammal. Their bodies must be kept warm, so that they can produce enough energy to fly.

Secondly, the feathers on the tail and wings give a large surface area, which helps to keep the bird in the air.

The body is streamlined When a bird is flying, its body is shaped to cut cleanly through the air. The feathers lie smoothly against its body, so that the air can easily flow over them.

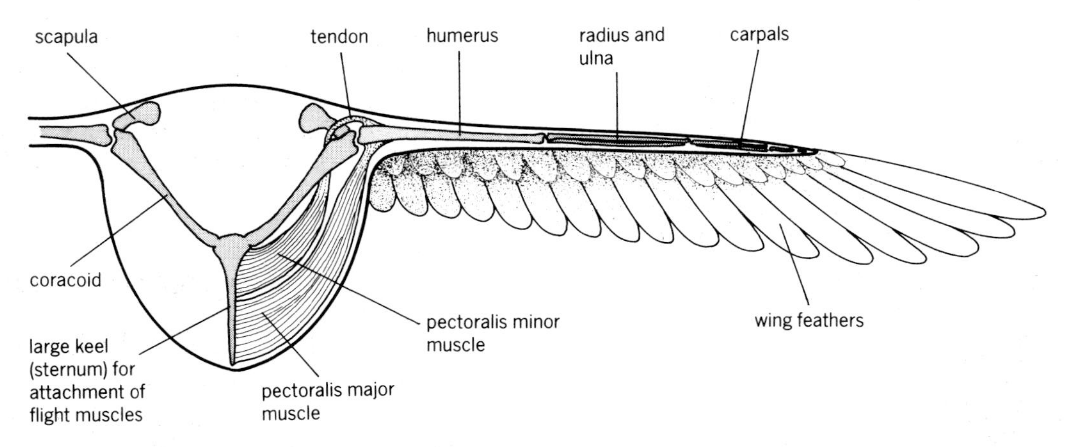

11.15 Transverse section through the thorax of a bird

Support and movement in fish

11.13 Fish are adapted for swimming.

Swimming in water presents very different problems from walking on land like a human, or flying in air like a bird. Herrings, which are very active marine fish, have several adaptations which help them to swim.

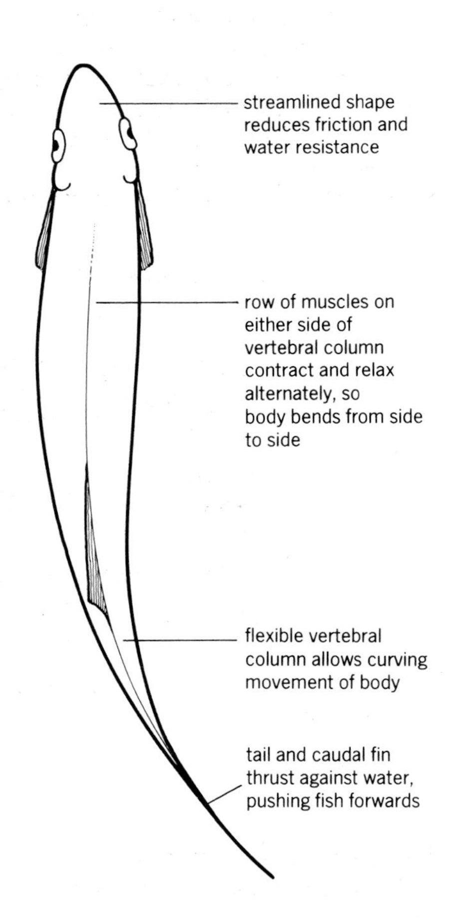

11.16 Movement of a herring

They have a flexible vertebral column Most fish swim by moving their bodies from side to side. Fig 11.16 illustrates how this produces a forward movement. The vertebral column has to be able to bend, to allow the fish's body to curve like this.

They have a swim bladder This is an air filled sac just below the vertebral column. It is found in all bony fish, such as herrings. The fish can adjust the amount of air in it. This helps to keep them afloat at the right depth in the water. The more air in the swim bladder, the nearer they will float to the surface.

Because the buoyancy of the water supports them, fish do not need such strong skeletons as mammals and birds.

They are streamlined This allows them to slide easily through the water. The scales overlap, and point backwards. The scales are covered by a thin, transparent skin. The skin secretes mucus, which makes a fish feel slimy to touch, and which reduces drag when swimming quickly.

They have fins which help them to balance in the water.

Support and movement in insects

11.14 Insects have an exoskeleton of chitin.

An insect's skeleton is built on a completely different plan from a vertebrate's skeleton. Instead of being inside the body, it is on the outside of it. It is called an exoskeleton.

An insect's exoskeleton is mostly made of **chitin.** Chitin is quite a flexible substance, but when it combines with protein it becomes hard. Where the exoskeleton needs to be strong and rigid, it is made of chitin and protein. Where it needs to be flexible, such as at joints, or on the wings, it is made only of chitin.

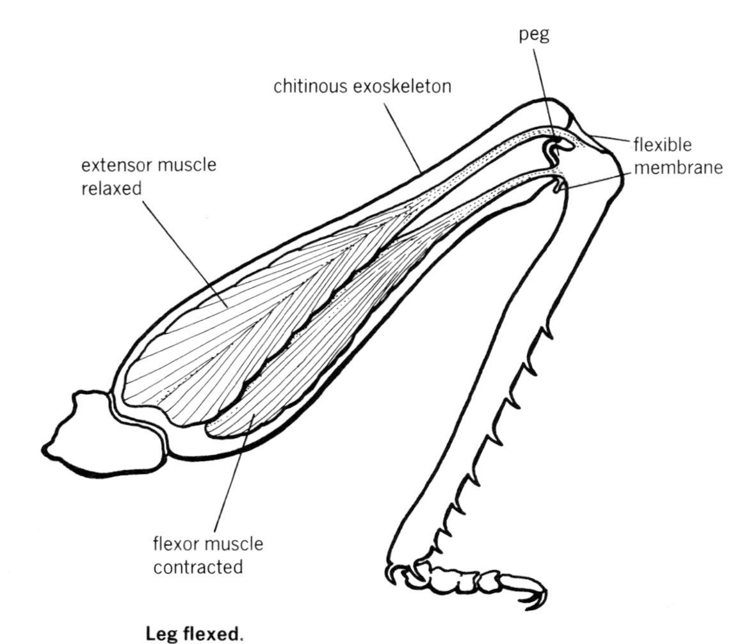

Leg flexed.

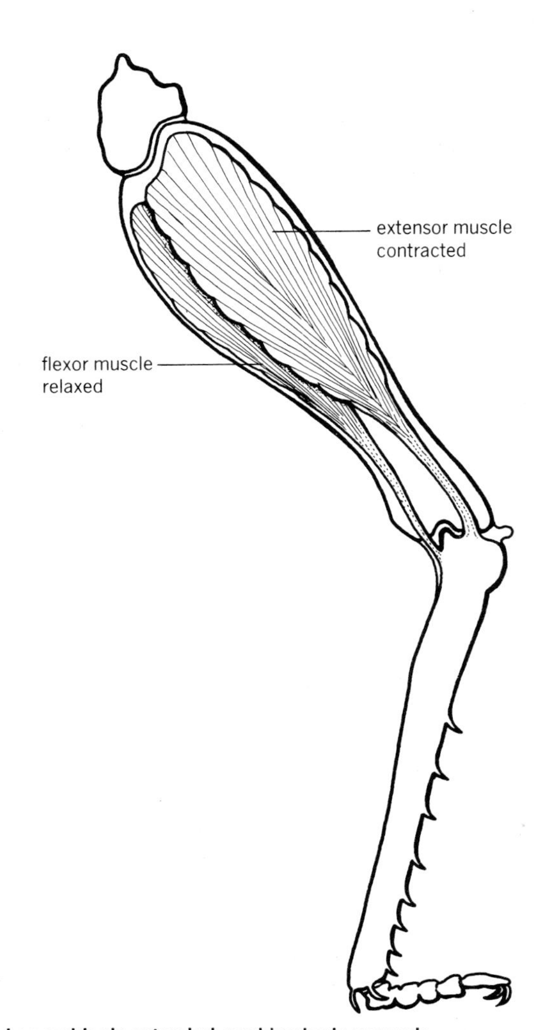

Leg suddenly extended, pushing body upwards.

11.17 How a locust hops

The exoskeleton is covered with a layer of wax, called a **cuticle**. The cuticle stops water evaporating from the insect's body.

Fig 11.17 illustrates how an insect's muscles are attached to its exoskeleton, and how they cause movement.

Support and movement in earthworms

11.15 Earthworms have hydrostatic skeletons.

Earthworms have no hard parts to support their bodies. They are held in shape by the fluid inside them. The fluid presses out on the muscular body wall, keeping the earthworm's body firm. This is called a **hydrostatic skeleton.**

11.16 How an earthworm moves.

Earthworms are adapted for squeezing through burrows in the soils. Their long, cylindrical shape helps them to do this. The mucus on their skin helps them to slide past soil particles without damaging themselves. They have small bristles called **chaetae** on their underside, which help them to grip the sides of their burrow.

Earthworms have two sets of muscles outside their fluid skeleton. **Circular muscles** run round the earthworm's body. When they contract, they make its body long and thin. **Longitudinal muscles** run lengthways. When they contract, they make the earthworm's body short and fat. These two sets of muscles work antagonistically (see section 11.8).

Fig 11.18 shows how an earthworm moves.

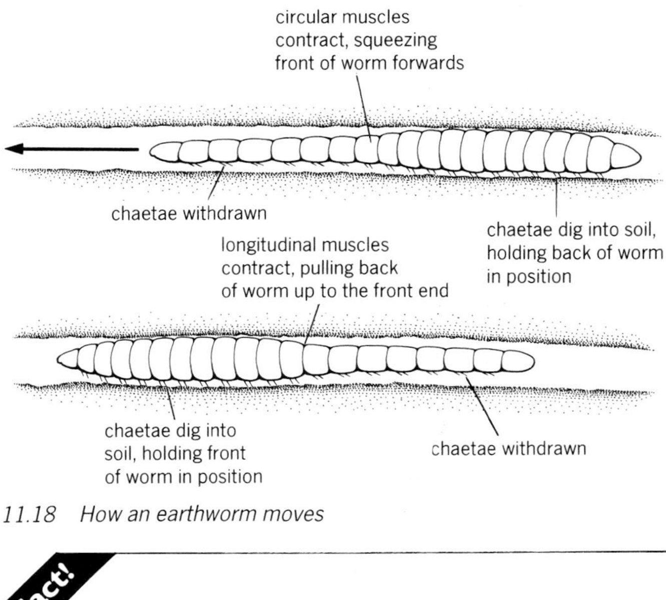

11.18 How an earthworm moves

Fact!

The longest species of earthworm lives in South Africa. It can grow up to 6.7 m long.

Support in plants

11.17 Plant skeletons.

All the skeletons so far described in this chapter have two main functions – support and movement. Plants, though, do not move very much. They only need to be supported, so they do not need elaborate skeletons.

11.18 Xylem forms wood and supports plants.

Plant stems, roots and leaves contain xylem (see section 6.27). Xylem is made of cells which have very strong walls containing lignin. These lignified xylem vessels help to support the stem.

The larger and taller the plant, the more support it needs. Trees are supported by the wood in their trunks and branches, which is made almost entirely of xylem.

11.19 Cell turgor supports herbaceous plants.

In parts of plants where there is not much xylem, another means of support is needed.

When a plant has plenty of water, the contents of each cell press outwards on the cell wall (see Fig 4.6). This makes the cells firm, or **turgid.** They press against each other, holding the plant firm and upright. This is particularly important in leaves, and in **herbaceous plants** which do not have woody stems.

11.19 A wilting geranium shoot

Questions

1 List four ways in which birds are adapted for flight.
2 Look at Fig 11.15. What will happen to the wing when (a) the pectoralis major contracts, and (b) the pectoralis minor contracts?
3 What is a swim bladder, and how is it used?
4 What is meant by a hydrostatic skeleton?
5 What is wood made of?
6 How does turgor help to support plants?

Table 11.2 Summary of the skeletons of different organisms.

	Type of skeleton	*Material skeleton is made of*	*Main functions*
Mammal	endoskeleton	bone and cartilage	support, movement and protection
Bird	endoskeleton	bone and cartilage	support, movement and protection
Fish	endoskeleton	bone and cartilage	support, movement and protection
Insect	exoskeleton	chitin and protein	support, movement and protection
Earthworm	hydrostatic endoskeleton	fluid	support and movement
Plant	(a) hydrostatic skeleton and (b) endoskeleton	(a) fluid in cells (b) cellulose and lignin in cell walls	support

Chapter revision questions

1 Match each word below with its definition.
collagen, cartilage, ligament, tendon, flexor, extensor, keel, chitin, chaetae, lignin
(a) a structure which holds two bones together at a joint, allowing movement between them
(b) a muscle which pulls two bones closer together when it contracts
(c) a tough material from which an insect's exoskeleton is made
(d) the very large sternum of a bird, to which its flight muscles are attached
(e) small hair-like structures, which help an earthworm to grip the sides of its burrow
(f) the protein found in bone
(g) a muscle which pulls two bones away from one another when it contracts
(h) the strong substance found in the walls of xylem vessels
(i) a strong but flexible substance, found on the ends of moveable bones
(j) a structure which attaches a muscle to a bone

2 (a) What is meant by antagonistic muscles? Give one example.
(b) Make a large, labelled diagram to show the position of one pair of antagonistic muscles in a named mammal, including the structure of the joint whose movement they cause.
(c) Describe how these muscles cause flexing of the joint you have shown.

12 Sensitivity and coordination

12.1 Organisms detect changes around them.

All living organisms are sensitive to their environment. This means that they can detect changes in their environment. The changes they detect are called **stimuli.**

The parts of the organism's body that detect stimuli are called **receptors.** In animals, the receptors are often part of a **sense organ.** For example, your eye is a sense organ and the rod and cone cells in the retina of your eye are receptors. They are sensitive to light.

Touch and taste

12.2 Skin contains receptive nerve endings.

One of the functions of skin is to pick up various kinds of information about your environment. If you look back at Fig 9.5 you will see that there are several sorts of nerve endings in the dermis. These nerve endings are receptors for touch, heat, cold, pressure and pain.

If you do Investigation 12.1, you can find out which parts of your skin contain the most touch receptors.

Investigation 12.1 To find which part of the skin contains the most touch receptors

This experiment tests the skin on the back of the hand, the palm of the hand, and the back of the neck. You could try different parts of the body if you like.

1. Copy out the results table, ready to fill in your results.
2. Set your two pins at exactly 2 cm apart. Keep checking that the points stay 2 cm apart all the time you use them.
3. Ask your partner to close his or her eyes. Touch your partner gently on the back of the hand with either one or two pins. Ask your partner to tell you how many pins are touching the hand. Put a tick in the first space in your table if your partner is right, and a cross if your partner is wrong.
4. Repeat this nine more times. You should now have ten ticks or crosses in the first space in the table.
5. Now repeat steps 3 and 4, still keeping the pins exactly 2 cm apart, but this time touching the skin on the palm of the hand.
6. Repeat on the forehead.
7. Now adjust the pins to 1 cm apart. Test the back of hand, the palm of the hand and the back of the neck as before.
8. Adjust the pins to 0.5 cm apart, and test as before.
9. If there is time, test again with the pins 0.2 cm apart.
10. Use your results to draw a histogram. Fig 12.1 shows how you can do this.

Results table	Distance apart of pins 2cm	1cm	0·5cm	0·2cm
back of hand				
palm of hand				
forehead				

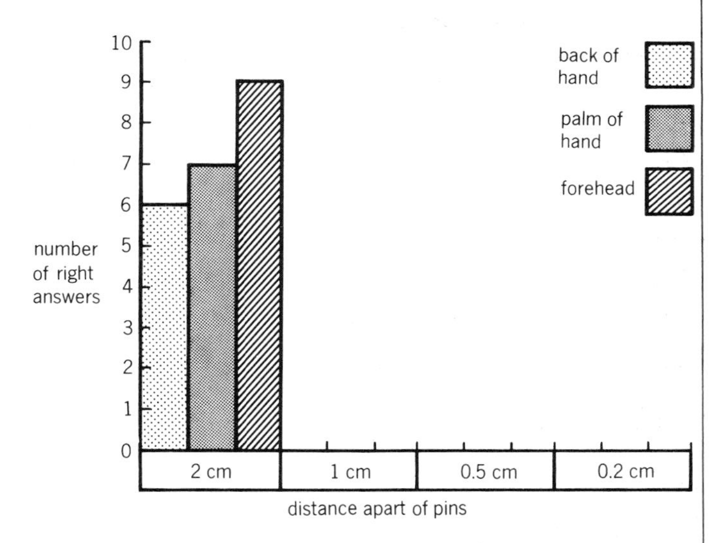

12.1 Results graph from experiment to find which part of the skin contains the most touch receptors

Questions

1. Why does it get more difficult to tell how many pins are touching your skin, as the pins get closer together?
2. Which of the three parts of the body you tested had the most touch receptors?
3. Why do you think this part of the body needs to be so sensitive?

12.3 Some receptors detect chemicals.

The nose and tongue both contain receptors which respond to chemical stimuli. They are sensitive to chemicals in the air, or in food.

On the tongue, these receptor cells are in small groups, called **taste buds** (see Fig 12.2). The taste buds do not all respond to the same kinds of chemical. Try Investigation 12.2, to find out which parts of your tongue can taste which kinds of flavour.

> **Fact!** The animal with the most acute sense of smell is the male Emperor moth. Using its antennae, it can detect a female Emperor moth 11 km upwind.
>
> The biggest mammalian tongue that has ever been weighed belonged to a Blue whale caught by Russian trawlers in 1947. Its tongue weighed 4.3 tonnes.

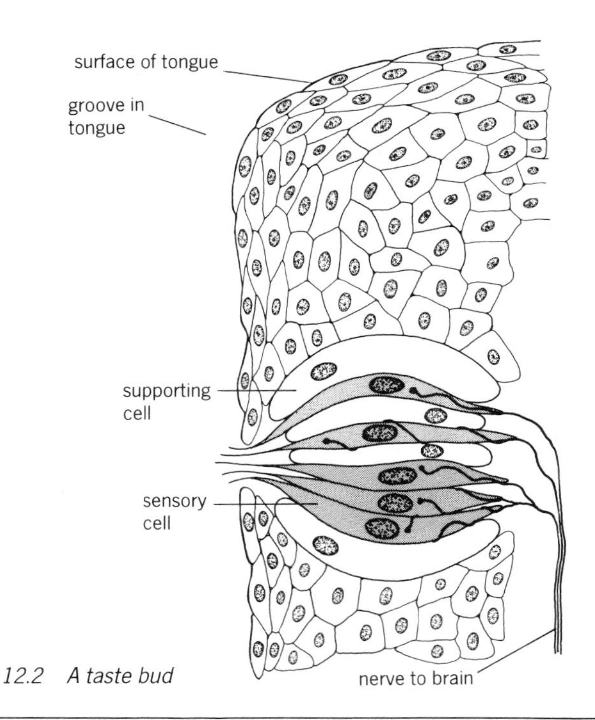

12.2 *A taste bud*

Investigation 12.2 To see which parts of the tongue can taste which flavours

The four tastes you are going to test are sweet, sour, bitter and salty.

1. Copy out the results table, ready to fill in your results.
2. Arrange a communications system between yourself and your partner, so that he or she can tell you which flavour can be tasted without moving the tongue.
3. Put a cotton wool stick into each of the four kinds of solution. Each piece of cotton wool must be put back into its own solution as soon as you finish using it.
4. Ask your partner to shut his or her eyes, and put out his or her tongue. Choose one piece of cotton wool, and touch it onto one of the areas of the tongue shown in Fig 12.3. *Do not* let your partner put his or her tongue back into the mouth yet.
5. Ask your partner to indicate what flavour has been put on their tongue. Your partner can now put his or her tongue back into his or her mouth and moisten it ready for the next test. Fill in the result in the appropriate space in the table, with a tick or a cross.
6. Repeat with each flavour on each part of the tongue, in a random order. If there is time, do each test twice.

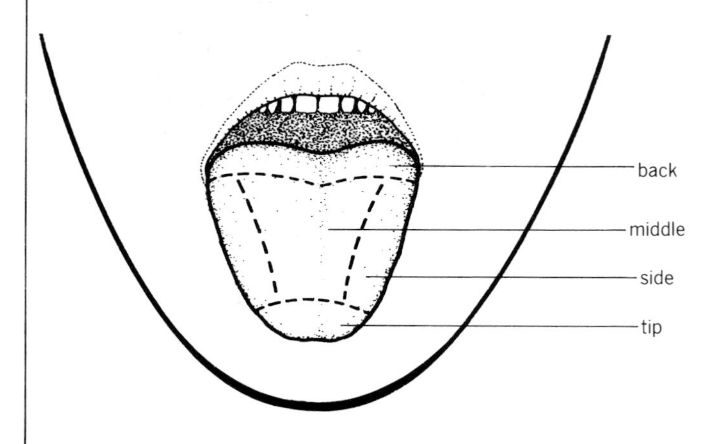

12.3 *Testing the sensitivity of the tongue*

Results table

	Part of tongue			
Flavour	front	centre	sides	back
sweet				
sour				
bitter				
salty				

Questions

1. Why must your partner not be allowed to put his or her tongue into his or her mouth before telling you what flavour they can taste?
2. Why is it best to do the tests in a random order?
3. Copy the drawing of the tongue, and show on it which parts of the tongue are most sensitive to sweet, sour, bitter and salty tastes.

The eye

12.4 The eye is well protected.

The part of the eye which contains the receptor cells is the **retina** (see Fig 12.4). This is the part which is actually sensitive to light.The rest of the eye simply helps to protect the retina, or to focus light onto it.

Each eye is set in a bony socket in the skull, called the **orbit.** Only the very front of the eye is not surrounded by bone.

The front of the eye is covered by a thin, transparent membrane called the **conjunctiva,** which helps to protect the parts behind it. The conjunctiva is always kept moist by a fluid made in the **tear glands.** This fluid contains an enzyme called **lysozyme,** which can kill bacteria.

The fluid is washed across your eye by your eyelids when you blink. The eyelids, eyebrows and eyelashes also help to stop dirt from landing on the surface of your eyes.

Even the part of the eye inside the orbit is protected. There is a very tough coat surrounding it called the **sclera.**

12.5 Cells in the retina are receptive to light.

The retina is at the back of the eye. It contains two sorts of receptor cell. **Rods** are sensitive to quite dim light, but only let you see in black and white. **Cones** give colour vision, but only in bright light.

When light falls on a receptor cell in the retina, the cell sends a message along the **optic nerve** to the brain. The brain sorts out all the messages from each receptor cell, and builds up an **image.**

The closer together the receptor cells are, the clearer the image the brain will get. The part of the retina where the receptor cells are packed most closely together is called the **fovea.** This is the part of the retina where light is focused when you look straight at an object. All the receptor cells in the fovea are cones. The rods are scattered further out on the retina.

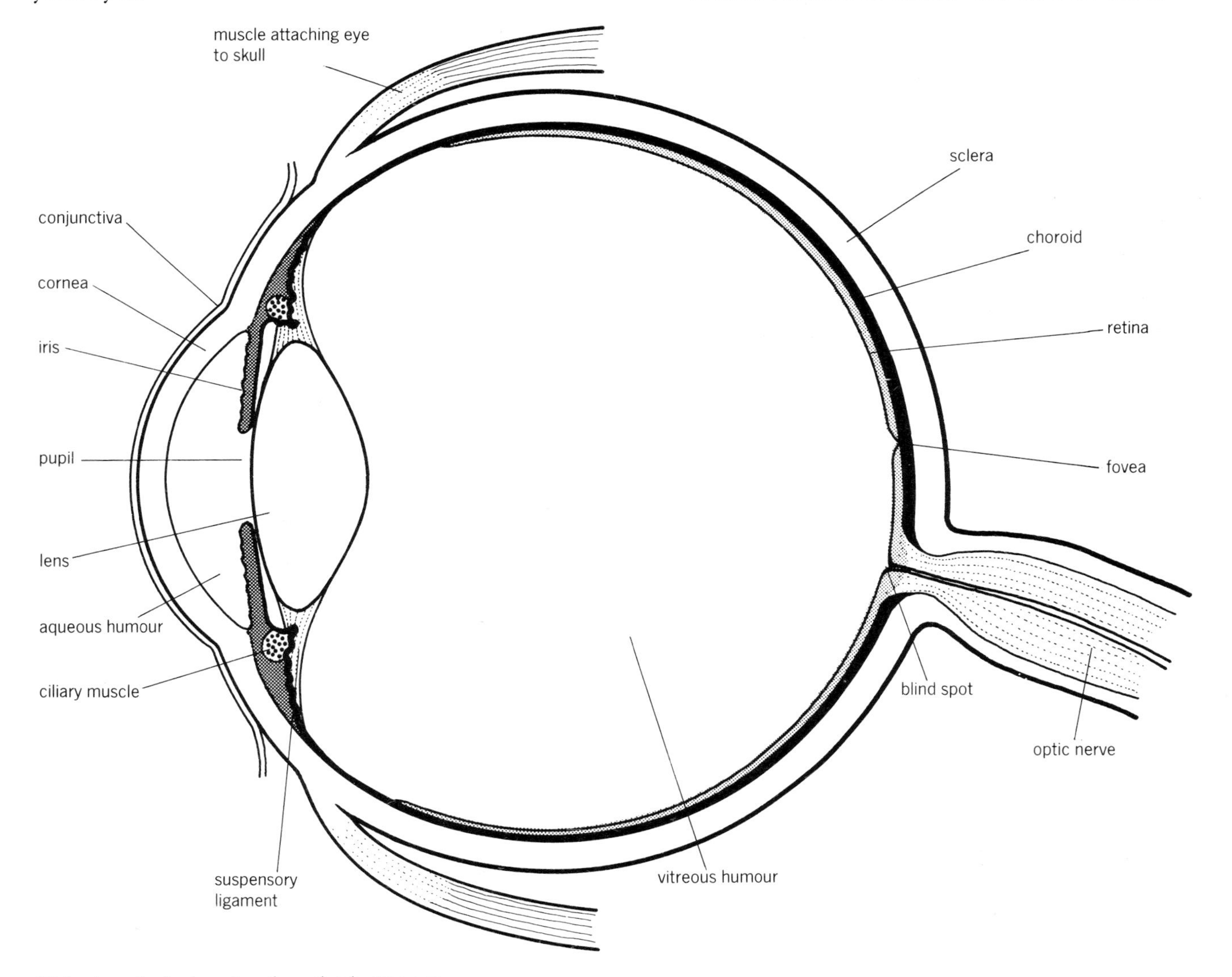

12.4 Longitudinal section through a human eye

There are no receptor cells where the optic nerve leaves the retina. This part is called the **blind spot.** If light falls on this place, no messages will be sent to the brain.

Behind the retina is a black layer called the **choroid.** The choroid absorbs all the light after it has been through the retina, so it does not get scattered around the inside of the eye.

12.6 How the eye focuses light.

For the brain to see a clear image, there must be a clear image focused on the retina. Light rays must be bent, or **refracted,** so that they focus exactly onto the retina.

The **cornea** is responsible for most of the bending of the light. The **lens** makes fine adjustments.

Fig 12.5 shows how these two parts of the eye focus light onto the retina.

The image on the retina is upside down. The brain interprets this so that you see it the right way up.

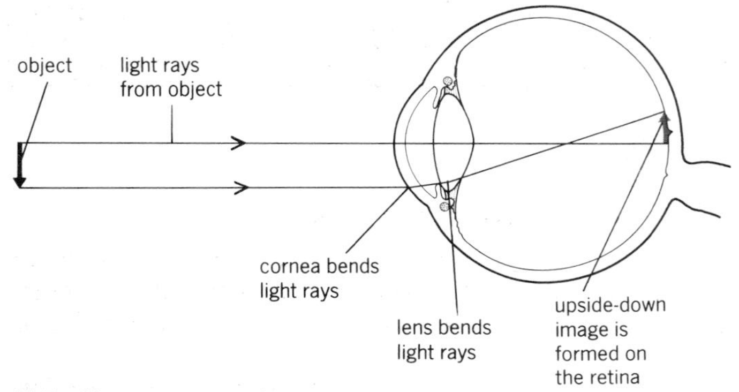

12.5 How an image is focused onto the retina

12.7 The lens adjusts the focusing.

Not all light rays need bending the same amount to focus them onto the retina. Light rays coming from a nearby object are going away from one another, or diverging. They will need to be bent inwards quite strongly.

Light rays coming from an object in the distance will be almost parallel to one another. They will not need bending so much.

The shape of the lens can be adjusted to bend light rays more. The fatter it is, the more it will bend them. The thinner it is, the less it will bend them. The adjustment in the shape of the lens, to focus light coming from different distances, is called **accommodation.**

Figs 12.6 and 12.7 show how the shape of the lens is changed. It is held in position by a ring of **suspensory ligaments.** The tension on the suspensory ligaments, and thus the shape of the lens, is altered by means of the **ciliary muscle.** When it contracts, the suspensory ligaments are loosened. When it relaxes, they are pulled tight (see Fig 12.8). When the suspensory ligaments are tight, the lens is pulled thin. When they are loosened, the lens get fatter.

12.8 The iris adjusts how much light enters the eye.

In front of the lens is a circular piece of tissue called the **iris.** The iris contains pigments, which absorb light and stop it getting through to the retina.

In the middle of the iris is a gap, called the **pupil.** The size of the pupil can be adjusted. The wider the pupil is, the more light can get through to the retina. In strong

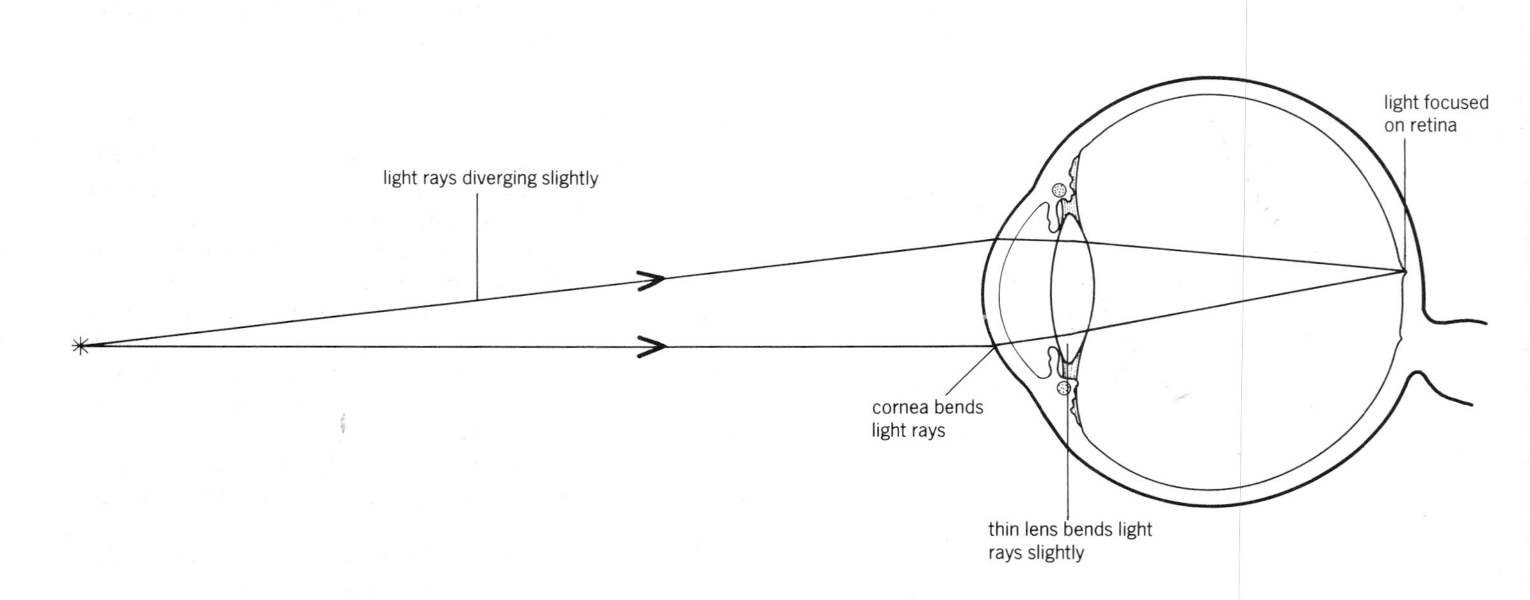

12.6 Focusing on a distant object

light, the iris closes in, and makes the pupil small. This stops too much light getting in and damaging the retina.

To allow it to adjust the size of the pupil, the iris contains muscles. **Circular muscles** lie in circles around the pupil. When they contract, they make the pupil constrict, or get smaller. **Radial muscles** run outwards from the edge of the pupil. When they contract, they make the pupil dilate, or get larger.

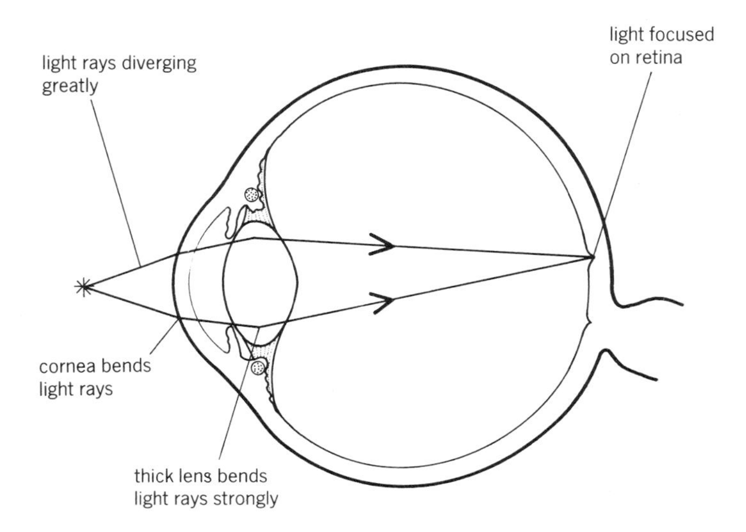

12.7 Focussing on a nearby object

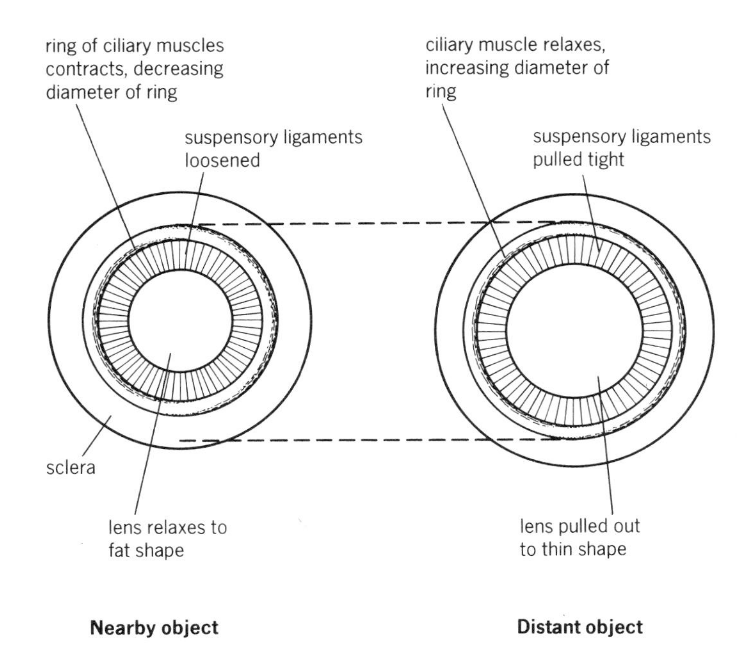

12.8 How the shape of the lens is changed

Questions

1. What is a stimulus?
2. Name two parts of the body which contain receptors of chemical stimuli.
3. Which part of the eye contains cells which are sensitive to light?
4. Your brain can build up a very clear image when light is focused onto the fovea. Explain why it can do this.
5. If you look straight at an object when it is nearly dark, you may find it difficult to see it. It is easier to see if you look just to one side of it. Explain why this is.
6. What is the choroid, and what is its function?
7. List, in order, the parts of the eye through which light passes to reach the retina.
9. Name two parts of the eye which refract light rays.
10. What is meant by accommodation?
11. (a) What do the ciliary muscles do when you are focussing on a nearby object?
(b) What effect does this have on (i) the suspensory ligaments, and (ii) the lens?

The ear

12.9 The structure of the ear.

The ear has two functions. Firstly, it is sensitive to sound. It also contains receptor cells which are sensitive to the position and movement of your head. These cells help with balance.

Fig 12.9 illustrates the structure of the human ear. The outer ear and middle ear both contain air. The inner ear though, is filled with two sorts of fluid, **perilymph** ('peri' means round the outside) and **endolymph** ('endo' means inside).

12.10 Hearing.

The cells which are sensitive to sound waves are inside the **cochlea.** Sound waves need to be made stronger, or amplified, before these cells will respond to them.

Firstly, the sound waves make the air in the outer ear vibrate. This makes the eardrum vibrate. Touching the eardrum is a tiny bone or ossicle, called the hammer or **malleus.** As the eardrum vibrates, the malleus vibrates too. The vibrations pass along the chain of ossicles. The malleus vibrates on the anvil or **incus,** and the incus vibrates on the stirrup or **stapes.**

The stapes lies against a membrane over the **oval window.** The oval window transmits the vibrations to the perilymph.

This chain of events helps to amplify the size of the vibrations. The vibrations in the perilymph are then passed into the cochlea. Fig 12.10 shows how the inside of the cochlea would look if it was uncoiled. Fig 12.11 shows a cross section of the cochlea. As the

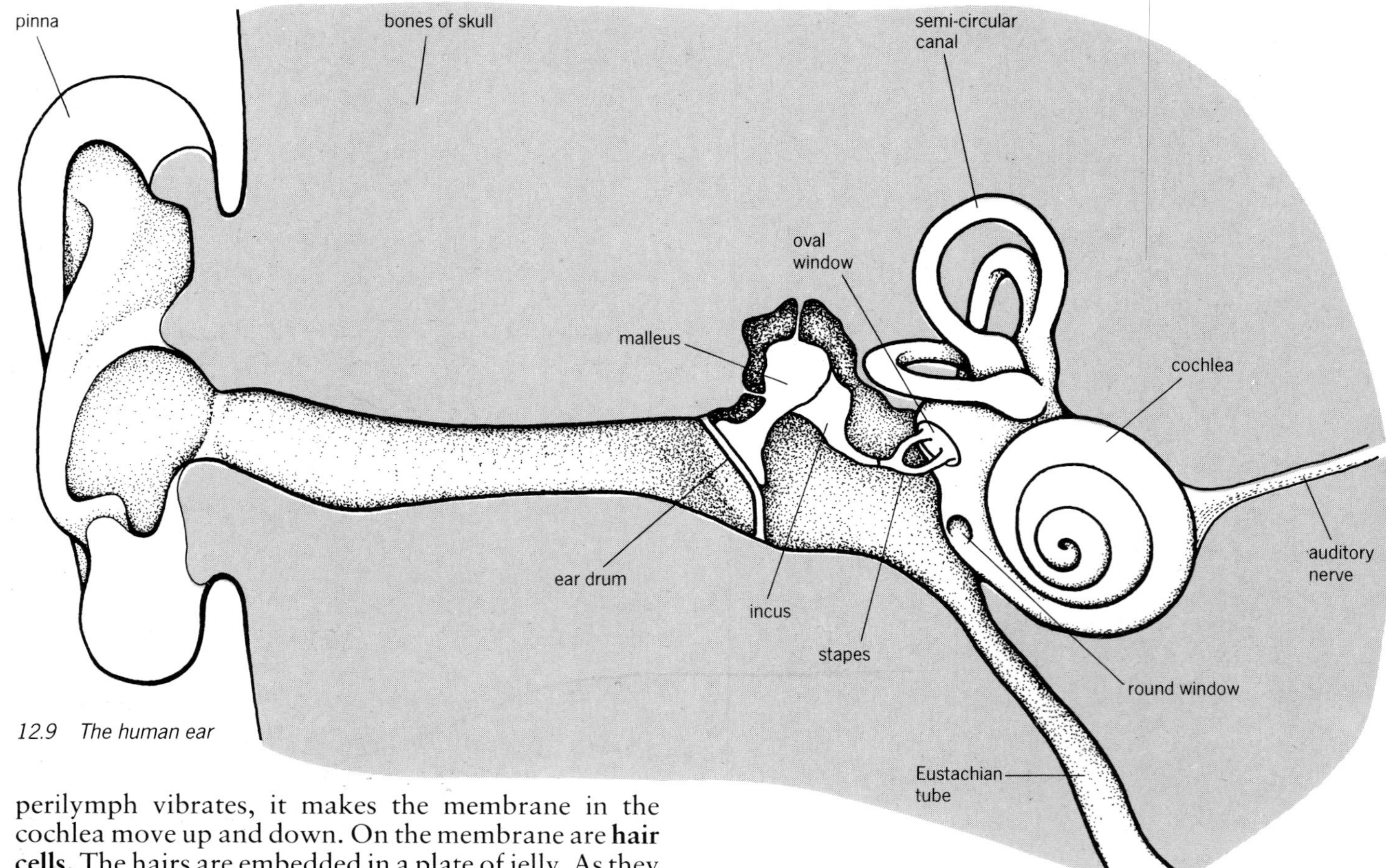

12.9 The human ear

perilymph vibrates, it makes the membrane in the cochlea move up and down. On the membrane are **hair cells.** The hairs are embedded in a plate of jelly. As they move up and down, the hairs are pulled and pushed against the jelly plate. This makes the hair cells send messages along the auditory nerve to the brain.

The hair cells in different parts of the cochlea respond to different frequencies of vibration. The ones nearest to the oval window respond to high frequencies (high pitched sounds). Low frequency sounds are picked up by the cells nearest the middle of the coil.

12.11 Balance.

The three **semi-circular canals** are sensitive to movements of the head. They are filled with endolymph. Each semi-circular canal has a swelling near one end of it, called an **ampulla.**

Fig 12.12 shows part of the inside of an ampulla. Like the cochlea, it contains hair cells with their hairs embedded in a plate of jelly. The jelly is called the **cupula.** When you move your head, the cupula moves in the endolymph. It pulls on the hairs, so the hair cells send messages along a nerve to the brain.

Each ear has three semi-circular canals, all at right angles to one another. By comparing the messages from each one, your brain can tell exactly how your head is moving, and so enable you to keep your balance. Usually, messages from your eyes help with this as well. If the messages from your eyes and ears do not match, such as when you are reading a book while travelling in a car, then you may feel sick.

> **Fact!**
> High-pitched sounds are rapid vibrations of the molecules in air. Humans can hear sounds which vibrate at about 19 kHz. Some bats can hear ultrasonic sounds up to about 160 kHz.

12.12 The Eustachian tube prevents eardrum damage.

Both the outer and the middle ear are filled with air. They are separated by the eardrum. If there is a big difference in the air pressure on the two sides of the eardrum, it may burst.

To stop this happening, there is a tube leading from the middle ear to the back of the throat. It is called the **Eustachian tube.** Usually, it is kept closed by a sphincter muscle at the bottom. But when you swallow, it opens, so that air taken into your mouth can enter the Eustachian tube. This allows the air pressure in the middle ear to equalise with the air pressure outside.

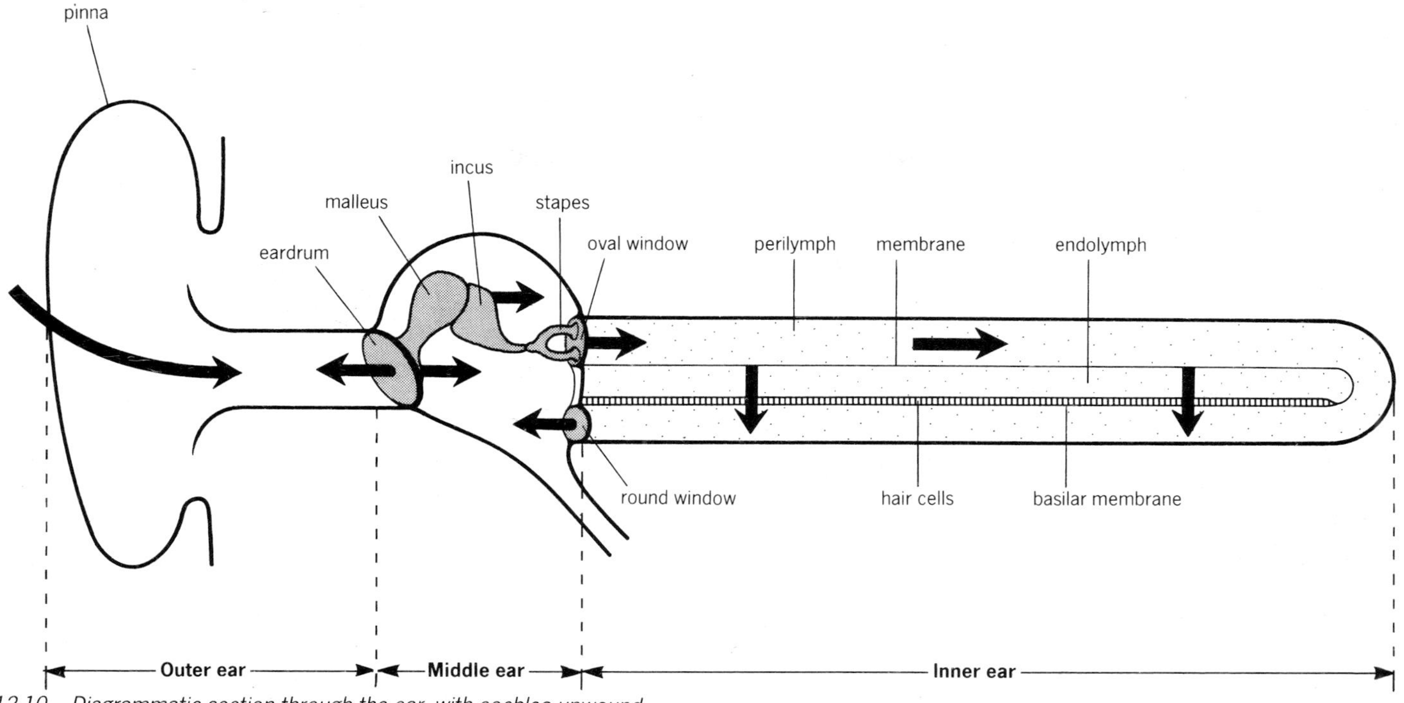

12.10 Diagrammatic section through the ear, with cochlea unwound

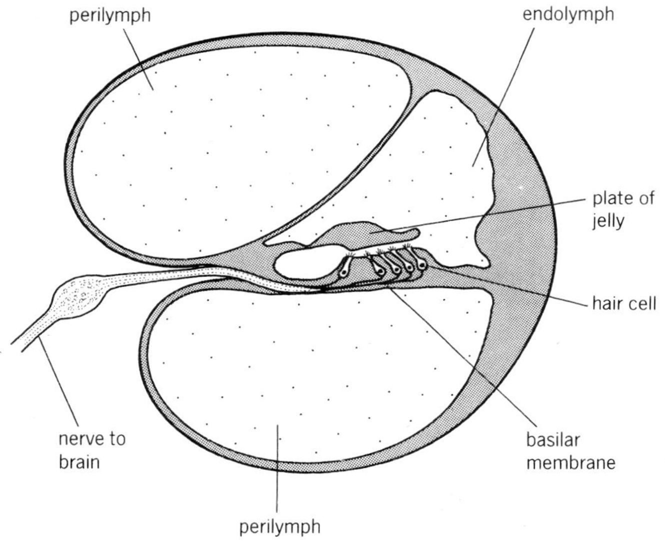

12.11 Transverse section through part of cochlea

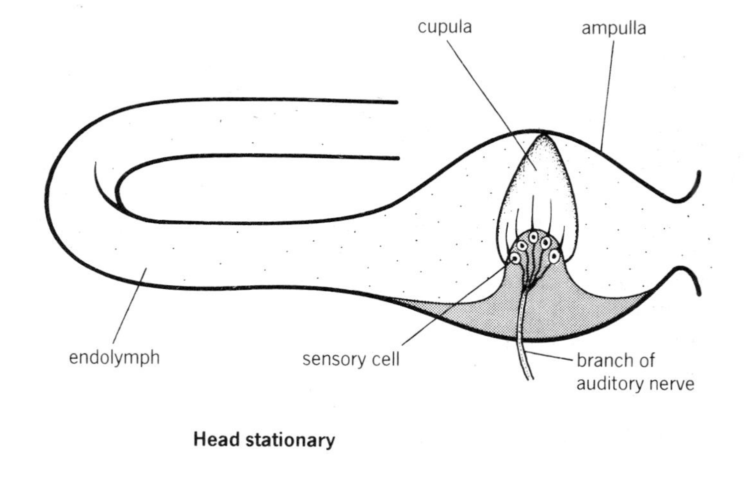

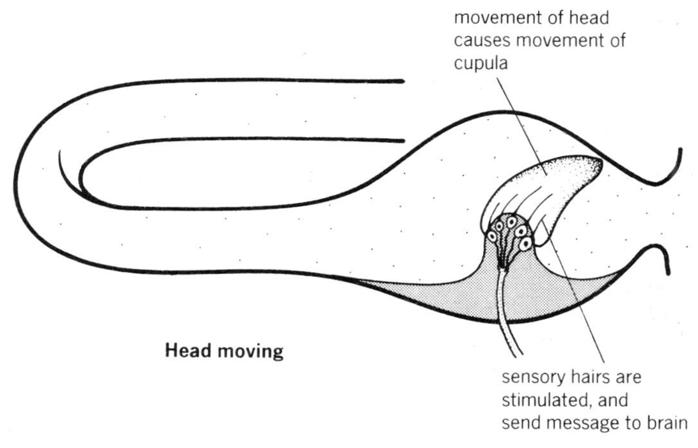

12.12 Section through part of a semi-circular canal

Questions

1. In which part of the ear are the cells which are sensitive to sound?
2. Explain how sound brings about a response by these cells.
3. Which part of the ear helps with balance?
4. Name one other sense organ which helps with balance.
5. Where is the Eustachian tube?
6. As an aeroplane takes off, the air pressure inside it drops. Your ears will feel less uncomfortable if you suck a sweet. Explain why this is.

Co-ordination and response

12.13 Effectors respond when the body is stimulated.

Being able to detect stimuli is not much use to an organism unless it can respond to them in some useful way. The part of the body which responds to a stimulus is called an **effector.**

Muscles are effectors. For example, if you touch something hot, the muscles in your arm contract, so that your hand is quickly pulled away. Glands can also be effectors. If you smell food cooking, your salivary glands may react by secreting saliva.

To make sure that the right effectors respond at the right time, there needs to be some kind of communication system between receptors and effectors. The pain receptors on your fingertips need to send a message to your arm muscles to tell them to contract. The chemical receptors in your nose must communicate with your salivary glands, to make them secrete saliva. The way in which receptors pick up stimuli, and then pass messages on to effectors, is called **co-ordination.**

12.14 Hormones and nerves allow communication.

Animals need fast and efficient communication systems between their receptors and effectors. This is partly because most animals move in search of food. Many animals need to be able to respond very quickly to catch their food, or to avoid predators.

Most animals have two methods of sending messages from receptors to effectors. The fastest is by means of **nerves.** The receptors and nerves make up the animal's **nervous system.** A slower method, but still a very important one, is by means of chemicals called **hormones.** Hormones are part of the **endocrine system.**

Nervous systems

12.15 Neurones carry nerve impulses.

Nervous systems are made of special cells called **neurones.** Fig 12.13 illustrates a neurone from a mammal's body.

Neurones contain the same basic parts as any animal cell. Each has a nucleus, cytoplasm, and a cell membrane. But their structure is specially adapted to be able to carry messages very quickly.

To enable them to do this, they have long, thin fibres of cytoplasm stretching out from the cell body. They are called **nerve fibres.** The longest fibre is usually called an **axon.** Axons can be more than a metre long. The shorter fibres are called **dendrons** or **dendrites.**

The dendrites pick up messages from other neurones lying nearby. They pass the message to the cell body, and then along the axon. The axon might then pass it on to another neurone.

12.16 Myelinated neurones carry impulses quickly.

The nerve fibres of active animals like mammals are wrapped in a layer of fat and protein called **myelin.** Every now and then, there are narrow gaps in the myelin sheath.

The messages that neurones transmit are in the form of electrical impulses. Myelin insulates the nerve fibres, so that they can carry impulses much faster. A myelinated nerve fibre in a cat's body can carry impulses at up to 100 metres per second. A fibre without myelin can only carry impulses at about 5 metres per second.

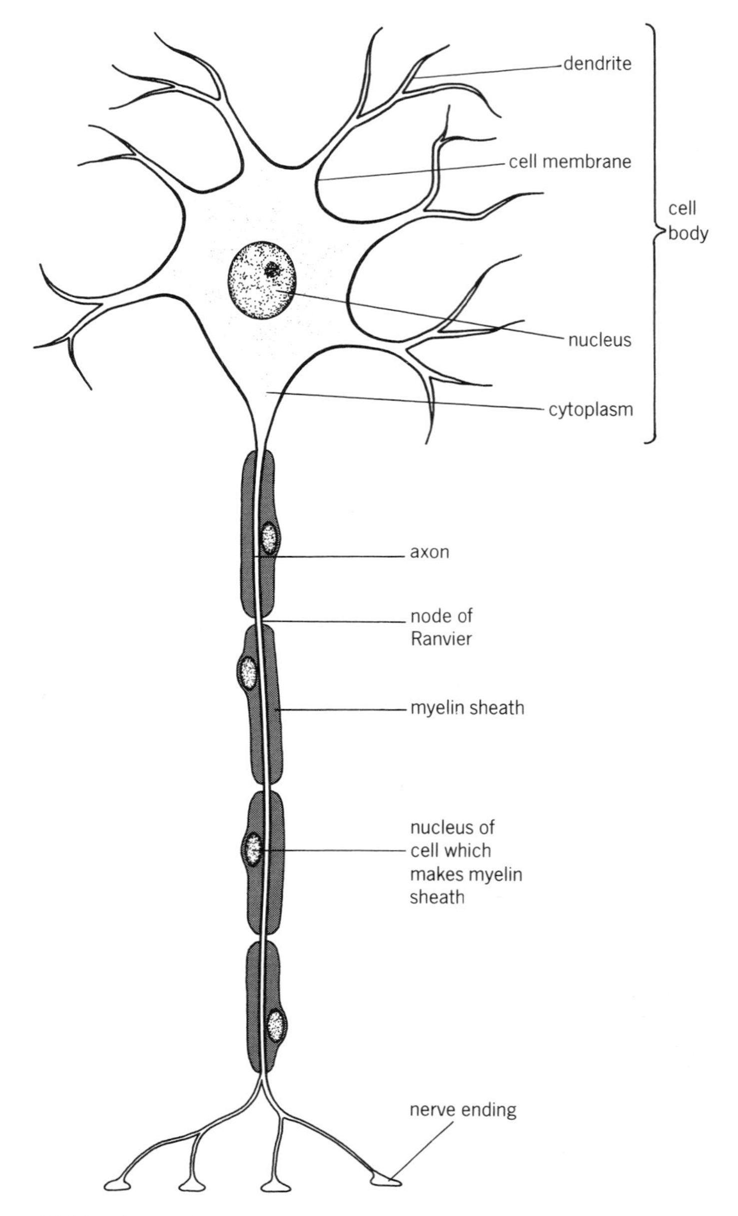

12.13 A neurone

12.17 A nerve net is a simple nervous system.

One of the simplest arrangements of neurones is found in a small animal called *Hydra*. *Hydra* belongs to the jellyfish or Coelenterate phylum (see Chapter 17). It lives in fresh water, where it attaches itself to weeds. It feeds by catching small organisms with its tentacles.

Hydra does not need to move very quickly, and its behaviour is not at all complicated. It has a simple nervous system called a **nerve net** (see Fig 12.14).

Each neurone in a nerve net has several fibres leading out of the nerve cell body. Each fibre connects with one or more fibres from a nearby neurone. So if one part of the nervous system receives a stimulus, an electrical message spreads all over the *Hydra*'s body, from neurone to neurone.

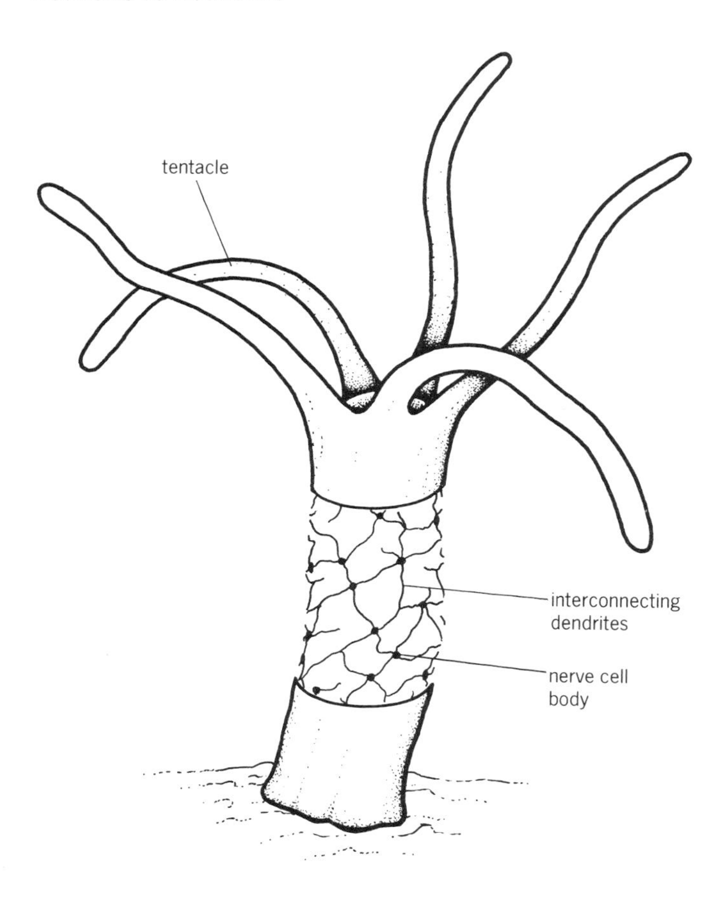

12.14 A Hydra with its outer layer of cells removed, to show its nerve net

Fact!

Your brain contains about 100 000 000 000 neurones. After the age of eighteen, you will lose about 1000 of these every day.

A nerve impulse can travel through some parts of your nervous system at a speed of 288 km/hr.

12.18 A nerve is a bundle of nerve fibres.

A simple nerve net like this works perfectly well for *Hydra*. But an active animal, particularly one with such complex behaviour as a human, needs a more elaborate system. One difference between the human nervous system and that of *Hydra* is that most human nerve fibres have myelin around them, whereas *Hydra*'s do not. This means that messages can travel more quickly along human nerve fibres.

Another difference is that human nerve fibres do not run through the body on their own. They are usually in groups of several hundred, called a **nerve.** Fig 12.15 shows a cross section through a nerve.

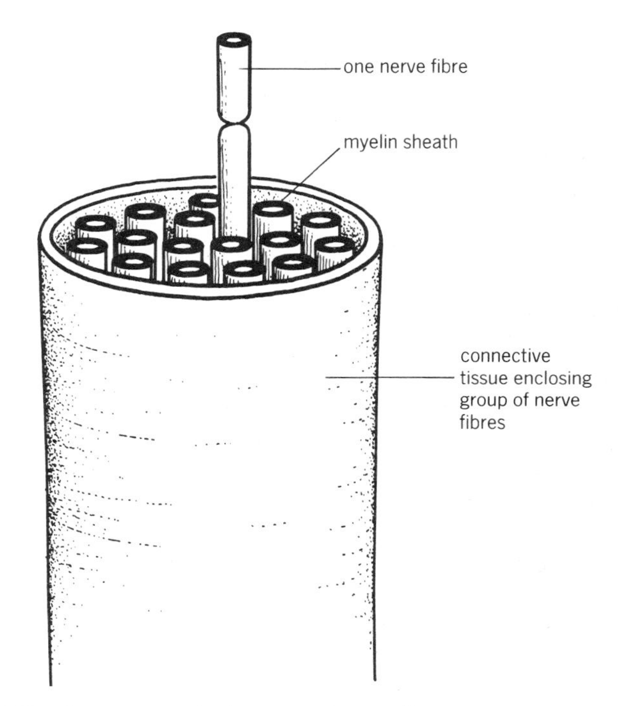

12.15 A nerve consists of a group of nerve fibres

12.19 Most animals have a central nervous system.

Perhaps the most important difference between the human nervous system and *Hydra*'s simple nerve net is that humans have a **brain** and **spinal cord.** These make up the **central nervous system,** or **CNS** (see Fig 12.16). Like the rest of the nervous system, the CNS is made up of neurones. Its job is to co-ordinate the messages travelling through the nervous system.

When a receptor detects a stimulus, it sends a message to the brain or spinal cord. The brain or spinal cord receives the message, and 'decides' which effectors need to react to the stimulus. It then sends the message on, along the appropriate nerve fibres, to the appropriate effector.

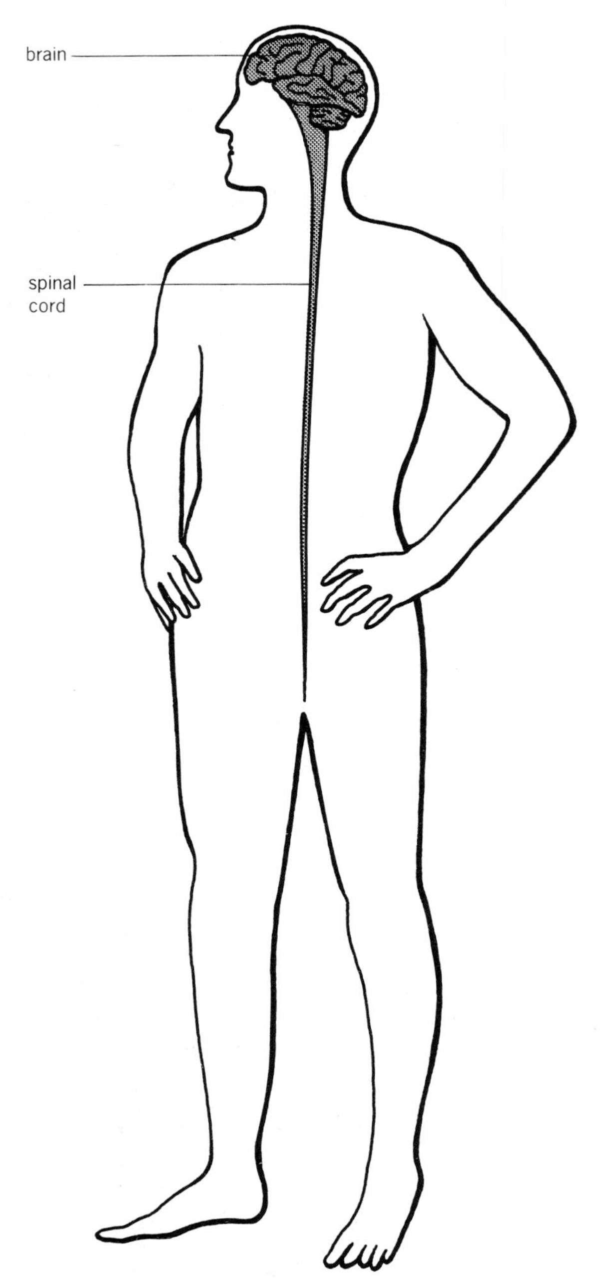

12.16 The human central nervous system

Investigation 12.3 To measure reaction time

The time taken for a message to travel from a receptor, through your CNS and back to an effector is very short. It can be measured, but only with special equipment. However you can get a reasonable idea of the time it takes if you use a large number of people, and work out an average time.

1. Get as many people as possible to stand in a circle, holding hands.
2. One person lets go of his or her neighbour with the left hand, and holds a stop watch in it. When everyone is ready, this person simultaneously starts the stopwatch, and squeezes their neighbour's hand with the right hand.
3. As soon as each person's left hand is squeezed, he or she should squeeze his or her neighbour with the right hand. The message of squeezes goes all round the circle.
4. While the message is going round, the person with the stopwatch puts it into the right hand, and holds his or her neighbour's hand with the left hand. When the squeeze arrives, he or she should stop the watch.
5. Keep repeating this, until the message is going round as fast as possible. Record the time taken, and also the number of people in the circle.
6. Now try again, but this time make the message of squeezes go the other way around the circle.

Questions

1. Using the fastest time you obtained, work out the average time it took for one person to respond to the stimulus they received.
2. Did people respond faster as the experiment went on? Why might this happen?
3. Did the message go as quickly when you changed direction? Explain your answer.

12.20 Reflex arcs allow rapid response.

Fig 12.17 shows how these messages are sent. If your hand touches a hot plate, a message is picked up by a sensory receptor in your finger. It travels to the spinal cord along the axon from the receptor cell. This cell is called a **sensory neurone,** because it is carrying a message from a sensory receptor.

In the spinal cord, the neurone passes its message on to several other neurones. Only one is shown in Fig 12.17. These neurones are called **relay neurones,** because they relay the message on to other neurones. The relay neurones pass the message on to the brain. They also pass it on to an effector.

In this case, the effectors are the muscles in your arm. The message travels to the muscle along the axon of a **motor neurone.** The muscle then contracts, so that your hand is pulled away.

This sort of reaction is called a **reflex action.** You do not need to think about it. Your brain is made aware of it, but you only consciously realise what is happening after the message has been sent on to your muscles.

Reflex actions are very useful, because the message gets from the receptor to the effector as quickly as possible. You do not waste time in thinking about what to do.

The arrangement of sensory neurone, relay neurones and motor neurone is called a **reflex arc.**

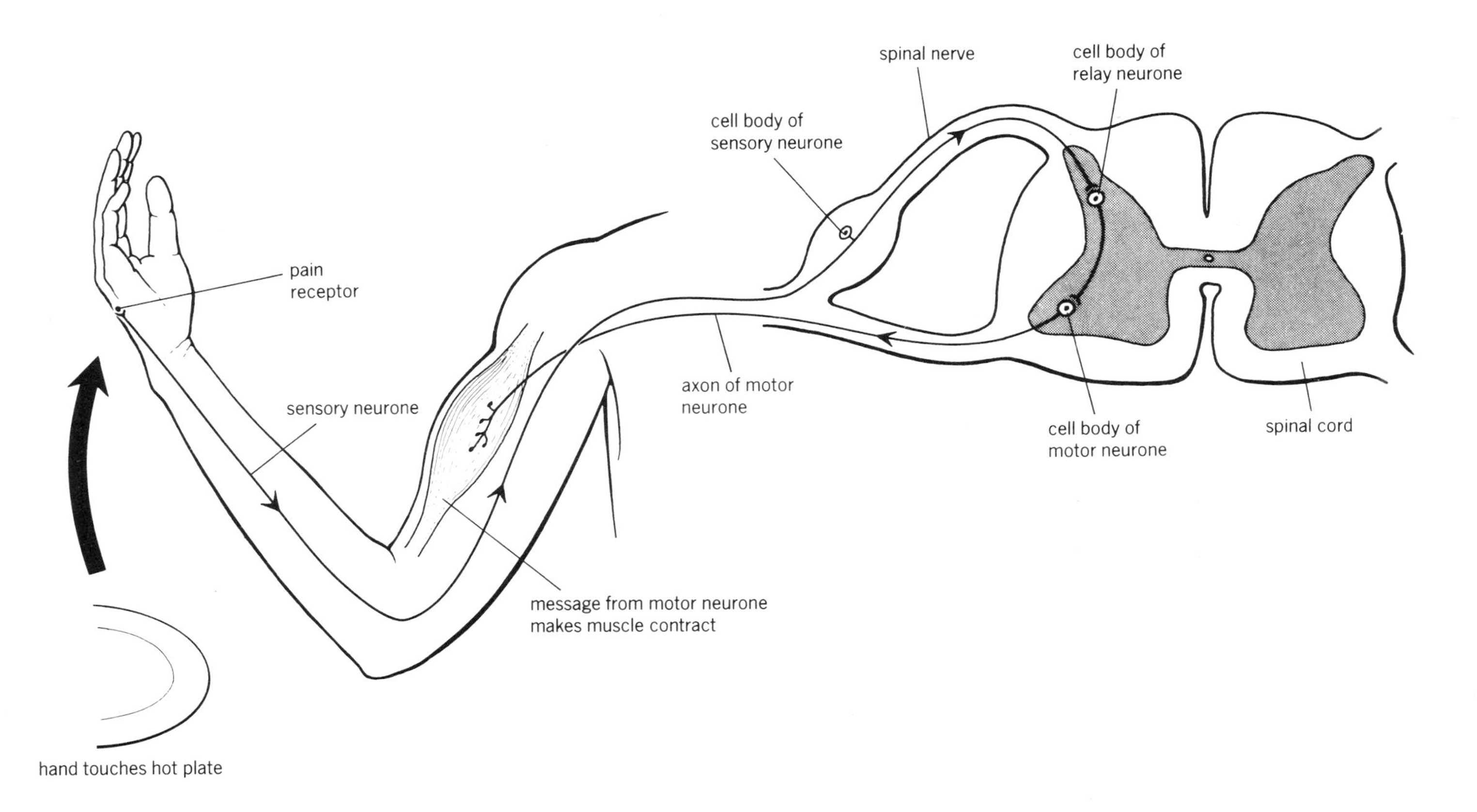

12.17 A reflex arc

12.21 Synapses connect neurones.

If you look carefully at Fig 12.17, you will see that the three neurones involved in the reflex arc do not quite connect with one another. There is a small gap between each one. These gaps are called **synapses.**

Fig 12.18 shows a synapse in more detail. Inside the sensory neurone's axon are hundreds of tiny vacuoles, or **vesicles.** These each contain a chemical, called **transmitter substance.**

When an impulse comes along the axon, it makes these vesicles empty their contents into the space between the two neurones. The transmitter substance quickly diffuses across the tiny gap, attaches to the membrane of the relay neurone and triggers an impulse in the relay neurone. The relay neurone then sends the message onwards.

Synapses act like one-way valves. There is only transmitter substance on one side of the gap, so messages can only go across from that side. Synapses ensure that nervous impulses only travel in one direction.

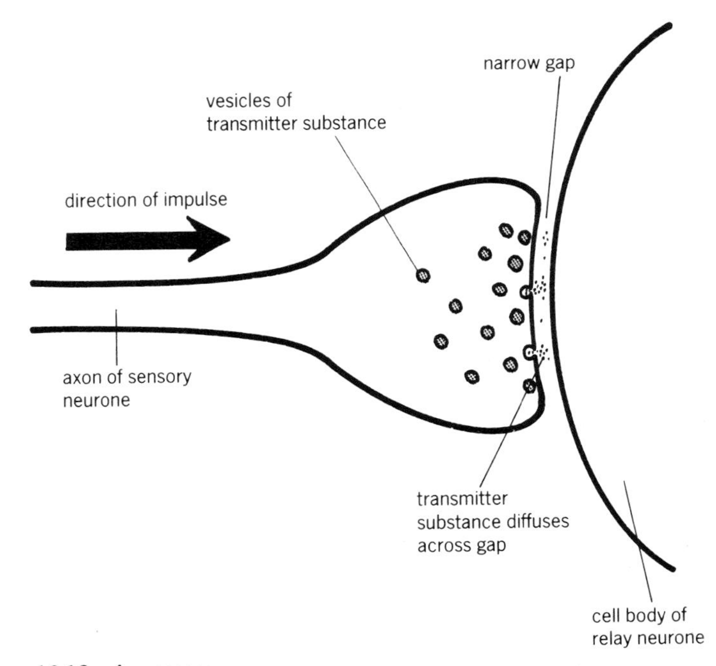

12.18 A synapse

12.22 A complex CNS allows complex behaviour.

Why is the central nervous system needed? Would it not be much quicker if pain receptors in your hand could just send a message straight to your arm muscles to tell them to move your hand away from the hot plate, rather than all the way to the spinal cord and back? Yes, it would, but that system would not be good enough for animals which need to be able to vary their behaviour under different circumstances.

With a central nervous system it is possible to give a modified, more 'intelligent' response. Say, for example, that you started to pick up the hot plate before you knew it was hot. If you just pulled your hand away, you would drop the plate and break it.

When the message from your fingers saying 'hot plate' arrives at your CNS, there is already another message there saying 'but don't drop it'.

The CNS will 'consider' the two messages together. It will probably send a message to your muscles to tell them to put the plate down gently, not to drop it.

The job of the CNS is to collect up all the information from all the receptors in your body. This information will be added together before messages are sent to effectors. In this way, the best action can be taken in a particular set of circumstances.

12.23 Brain functions are localised.

The brain and spinal cord both help to receive impulses from receptors, and pass them on to effectors. But the brain does much more than this.

Figs 12.19 and 12.20 show the structure of the human brain. It is surrounded by three membranes or **meninges,** which help to protect it.

The **cerebrum** is the largest part of the brain. It is made of two **cerebral hemispheres.** Mammals have much larger cerebral hemispheres than any other kind of animal. Humans have the largest ones of all, compared with the size of the rest of the brain.

Conscious thought and memory take place in the cerebrum. Different parts of the cerebrum have different functions. For example, some areas deal with sight, others with speech. An area near the front determines your personality.

The **hypothalamus** lies underneath the front part of the cerebrum. This is the part of the brain which controls osmoregulation and temperature regulation. The **cerebellum** is in control of co-ordination of body movements, and posture. The **medulla oblongata** controls heart beat and breathing.

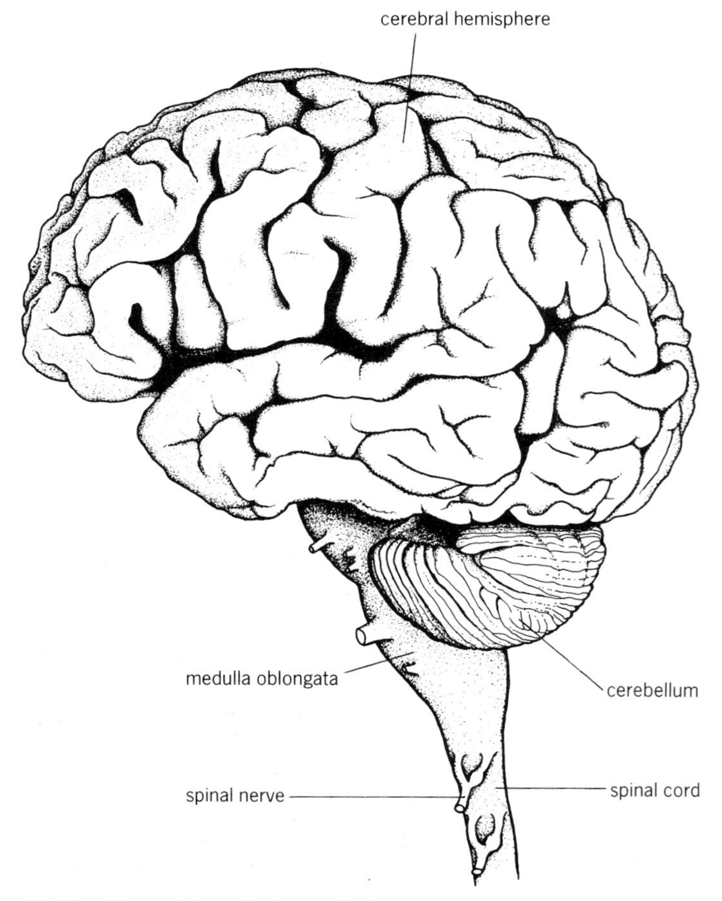

12.19 External view of a human brain

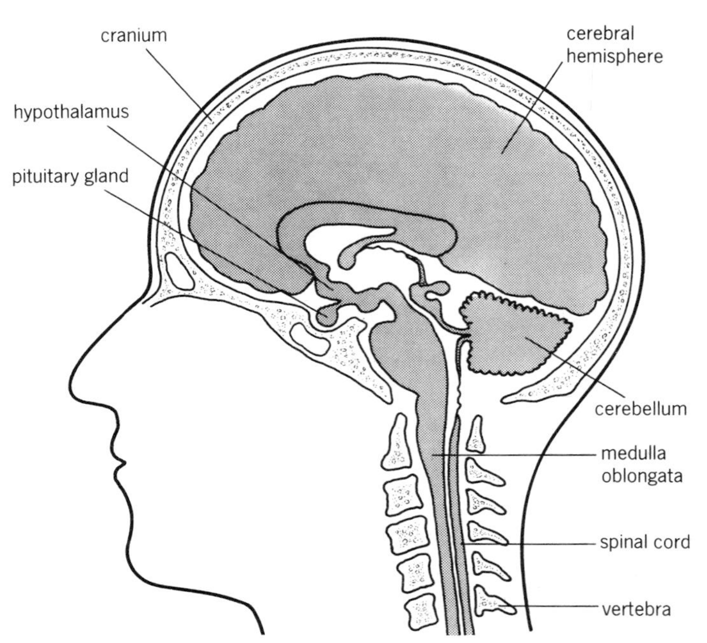

12.20 Section through a human head, to show the brain

Questions

1. Give two examples of effectors.
2. What are the two main communication systems in an animal's body?
3. List three ways in which neurones are similar to other cells.
4. List three ways in which neurones are specialised to carry out their function of transmitting messages very quickly.
5. What is a nerve?
6. What is the function of the central nervous system?
7. Where are the cell bodies of each of these types of neurone found; (a) sensory neurone, (b) relay neurone, and (c) motor neurone?
8. What is the value of reflex actions?
9. Describe two reflex actions, other than the one described in section 12.20.
10. How many synapses are there in the reflex arc shown in Fig 12.17?

The endocrine system

12.24 Endocrine glands make hormones.

Nerves can carry electrical messages very quickly from one part of an animal's body to another. But animals also use chemical messages.

The chemicals are called **hormones.** Hormones are made in special glands called **endocrine glands.** Fig 12.21 shows the positions of the most important endocrine glands in the human body.

Endocrine glands have a good blood supply. They have blood capillaries running right through them. When the endocrine gland makes a hormone, it releases it directly into the blood (see Fig 12.22).

Other sorts of gland do not do this. The salivary glands, for example, do not secrete saliva into the blood. Saliva is secreted into the salivary duct, which carries it into the mouth. Endocrine glands do not have ducts, so they are sometimes called ductless glands.

Once the hormone is in the blood, it is carried to all parts of the body, dissolved in the plasma. Each kind of hormone only affects certain parts of the body.

12.25 Adrenalin prepares the body for action.

There are two adrenal glands, one above each kidney. They make a hormone called **adrenalin.** When you are frightened, excited or keyed up, your brain sends messages along a nerve to your adrenal glands. This makes them secrete adrenalin into the blood.

Adrenalin has several effects which are designed to help you to cope with danger. For example, it makes your heart beat faster, supplying oxygen to your brain and muscles more quickly. This gives them more energy for fighting or running away.

The blood vessels in your skin and digestive system contract right down so that they carry very little blood. This makes you go pale, and gives you 'butterflies in your stomach'. As much blood as possible is needed for your brain and muscles in the emergency.

All of this is very useful if you really have to fight an enemy. It is also useful if you are an athlete at the start of a race. But it does not help at all if you are on your way to the dentist, or watching a horror film.

Like most hormones, adrenalin breaks down very quickly after it is released, so its effects do not last long. If you need to go on feeling frightened, then your brain will keep telling the adrenal glands to secrete more adrenalin.

Fact!

The animal with the heaviest brain is the sperm whale. Its brain weighs about 9.2 kg.

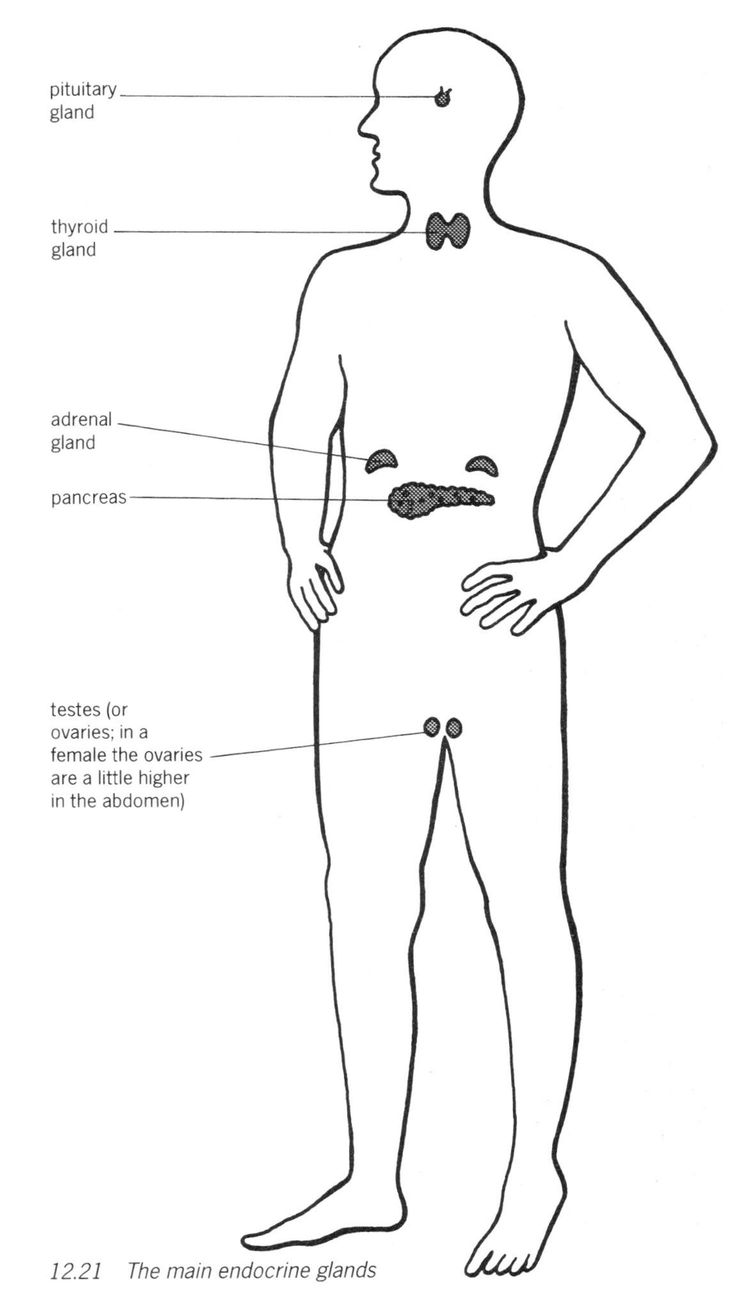

12.21 The main endocrine glands

12.26 The thyroid regulates metabolism and growth.

The thyroid gland secretes **thyroxine.** Thyroxine is secreted almost all the time, but in quite small amounts. Thyroxine helps to control the metabolic rate, mostly by regulating the speed at which mitochondria break down glucose in respiration. This is particularly important in children. If a child does not have enough thyroxine, it will not grow properly, and its brain will not develop. A child like this is called a **cretin.** Cretinism can be cured by giving injections of thyroxine.

Adults who are short of thyroxine are sluggish, and tend to be overweight. Too much thyroxine makes a person overactive, thin and edgy.

Thyroxine contains iodine. A lack of iodine in a diet will mean that the thyroid gland cannot make enough thyroxine. To compensate for this, the thyroid gland may get bigger, forming a swelling or **goitre.**

12.27 Insulin regulates blood sugar levels.

The pancreas is two glands in one. Most of it is an ordinary gland with a duct. It makes pancreatic juice, which flows along the pancreatic duct into the duodenum (see section 2.36).

Scattered through the pancreas, however, are groups of cells called **islets of Langerhans.** These cells do not make pancreatic juice. They make a hormone called **insulin.** The insulin is secreted directly into the blood, so the islets of Langerhans are an endocrine gland.

Insulin helps the liver to control the amount of glucose in the blood. If you eat a meal containing a lot of

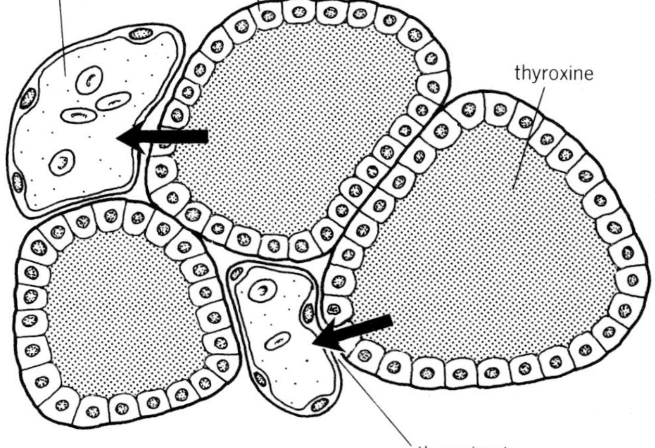

12.22 Section through the thyroid gland

Table 12.1 Mammalian endocrine glands and hormones.

Hormone	*Gland which secretes it*	*When secreted*	*Function*	*Other points*
Adrenalin	adrenal gland	in small amounts all the time; in large amounts when frightened.	prepares the body for fight or flight	
Thyroxine	thyroid gland	throughout life	controls metabolic rate, especially respiration in mitochondria	thyroxine contains iodine; lack of thyroxine in childhood causes cretinism.
Insulin	islets of Langerhans in pancreas	when blood glucose level rises above normal	causes liver and muscles to take up glucose, so restoring blood glucose level to normal	lack of insulin causes diabetes
Testosterone	testes	in small quantities throughout life; in larger quantities from puberty onwards	controls development of male sex organs and secondary sexual characteristics	
Oestrogen	ovaries	in small quantities throughout life; in larger quantities from puberty onwards, particularly when follicle is developing in ovary.	controls development of female sex organs and secondary sexual characteristics; causes lining of uterus to get thick and spongy	
Progesterone	corpus luteum	after ovulation	maintains lining of uterus	if placenta does not secrete enough progesterone, a miscarriage may occur
	placenta	throughout pregnancy		
ADH (Anti-diuretic hormone)	pituitary gland	when quantity of water in blood gets too low	causes kidneys to reabsorb water from urine	
Thyroid stimulating hormone	pituitary gland	throughout life	causes thyroid gland to secrete thyroxine	
Growth hormone	pituitary gland	throughout life, especially during growing period	stimulates growth	lack of growth hormone causes dwarfism; too much causes gigantism

glucose, the level of glucose in the blood goes up. The islets of Langerhans in the pancreas detect this, and secrete insulin into the blood.

When the insulin arrives at the liver and muscles, it makes them absorb glucose from the blood, and use it up. Some of the glucose is used in respiration. Some of it is built up into glycogen molecules, which are stored in the liver and muscle cells.

When the blood glucose level has dropped to the right level, the pancreas stops secreting insulin. The liver and muscles stop using up glucose from the blood so quickly, so the level returns to normal (see Fig 12.23).

The control of blood glucose is an important part of homeostasis (see section 10.19).

12.28 Male sex hormones.

Male sex hormones are called **androgens.** The most important androgen is **testosterone.** Testosterone is made in the testes.

Testosterone and other androgens regulate the development of the male sex organs. They also control the development of the male secondary sexual characteristics (see section 8.23).

12.29 Female sex hormones.

Female sex hormones are called **oestrogens.** They regulate the development of the female sex organs, and the female secondary sexual characteristics.

Whereas male mammals make sperms all the time, females only produce eggs at certain times. In humans, ovulation (see section 8.12) happens once a month. Ovulation is part of the **menstrual cycle.** The menstrual cycle is controlled by hormones (see Fig 12.24).

Fig 8.25 illustrates what happens during the human menstrual cycle. First, a follicle develops inside an ovary. The developing follicle secretes a hormone called **oestrogen.** The oestrogen makes the lining of the uterus grow thick and spongy.

When the follicle is fully developed, ovulation takes place. The follicle stops secreting oestrogen. It becomes a **corpus luteum.** The corpus luteum starts to secrete another hormone, called **progesterone.**

Progesterone keeps the uterus lining thick, spongy, and well supplied with blood, in case the egg is fertilised. If it is not fertilised, then the corpus luteum gradually disappears. Progesterone is not secreted any more, and so the lining of the uterus breaks down.

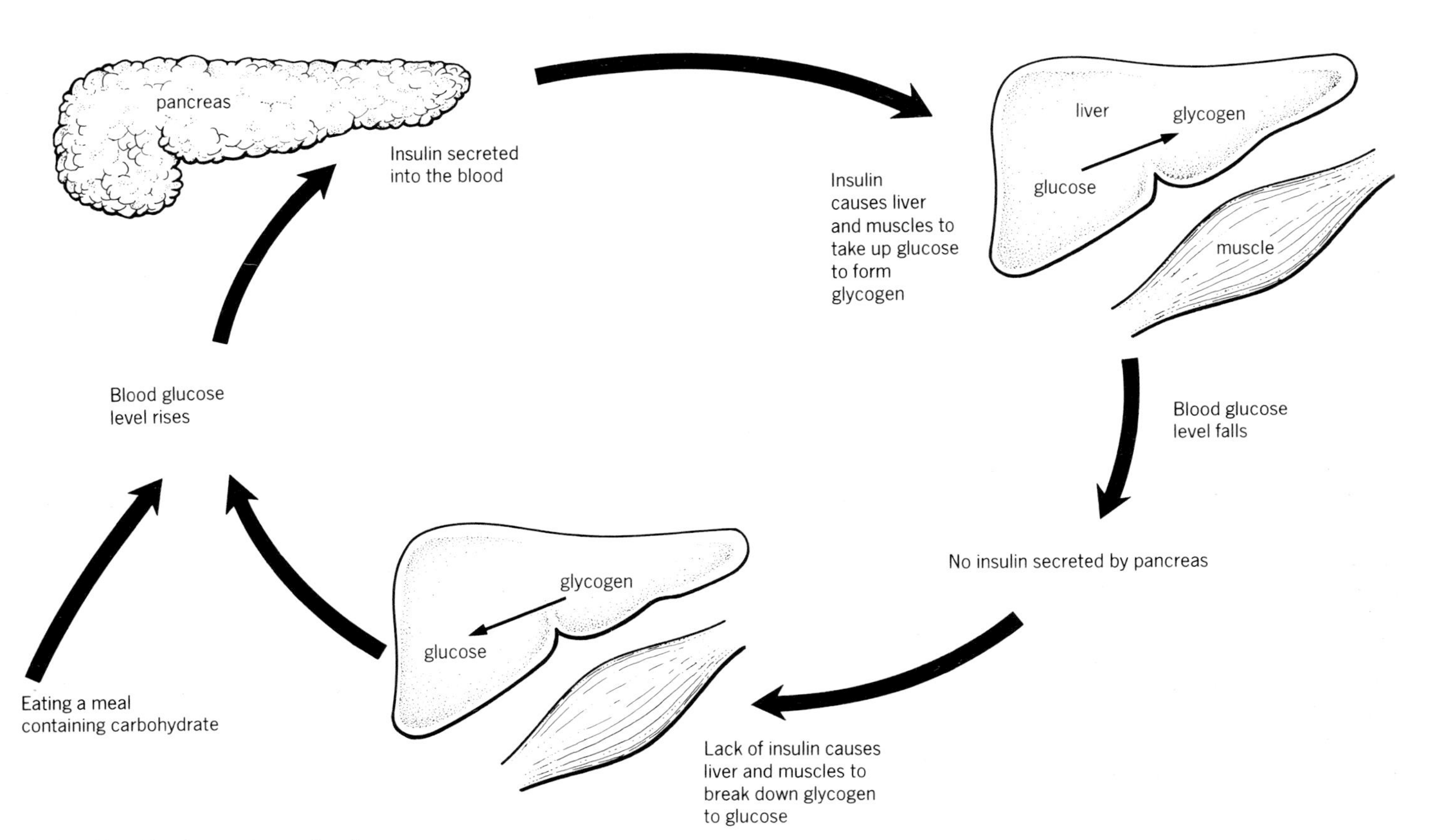

12.23 How blood sugar is regulated

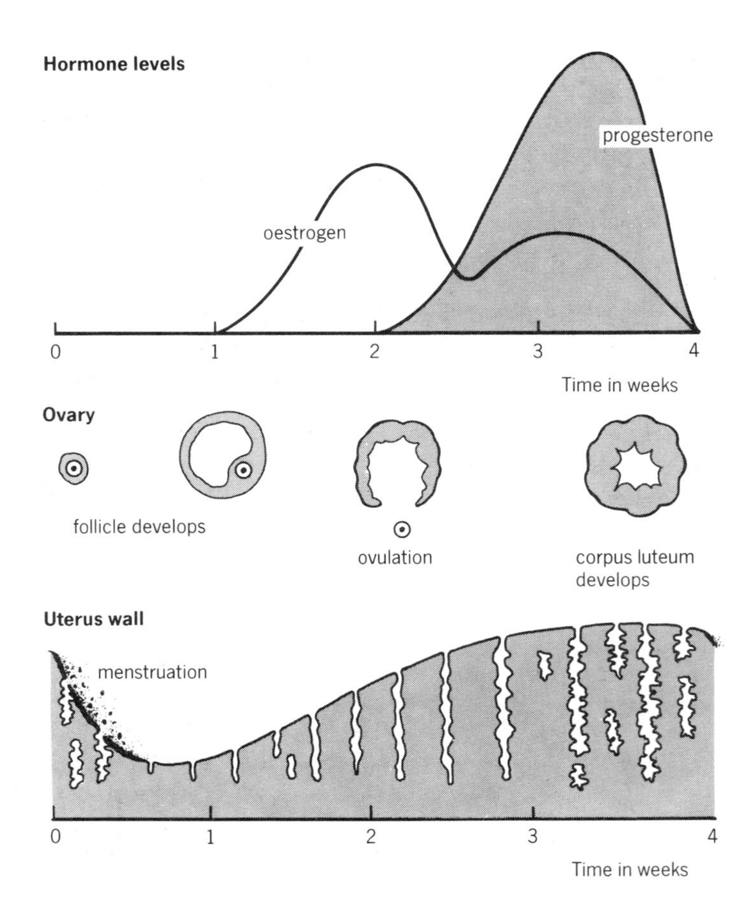

12.24 Hormones and the menstrual cycle

Menstruation happens. A new follicle starts to develop in the ovary, and the cycle begins again.

But if the egg is fertilised, the corpus luteum does not degenerate so quickly. It carries on secreting progesterone until the embryo sinks into the uterus wall, and a placenta develops. Then the placenta secretes progesterone, and carries on secreting it all through the pregnancy. The progesterone maintains the uterus lining, so that menstruation does not happen during the pregnancy.

12.30 The pituitary gland controls other glands.

The pituitary gland is in the centre of the head. It is attached to the part of the brain called the **hypothalamus.**

The pituitary gland secretes a large number of hormones. For example, when receptor cells in the hypothalamus sense that there is not enough water in the blood, they send messages along nerves to the pituitary gland. The pituitary gland then secretes ADH. ADH stops the kidneys allowing too much water to leave the body in the urine (see section 10.17).

Many of the hormones secreted by the pituitary gland control the other endocrine glands. **Thyroid stimulating hormone,** for example, makes the thyroid gland secrete thyroxine.

The pituitary gland also secretes **growth hormone.** Growth hormone stimulates the growth of the body, partly by causing proteins to be built up in cells. Sometimes, not enough growth hormone is secreted during childhood. If this happens, the child may develop into a dwarf.

Table 12.2 A comparison of the nervous and endocrine systems in a mammal.

Nervous system	*Endocrine system*
Made of neurones	Made of secretory cells
Messages transmitted in the form of electrical impulses	Messages transmitted in the form of chemicals called hormones
Messages transmitted along nerve fibres	Messages transmitted through the blood system
Messages travel very quickly	Messages travel more slowly
Effect of message usually only lasts a very short while	Effect of messages usually lasts longer

Questions

1 How do endocrine glands differ from other glands?
2 Describe three effects of adrenalin, and explain the value of each one.
3 Why is it important that adrenalin is broken down very quickly in your body?
4 Why do you need iodine in your diet?
5 Where are the islets of Langerhans?
6 Explain what happens when you eat a meal containing a lot of carbohydrate.
7 Which female hormone is secreted by a follicle as it develops inside an ovary?
8 What effect does this hormone have?
9 Which hormone is secreted by a corpus luteum?
10 What effect does this hormone have?
11 Name one other structure which secretes this hormone.
12 The pituitary gland is sometimes called 'the master gland'. Suggest why this is.

Fact!

An average female human may ovulate as many as 500 times during her life.

Co-ordination in plants

12.31 Tropisms are directional growth responses.

Most plants cannot respond to stimuli as quickly as animals can. They respond more slowly, usually by growing. They grow either towards or away from the stimulus. This sort of response is called a **tropism.**

Two important stimuli for plants are light and gravity. For example, the shoot of a plant grows towards light. This is called **phototropism** ('photo' means light). Because the shoot grows towards the light, the response is called **positive phototropism.**

Plants can respond to gravity by growing either towards or away from the centre of the Earth. This is called **geotropism.** Shoots tend to grow away from the pull of gravity. This is called **negative geotropism** because the shoot is growing away from the stimulus.

Roots are **positively geotropic** – they grow towards the pull of gravity (see Figure 12.26). Some roots also respond to light by growing away from it, which means that they are **negatively phototropic.**

Investigation 12.4 To find out how shoots respond to light

1. Label three petri dishes A, B and C. Line each with moist cotton wool or filter paper, and sprinkle on some mustard seeds.
2. Leave all three dishes in a warm place for a day or two, until the seeds begin to germinate. Check that they do not dry out.
3. Now put dish A into a light-proof box with a slit in one side, so that the seedlings get light from one side only.
4. Put dish B onto a **clinostat** (see Fig 12.25) in a light place. The clinostat will slowly turn the seedlings around, so that they get light from all sides equally.

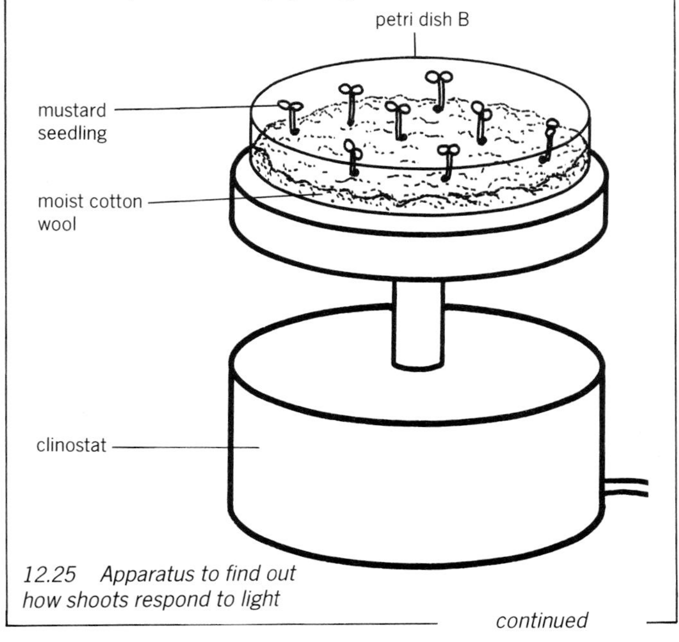

12.25 Apparatus to find out how shoots respond to light

continued

5. Put dish C into a completely light-proof box.
6. Leave all the dishes for a week, checking that they do not dry out.
7. Make labelled drawings of one seedling from each dish.

Questions

1. How had the seedlings in A responded to light from one side? What is the name for this response?
2. Why was dish B put onto a clinostat, and not simply left in a light place?
3. Explain what happened to the seedlings in dish C.
4. What was the control in this experiment?

Whichever way up a seed is planted, its radicle always grows downwards.

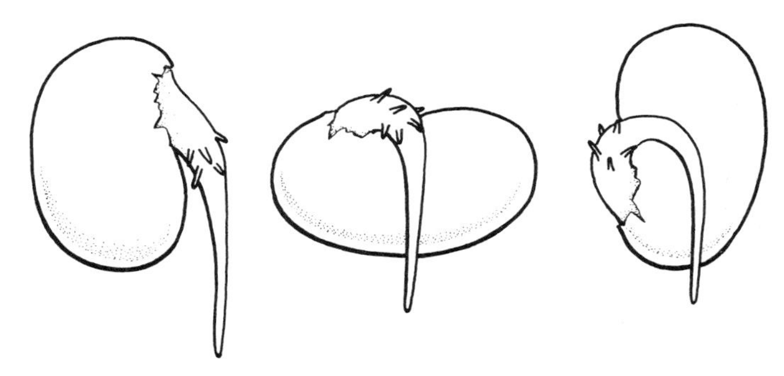

12.26 Positive geotropism in roots

Investigation 12.5 To find out how roots respond to gravity

1. Germinate several broad bean seeds, as in Investigation 7.2. Leave them until their radicles are about 2 cm long.
2. Line the containers of two clinostats with blotting paper. Dampen the paper.
3. Cover the cork discs on the clinostats with wet cotton wool.
4. Choose about eight bean seeds with straight radicles. Pin four onto each disc, with their radicles pointing straight outwards (see Fig 12.27). Put the containers lined with blotting paper over them.
5. Turn both clinostats on their sides. Switch on clinostat B.
6. Leave both clinostats for a few days. Check the beans occasionally to see that they have enough water.
7. Make labelled drawings of the bean seedlings from each clinostat.

continued

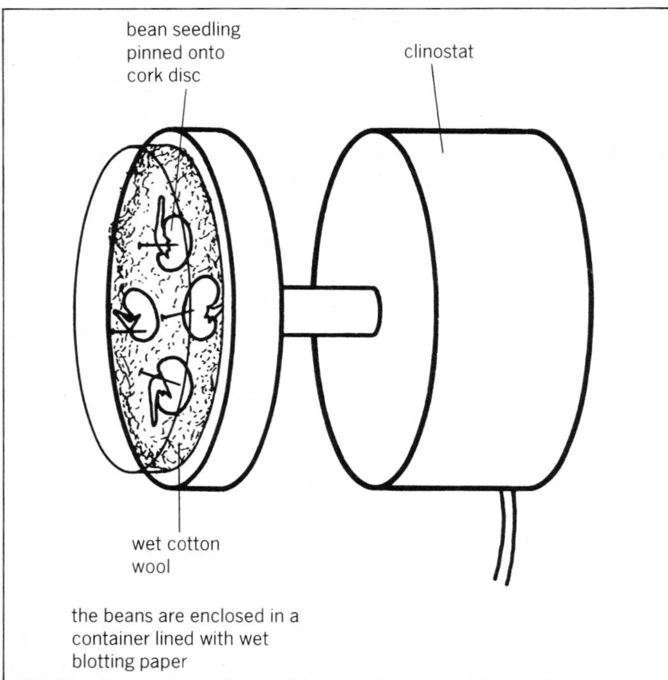

12.27 Apparatus to find out how roots respond to gravity

Questions

1 How had the seedlings in clinostat A responded to the pull of gravity? What is the name for this response?
2 What was the purpose of clinostat B?
3 Design an experiment to find out how shoots respond to gravity. Give full instructions for the method, including labelled diagrams. What results would you expect?

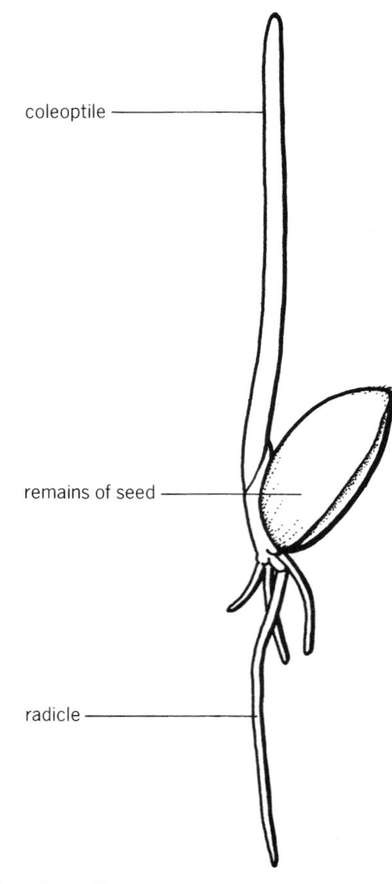

12.28 An oat seedling

Investigation 12.6 To find which part of a shoot is sensitive to light

1 Germinate several oat seeds in 3 pots, labelled A, B and C. Space the seeds well out from one another. The seeds will grow shoots called **coleoptiles** (see Fig 12.28).

Lightproof boxes, allowing light in from one side only.

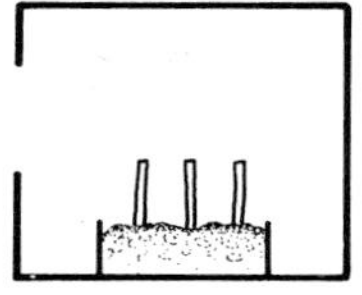

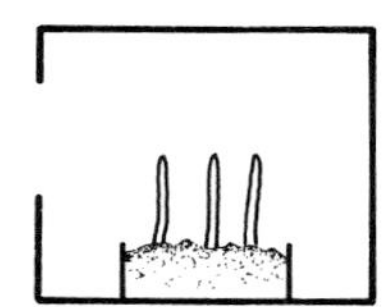

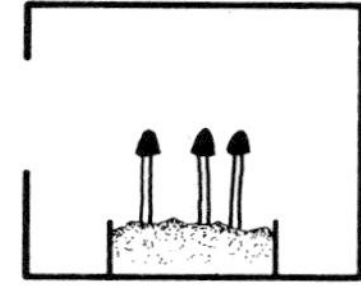

12.29 Apparatus to find which part of a shoot is sensitive to light

2 Cut the tips from each coleoptile in pot A.
3 Cover the tips of each coleoptile in pot B with foil.
4 Measure the length of each coleoptile in each pot. Find the average length of the coleoptiles in each pot, and record it.
5 Put pots A, B and C into light-proof boxes with light shining in from one side. Leave them for several days.
6 Find the new average length of the coleoptiles in each pot. Compare it with the original average length, to see whether the coleoptiles have grown or not.
7 Copy out the results table and fill it in.

Results table

Pot	A	B	C
Treatment	tips removed	tips covered	tips left alone
Did they grow?			
Did they grow towards the light?			

Questions

1 Explain why some coleoptiles grew, and some did not.
2 Which coleoptiles grew towards the light, and which did not? Explain why.

12.32 Tropisms aid plant survival.

It is very important to the plant that its roots and shoots behave like this. Shoots must grow upwards, away from gravity and towards the light, so that the leaves are held out into the sunlight. The more light they have, the better they can photosynthesise. Flowers, too, need to be held up in the air, where insects or the wind can pollinate them.

Roots, though, need to grow downwards, into the soil in order to anchor the plant in the soil, and to absorb water and minerals from between the soil particles.

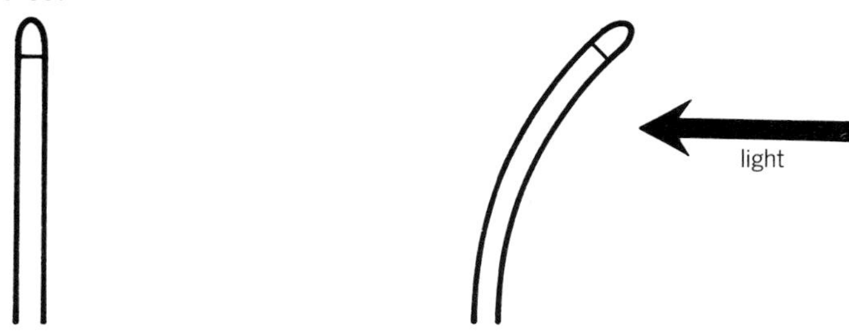

If the tip of a coleoptile is cut off and then replaced, the coleoptile will still grow towards the light.

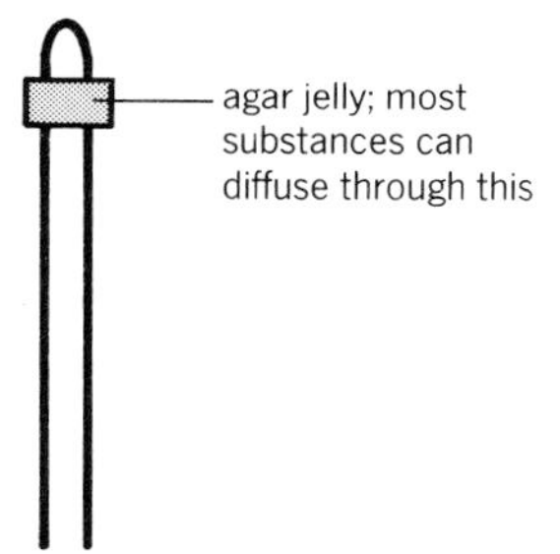

If the tip is cut off, and separated from the rest of the coleoptile by a piece of agar jelly, the coleoptile still grows towards the light.

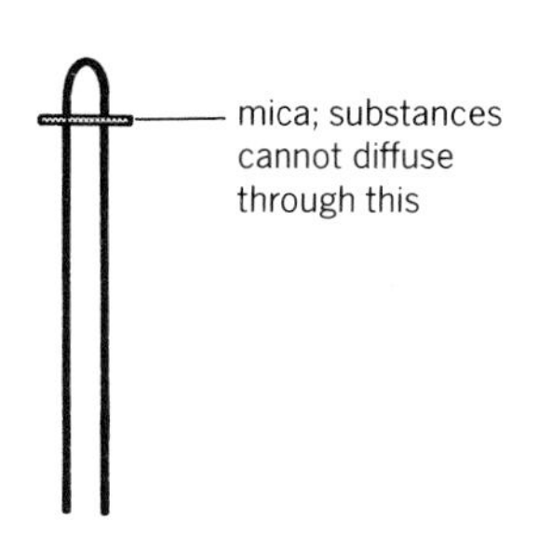

But if a piece of mica separates the tip from the rest of the coleoptile, then it does not grow towards the light.

This suggests that the response to light is caused by a substance which is made in the tip, and diffuses down the coleoptile.

12.30 An experiment investigating the method by which shoots respond to light

12.33 How a shoot responds to light.

In section 12.13, we saw that for an organism to respond to a stimulus, there must be a **receptor** to pick up the stimulus, an **effector** to respond to it, and some kind of **communication system** in between. In mammals, the receptor is often part of a sense organ, and the effector is a muscle or gland. Messages are sent between them along nerves, or sometimes by means of hormones.

Plants, however, do not have complex sense organs, muscles or nervous systems. So how do they manage to respond to stimuli like light and gravity?

Fig 12.30 shows an experiment that can be done to find out which part of a shoot picks up the stimulus of light shining onto it. The sensitive region is the tip of the shoot. This is where the receptor is.

The results of Investigation 12.6 will probably show that the part of the shoot which responds to the stimulus is the part just below the tip. This is the effector.

These two parts of the shoot must be communicating with one another somehow. They do it by means of **hormones.**

12.34 Changes in auxin concentration cause phototropisms.

One kind of plant hormone is called **auxin.** Auxin is being made all the time by the cells in the tip of a shoot. The auxin diffuses downwards from the tip, into the rest of the shoot.

Auxin makes the cells just behind the tip get longer. The more auxin there is, the faster they will grow. Without auxin, they will not grow.

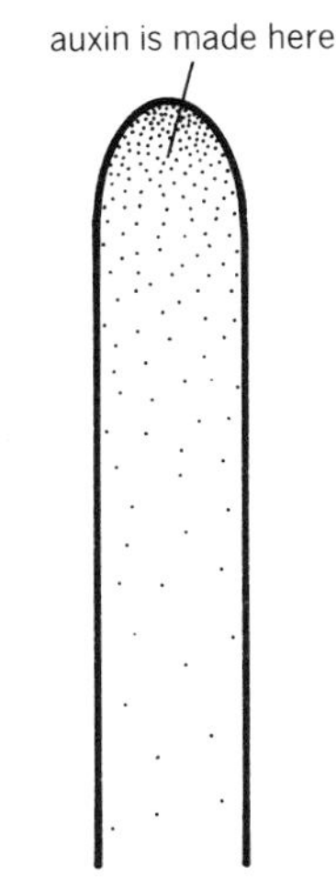

Auxin made in the tip diffuses unevenly down the shoot, concentrating on the shady side.

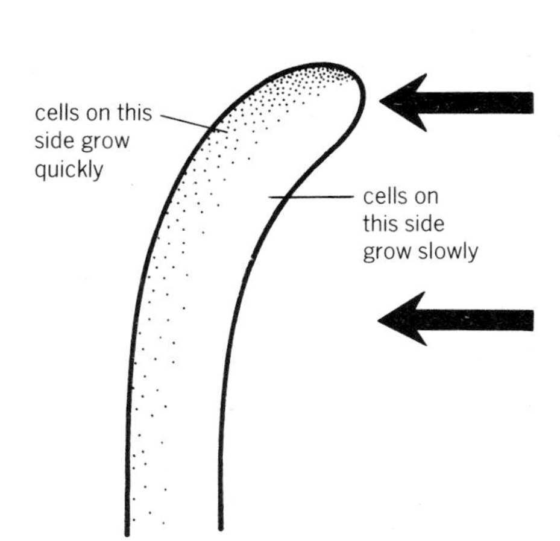

The uneven concentration of auxin causes the shady side to grow faster than the light side, so the shoot bends towards the light.

12.31 Auxin and phototropism

When light shines onto a shoot from one side, the auxin at the tip concentrates on the shady side (see Fig 12.31). This makes the cells on the shady side grow faster than the ones on the bright side, so the shoot bends towards the light.

Investigation 12.7 To find how auxin affects shoots

In this experiment, you will use a kind of auxin called indole acetic acid, or IAA. When you put it onto a shoot, you need to mix it with lanolin, so that it will stick on.

1. Germinate some oat coleoptiles in three pots, labelled A, B and C.
2. Mix some IAA with a little warm lanolin. Gently smear the mixture down one side only of each coleoptile in pot A. Put the IAA on the same side of each coleoptile. Put a label in the pot to show which side of the coleoptiles the IAA was put on.
3. Do the same with the coleoptiles in pot B, but use pure lanolin, with no IAA in it.
4. Put all three pots onto clinostats in a light place, and leave them for about a day.

Questions

1. What has happened to the coleoptiles in pots A, B and C? Explain why.
2. What was the reason for smearing the coleoptiles in pot B with lanolin?
3. Why were all the pots put onto clinostats?

12.35 Plants become etiolated in the dark.

If you did Investigation 12.4, you will probably have found that the seedlings which had no light looked very different from the others. Plants without light become yellow and spindly. They grow very tall and thin, and have smaller leaves, which are further apart than in a normal plant. Plants like this are said to be **etiolated.**

Plants in darkness grow like this because fast, upwards growth gives them the best chance of reaching light. As they have no light, they cannot photosynthesise, so they lose their chlorophyll. If they do not succeed in reaching light, they will eventually die.

Questions

1. What is a tropism?
2. Which parts of a plant show positive phototropism?
3. Which parts of a plant show positive geotropism?
4. Which part of a shoot is sensitive to light?
5. Which part of a shoot responds to light?
6. How do these parts communicate with one another?
7. Describe three features of an etiolated shoot.

Chapter revision questions

1. Explain the difference between each of the following pairs of terms, giving examples whenever they make your answer clearer.
   (a) cornea, conjunctiva;
   (b) choroid, sclera;
   (c) neurone, nerve;
   (d) receptor, effector;
   (e) endolymph, perilymph;
   (f) sensory neurone, motor neurone;
   (g) cerebrum, cerebellum;
   (h) thyroxine, thyroid stimulating hormone;
   (i) oestrogen, progesterone;
   (j) negative geotropism, positive geotropism.
2. If you walk from a brightly lit street into dark room, your pupil will rapidly dilate.
   (a) What type of action is this?
   (b) Using each of these words at least once, but not necessarily in this order, explain how this reaction is brought about.
   synapse, receptor, motor neurone, sensory neurone, relay neurone, radial muscles.
   (c) As well as the muscles in the iris, the eye also contains muscles in the ciliary body. What is their function?
3. (a) Make a large, labelled diagram of a nerve cell or neurone.
   (b) List three ways in which this cell is similar to other animal cells.
   (c) List three ways in which this cell differs from other animal cells, explaining how each of these differences enable it to perform its function efficiently.
4. (a) Why do body cells need glucose?
   (b) In healthy humans, the blood contains 60–110 mg of glucose per 100 ml of blood. Which gland secretes the hormone which is responsible for keeping this level fairly constant?
   (c) The graph in Fig 12.32 shows the changes in blood glucose level after a meal. Explain the shape of the graph between (i) A and B, (ii) B and C, and (iii) C and D.
   (d) People with the disease diabetes mellitus cannot make insulin. Why is it dangerous for diabetics to eat a meal containing a lot of sugar?
   (e) The regulation of blood glucose levels is one example of homeostasis. What is homeostasis, and why is it important?

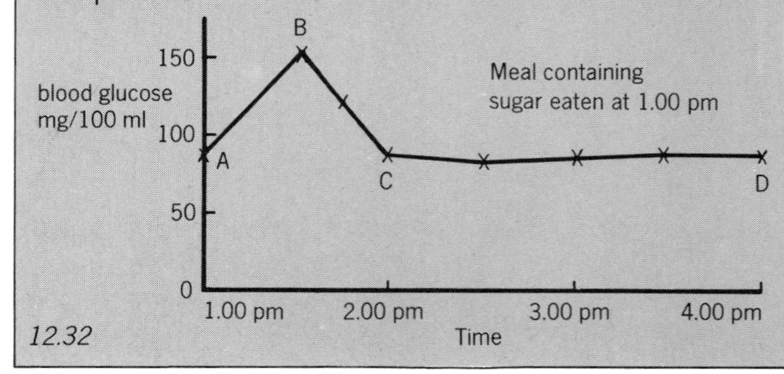

12.32

13 Living organisms in their environment

13.1 Organisms interact with their environment.

One very important way of studying living things is to study them where they live. Animals and plants do not live in complete isolation. They are affected by their surroundings, or **environment.** Their environment is also affected by them. The study of the interaction between living organisms and their environment is called **ecology.**

You cannot really learn about ecology without doing a lot of practical work outside. The information in this chapter and the next should help you to interpret what you find out from your practical work.

13.2 Some important words.

There are many words used in ecology with which you need to be familiar. The area where an organism lives is called its **habitat.** The habitat of a tadpole might be a pond. There will probably be many tadpoles in the pond, forming a **population** of tadpoles. A population is a group of organisms of the same species.

But tadpoles will not be the only organisms living in the pond. There will be many other kinds of animals and plants making up the pond **community.** A community is all the organisms, of all the different species, living in the same habitat.

The living organisms in the pond, the water in it, the stones and the mud at the bottom, make up an **ecosystem.** An ecosystem consists of a community and its environment.

Within the ecosystem, each living organism has its own life to live and role to play. The way in which an organism lives its life in an ecosystem is called its **niche.** Tadpoles, for example, eat algae and other weeds in the pond; they disturb pebbles and mud at the bottom of shallow areas in the pond; they excrete ammonia into the water; they breathe in oxygen from the water, and breathe out carbon dioxide. All these things, and many others, help to describe the tadpoles' role, or niche, in the ecosystem.

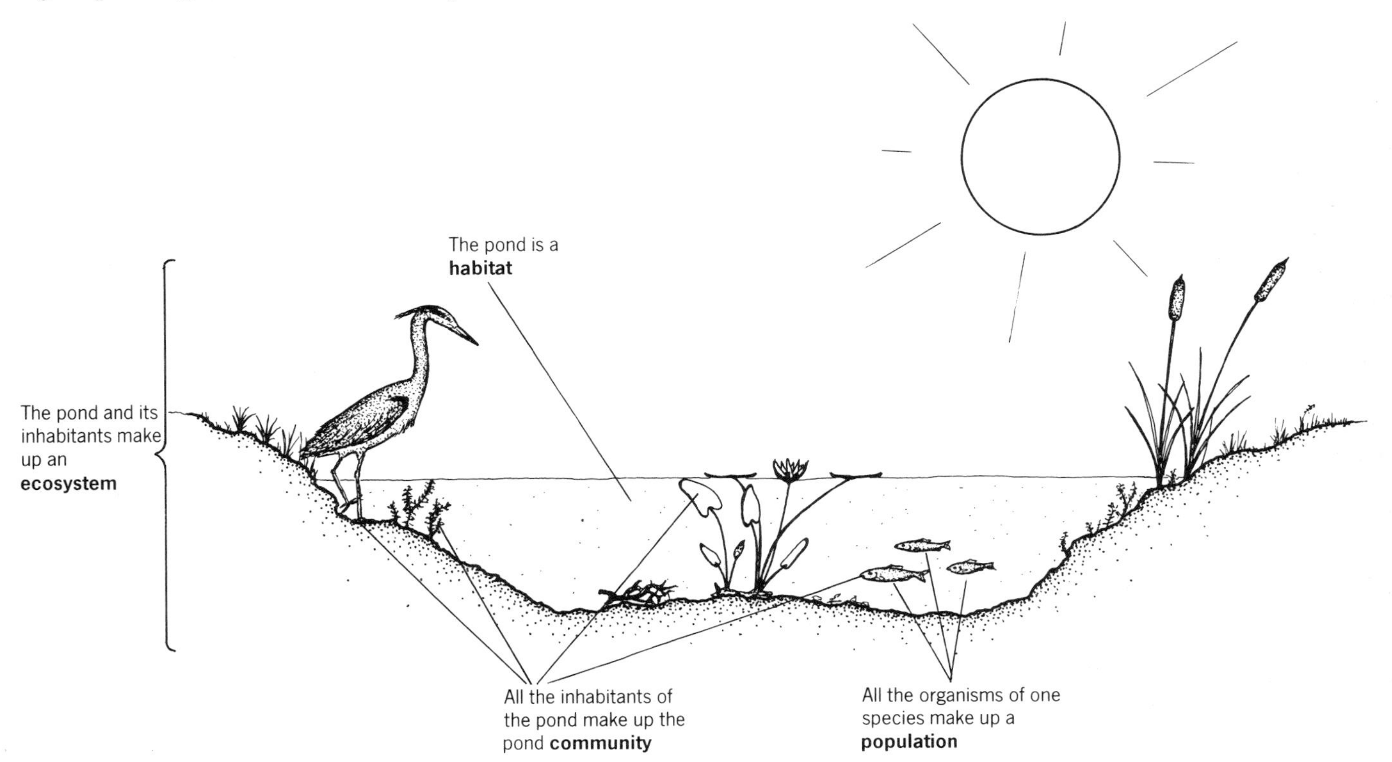

13.1 A pond and its inhabitants–an example of an ecosystem

Use this key to identify these five animals

1	Aquatic animal, with flippers	**Seal**
	Terrestrial animals	**2**
2	Dog-like; back not very supple	**Fox**
	Not dog-like; supple body	**3**
3	Flank light; dark underneath	**Polecat**
	Upper side brown; underside yellow	**4**
4	Black tip to tail	**Stoat**
	No black tip to tail	**Weasel**

Questions

1 What is ecology?
2 What is a population?
3 Give two examples of an ecosystem, other than a pond.
4 What is a niche?

Studying ecosystems

13.3 Keys are used for identification.

There are many different ways of studying ecosystems. But whatever ecosystem you study, and however you decide to study it, you will have to begin by identifying the living organisms in it.

Some of them, particularly the bigger ones, you may be able to identify quite quickly from pictures in books. But there will almost certainly be many that you cannot find pictures of, or where you are not sure if the picture really is of your plant or animal.

When this happens, you will need to use a **key.** A key is a way of leading you through to the name of your organism by giving you two descriptions at a time, and asking you to choose between them. Each choice you make then leads you onto another pair of descriptions, until you end up with the name of your organism.

An example of a key like this is shown in Fig 13.2. It is called a **dichotomous key,** because each time you choose between *two* descriptions ('di' means two).

13.4 Sampling can estimate abundance.

When you have identified as many of the organisms as possible in your ecosystem, you can make a list of them, called a **species list.** The next job will probably be to try to find out how many of each species live there.

Sometimes, you can simply count them. You could

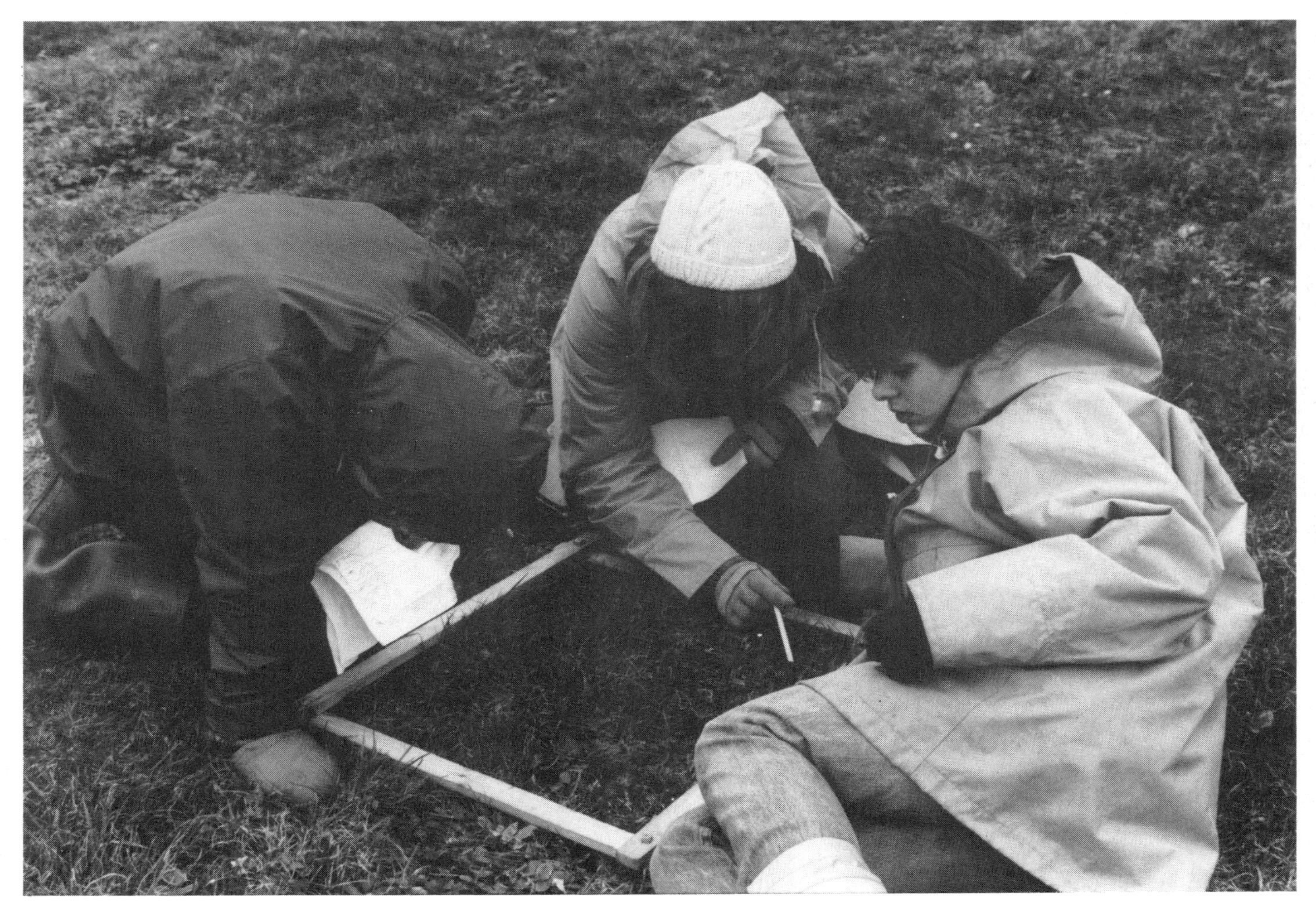

13.3 A quadrat being used to estimate the relative abundance of different plant species on a lawn

find how many oak trees there were in a small wood just by counting them. Often it is not quite that easy. If you are studying part of a field, for example, you could not possibly count every buttercup plant. You will have to take a **sample** of the field, and count the numbers of each species in that sample. If you work out the scale factor of the sample to the whole field, you could then get an indication of the numbers of each species in the field.

13.5 Quadrats are used to sample plant cover.

One very useful way of taking a sample is to use a **quadrat.** A quadrat is a square. It can be any size, but one with sides of about 0.5 m is a convenient size to use in a field.

The quadrat is put down onto the ground, and the numbers of each species of plant inside it are counted. With some species, like grasses, though, it is impossible to say where one plant stops and another begins. In this case, you can estimate what percentage of the quadrat area is covered by grass, and by other plants.

If the plants in your quadrat are quite tall, there may be more than one layer of plants. In this case, the total of all your percentages may be more than 100%.

13.6 Sampling should be random.

Your quadrat sample only gives you an idea of the numbers of plants in one small area. You cannot guarantee that that area is representative of the whole field. You will need to do many quadrats, and average your results from each, to be sure of getting a representative sample.

The placing of your quadrat is very important. If you just choose where to put it, the part of the field full of gorse bushes and nettles, with a bull standing behind them, will probably not get sampled very often! So you must use some way of placing your quadrats randomly in the field.

There are several ways of doing this. One way is to divide the piece of ground into squares and use pairs of random numbers as co-ordinates. These you can get from tables. If your numbers are 12,8 for example, you could go twelve squares forward from a corner along an edge, and then eight squares out into the field, and put your quadrat down at that point.

13.7 Transects sample changes between habitats.

Another way of sampling the distribution of organisms in your field is to use a **transect.** A transect is a line crossing the field. You can use a long tape measure to mark the transect. You then record the species of plants touching the tape.

Often, it would take far too long to record all the plants touching the tape. Instead, you might record them at intervals, say every 10 cm.

Transects are particularly useful where one kind of habitat is changing into another. You could use one, for example, where a grassy field merged into a wood, or into a stream. A transect will give you information about how the numbers and kinds of species change, as the environment changes.

13.8 Mark, release, recapture estimates numbers.

Quadrats and transects are very useful ways of finding out how many organisms of different species are living in a habitat. But they can only be used with organisms which stay in one place for most of the time. This usually means plants, though on a seashore you can also count limpets, barnacles, sea anemones and many other animals in this way.

You need a different method for estimating the numbers of animals that move around a lot. One method is the **mark, release, recapture technique.** It works so long as there are reasonable numbers of each kind of animal, and so long as they move around quite freely.

Suppose that you wanted to estimate the size of a population of woodlice. First, you need to capture a sample of perhaps 30 woodlice. Each woodlouse is marked with a small spot of waterproof paint, and then released.

The woodlice are then left alone for about a day, to give the marked ones a chance to become mixed up with any unmarked ones. You then capture a second sample, of as many woodlice as you can. Count the total number, and the number of marked ones.

Suppose that you caught 100 woodlice in your second sample, and 10 of them had been marked. You have recaptured 10 of the 30 you originally marked, or ⅓ of them. So it is probable that you have caught about ⅓ of the whole woodlouse population. The size of the population will therefore be about 3 × 100 woodlice, that is 300.

In general, the formula for working this out is

$$\text{number of animals caught the first time} \times \frac{\text{number of animals caught the second time}}{\text{number of marked animals caught the second time}}.$$

Questions

1. When using a quadrat, how can you estimate the amount of a plant such as grass, in your sample?
2. Why is it important to place quadrats randomly?
3. When might you use a transect?
4. 50 water beetles were caught and marked, before being returned to their pond. The next day, another 50 water beetles were caught, 10 of which had been marked. About how many water beetles were in the pond altogether?

13.4 Transects being taken up a shingle shore to measure the changes in the composition of plant communities away from the sea

Diagram of area being studied

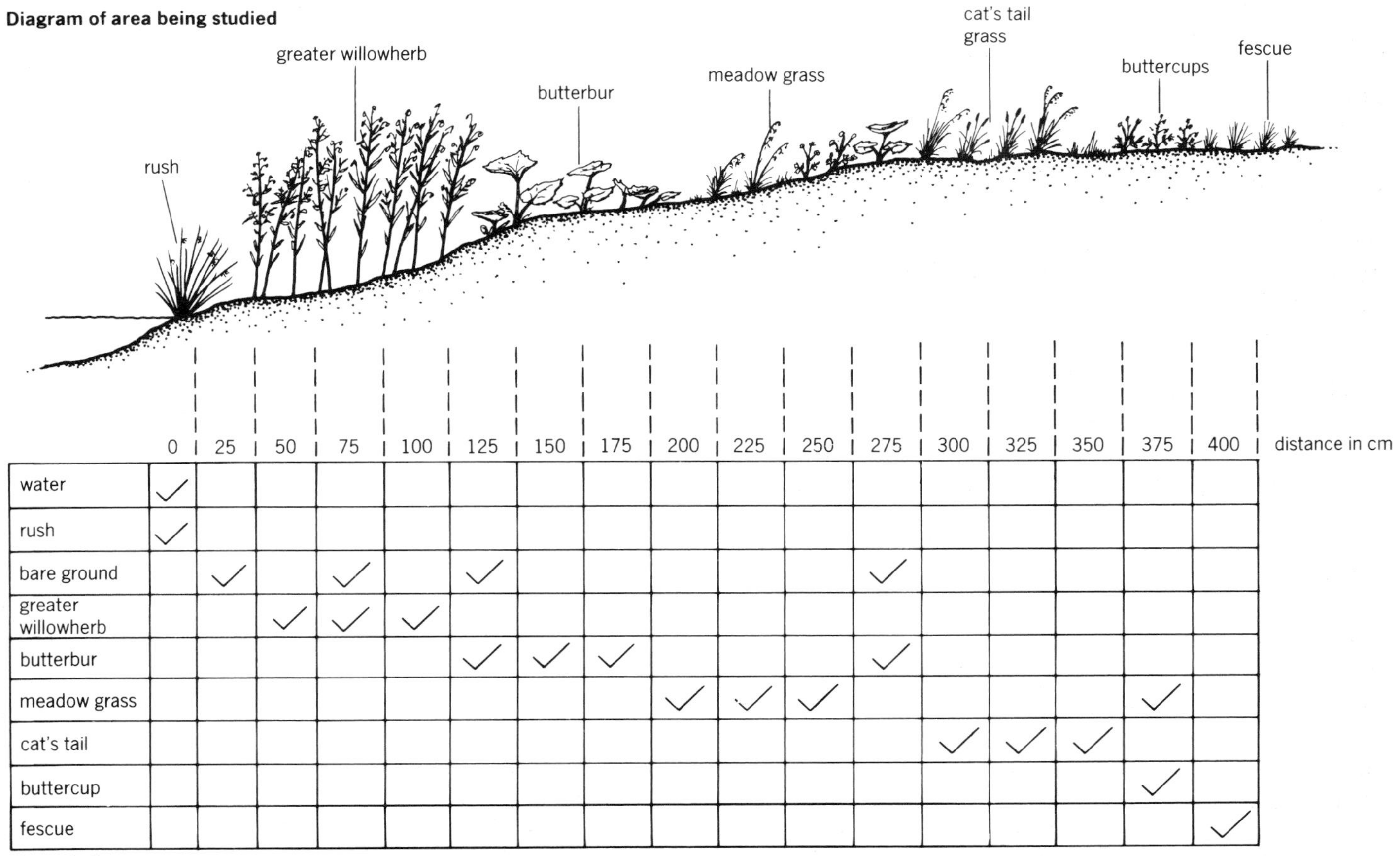

	0	25	50	75	100	125	150	175	200	225	250	275	300	325	350	375	400
water	✓																
rush	✓																
bare ground		✓		✓		✓						✓					
greater willowherb			✓	✓	✓												
butterbur						✓	✓	✓				✓					
meadow grass									✓	✓	✓					✓	
cat's tail													✓	✓	✓		
buttercup																✓	
fescue																	✓

Record of transect

13.5 A transect recording changes in vegetation from wet to dry ground

Investigation 13.1 Estimating the size of a bead population, using the mark, release, recapture technique.

1 Fill a bucket or large tray with a large number of beads, all the same colour and size.
2 Capture a sample of about 50 beads.
3 'Mark' the captured beads, by exchanging them for beads of a different colour. Return the marked beads to the population.
4 Thoroughly mix the marked beads with the rest of the population.
5 Capture a second, quite large but random, sample of beads. It may be best to use a blindfold for this sample to ensure that it is random.
6 Count (a) the number of marked beads in your second sample, and (b) the total number of beads in this sample.
7 Work out the estimated population of beads using the formula

$$\text{number of beads in first sample} \times \frac{\text{number of beads in second sample}}{\text{number of marked beads in second sample}}$$

8 Now count the actual number of beads in the population.

Questions

1 How close was your estimate to the actual number of beads in the population? Do you consider it was close enough for this to be a useful technique?
2 When using the technique in the field, how could you ensure that your estimate came as close as possible to the real size of the population?
3 For which of these populations would this method be suitable?
(a) snails in a small garden
(b) rabbits in a hedgerow
(c) dandelions in a lawn
(d) killer whales in the Atlantic Ocean
(e) lichens on a tree trunk
Give reasons for your answers, and suggest alternative methods if you do not think that mark, release, recapture would be suitable.

Food and energy in an ecosystem

13.9 Energy passes along food chains.

All living organisms need energy. They get energy from food, by respiration. All the energy in an ecosystem comes from the sun. Some of the energy in sunlight is captured by plants, and used to make food – glucose, starch and other organic substances such as fats and proteins. These contain some of the energy from the sunlight. When the plant needs energy, it breaks down some of this food by respiration.

Animals get their food, and therefore their energy, by eating plants, or by eating animals which have eaten plants.

The sequence by which energy, in the form of food, passes from a plant to an animal and then to other animals, is called a **food chain.** Fig 13.6 shows one example of a food chain.

13.10 Consumers use food made by producers.

Every food chain begins with green plants because only they can capture the energy from sunlight. They are called **producers,** because they produce food.

Animals are **consumers.** An animal which eats plants is a **primary consumer,** because it is the first consumer in a food chain. An animal which eats that animal is a **secondary consumer,** and so on along the chain.

13.11 Food chains are usually short.

As the energy is passed along the chain, each organism uses some of it. So the further along the chain you go, the less energy there is. There is plenty of energy available for producers, so there are usually a lot of them. There is less energy for primary consumers, and less still for secondary consumers. This means that towards the end of the food chain, the organisms get fewer in number, or smaller in total size.

The loss of energy along the food chain also limits the length of it. There are rarely more than five links in a chain, because there is not enough energy left to supply the next link. Many food chains only have three links.

13.12 Consumers feed at different trophic levels.

In Fig 13.8, the number of organisms in the food chain is shown as a pyramid. The size of each block in the pyramid represents the number of organisms. It is called a **pyramid of numbers.** Each level in the pyramid is called a **trophic level** ('trophic' means feeding).

Many organisms feed at more than one trophic level. You, for example, are a primary consumer when you eat vegetables, a secondary consumer when you eat meat or drink milk, and a tertiary consumer when you eat a predatory fish such as a salmon.

13.13 Understanding energy flow helps agriculture.

Understanding how energy is passed along a food chain can be useful in agriculture. We can eat a wide variety of foods, and can feed at several different trophic levels. Which is the most efficient sort of food for a farmer to grow, and for us to eat?

The nearer to the beginning of the food chain we

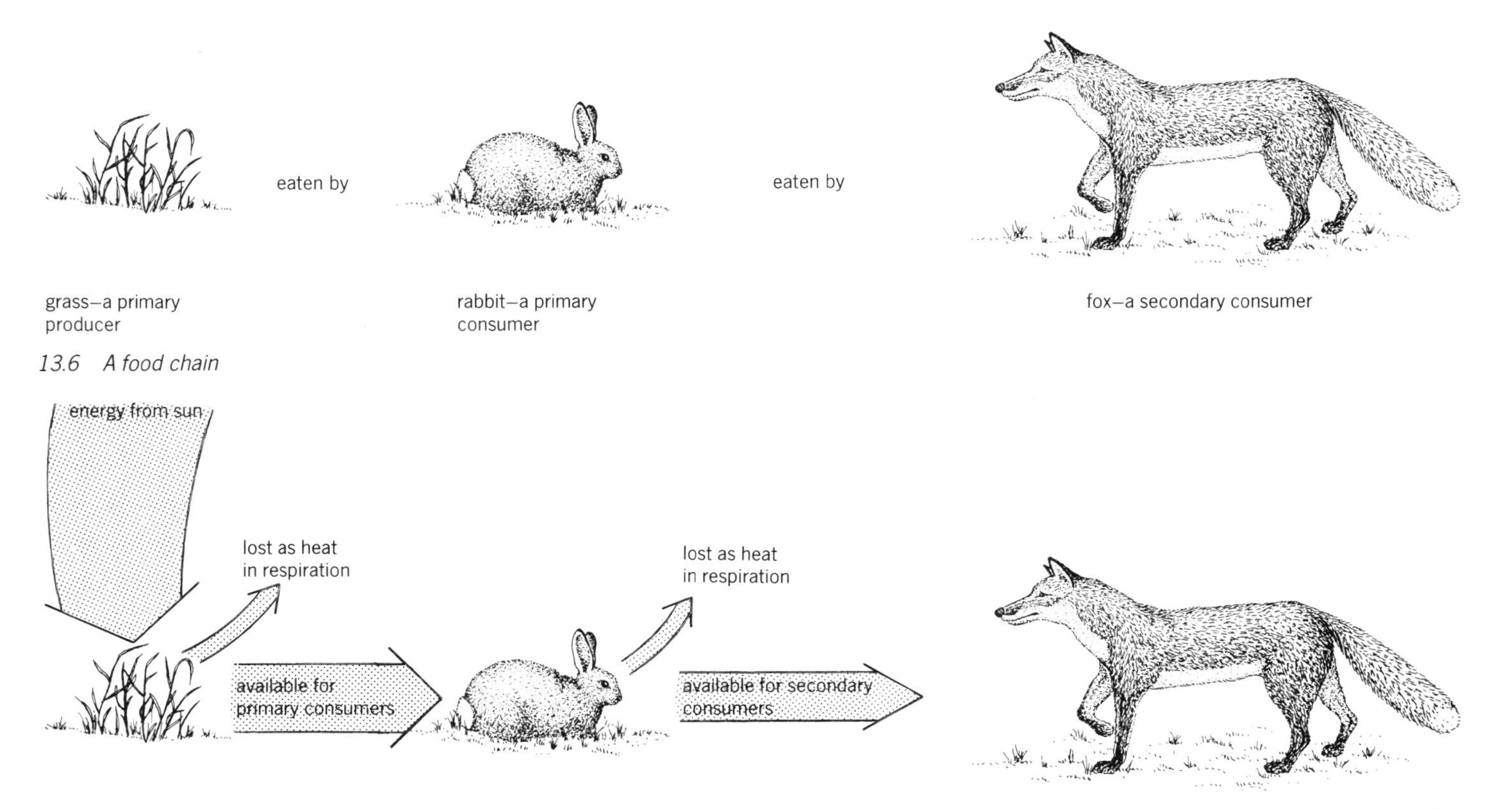

13.6 A food chain

13.7 Energy losses in food chain

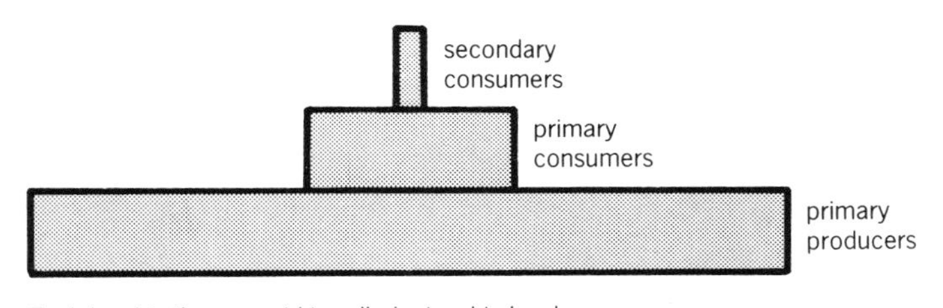

Each level in the pyramid is called a trophic level. The size of each level represents the numbers of organisms feeding at that level.

13.8 A pyramid of numbers

feed, the more energy there is available for us. This is why our staple foods, wheat, rice, potatoes, are plants.

When we eat meat, eggs or cheese or drink milk, we are feeding further along the food chain. There is less energy available for us from the original energy provided by the sun. It would be more efficient in principle to eat the grass in a field, rather than to let cattle eat it, and then eat them.

In fact, however, although there is far more energy in the grass than in the cattle, it is not available to us. We simple cannot digest the cellulose in grass, so we cannot release the energy from it. The cattle can; they turn the energy in cellulose into energy in protein and fat, which we can digest.

However, there are many plant products which we can eat. Soya beans, for example, yield a high amount of protein, much more efficiently and cheaply than cattle or other animals. A change towards vegetarianism would enable more food to be produced on the Earth, if the right crops were chosen.

Questions

1. Where does all the energy in living organisms originate from?
2. Write down a food chain
   (a) which ends with humans,
   (b) in the sea,
   (c) with five links in it.
3. Why are green plants called producers?
4. Why are there rarely more than five links in a food chain?
5. At which trophic level are you feeding when you eat (a) roast beef, (b) bread, (c) eggs, (d) an apple, (e) strawberries?

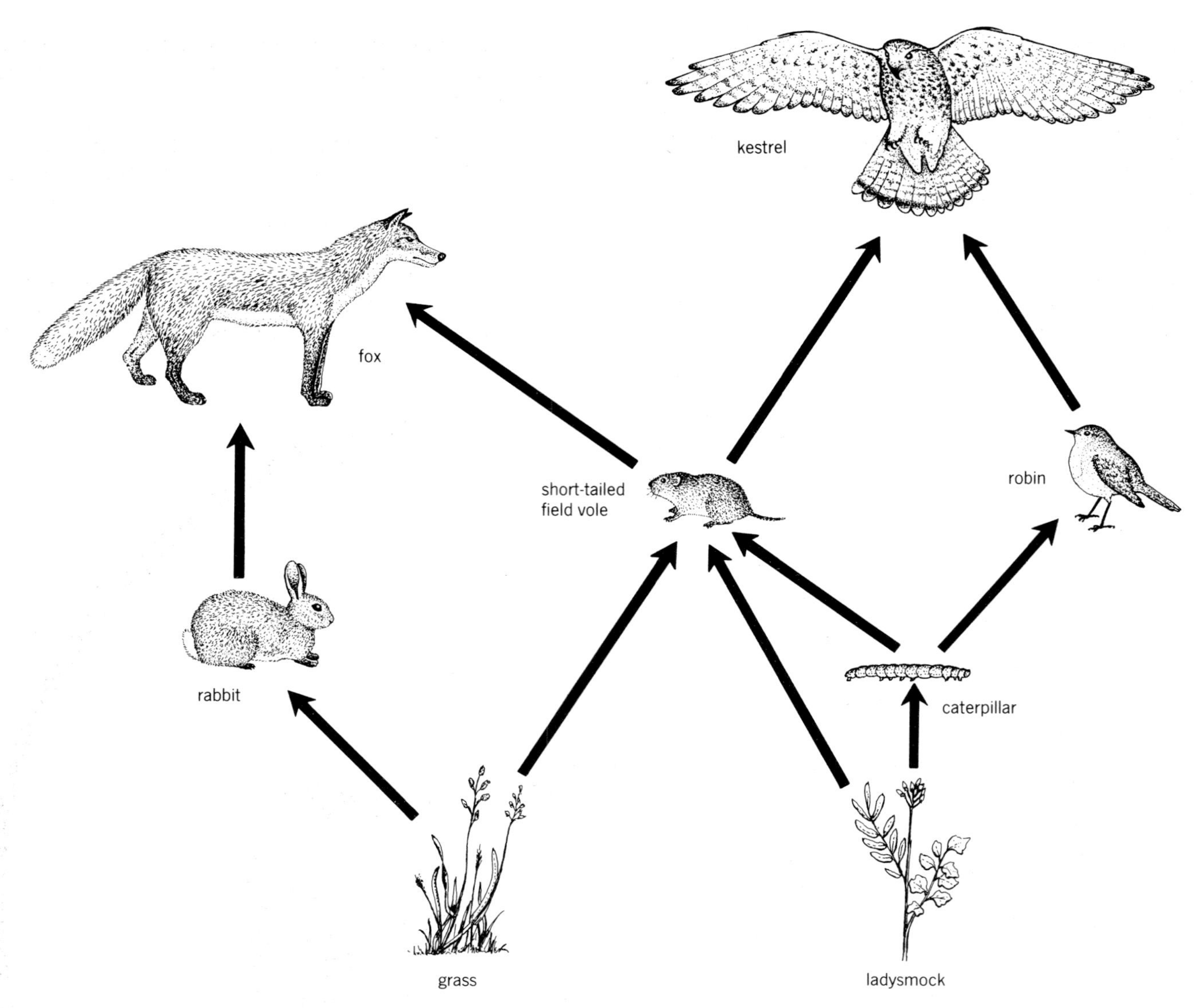

13.9 A food web

Nutrient cycles

13.14 Decomposers release minerals from dead organisms.

One very important group of organisms which it is easy to overlook when you are studying an ecosystem, is the **decomposers.** They feed on waste material from animals and plants, and on their dead bodies. Many fungi and bacteria are decomposers.

Decomposers are extremely important, because they help to release substances from dead organisms, so that they can be used again by living ones. Two of these substances are carbon and nitrogen.

13.15 Carbon is recycled.

Carbon is a very important component of living things, because it is an essential part of carbohydrates, fats and proteins.

Fig 13.10 shows how carbon circulates through an ecosystem. The air contains about 0.03% carbon dioxide. When plants photosynthesise, carbon atoms from carbon dioxide become part of glucose or starch molecules in the plant.

Some of the glucose will be broken down by the plant in respiration. The carbon in the glucose becomes part of a carbon dioxide molecule again, and is released back into the air.

Some of the carbon in the plant will be eaten by animals. The animals respire, releasing some of it back into the air as carbon dioxide.

When the plant or animal dies, decomposers will feed on them. The carbon becomes part of the decomposers' bodies. When they respire, they release carbon dioxide into the air again.

13.16 Few organisms can use nitrogen gas.

Living things need nitrogen to make proteins. There is plenty of nitrogen around. The air is about 79% nitrogen gas. Molecules of nitrogen gas, N_2, are made of two nitrogen atoms joined together. These molecules are very inert, which means that they will not readily react with other substances.

So, although the air is full of nitrogen, it is in such an unreactive form that plants and animals cannot use it at all. It must first be changed into a more reactive form, such as ammonia, (NH_3), or nitrates, (NO_3^-).

Changing nitrogen gas into a more reactive form is called **nitrogen fixation.** There are several ways that it can happen.

Lightning Lightning makes some of the nitrogen gas in the air combine with oxygen, forming nitrogen oxides. They dissolve in rain, and are washed into the soil, where they form nitrates.

Artificial fertilisers Nitrogen and hydrogen can be made to react in an industrial chemical process, forming ammonia. The ammonia is used to make am-

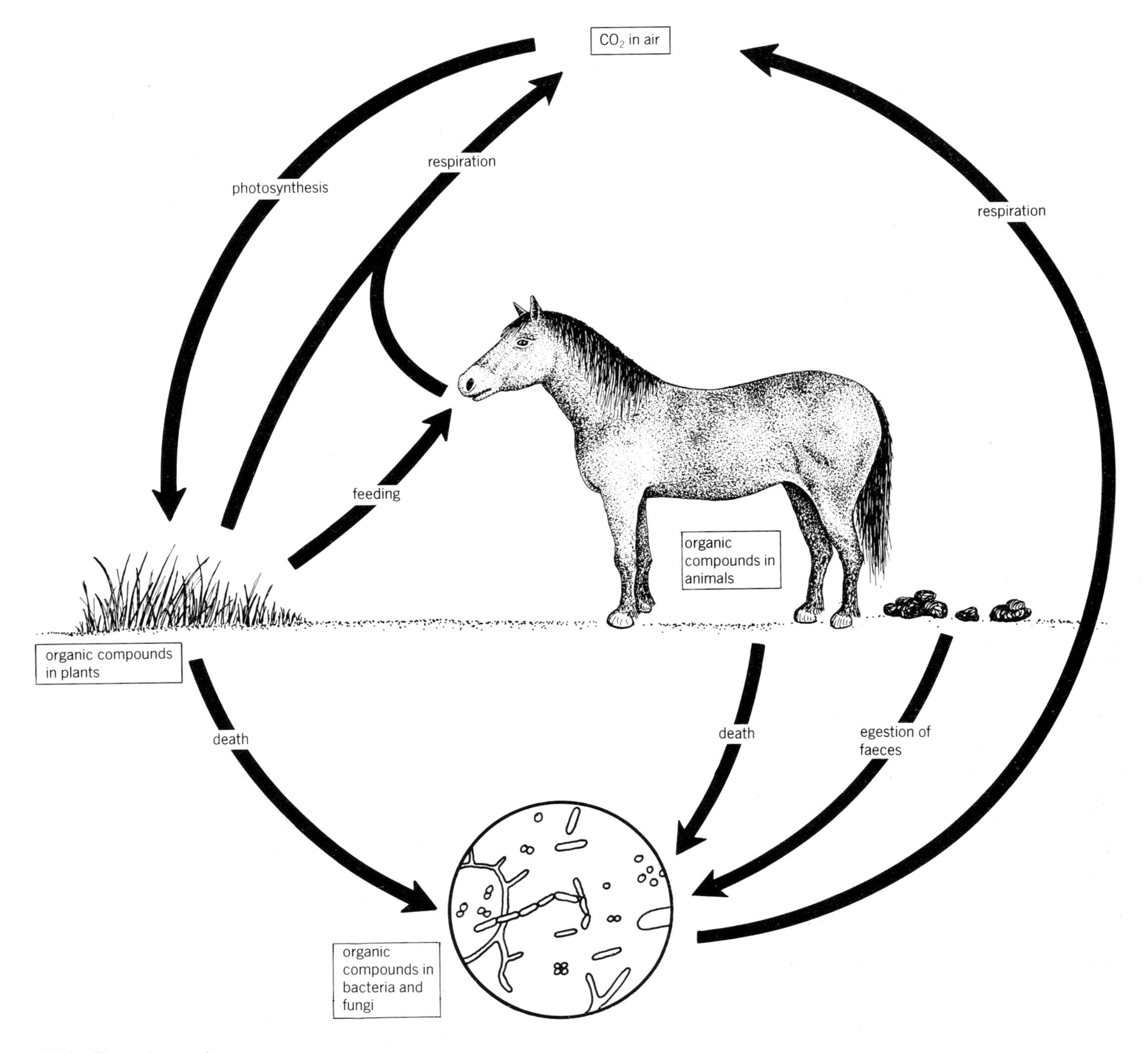

13.10 The carbon cycle

monium compounds and nitrates, which are sold as fertilisers.

Nitrogen fixing bacteria These bacteria live in the soil, or in root nodules (small swellings) on plants like peas, beans and clover. One kind is called *Rhizobium* ('rhizo' means root, 'bium' means living). They use nitrogen gas from the air spaces in the soil, and combine it with other substances to make nitrates and other compounds.

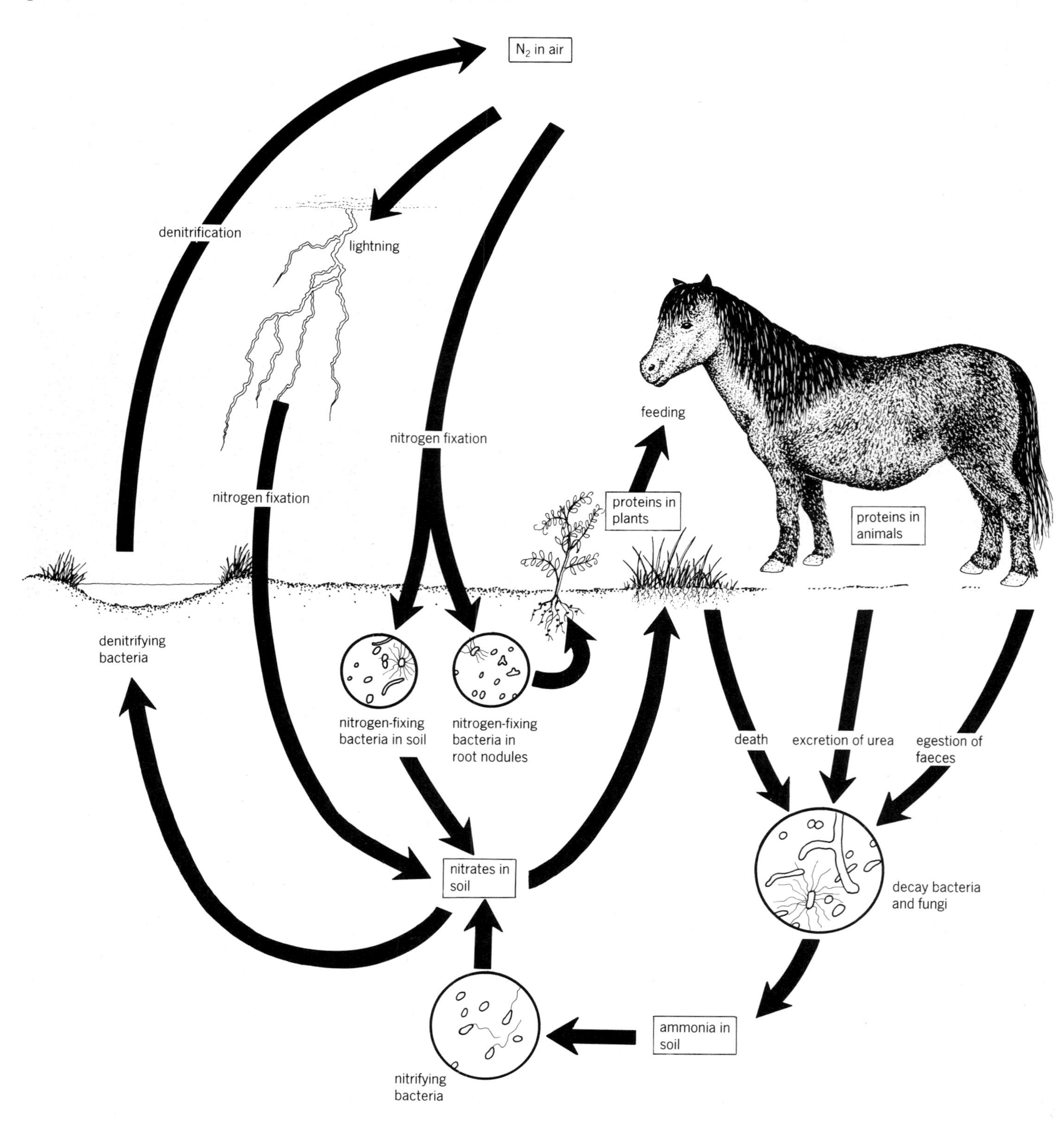

13.11 The nitrogen cycle

13.17 Fixed nitrogen moves round the nitrogen cycle.

Once the nitrogen has been fixed, it can be absorbed by the roots of plants, and used to make proteins. Animals eat the plants, so animals get their nitrogen in the form of proteins.

When an animal or plant dies, bacteria and fungi decompose the bodies. The protein, containing nitrogen, is broken down to ammonia and this is released. Another group of bacteria, called **nitrifying bacteria,** turn the ammonia into nitrates, which plants can use again.

Nitrogen is also returned to the soil when animals excrete nitrogenous waste material. It may be in the form of ammonia or urea. Again, nitrifying bacteria will convert it to nitrates.

13.18 Denitrifying bacteria make nitrogen gas.

A third group of bacteria complete the nitrogen cycle. They are called **denitrifying bacteria,** because they undo the work done by nitrifying bacteria. They turn nitrates and ammonia in the soil into nitrogen gas, which goes into the atmosphere.

13.19 Carnivorous plants get nitrogen from insects.

Nitrogen fixing bacteria can only use nitrogen gas if there is plenty of air in the soil where they live. They need to get nitrogen as a gas from the air spaces in the soil.

If the soil is waterlogged, nitrogen fixing bacteria cannot live there, but denitrifying ones can. So boggy soil is usually very short of nitrates. Plants living in

13.12 The insect-eating sundew which lives in boggy soils

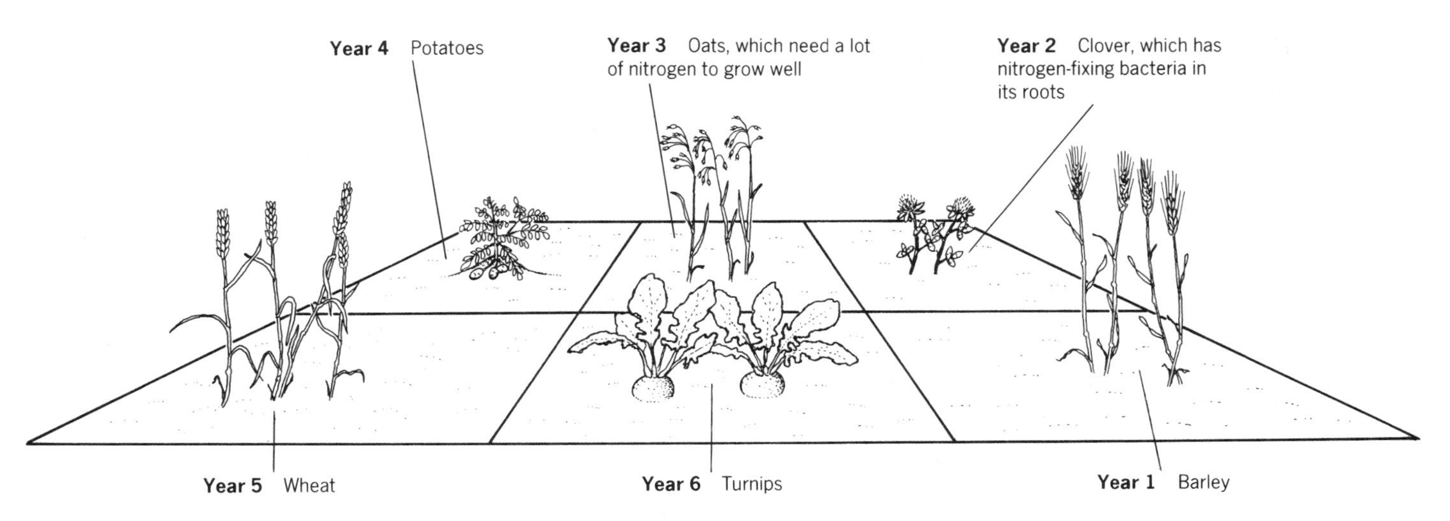

By growing different crops on the same piece of land in successive years, a farmer can gain several benefits. For example, crops like clover may provide nitrogen for the following crop. Also, a disease of one crop will not get the chance to infect that crop in the following year, and may die out before the same crop comes round again.

13.13 One type of crop rotation

these places either have to manage with very little nitrogen, or get it from somewhere else. Some of them have become carnivorous. Plants like the Venus fly trap, or the sundews, supplement their diet with insects. They digest them with enzymes, and get extra nitrogen from the protein in the insects' bodies.

13.20 Crop rotation reduces the need for fertiliser.

Farmers can save themselves money on fertilisers containing nitrogen, if they grow plants which actually increase the amount of nitrates in the soil. Peas, beans and clover do this, because they have *Rhizobium* living in their root nodules. When the crop is harvested, the roots are left in the soil. The next crop can then absorb nitrogen from them.

Fig 13.13 shows one possible sequence of crop rotation. In many places, however, farmers do not use crop rotation any more. They find it more convenient to keep growing the same crop in one field for several years, even though this means that they have to buy expensive fertilisers.

Research is now taking place to try to breed varieties of wheat and other crops which have their own nitrogen fixing bacteria in their roots. This would cut down considerably the amount of nitrogenous fertilisers needed to produce high yields of these crops.

Questions

1 What is a decomposer?
2 Why are decomposers important?
3 Why do living organisms need carbon?
4 How do carbon atoms become part of a plant?
5 What happens to some of these carbon atoms when a plant respires?
6 How do decomposers help in the carbon cycle?
7 Why do living organisms need nitrogen?
8 Why can plants and animals not use the nitrogen in the air?
9 What is nitrogen fixation?
10 Where do nitrogen-fixing bacteria live?
11 How do animals obtain nitrogen?
12 What do nitrifying bacteria do?
13 Which type of bacteria return nitrogen to the air?
14 Why do you often find carnivorous plants growing in bogs?

Chapter revision questions

1 Explain the difference between each of the following pairs, giving examples where you can.
(a) habitat, niche
(b) community, population
(c) quadrat, transect
(d) primary consumer, secondary consumer
(e) nitrogen fixing bacteria, nitrifying bacteria

2 Construct a dichotomous key, to enable someone to identify six people in your class.

3 Using the following list of words, in order, explain how
(a) a carbon atom in the air becomes part of a glucose molecule in your biceps muscle, and
(b) how that carbon atom might return to the air again.
broad bean plant, stomata, photosynthesis, glucose, sucrose, phloem vessel, bean seed, starch, feeding, amylase, maltose, maltase, glucose, ileum, hepatic portal vein, liver, hepatic vein, heart, aorta, subclavian artery, capillary, diffusion, muscle, respiration, carbon dioxide, diffusion, capillary, subclavian vein, heart, pulmonary artery, capillary, diffusion, alveolus, expiration

4 (a) Why is nitrogen important to living organisms?
(b) In what form do each of the following obtain their nitrogen?
(i) a green plant
(ii) nitrogen fixing bacteria
(iii) a mammal
(c) In the sea, the main nitrogen fixing organisms are blue-green algae, which float near the top of the water in the plankton. Construct a diagram or chart similar to Fig 13.11, showing how nitrogen is circulated amongst marine organisms.

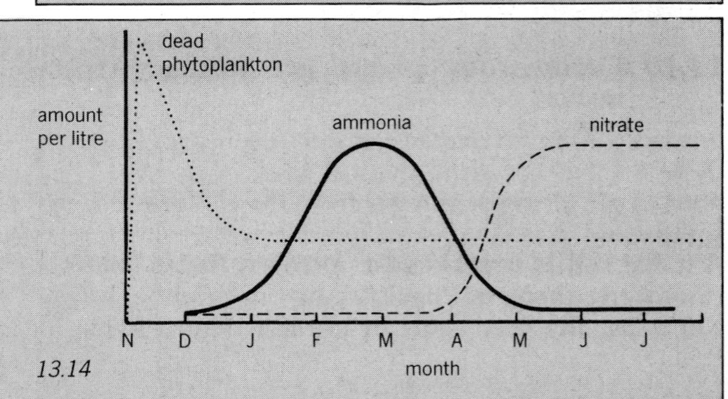

13.14

5 A fish tank was filled with water, and some bacteria were added. Some phytoplankton (microscopic plants) were then introduced. The tank was put into a dark place, and left for eight months.

At intervals, the water was tested to find out what it contained. The results are shown in the graph in Fig 13.14.
(a) Why did the phytoplankton die so quickly?
(b) The phytoplankton contain nitrogen in their cells. In what form is most of this nitrogen?
(c) Why does the quantity of dead phytoplankton decrease during the first two months of the experiment?

After one month, ammonia begins to appear in the water.
(d) Where has this ammonia come from?
(e) What kind of bacteria are responsible for its production?
(f) When does nitrate begin to appear in the water?
(g) What kind of bacteria are responsible for its production?

14 The distribution of living organisms

14.1 Environmental factors partly determine organism distribution.

Why do living organisms live where they do? Why do polar bears live in the Arctic, and not in Africa? Why do poppies grow on recently disturbed grass verges or in cornfields, but not on lawns?

The simple answer is that organisms tend to live where the environment is suitable for them to live. Any feature of the environment which affects a living organism is called an **environmental factor.** Each kind of living organism is especially equipped, or adapted, to cope with a particular set of environmental factors.

Polar bears, for example, are adapted to live in the intense cold of the Arctic. They have thick fur and a thick layer of fat beneath their skin to insulate their bodies. Poppies grow where the ground has recently been disturbed, because this is where their seeds can germinate easily. They cannot cope with the constant mowing on a lawn. The cold of the Arctic, the disturbance of ground and the mowing of a lawn are all examples of environmental factors.

Environmental factors alone, however, cannot completely explain the distribution of living organisms. Sometimes, an environment may seem just right for an

14.1 A polar bear has a thick coat of insulating fur, which helps it to maintain its body temperature

organism, and yet it is not found there. This may be because it has never been able to spread to that area.

14.2 Biotic and abiotic factors.

Because there are so many different environmental factors, it is useful to try to group them in some way. The two main groups are **biotic factors,** which are the influences of other living things, and **abiotic factors,** which are the influences of non-living parts of the environment.

Each of these main groups includes several kinds of environmental factor.

Abiotic factors These include **climatic** factors, such as sunlight, rainfall, humidity and temperature. Also important are **chemical** and **physical** factors, such as the amount of oxygen dissolved in a pond or stream, the amount of hydrogen sulphide gas in the air, or the pH of pond water. Factors caused by the soil are also very important. They are sometimes called **edaphic** factors.

Biotic factors These include availability of **food,** and how many **predators** there are. **Parasites** and **pathogens** (disease-causing organisms) are also important biotic factors. Another is the amount of **competition** with other organisms for food, shelter, or anything which an organism needs.

Environmental factors affect living organisms in many ways. For example, they affect their distribution, their size, their numbers, and their ability to reproduce. You will find many examples when studying any ecosystem. This chapter can describe only a few factors, and some of the effects they have on some organisms.

Questions

1. What is an environmental factor?
2. Describe at least four ways in which (a) a herring, and (b) a locust are adapted to their way of life.
3. What is meant by (a) a biotic factor, and (b) an abiotic factor?
4. What are edaphic factors?

Abiotic factors – climate

14.3 Climate influences natural vegetation.

On a world scale, climate has a great influence on the kinds of plants and animals which can live in different areas (see Fig 14.2). The two most important factors are temperature and rainfall.

For example, in hot places with plenty of rainfall, a large variety of plants can thrive, and **tropical rain forest** is formed. This provides a very rich and varied environment for animals, and so the number and variety of animals is also large.

14.2a World vegetation

Dry or **desert** areas, whether hot or cold, are more difficult for both plants and animals, because they lose water by evaporation and cannot easily replace it. Only a very few plants and animals are adapted to live in these conditions, so deserts are sparsely populated.

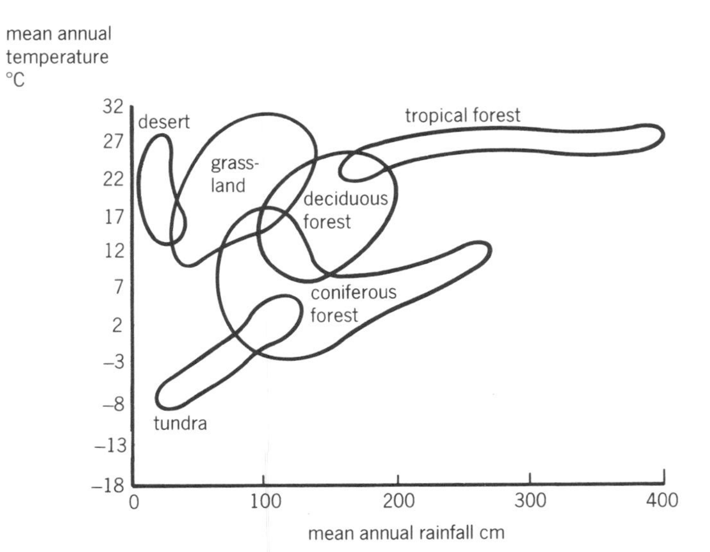

14.2b Temperature and rainfall affect the type of vegetation

14.4 In deserts such as the Namib, plants are few and far between

14.3 A great variety of plants grows densely where the climate is always warm and moist, as here on the Seychelles islands

14.4 Microclimate is important to many organisms.

Climate is also important to living organisms on a much smaller scale. To a woodlouse, for example, the climate which immediately affects it is the climate where it lives, perhaps under a rotting log. The climate in a small space like this is called a **microclimate**.

Microclimates may be quite different from the general climate in that area. Beneath the log, for example, humidity will probably be nearly 100%, whereas the air outside might be quite dry. Woodlice are not well adapted to conserve water, so they tend to stay under cover during the day, and come out at night when the air is cooler and more humid. They are **cryptozoic** animals ('crypto' means hidden).

Abiotic factors – chemical and physical factors

14.5 Oxygen.

Most living organisms need oxygen, for respiration. Usually, there is plenty of oxygen available in the air.

Aquatic organisms rely on oxygen which is dissolved in the water. Some of this will come from the air, and some from water plants which give off oxygen during photosynthesis.

Oxygen can quite often be in short supply in water. This is because oxygen is not very soluble in water, and does not diffuse through it very quickly. So the bottom of a deep lake may have little or no oxygen, especially as it will be too dark for plants to grow there.

14.5 Salmon will only breed in unpolluted fast-flowing rivers with plenty of dissolved oxygen

Shallow, fast flowing streams, however, always have plenty of oxygen. Trout and salmon, which swim very actively and need plenty of oxygen, may be found here. They are not found in deep, poorly oxygenated water.

Pollution of streams and rivers by sewage causes the amount of oxygen dissolved in them to decrease. This is because the sewage provides food for bacteria. A large population of bacteria builds up. This uses up the oxygen in the water so fish and other organisms cannot live there (see Fig 14.6).

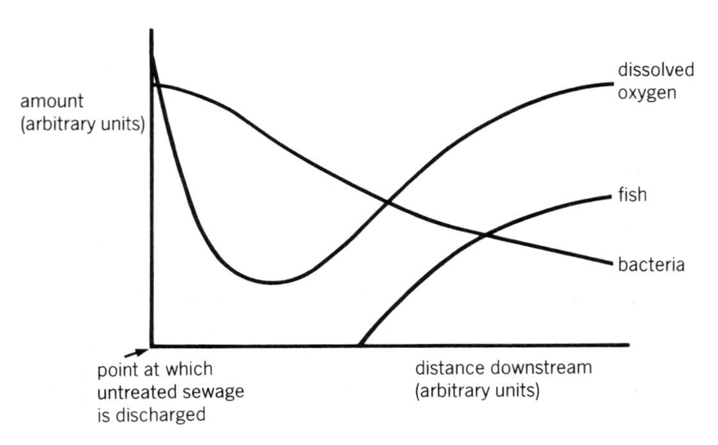

14.6 The effect of sewage pollution on a stream

14.7 Rivers running through cities are often severely polluted

14.6 Light.

Light is a very important environmental factor for plants, because they need it for photosynthesis. Many plants, such as poppies, need as much light as they can get. Others, such as dog's mercury (see Fig 14.8) are adapted to live in more shady places. Dog's mercury is usually found in woodland, where it can tolerate the shade of the trees. By growing in the shade, it avoids competition with other plants which need more light.

14.8 Dog's mercury can grow well, even in the shade of an oak wood

Questions

1. What are the two most important factors influencing the type of vegetation found in an area?
2. What type of vegetation would you expect to find in an area with a mean annual temperature of 20 °C and a rainfall of 80 cm?
3. What is a microclimate?
4. Why are woodlice more active at night than in the day-time?
5. The river Thames used to contain salmon, but up until the 1980s there have been no salmon in it for many years. Why is this?

Abiotic factors – soil

14.7 Soil affects plants, and therefore animals too.

Soil is a very important environmental factor, because plants rely on it for many of their requirements.

Anchorage Soil provides an anchorage for plant roots. A thin or very loosely structured soil will not support many plants, because their roots will not be able to get a good grip.

Nutrient minerals Soil provides nutrients for plants, particularly minerals such as nitrates, potassium salts etc.

Water Plants obtain water from the soil.

Air Plant roots and other soil organisms need air, to provide them with oxygen for respiration. A good soil has plenty of air spaces.

So the type of soil in a particular area has a large effect on the plants growing in it. This in turn will affect the animals which live there.

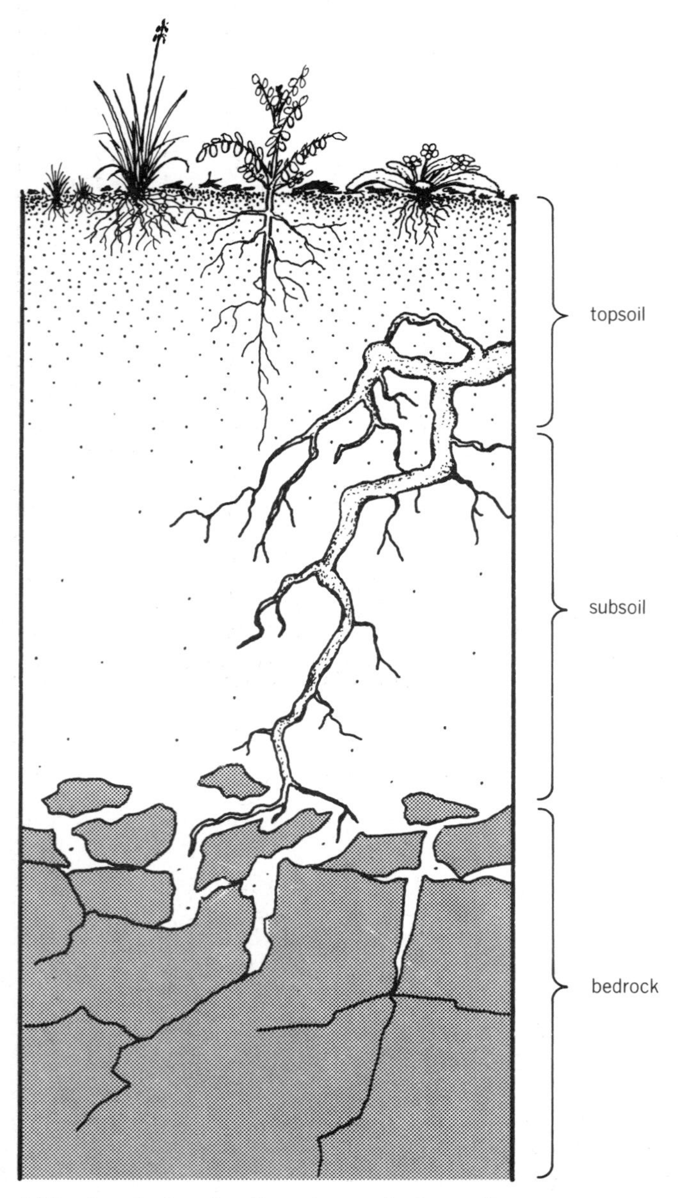

14.9 A vertical section through an agriculture soil

14.8 Soil is slowly formed from rock.

Soil is formed from rock. When rocks are weathered by wind, freezing and thawing, or by water flowing over them, they are broken down into small particles, called **rock waste.**

These particles are gradually colonised by a few flowering plants, and then lichens and mosses. As the plants die and decay, their remains add organic materials to the mineral particles of the rock waste. Other plants and animals can then begin to colonise the soil. This takes a very long time. It probably takes thousands of years to form a good, deep soil suitable for agriculture.

14.9 Soil has several components.

Fig 14.9 shows a vertical section through a good agricultural soil. The top layers are called **topsoil.**

(a)

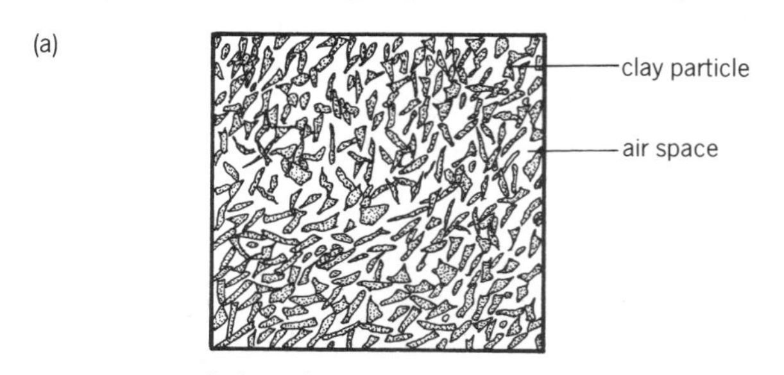

A clay soil contains small particles, which pack closely together.

(b)

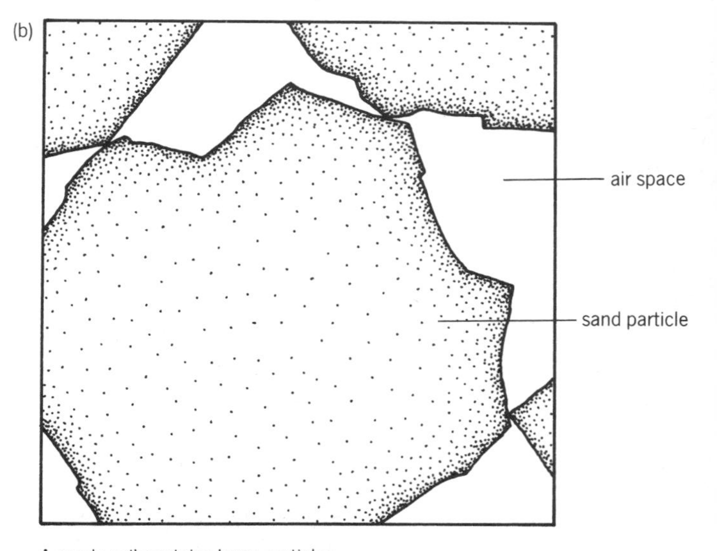

A sandy soil contains large particles, with large air spaces between them.

14.10 Soil particles

Topsoil has six main constituents. They are mineral particles, humus, water, nutrient ions, air, and living organisms.

14.10 The size of soil particles is important.

Mineral particles are formed from rocks, by weathering. The size of the mineral particles in a soil is very important. Very small particles form a soil called **clay,** while larger ones form **sand.**

Clay soils A clay soil contains very small soil particles, which can pack tightly together (see Fig 14.10(a)). Because they are so closely packed, they tend to hold water between them, by **capillarity.** Capillarity is the tendency for water to move into very narrow spaces. Clay soils do not dry out quickly in dry weather.

However, this can be a disadvantage. In wet conditions, the small spaces between the soil particles fill up with water, so there is no room for air. The soil becomes waterlogged (see section 14.14).

Clay particles have a slight electrical charge, and so mineral ions like potassium (K^+) and calcium (Ca^{2+}) are attracted to them. This is useful, because the clay particles hold the ions, stopping them from being washed or **leached** out of the soil by rain water.

Sandy soils A sandy soil contains larger soil particles (see Fig 14.10(b)). The large particles cannot pack very closely together, so there are large air spaces between them. Sandy soils are usually well aerated.

The large spaces, however, mean that water is not held by capillarity. Sandy soils drain very quickly. Sand particles do not hold mineral ions in the same way that clay particles do. So minerals are leached out of a sandy soil more quickly.

Loam A loam is a soil which contains a good mixture of sand and clay particles. If the balance is right, it will hold water and mineral ions, but will not get waterlogged too easily.

Investigation 14.1 Making a rough estimate of the proportions of particles of different sizes in a soil sample

1 Put the soil sample into a gas jar.
2 Fill the gas jar to within 5 cm of the top with tap water.
3 Stir or shake the jar, to mix the soil and water completely.
4 Leave the jar undisturbed, until the particles have settled into layers (see Fig 14.11).

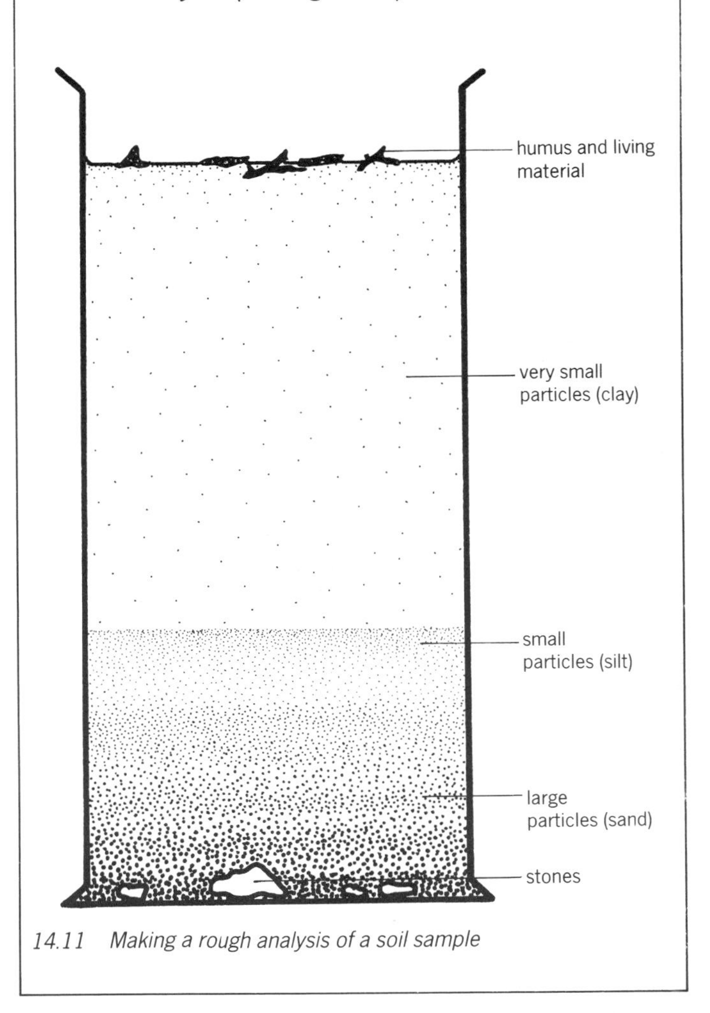

14.11 Making a rough analysis of a soil sample

14.11 Remnants of decayed organisms form humus.

The dead bodies of animals and plants, and any other organic waste such as faeces, are decomposed by bacteria and fungi in the soil. They are slowly broken down to a dark, sticky material called **humus.**

Humus forms a coating over the mineral particles in soil. It sticks the soil particles together into small groups, called **crumbs.** A soil with a good crumb structure tends to be well drained and aerated, and yet holds water and minerals.

It takes a long time before bacteria and fungi can break humus down completely. So humus provides a long-term store of useful substances such as nitrogen, which plants can eventually use.

Humus also provides food for other living organisms in the soil, such as earthworms.

14.12 Water coats soil particles.

Plants obtain all their water from the soil. The water is absorbed by osmosis, through root hairs (see section 6.32).

Even in dry weather, a good loam with plenty of humus will hold water. The water forms a film around each soil particle. Too much water in soil, however, causes waterlogging.

Investigation 14.2 To estimate the percentage of water in a soil sample

1 Weigh an evaporating dish.
2 Put your soil sample into the dish, and reweigh. Work out the weight of the soil sample, and record it.
3 Put the dish and soil into a cool oven. The warmth will dry out the soil. The oven must not be too hot, or the organic material in the soil will break down.
4 After a day or so, reweigh the dish and soil. Replace in the oven, and leave for a few hours more. Reweigh. If the two weights are the same, the soil is dry. If not, replace in the oven. This is called drying to constant weight.
5 Work out the weight of water in the soil, by subtracting the weight of the dried soil plus dish from the weight of the wet soil plus dish.
6 Work out the percentage of water in the soil like this.

$$\text{\% water in soil sample} = \frac{\text{weight of water in sample}}{\text{original weight of soil sample}} \times 100.$$

Questions

1 Why must the oven not be hot enough to break down organic material in the soil?
2 What type of soil would you expect to contain the highest percentage of water?

Investigation 14.3 To estimate the percentage of humus in a soil sample

1 Take the dried soil sample in its dish from Investigation 14.2. Reweigh to check its weight. If necessary, dry to constant weight again.
2 Heat the soil strongly, either in a very hot oven, or over a bunsen burner. The high temperature will oxidise the humus.
3 Allow the sample to cool, and reweigh.
4 Heat again, cool, and reweigh, until the weight is constant.
5 Work out the weight of humus in the soil, by subtracting the weight of heated soil plus dish, from the weight of dried soil plus dish.
6 Work out the percentage of humus in the soil like this.

$$\text{\% humus in soil sample} = \frac{\text{weight of humus in sample}}{\text{original weight of soil sample (before drying)}} \times 100.$$

Questions

1 Why was dried soil used for this experiment?
2 If the dried soil has been left for a while since the last experiment, you may find that its weight has increased slightly when you reweigh it at the beginning of this experiment. Explain this.
3 How could you use your results from Investigations 14.2 and 14.3 to calculate the % of minerals in the soil?

14.13 Minerals are dissolved in soil water.

Plants obtain nutrient or mineral ions from soil. They are absorbed by diffusion or active transport through root hairs. Table 3.2 lists the main mineral ions required by plants.

The mineral ions in soil are dissolved in the soil water. The kind and amount of ions depends partly on the kind of rock from which the soil was made, and partly on the activities of bacteria in the soil. For example, a soil with plenty of nitrogen-fixing and nitrifying bacteria will contain plenty of nitrate ions.

14.14 Organisms in soil need air.

Some of the spaces between the soil particles in soil are filled with air. If all the spaces are filled with water, so that there is no room for air, then the soil is said to be **waterlogged.**

Plant roots need oxygen from the air spaces for respiration. In a waterlogged soil, plant roots respire anaerobically. They make alcohol as a waste product (see section 5.9), which may kill them. Some plants, though, such as the marsh ragwort (see Fig 14.12) can tolerate the alcohol, and so are well adapted to live in boggy soils.

Soil animals also need oxygen for respiration. Nitrogen-fixing bacteria need nitrogen from the air in the soil.

14.12 Marsh ragwort tolerates boggy soils which kill other plants

14.15 Soil is a habitat for many organisms.

A good soil contains a large variety of living organisms. They include plant roots, earthworms, fungi and bacteria.

Earthworms can help to improve soil for plant growth. Their burrows help to aerate soil, and improve drainage. They also add humus to the soil by pulling in dead leaves, by excreting waste material, and from their decaying bodies after death. Earthworms feed by eating soil particles and extracting organic material from them, and the remains which they egest improve the texture of the soil.

Bacteria occur in soil in huge numbers. These bacteria, along with other decomposers such as fungi, feed on the dead bodies and faeces of animals and plants. They form humus. They help to release nutrients into the soil so that these can be reused by other living organisms (see sections 13.15 and 13.17).

14.16 Acid soils can form peat.

Acid soils form where rainfall is very high and drainage is poor, such as on high ground in the north and west of Britain. Because there is so much water in the soil, bacteria are unable to get enough air from the soil to allow them to decay dead plants and animals. The partly decomposed remains of the plants and animals build up as **peat.** Peat has a very high capacity for holding water, so the ground becomes even more waterlogged.

The partly decomposed plant remains release acids, which lower the pH in the soil. Few bacteria can live in these acidic, airless conditions. Because there are not many bacteria there are not many nitrates available to plants.

It is not surprising that only a few types of plant, such as heather, sedges and sphagnum moss (see Fig 14.14) are adapted to live in wet, acid, peat soils.

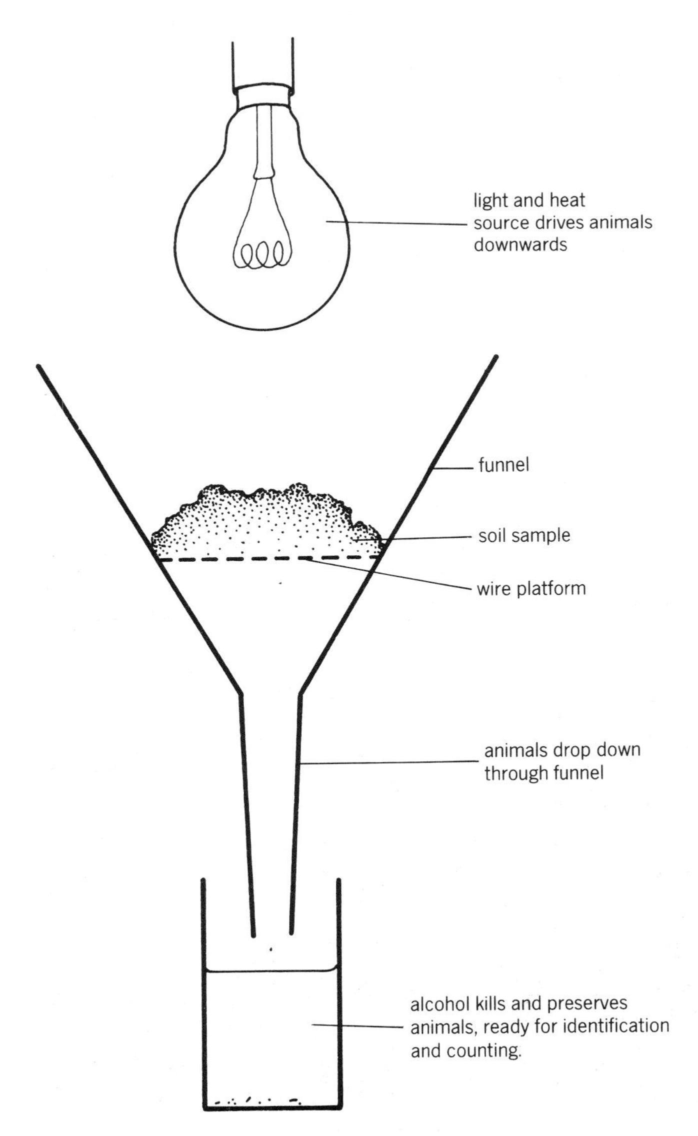

14.13 A Tullgren funnel–a method of collecting small soil animals

14.14 Heather-covered moorland occurs where the soil is wet and acid

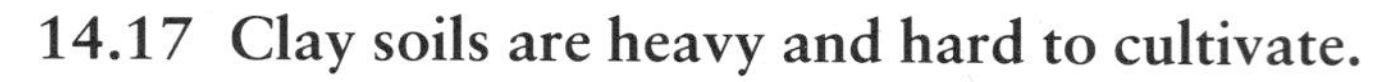

14.15a A wet clay soil can be compacted by heavy machinery

14.15b Clay soils may crack badly when dry

14.17 Clay soils are heavy and hard to cultivate.

A clay soil has the advantage that it holds water and mineral ions well. However, drainage and aeration are poor.

A clay soil is difficult to dig or plough. When wet, it is soft and sticky, and heavy farm machinery will compact it, squeezing out air from between the soil particles. When dry, it becomes very hard, and difficult to break up.

A clay soil may be made easier to work, and better for plants to grow in, by improving its crumb structure. There are two main ways that this can be done.

One way is by adding **lime** to the soil. Lime is calcium hydroxide. It reacts with the clay particles, making them clump together, or **flocculate.** The large crumbs improve the drainage and aeration of the soil.

Another way of improving the crumb structure is by adding humus to the soil, in the form of manure or compost. This also sticks the clay particles together to form crumbs. Other advantages of adding humus are that it adds nutrients to the soil, and encourages soil organisms.

A clay soil, if properly managed, can become an excellent agricultural soil.

14.18 Sandy soils are dry and poor in nutrients.

The natural advantage of a sandy soil is its good drainage. This makes it easy to work in winter or summer, because the large particles separate easily, and do not become compacted.

Sandy soils, though, lack nutrients and have poor water retentive properties. Both of these can be improved by the addition of humus.

14.19 Drainage improves acid soils.

It is not easy to convert a wet, acid peat soil into good agricultural land. The first step is to drain it. As the land becomes drier, it becomes better aerated, and bacteria and other organisms can colonise it. It may be ploughed, and sown with grass seed.

The main problem is that rainfall will still be high, and so it is very difficult to keep the land well drained.

Table 14.1 A comparison of clay and sandy soils

	Clay soil	*Sandy soil*
Particle size	Small	Large
Aeration	Spaces between particles are small, so soil is often poorly aerated	Spaces between particles are large, so soil is usually well aerated
Water holding capacity	Water is held in the small spaces by capillarity	Water is not held by capillarity, because the spaces are too large
Drainage	Water only drains slowly through the small spaces	Water drains quickly through the large spaces
Mineral ions	Many, because they are bound to the clay particles, slow drainage prevents them being leached out	Few, because they are quickly leached out as water drains through

Investigation 14.4 To find the effect of lime on clay particles

1 Put a small amount of powdered clay into a container, and mix it up thoroughly with plenty of tap water. The fine clay particles will form a cloudy suspension in the water.
2 Now add a small amount of lime (calcium hydroxide or calcium oxide) to the clay suspension. Watch carefully to see what happens (a) immediately, and (b) after a few minutes.

Questions

1 What effect did the lime have on the clay particles?
2 What are the main problems that farmers have when trying to cultivate a clay soil?
3 How might the addition of lime improve such a soil?

Questions

1 List four reasons why soil is important to plants.
2 What is topsoil?
3 List the six constituents of topsoil.
4 Explain why a clay soil usually contains plenty of mineral ions.
5 Why are sandy soils better aerated than clay soils?
6 What is a loam?
7 What is humus?
8 List three advantages of having plenty of humus in a soil.
9 What is meant by a waterlogged soil?
10 Why do many plants die if their soil becomes waterlogged?
11 How does a large earthworm population benefit a soil?
12 What is peat?
13 Why are peat soils acidic?
14 What are the problems of trying to cultivate a clay soil?
15 How may a sandy soil be improved?

Biotic factors

14.20 Relationships between living organisms.

Biotic factors are ones caused by other living things. Every living organism is affected by others in some way.

There are many kinds of relationships between living organisms. The relationship may be close or casual, beneficial or harmful. They include the relationships between:

1 **predators** and their prey
2 **parasites** and their hosts

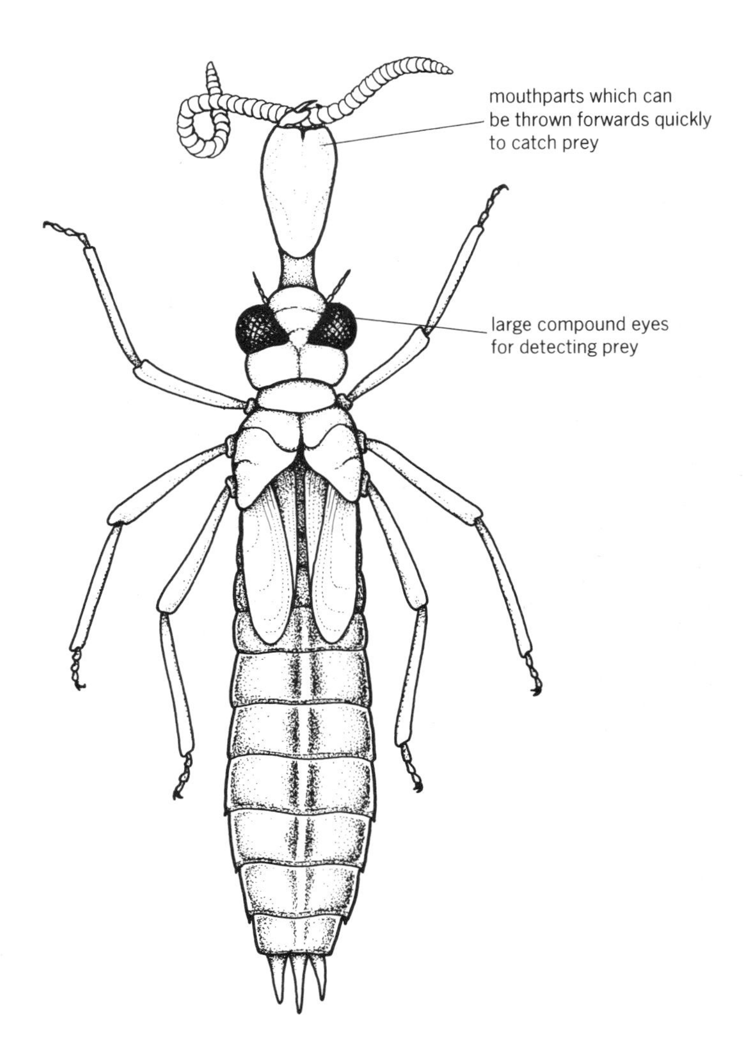

14.16 A dragonfly nymph–an example of a predator

3 **pathogens** (disease causing organisms) and their hosts
4 **mutualistic partners** (organisms which live together for mutual benefit)
5 **competitors** – organisms which need the same things, and so may compete with each other.

14.21 Predators kill prey for food.

A **predator** is an animal which kills another living organism, called its **prey**, for food. An example of a predator is the nymph of the dragonfly (see Fig 14.16). The nymph lives at the bottom of freshwater ponds, and feeds on any small living organisms that it can catch.

Predators need to be adapted to catch and kill their prey. Some of the adaptations of the dragonfly nymph are shown in Fig 14.16.

Animals which may be preyed on must also be adapted to protect themselves from their predators. Caddis larvae, for example, build protective cases around their soft bodies (see Fig 14.18).

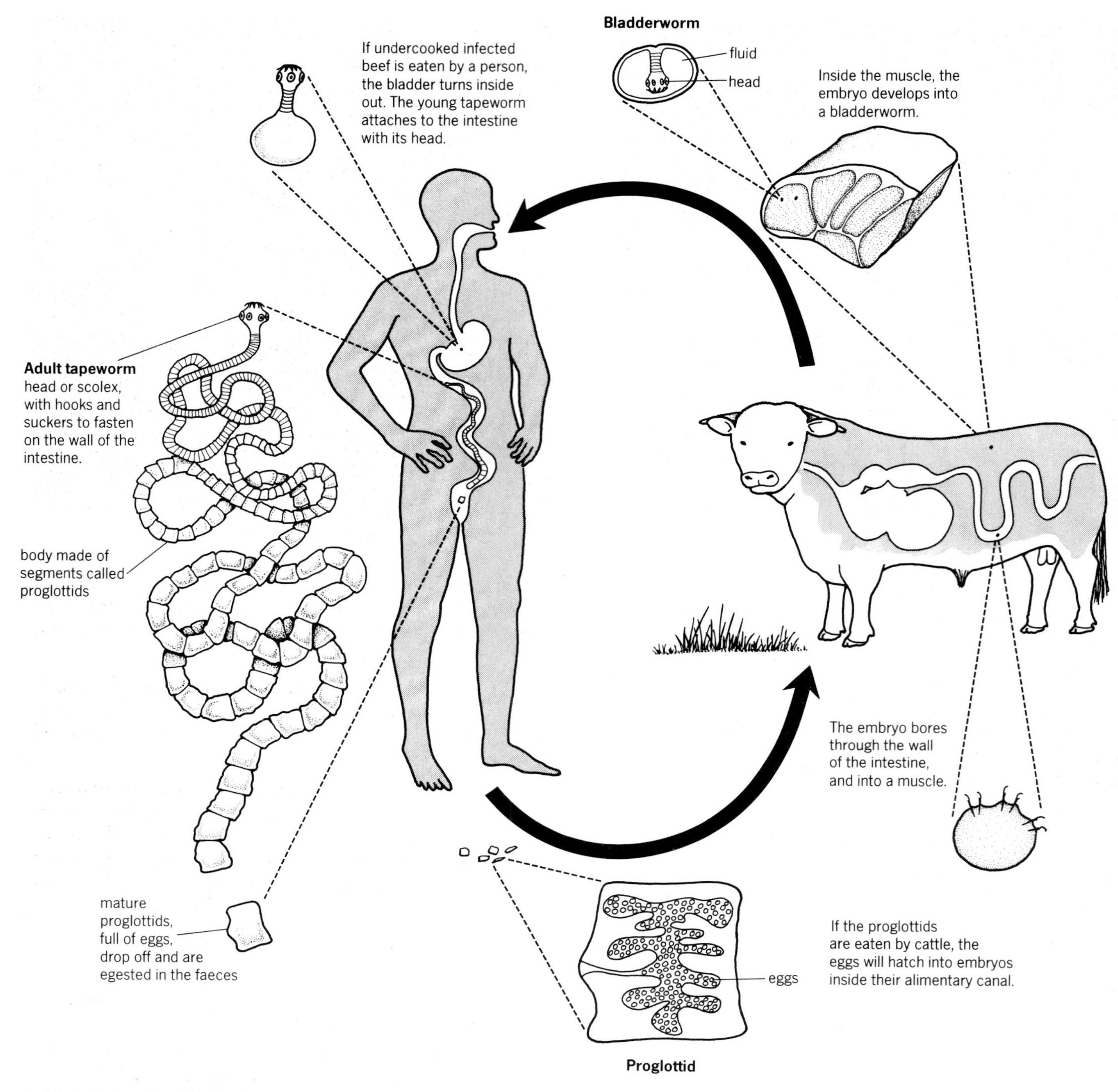

14.17 Life cycle of the beef tapeworm

14.22 Parasites harm their host.

A **parasite** is an organism which lives in close association with another organism, called its **host.** The parasite usually harms the host.

A tapeworm (see Fig 14.17) is an example of parasite. It lives inside the body of its host, so it is an **endoparasite.** The head louse (see Fig 14.19) lives on the outside of a person's body. It is an **ectoparasite.**

14.23 Parasites have special adaptations.

A parasite's life is not always an easy one. Although it has plenty of food, it has many other problems.

One problem is staying in position. Tapeworms have hooks and suckers to fix them firmly to the wall of the alimentary canal. If they did not, then peristalsis would push them through the digestive system.

Head lice can grip hair firmly when it is combed, so

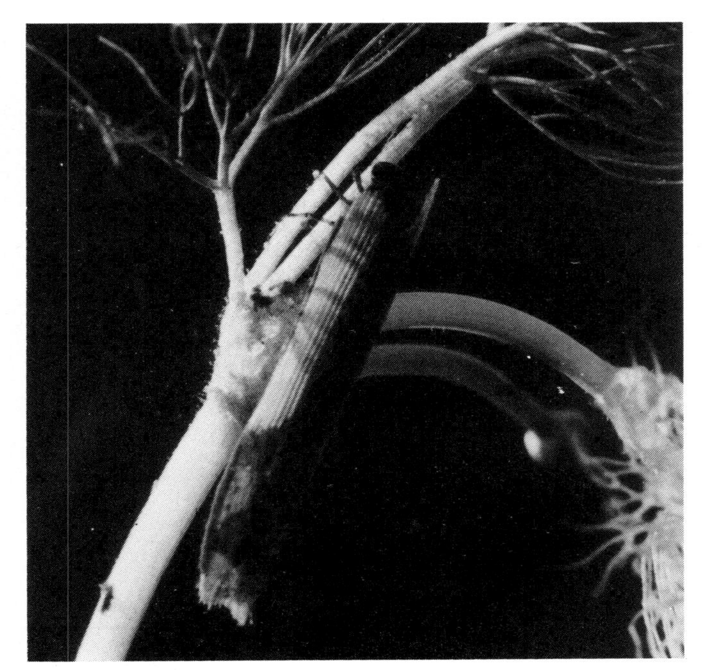

14.18 A case made of leaves camouflages this caddis larva

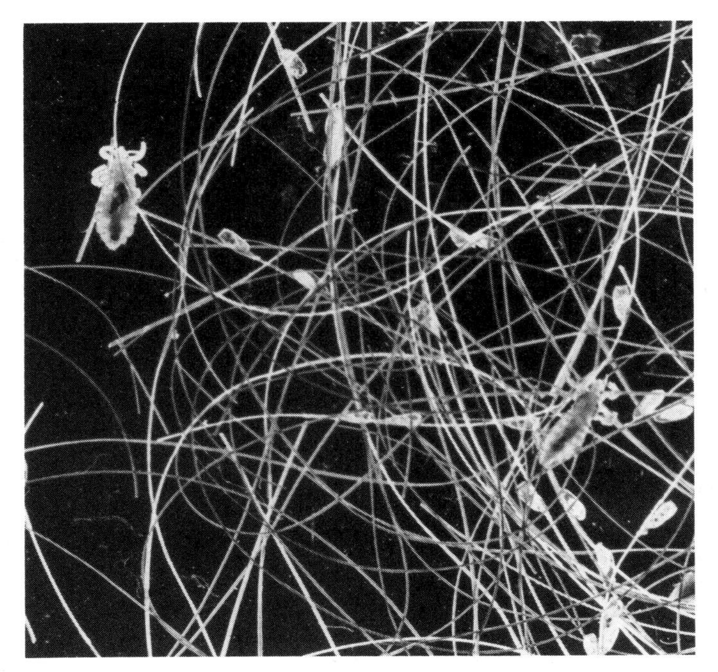

14.19 Head lice and their eggs on human hair

they do not get dislodged. Their flattened shape lets them lie close against the scalp. Their eggs are cemented firmly to the hairs.

A parasite is completely dependent on its host. If the host dies, the parasite will probably die too. Well adapted parasites do not kill their hosts.

One problem for all parasites is for their offspring to find a new host. Many parasites, such as the tapeworm, have complicated life cycles which help them to do this.

Because finding a host is such a risky business, many parasites produce large numbers of eggs and young. Even though most of them will not find a host, there is then a better chance that at least some of them will.

14.24 Pathogens cause disease.

A **pathogen** is an organism which causes disease. The most important pathogens are bacteria (see Fig 14.20) and viruses (see Fig 14.21). Some fungi and protista may also be pathogens.

14.25 Pathogens enter the body in different ways.

Pathogens cause disease by breeding in certain parts of the body. There are several ways in which they can get in.

Through the skin Some bacteria and viruses can get into the body through the skin, even when it is undamaged. Wart viruses can do this. Others are more likely to cause infections when the skin is damaged. *Staphylococcus* bacteria, which can infect wounds and turn them septic, or cause boils, enter through damaged skin.

Through the respiratory system Cold and influenza viruses are carried in the air in small droplets of moisture. If you breathe these in you may become infected. This is called droplet infection.

In food or water Bacteria, such as *Salmonella,* which cause food poisoning are taken into the alimentary canal in food. Poliomyelitis viruses may be transmitted in water.

By vectors A **vector** is an organism which transmits a pathogen to its host. Malaria is caused by a protistan called *Plasmodium.* It lives in the salivary glands of some types of mosquito, and is injected into a person's blood when the mosquito feeds (see Fig 14.24). Mosquitoes are therefore a vector for *Plasmodium.*

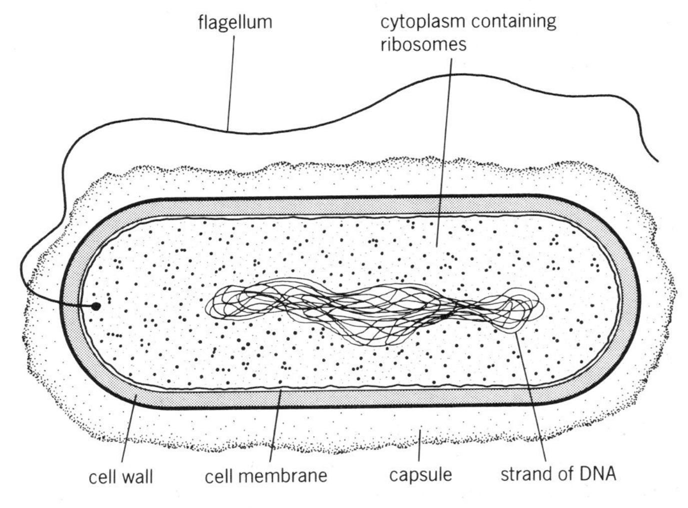

14.20 A bacterium

Table 14.2 How the body prevents infection

Method of entry	*Example*	*Natural defences*
Through skin	*Staphylococcus* bacterium	1 Epidermis is a barrier between pathogens and body 2 When skin is damaged, blood clots seal wound and prevent entry of pathogens 3 Tears contain lysozyme, which helps to prevent eye infections
Into respiratory system	Influenza virus	1 Cilia and mucus in respiratory passages trap dust particles which may carry pathogens, and sweep them upwards
In food or water, into alimentary canal	*Salmonella*	1 Distaste for food which looks or smells bad 2 Hydrochloric acid in stomach kills many bacteria
Injection into body by a vector	*Plasmodium*	None
By sexual intercourse	Bacterium causing gonorrhoea	None

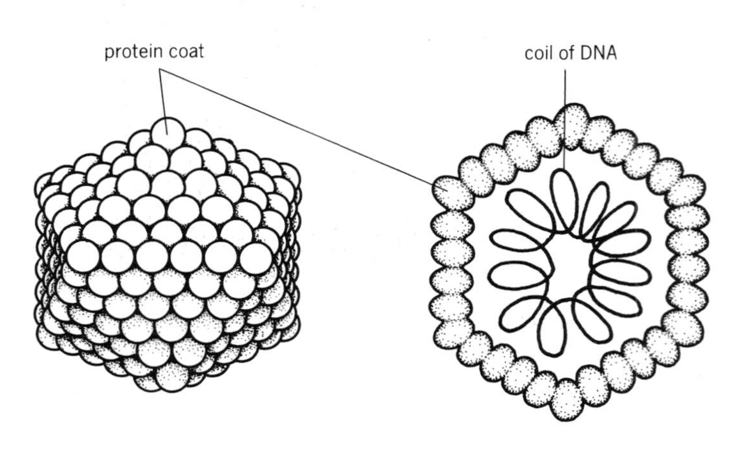

14.21 A virus, of the type which can cause sore throats

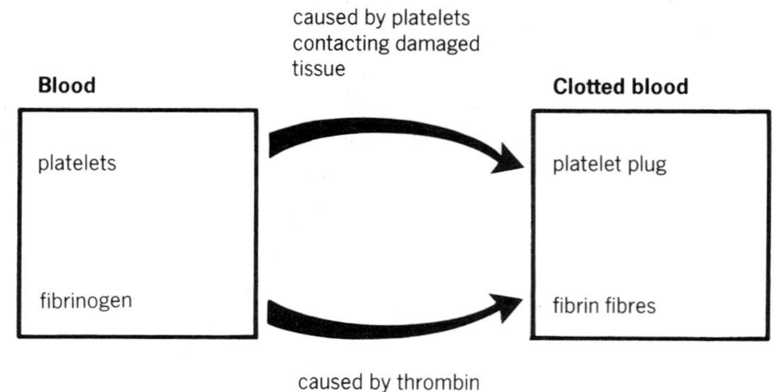

14.22 Blood clotting

14.26 Blood clotting protects the body.

The human body has many natural defences against the entry of pathogens. Some of them are listed in Table 14.2.

One way in which pathogens may enter the body is through a cut in the skin. Blood clotting helps to prevent this.

When blood platelets come into contact with a damaged tissue, they stick to the edges of the damaged area, and then to each other, forming a platelet plug (see Fig 14.23). If the wound is small, this will be enough to stop bleeding.

Larger wounds, however, need a larger barrier than this. Blood plasma contains several substances which are involved in blood clotting. There are thirteen of these **blood clotting factors.** If any one of them is defective, then blood will not clot. For example, if factor VIII is missing, the blood will not clot and bleeding continues. Continued bleeding from a small wound due to missing factor VIII is a disease called haemophilia.

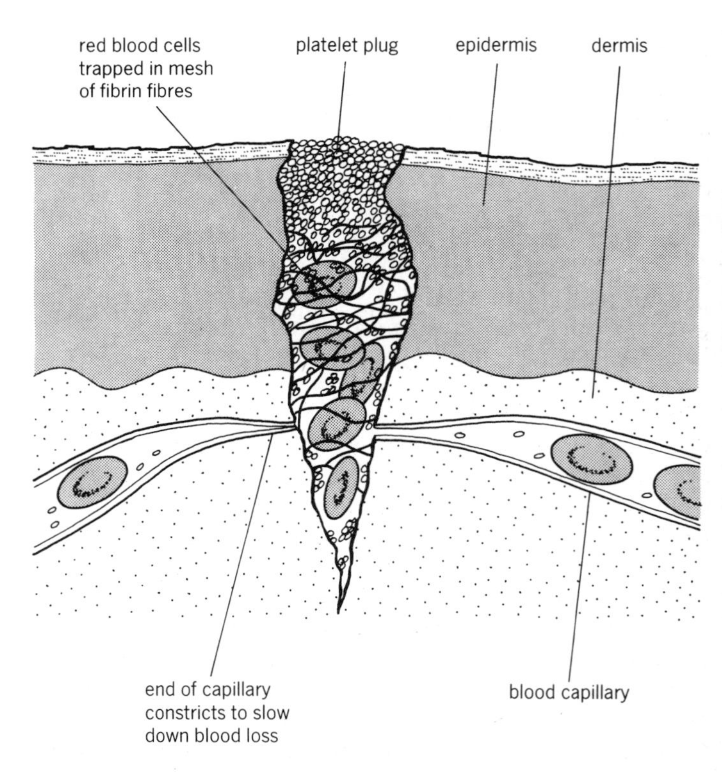

14.23 Vertical section through a blood clot

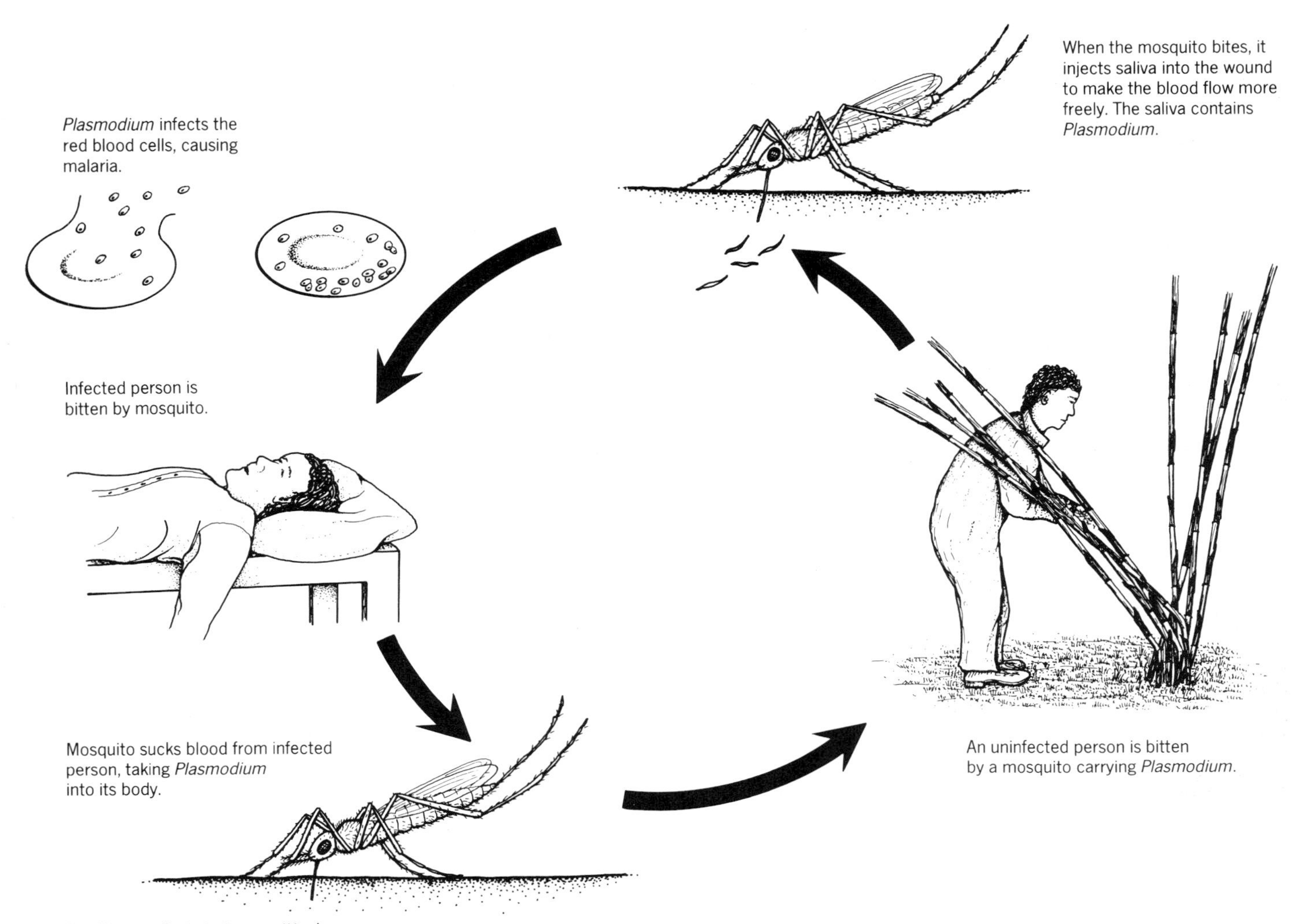

14.24 How malaria is transmitted

Two of these blood clotting factors are **prothrombin** and **fibrinogen,** which are soluble proteins dissolved in the blood plasma. If a tissue is damaged, it releases a chemical called **thromboplastin.** This converts prothrombin to **thrombin.** Thrombin acts on fibrinogen, converting it to the protein **fibrin.** Fibrin is insoluble, and forms fibres across the wound (see Fig 14.25). Blood cells and platelets get caught up in the fibres, forming a clot.

14.27 Bacteria release fever-causing toxins.

Once a pathogen, for example a bacterium, has successfully entered the body, it begins to reproduce. It usually takes some time before the colony has become large enough to have an effect on the body. This length of time is called the **incubation period.** It may be a few hours, or several days.

If they are able to breed successfully, then the bacteria will eventually begin to affect the body. The symptoms of the infection appear. Many symptoms are caused by the remains of dead bacteria, or substances released by living bacteria. These are called **toxins.**

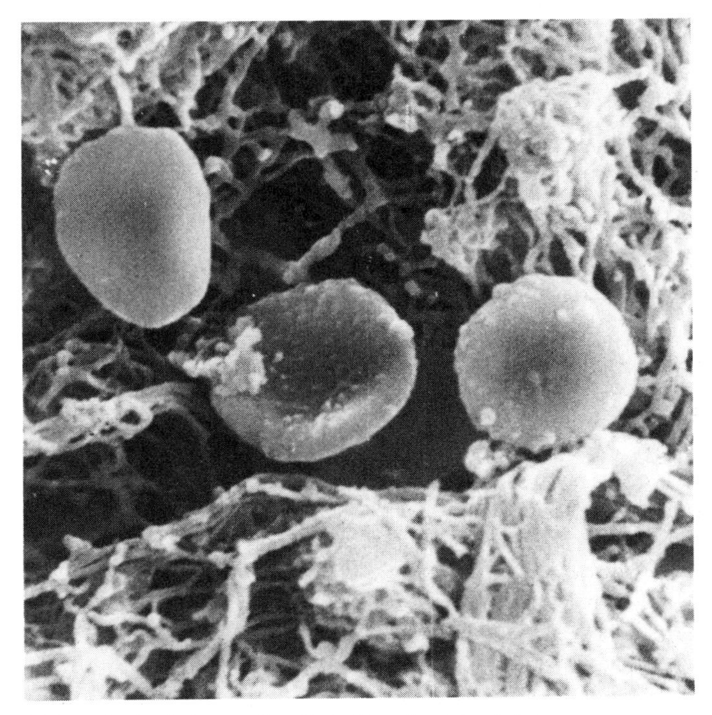

14.25 Red cells trapped by fibrin fibres in a blood clot

Some of these substances may affect the hypothalamus in the brain. Because the hypothalamus is responsible for regulating body temperature, your temperature may become much higher than usual. This is a fever.

A fever is usually only one symptom of a disease. Each kind of pathogen produces its own set of symptoms. Some of them are listed in Table 14.3.

Questions

1. How does a parasite differ from a predator?
2. List three problems that a tapeworm has in trying to survive and reproduce.
3. What is a pathogen?
4. How do the following pathogens enter the body? (a) *Staphylococcus*, (b) *Salmonella*, (c) *Plasmodium*
5. What is an incubation period?
6. Why do you often have a high temperature when you get an infection?

Fact!

The commonest disease in the world is caused by bacteria. It is tooth decay, or dental caries. In Great Britain, 13% of people have none of their own teeth left by the time they are 21 years old.

14.28 Antibodies act against invaders.

Your body does not just sit back and let the pathogen take over. White blood cells work to try to destroy the invading bacteria or viruses.

The white cells recognise the invading pathogen as 'foreign'. This is because it has chemicals on its cells which are not found on the cells in your body. These chemicals are called **antigens.**

One group of white cells is able to make another set of chemicals in response to the antigens. These chemicals are called **antibodies.**

The antibody molecules bind onto the antigen molecules. Each type of antibody will only fit onto one kind of antigen.

The binding of the antibody to the antigen can have several different effects. One effect is that the cells carrying the antigens clump together (see Fig 14.26). The white cells can then destroy them, by phagocytosis (see Fig 6.14).

It may take some time for the white cells to make enough of the right antibody. This gives the pathogen a chance to breed, and produce the symptoms of the disease. Eventually, though, the white cells usually manage to make enough antibodies to destroy the pathogen, and you recover from the illness.

14.29 People can be immune to certain diseases.

What happens if the same kind of pathogen invades your body again some time later? This time, the white cells are prepared. They recognise the antigen straight away, and quickly make large quantities of the appropriate antibody. The pathogen is destroyed before it has a chance to breed. You have become **immune** to the disease.

Having a disease and recovering from it is one way

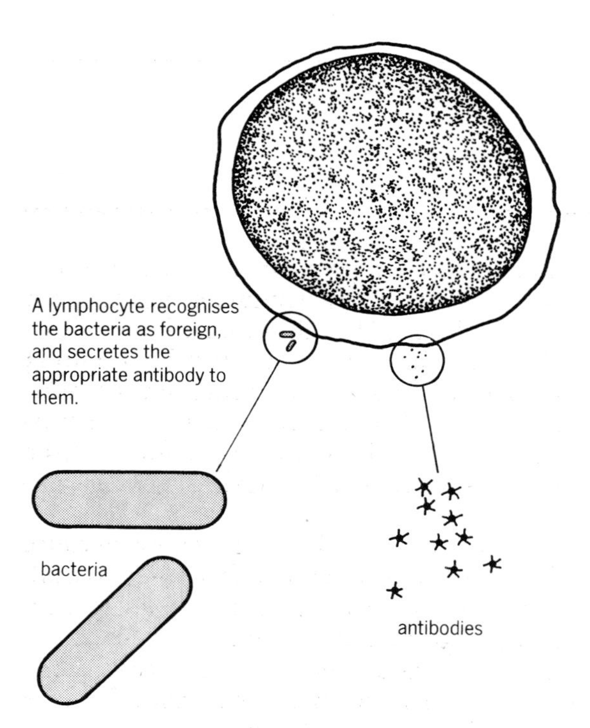

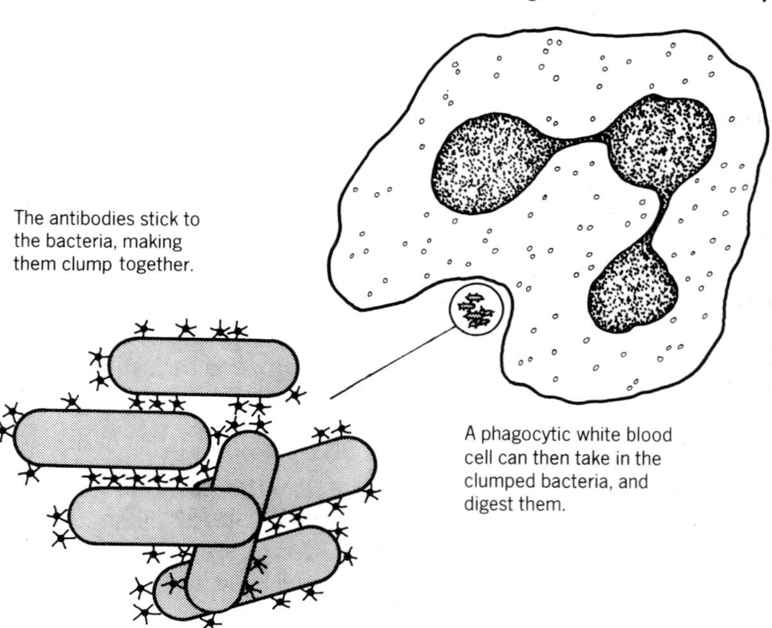

14.26 How white cells destroy bacteria

Table 14.3 Some important infectious diseases

Disease	*Organism causing it*	*How transmitted*	*Symptoms*	*Preventative measures*	*Other points*
Influenza	virus	droplets in air	fever, headache, muscular pains	vaccine	new strains of the virus keep appearing, so vaccines have limited use
Smallpox	virus	droplets in air, or contact with skin	fever, dark red spots which turn into scabs	vaccine	a world-wide campaign has now eradicated smallpox
Tuberculosis	bacterium	inhalation, or by drinking infected milk	usually no symptoms in early stages; later, fever, loss of weight, damage to lungs, sometimes a cough	eliminating poverty and overcrowding; pasteurization of of milk; mass radiography can diagnose infections early; BCG injections to immunise young people	if untreated, may last for several years
Gonorrhoea	bacterium	by sexual intercourse	in a man, a thick discharge from the urethra and pain when passing urine; often no symptoms in a woman	tracing and treatment of sexual contacts, so that they will not spread the disease any further	gonorrhoea is spreading rapidly, because more people now have more than one sexual partner
Dysentery	protistan – a kind of *Amoeba*	in water or food contaminated by faeces of infected person	ulcers in caecum, colon and rectum; diarrhoea, fever and stomach pain	good personal hygiene	most likely to occur in unhygienic, overcrowded conditions
Malaria	protistan – *Plasmodium*	in saliva of mosquitos	several types, with various symptoms, including fever	drainage of wet areas where mosquitos breed; wearing clothing to prevent bites, sleeping under nets; anti-malarial tablets	
Ringworm	fungus	by contact with infected people or animals, or with an infected person's hairbrushes, etc.	circular, bald patch on head, or circular scaly patch on body	good personal hygiene	

of becoming immune to it. This sort of immunity is called **active immunity,** because your white cells make the anitbodies themselves.

You can also acquire active immunity by having a vaccination. Some vaccines, for example the BCG vaccination for tuberculosis, contain bacteria which have been weakened. When they are injected into your body, they are too weak to reproduce. But the white cells recognise them as foreign, and 'learn' to make the antibodies to destroy them.

Another type of immunity is called **passive immunity.** Here your white cells do not make the antibodies. Instead, the antibody is put into your blood ready made. Breast fed babies get immunity to many diseases, because there are antibodies in breast milk.

Some vaccinations contain ready made antibodies. An anti-tetanus injection given by the doctor to patients with a cut, contains anti-tetanus antibodies from a horse. They are obtained by injecting the horse with a weakened form of the bacterium, and later taking some blood plasma from it and extracting the antibodies. It is however, also possible to be given an anti-tetanus vaccination containing weakened forms of the bacterium which will induce a longer-lasting active immunity.

Passive immunity does not last indefinitely, because the antibodies gradually disappear from your blood. This is why you are usually given an anti-tetanus injection whenever you go to a doctor with a cut.

Active immunity lasts much longer, because the white cells have 'learnt' to make the antibody. One BCG injection will last you a lifetime.

14.30 In mutualism, both partners benefit.

Mutualism means 'living together'. Mutualism happens when two organisms of different species live in a close association with one another, and both organisms benefit.

One example of mutualism is the relationship between the nitrogen-fixing bacterium *Rhizobium*, and clover. The bacteria live in swellings or nodules on the clover roots. The plant benefits because the bacteria take nitrogen gas from the air spaces in soil, and convert it into nitrogen compounds which the plant can use to make proteins. The bacteria benefit because they obtain sugars from the plant.

Lichens (see Fig 14.27) are another example of a very successful mutualistic partnership. A lichen consists of two types of organisms – a fungus and a green alga. The partnership is so successful that lichens can colonise places where no other plant can grow, such as roofs of buildings, or the soil in arctic regions. The alga photosynthesises, while the fungus extracts minerals from the roof or soil.

A third example of mutualism is the relationship between a rabbit and the bacteria which live in its caecum and appendix (see section 2.46). The rabbit benefits because the bacteria digest its cellulose. The bacteria obtain a plentiful food supply, and a place to live.

14.31 Organisms compete with each other.

Competition happens whenever two or more organisms need the same thing, which is in short supply. If the competition is between individuals belonging to the same species, it is called **intraspecific** competition. If it is between individuals belonging to different species, it is called **interspecific** competition.

Plants compete for light, root space, and sometimes for water and minerals from the soil. Animals compete for food, and a place to live and reproduce.

14.27 Lichens can grow in places that no other plant can tolerate; however, they grow slowly, as can be seen by the size of the patches on this 140 year old tombstone

Questions

1. Explain how white cells destroy bacteria.
2. How does a vaccination for (a) tuberculosis, and (b) tetanus prevent you from getting the disease?
3. What is mutualism?
4. Describe three examples of mutualism.
5. What is competition?
6. Using the illustration on the cover of this book, list (a) a predator and its prey, (b) two mutualistic partners, and (c) two organisms which may compete with each other.

Fact!

The most potentially dangerous insect in the world is the common housefly. It is thought that it can transmit as many as 30 diseases to man. These include cholera, typhoid, leprosy, dysentry, bubonic plague, scarlet fever, meningitis, diphtheria, smallpox and infantile paralysis.

14.32 Occupying different niches reduces competition.

Competition between living organisms only happens when their niches, or life styles, overlap. The more they overlap, the more likely it is that they will compete.

For example, black ants and yellow ants have quite similar niches. Both kinds of ant live in pastures in Britain. They both feed on aphids and other small insects. Because their niches overlap, competition occurs between the two species of ant. They compete for space and food. If the black ant colonies are destroyed, then the yellow ant colonies breed more rapidly than when the black ants are there. Therefore the competition has a harmful effect on the yellow ant colonies.

14.28 Ant hills in Wytham Wood, near Oxford; these hills are made by yellow ants. Black ants also live in this area, but under stones rather than in ant hills

Although the niches of the yellow and black ants overlap enough to cause them to compete with one another, they do not overlap completely. Yellow ants, for example, always look for food under the surface of the ground, while black ants forage on the surface. Yellow ants build ant hills to breed in, while black ants breed under stones.

Because their niches do not overlap completely, it is possible for yellow and black ants to live in the same place at the same time. This seems to be true for almost all species of living organism. Although niches may sometimes appear to be identical, each type of organism does in fact have its own niche which is different from every other. The more different an organism's niche is from those of other organisms, the less it will be in competition with them.

Competition is a very important factor in evolution. This is described in Chapter 16.

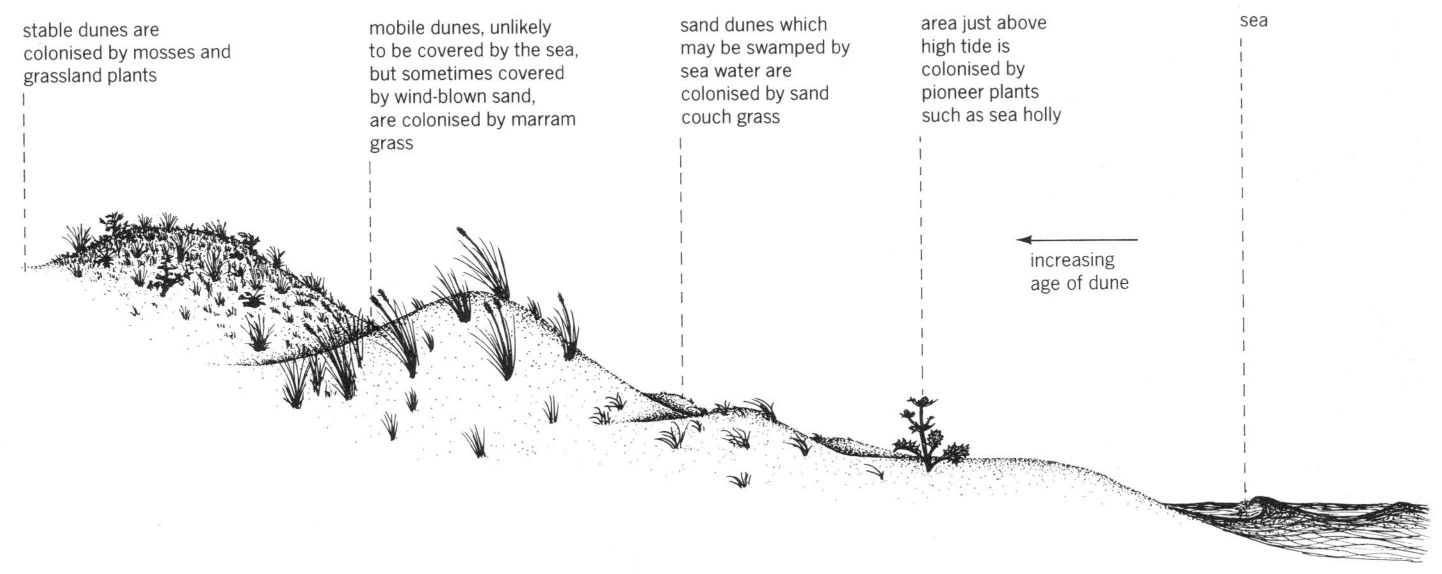

14.29 Succession on a sand dune – see section 14.34 overleaf

Colonisation and succession

14.33 Organisms cause changes in their environment.

This chapter has described some of the environmental factors affecting the distribution of living organisms. Each organism is adapted to live with a certain range of environmental factors. But environmental factors do not stay the same all the time. Over a period of time, they may change. Even the organism itself can change the environmental factors which are acting on it. They may be changed so much that the environment is no longer suitable for it. Other organisms may then move in to take its place.

The gradual change in the plant and animal community at a particular site is called **succession.** Finally, a community develops which does not change. This is called the **climax community.**

14.34 Sand dunes are examples of succession.

One place where it is quite easy to see succession happening is on a sand dune (see Fig 14.29).

The area near the sea just above high tide level is not an easy place for plants to live. Salt spray is blown onto them by the wind, and fresh water is in very short supply.

A few plants, though, are adapted to cope with these conditions, and they begin to colonise this area. They are called **pioneer plants.** One example is sea holly (see Fig 14.30). It has wax on its leaves which helps to prevent water loss by transpiration.

Once a few pioneer plants have colonised the area, wind blown sand begins to be trapped around them. A sand dune begins to form. One plant which can grow in these new conditions is sand couch grass (see Fig 14.31). It has rhizomes which run through the sand, helping to hold it in place. Sand couch grass can survive being swamped by sea water for a short time.

Gradually, the dune builds up until it is high enough to be out of reach of the sea. Now marram grass can move in. Marram grass could not grow on it before this, because it cannot tolerate immersion in sea water.

Marram grass has strong rhizomes which anchor it in the sand, and help to anchor the dune in position. It has leaves which can roll up (see Fig 14.32), enclosing the stomata and cutting down transpiration. Even if marram grass is covered to a depth of a metre by wind blown sand, it can grow up through it.

Marram grass only grows well on dunes where sand is periodically blown over it. These are sometimes called 'mobile' dunes. Another name for them is yellow dunes, because of the patches of sand between the grass.

Small animals can live amongst the marram grass. Their faeces and remains, and the remains of the grass, begin to build up the humus content of the dune. The soil can now retain moisture better, and has more nutrients in it. Other plants can now colonise the dune. Over a period of several hundred years, mosses, lichens and flowering plants grow on the dune, stabilising it so much that marram grass cannot grow on it any more. It is now a stable or grey dune.

Very gradually, the soil in the grey dunes becomes better, so that more and more plants can grow in it. What happens to it in the end partly depends on the

14.30 Sea holly can be an early coloniser of sand dunes

14.31 Sand couch grass on a young sand dune helps to hold the sand

14.32 Marram grass leaves roll up when dry, so less water is lost

kind of sand that formed it in the first place. If there was plenty of calcium carbonate, it will probably end up as grassland. If it is acidic, though, the climax community will be willows, birch and pine trees.

Questions

1. Why do plant and animal communities in some areas gradually change?
2. What is the name for this process?
3. What is a climax community?
4. Name a pioneer plant.
5. How is marram grass adapted to its way of life?

Fact!

The largest structure ever built by living organisms is the 2000 km long Great Barrier Reef. It covers an area of ¼ million km^2 and supports more life on its surface than any other habitat on Earth. It has been constructed by stony corals over a period of 600 million years.

Chapter revision questions

1 (a) List the components of topsoil.
(b) What problems are associated with the cultivation of (i) clay soil, (ii) sandy soil, and (iii) acid soil?
(c) Explain how each of the following can improve the properties of a clay soil.
(i) earthworms
(ii) addition of humus
(iii) addition of lime

2 (a) What is a parasite?
(b) List four problems faced by most parasites.
(c) For one named parasite, describe how it overcomes these difficulties.

3 (a) Name one disease which is transmitted in unclean water.
(b) What type of organism causes this disease?
(c) What natural defences does your body have against infection by this organism?
(d) How can your body fight the disease once infection has occurred?
(e) What measures can people take to prevent the spread of this disease?

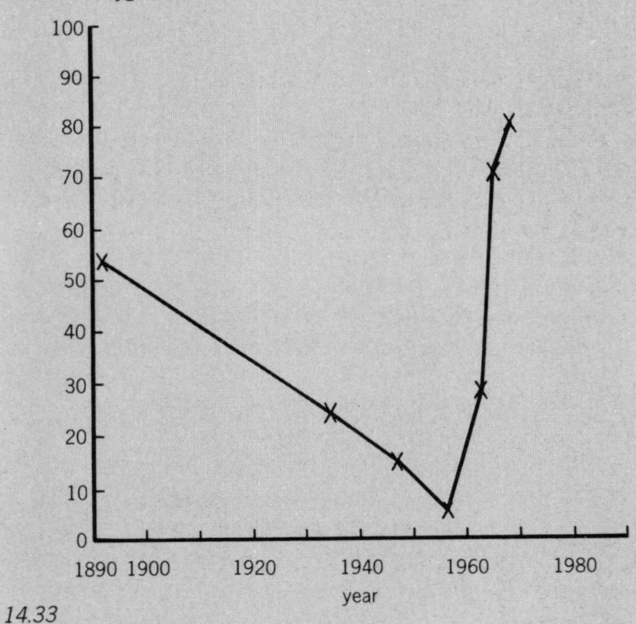

14.33

4 The graph in Fig 14.33 shows the amount of dissolved oxygen in the water of the river Thames 10 miles above London Bridge during the first part of this century.

In the 19th century, sewage from London houses drained directly into the river Thames. At the turn of the century, sewage treatment plants had been installed, which removed some of the organic material from the sewage before it entered the river. These plants have gradually become more efficient.

continued

(a) Give two ways in which water obtains dissolved oxygen.
(b) Explain how pollution by sewage causes dissolved oxygen levels to decrease.
(c) Suggest why dissolved oxygen levels in the Thames (i) decreased until 1948, and (ii) have increased since the 1950s.
(d) What effect would you expect a decrease in dissolved oxygen to have on the fish population in the Thames?
(e) Apart from affecting the levels of dissolved oxygen, what other harmful effects can the discharge of untreated sewage into rivers have?

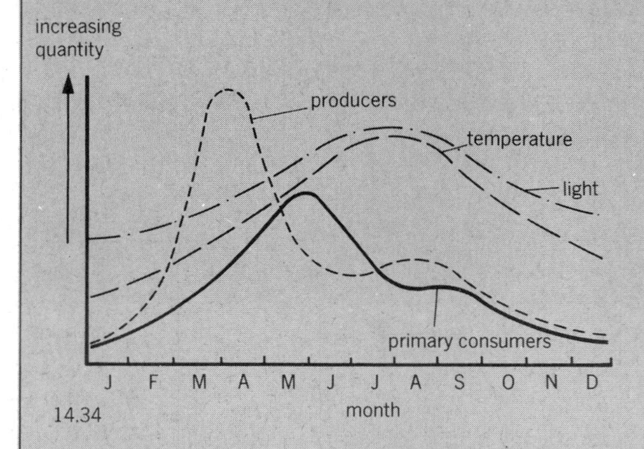

14.34

5 The graph in Fig 14.34 shows the changes in the size of a population of primary producers (plants) and primary consumers (herbivorous animals) in a lake, during one year.
(a) At what time of year is the amount of producers (i) lowest, and (ii) highest?
(b) How can the information about light and temperature given on the graph help to explain your answer to (a)?
(c) The population of primary consumers begins to increase about one month later than the increase in producers. Suggest why.
(d) Why does the population of producers begin to decrease in April?
(e) Why does the population of primary consumers begin to decrease in May and June?

Fact!

Of all the animal-borne diseases, the most devastating historically has been bubonic plague, or the black death, which has killed more than 50 million humans in recorded times. The disease is a bacterial infection injected into the blood by infected fleas from rats.

15 Genetics

15.1 Nuclei contain chromosomes carrying genes.

In the nucleus of every cell is a number of **chromosomes.** Chromosomes are long threads made of DNA (see section 15.23) and protein.

Most of the time, the chromosomes are too thin to see except with an electron microscope. But when a cell is dividing, they get shorter and fatter, so they can be seen with a light microscope (see Fig 15.1).

Chromosomes contain **genes.** It is the genes on the chromosomes which determine all sorts of things about you – what colour your eyes or hair are, whether you have a snub nose or a straight one, whether you can roll your tongue or not.

15.2 Each species has its own set of genes.

Each species of organism has its own number and variety of genes. This is what makes their body chemistry, their appearance and their behaviour different from those of other organisms.

Humans have a large number of genes. You have 46 chromosomes inside each of your cells, all with many genes on them. Every cell in your body has an exact copy of all your genes. But, unless you are an identical twin, there is no-one else in the world with exactly the same combination of genes that you have. Your genes make you unique.

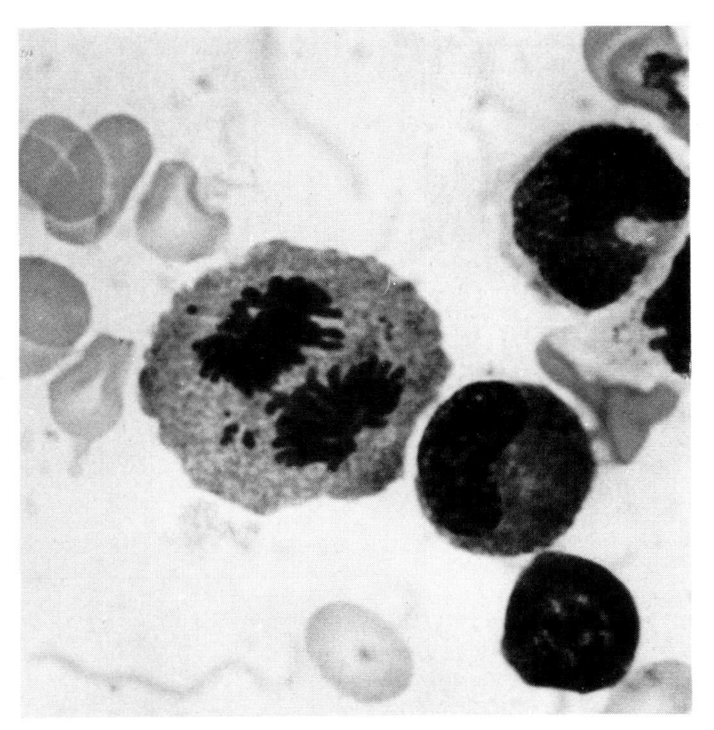

15.1 Chromosomes in a dividing cell from human bone marrow

15.3 Genes describe how to make particular proteins.

Genes provide information about making proteins. Sections 15.23 to 15.26 explain how they do this.

Every chemical reaction inside a living organism is catalysed by **enzymes.** Enzymes are proteins. So, by providing information for making enzymes, genes affect all the chemical reactions in an organism's body.

Each cell contains many genes which carry the information for making many proteins. But not all of these genes are used by any one cell. Just a few genes will be 'switched on' in any one cell at any one time. If you have red hair, for example you must have a red hair gene in all of your cells. But this gene will only have an effect in cells where hair grows, such as on your scalp. In heart cells, this gene will be switched off.

15.4 Most cells contain pairs of homologous chromosomes.

Fig 15.2(a) is a photograph of all the chromosomes from a human cell which is about to divide.

In Fig 15.2(b) the photographs of the chromosomes have been rearranged. You can see that there are, in fact, 23 pairs of chromosomes. The two chromosomes in a pair are the same size and shape.

The two chromosomes of a pair are called **homologous chromosomes.** One came from the person's mother, and the other from the father.

Each chromosome of a homologous pair carries genes for the same characteristic in the same place. For example, the two chromosome 1's might each carry a gene for tongue rolling. The tongue rolling gene will be at exactly the same position, or **locus,** on each chromosome 1 (see Fig 15.4).

Because there are two of each kind of chromosome, each cell contains two of each kind of gene. Let us look at one kind of gene, the tongue rolling gene, to see how it behaves, and how it is inherited.

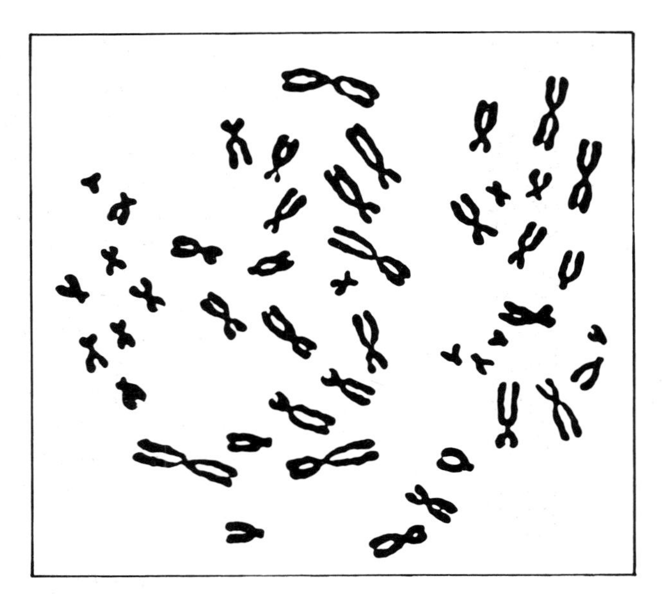

15.2(a) *Human chromosomes*

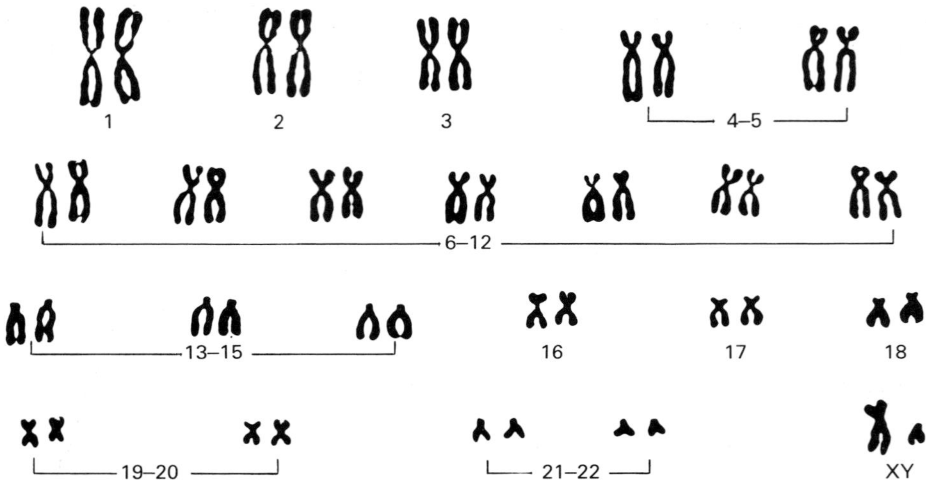

15.2(b) *Chromosomes from a normal male, arranged in order*

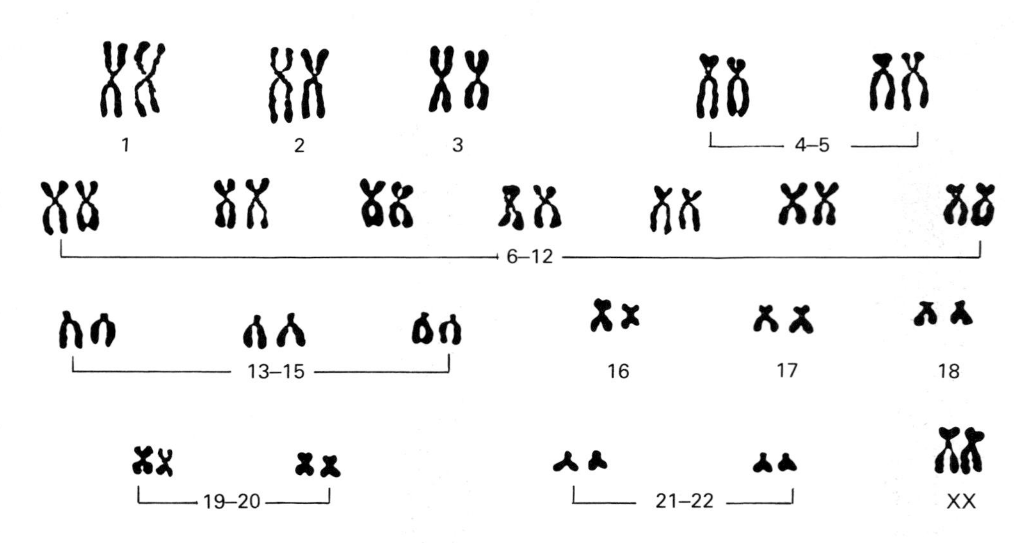

15.3 *Chromosomes from a normal female, arranged in order*

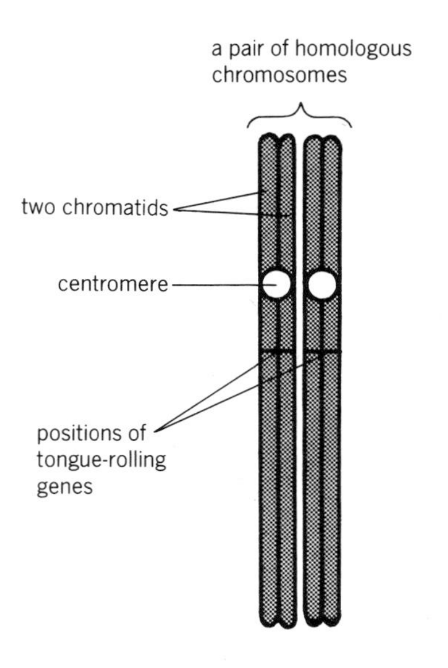

15.4 *Homologous chromosomes have genes for the same characteristic in the same position*

15.5 Each cell has two genes for any characteristic.

In each of your cells, there are two genes giving instructions about whether or not you can roll your tongue.

In humans, there are two kinds of tongue-rolling gene. One kind, *R*, allows you to roll your tongue. The other kind, *r*, does not. These two kinds of gene, defining a characteristic in different ways, are called **alleles.**

There are three possible combinations of genes for tongue rolling. You might have two *R* genes, *RR*. You might have two *r* genes, *rr*. Or you might have one of each, *Rr* (*rR* is just the same).

If the two alleles for tongue rolling in your cells are the same, that is *RR* or *rr*, then you are said to be **homozygous** for tongue rolling. If they are different, that is *Rr*, then you are **heterozygous** for tongue rolling.

15.6 Genotype can determine phenotype.

The genes that you have are your **genotype.** For tongue rolling, there are three possible genotypes – *RR*, *Rr* or *rr*.

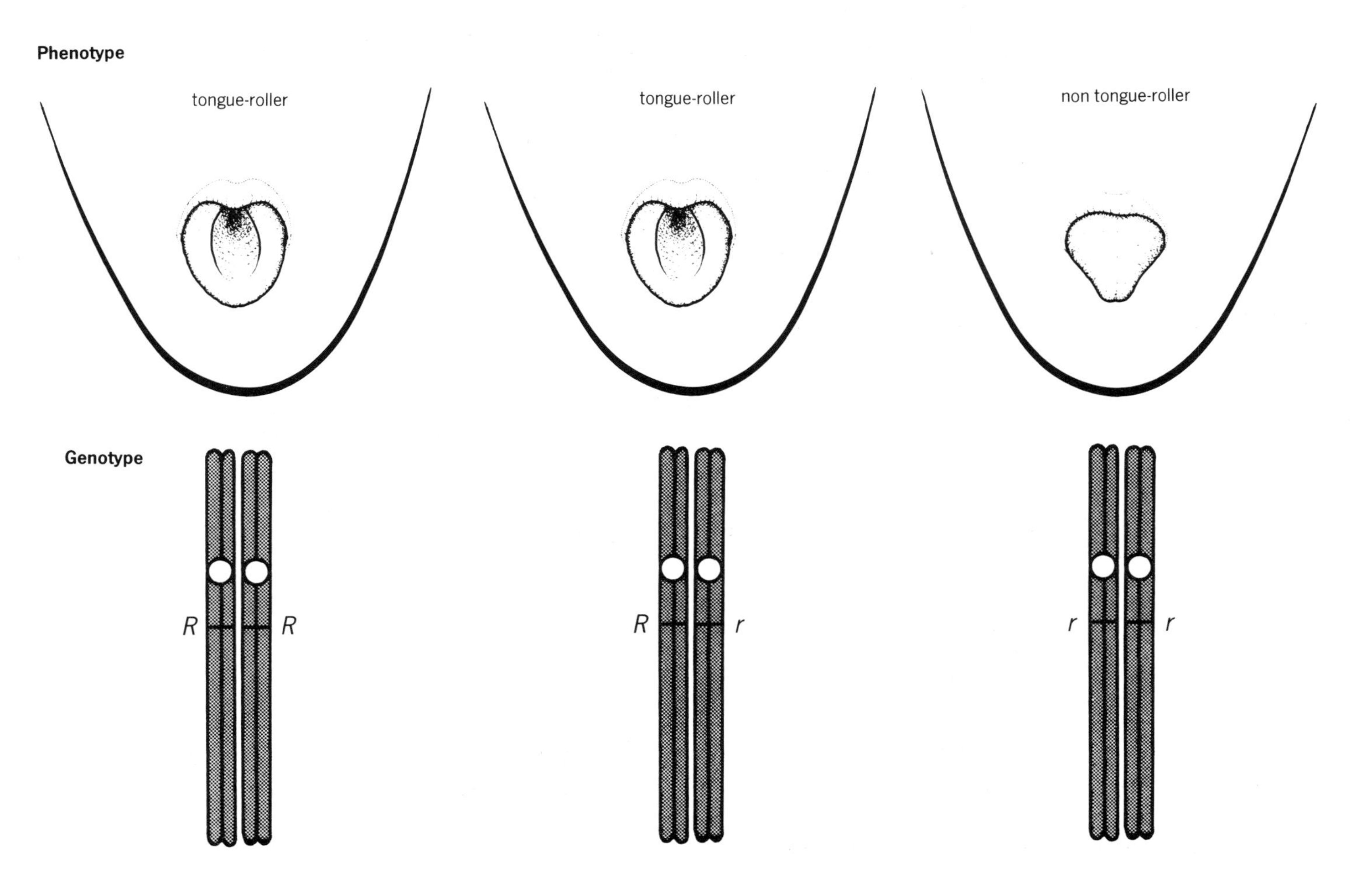

15.5 *Phenotypes and genotypes of tongue-rollers and non tongue-rollers*

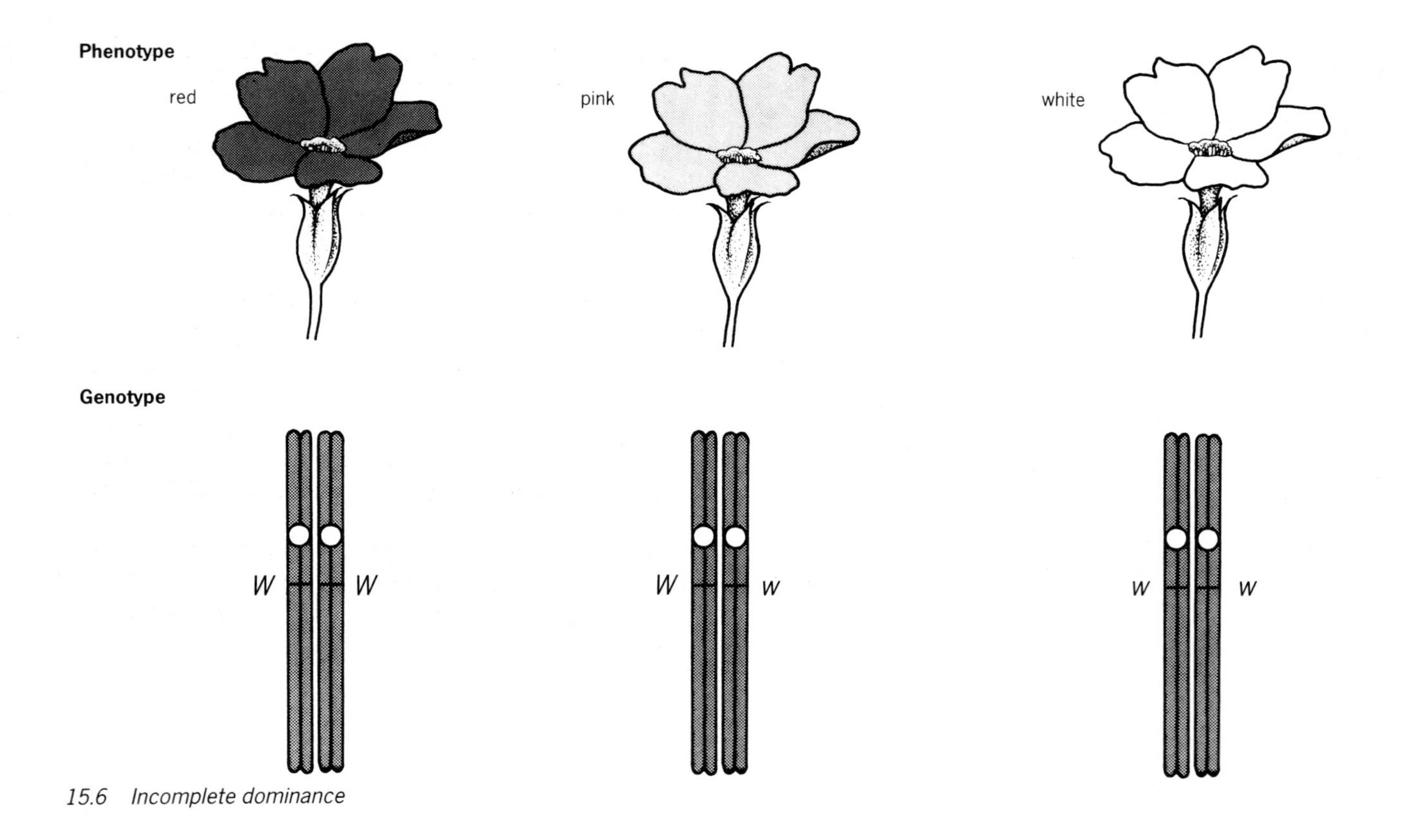

15.6 Incomplete dominance

The genotype determines whether or not you can roll your tongue. The effect that the genotype has is called your **phenotype.** Your phenotype for tongue rolling is either being able to roll your tongue, or not being able to do it.

15.7 Genes can be dominant or recessive.

So there are three kinds of genotype for tongue rolling, but only two kinds of phenotype. How does that happen?

It happens because the tongue rolling gene, *R*, is **dominant** over the non tongue rolling gene, *r*. If you are heterozygous for tongue rolling, *Rr*, then it is only the *R* gene which actually has any effect on the phenotype. You can roll your tongue. The effect of the *r* gene is hidden by the *R* one. The *r* gene is said to be a **recessive** gene.

This is summarised as follows.

genotype	*phenotype*
RR	tongue roller
Rr	tongue roller
rr	non tongue roller

15.8 Some alleles show codominance.

Sometimes, neither of a pair of alleles is dominant or recessive. Instead of one of them hiding the other in a heterozygote, they both show equally. This is called **incomplete dominance,** or **codominance.**

For example, imagine a kind of flower which has two alleles for flower colour. The gene *w* produces white flowers, while *W* produces red ones. If these alleles show incomplete dominance, then the genotypes and phenotypes are as below.

genotype	*phenotype*
ww	white flowers
Ww	pink flowers
WW	red flowers

Questions

1. What are chromosomes made of?
2. Why can you see chromosomes most easily when a cell is dividing?
3. Explain how genes affect all the chemical reactions in an organism's body.
4. What are homologous chromosomes?
5. What are alleles?
6. (a) The gene for brown eyes is dominant to the gene for blue eyes. Write down suitable symbols for these genes.
   (b) What will be the phenotype of a person who is heterozygous for this characteristic?
7. What is incomplete dominance?

15.9 Gametes have only one gene for any characteristic.

All of the cells in your body have come from one original cell. This first cell, the zygote, was made when a sperm fertilised an egg. The sperm and the egg each contained chromosomes with a certain set of genes on them, which have since been copied exactly into all your body cells.

To study inheritance, you need to understand how the sperm and the egg get their genes. Gametes, such as eggs and sperm, are made by meiosis. Fig 8.8 shows how meiosis happens. During the first division of meiosis, the chromosomes come together in their homologous pairs, and then separate from each other. In the second division, each of the chromosomes separates into its two chromatids, which are exact copies of each other. These identical chromatids are now called chromosomes from the moment they separate. They have a centromere, and are now independent of each other.

So, whereas normal cells have two of each kind of chromosome, gametes only have one of each kind. This also means that they only have one of each pair of alleles.

15.10 Alleles are separated in meiosis.

Fig 15.7 shows some of the stages in meiosis, to show what happens to one pair of genes when a sperm is made.

In this example, the person is a tongue roller, genotype *Rr*. For simplicity, only the chromosomes carrying the tongue rolling genes are shown. The other 44 have been left out.

During the first division of meiosis, these chromosomes come together and then separate. Two cells are made, one carrying the *R* gene and the other carrying the *r* gene.

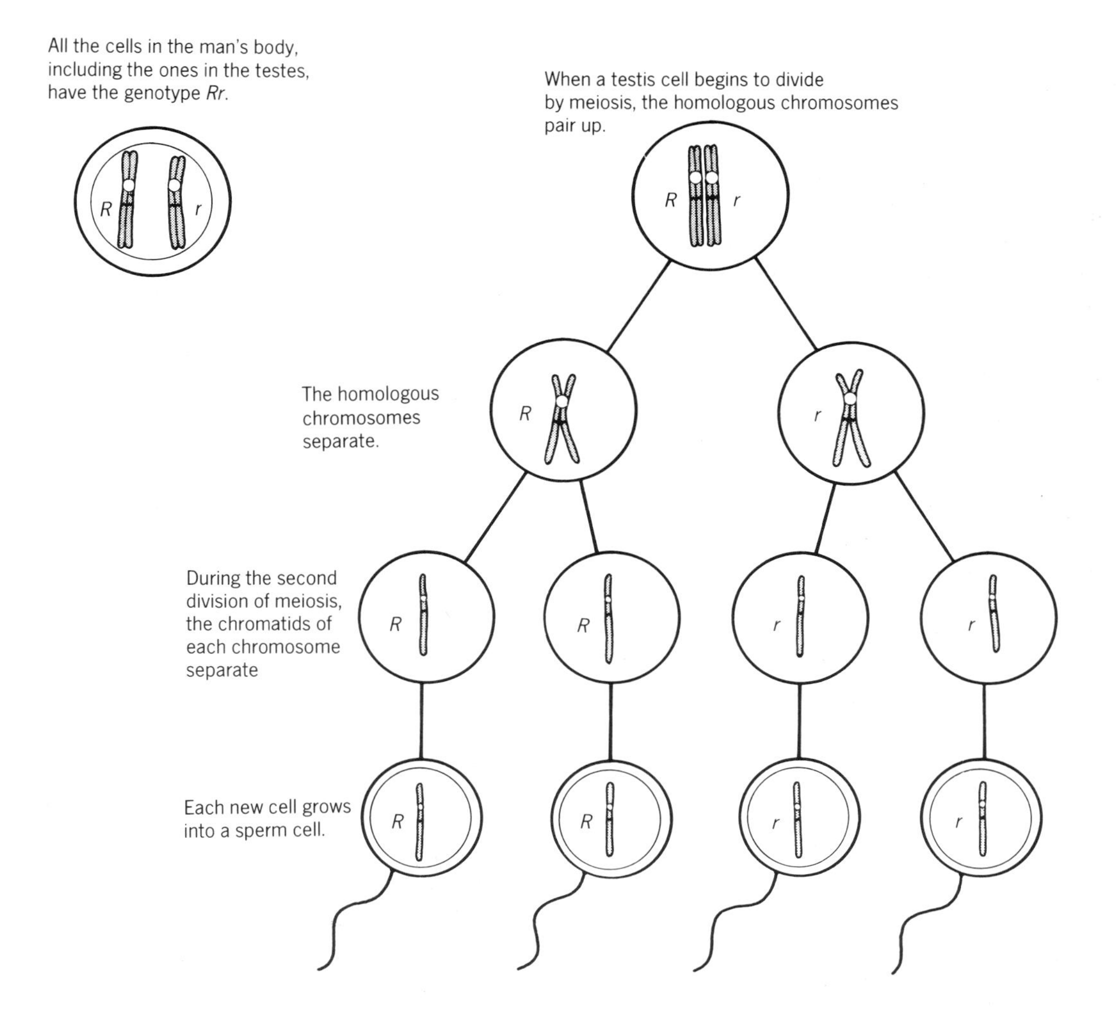

15.7 What happens to the genes of a heterozygous tongue-roller during meiosis

In the second division of meiosis, the chromosome in each cell splits into its two identical chromatids. These identical chromatids become full chromosomes during the development of the sperms. So at the end of meiosis, there are four cells, which will all grow into sperm cells. Half of them have the *R* gene and half have the *r* gene.

15.11 Genes and fertilisation.

If this man maries a woman with the genotype *rr*, will their children be able to roll their tongues or not?

The eggs that are made in the woman's ovaries are also made by meiosis. If you use Fig 15.7 to see what happens to her chromosomes during meiosis, you will see that she can only make one kind of egg. All of the eggs will carry an *r* gene.

During sexual intercourse, hundreds of thousands of sperms will begin a journey towards the egg. About half of them will carry an *R* gene, and half will carry an *r* gene. If there is an egg in the woman's oviduct, it will probably be fertilised. There is an equal chance of either kind of sperm getting there first.

If a sperm carrying an *R* gene wins the race, then the zygote will have an *R* gene from its father and an *r* gene from its mother. Its genotype will be *Rr*. After nine months, a baby will be born with the genotype *Rr*.

But if a sperm carrying an *r* gene manages to fertilise the egg, then the baby will have the genotype *rr*, like its mother (see Fig 15.8).

A man of genotype *Rr* produces equal numbers of *R* sperm and *r* sperm.

A woman of genotype *rr* produces eggs of genotype *r*.

There is an equal chance of either type of sperm fertilising an egg.

So the zygote formed has an equal chance of having the genotype *Rr* or *rr*.

15.8 Fertilisation between a heterozygous tongue-roller and a non tongue-roller

15.12 Genetic crosses must be written clearly.

There is a standard way of writing out all of this information. First, write down the phenotypes and genotypes of the parents. Next, write down the different types of gametes they can make, like this.

parents' phenotypes	tongue roller	non tongue roller
parents' genotypes	*Rr*	*rr*
gametes	*R* or *r*	*r*

The next step is to write down what might happen during fertilisation. Either kind of sperm might fuse with an egg. Lines can be drawn to join up each of the two kinds of sperm with the egg, like this.

gametes	*R* or *r*	*r*
offspring genotypes	*Rr*	*rr*
offspring phenotypes	tongue roller	non tongue roller

To finish your summary of the genetic cross, write out in words what you would expect the offspring from this cross to be.

'Approximately half of the children would be heterozygous tongue rollers, and half would be homozygous non tongue rollers.'

The offspring are sometimes called the first filial generation, or **F_1 generation.**

15.13 Another example.

What happens if both parents are heterozygous tongue rollers?

parents' phenotypes — tongue roller — tongue roller

parents genotypes — *Rr* — *Rr*

gametes

F_1 genotypes

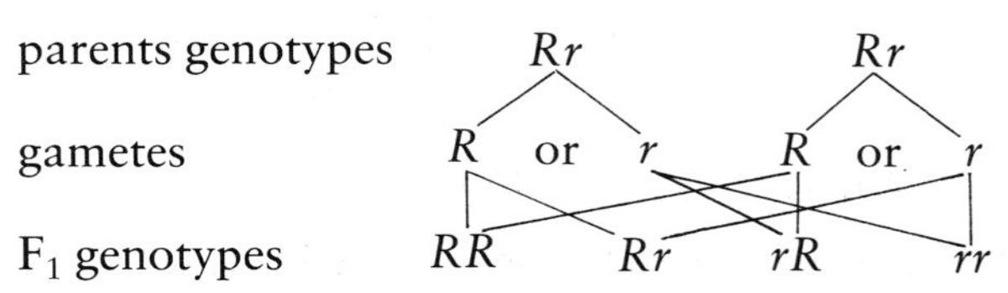

F_1 phenotypes — tongue roller — tongue roller — tongue roller — non tongue roller

Approximately 1/4 of the children would be homozygous tongue rollers, 1/2 would be heterozygous tongue rollers, and 1/4 would be homozygous non tongue rollers.

An alternative to drawing lines between the male and female gametes is to draw a square. Sperm genotypes are written down the side, and the egg genotypes along the top, like this.

		eggs	
		R	*r*
sperm	*R*	*RR*	*Rr*
	r	*rR*	*rr*

The offspring genotypes can then be filled in inside the squares.

15.14 Probability in genetics.

In the last example, there were four offspring at the end of the cross. This does not mean that the man and woman will have four children. It simply means that each time they have a child, these are the possible genotypes that it might have.

When they have a child, there is a 1 in 4 chance that its genotype will be *RR*, and a 1 in 4 chance that its genotype will be *rr*. There is a 2 in 4, or rather 1 in 2, chance that its genotype will be *Rr*.

However, as you know, probabilities do not always work out. If you toss a penny up four times you might expect it to turn up heads twice and tails twice. But does it always do this? Try it and see.

With small numbers like this, probabilities do not always match reality. If you had the patience to toss your coin up a few thousand times, though, you will almost certainly find that you get much more nearly equal numbers of heads and tails.

The same thing applies in genetics. The offspring genotypes which you work out are only probabilities. With small numbers, they are unlikely to work out exactly. With very large numbers of offspring from one cross, they are more likely to be accurate.

So, if the man and woman in the last example had eight children, they might expect six of them to be tongue rollers and two to be non tongue rollers. But they should not be too surprised if they have three tongue rollers and five non tongue rollers!

15.15 Solving genetics problems.

Here is a typical question that you might be given to answer. Follow the steps you would use to answer it.

> 'The seeds resulting from a cross between a tall pea plant and a dwarf one produced plants all of which were tall. When these plants were allowed to self-pollinate, the resulting seeds produced 908 tall plants and 293 dwarf plants. Account for these results. What would be the results of interbreeding the dwarf plants?'

Firstly, work out whether or not one gene is dominant over the other. Next, decide on symbols. Then write a list of possible genotypes and phenotypes. Your answer might go like this.

> 'There must be two alleles, one producing tallness and the other dwarfness. The plants are all either tall or dwarf, and no medium sized ones are mentioned. So one gene must be completely dominant over the other.
>
> When a tall plant is crossed with a dwarf plant, all the offspring are tall. This shows that it is the tall gene which is dominant.
>
> Therefore, let us use the symbol *T* for the gene for tall plants, and the symbol *t* for the gene for dwarf plants. The gene *T* is dominant over the gene *t*, which is recessive.
>
> The possible genotypes and phenotypes are these.
>
> | genotype | phenotype |
> |---|---|
> | *TT* | tall |
> | *Tt* | tall |
> | *tt* | dwarf' |

Now you are ready to begin explaining the first cross.

> 'The first cross is between a tall pea plant and a dwarf one. The possible genotypes of the tall pea plant are *TT* or *Tt*. The dwarf plant must have the genotype *tt*.
>
> If the tall plant is heterozygous, *Tt*, then half of its gametes would contain the *t* gene. During fertilisation, zygotes would be formed with the genotype *tt*, which would appear dwarf. But as none of the offspring of this cross are dwarf, the tall parent must be TT.
>
> The first cross is therefore as follows.
>
> | | | |
> |---|---|---|
> | parents' phenotypes | tall | dwarf |
> | parents' genotypes | *TT* | *tt* |
> | gametes | *T* | *t* |
> | F_1 genotypes | *Tt* | |
> | F_1 phenotypes | tall' | |

This has explained the first sentence of the question. Next, these tall plants self-pollinate, in other words, they cross with themselves. The offspring they produce are the second filial generation, or F_2 generation.

'The offspring from the first cross are all heterozygous, Tt. When they self-pollinate, the results would be expected to be as follows.

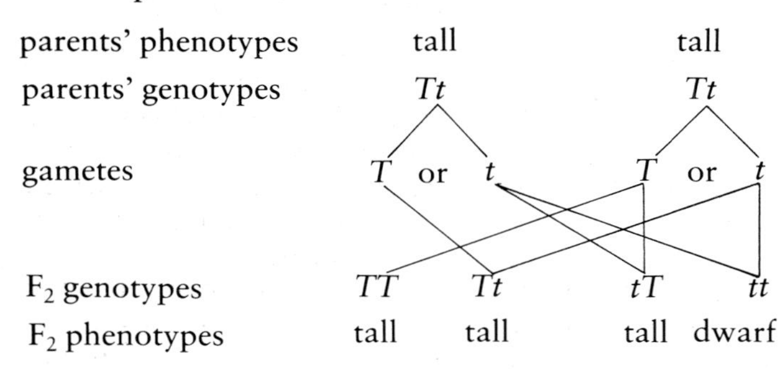

Approximately ¾ of the offspring would be expected to be tall and ¼ short. This fits in well with the figures of 903 tall, and 293 dwarf, which is a ratio of approximately 3 tall: 1 dwarf.'

Now all that is left to do is to answer the last sentence of the question.

'The dwarf plants all have the genotype tt. All of the gametes they produce will carry a t gene. Therefore all of their offspring will also be tt, that is homozygous dwarf plants.'

15.16 Back crosses help to determine genotype.

In the last example, there were two kinds of pea plant, tall and dwarf. Once you have decided that the gene for dwarfness, t, is recessive, then you know that the genotype of any dwarf plants must be tt. But a tall plant could be either Tt or TT.

If you had a tall pea plant, and wanted to know its genotype, how could you find out? The best way would be to cross the plant with a dwarf plant, and then see what kind of offspring you got from the cross. If your original plant was homozygous, TT, then all the offspring would have the genotype Tt, and be tall (see the first cross in section 15.15).

But if your plant was heterozygous, Tt, then not all the offspring will be tall. Try working it out. You will find that half of the offspring would have the genotype tt, and be dwarf.

By crossing a plant showing the dominant characteristic (in this case tallness) with a homozygous recessive plant (in this case, a dwarf one) you can find out the genotype of the plant with the dominant characteristic. This cross is called a **back cross.**

15.17 'Pure-breeding' means homozygous.

Some populations of animals or plants always have offspring just like themselves. For example, a rabbit breeder might have a strain of rabbits which all have brown coats. If he interbreeds them with one another, all the offspring always have brown coats as well. He has a **pure-breeding** strain of brown rabbits. Pure-breeding strains are always homozygous for the pure-breeding characteristic.

Questions

1. If a normal human cell has 46 chromosomes, how many chromosomes are there in a human sperm cell?
2. Using the symbols W for normal wings, and w for vestigial wings, write down the following.
   (a) the genotype of a fly which is heterozygous for this characteristic
   (b) The possible genotypes of its gametes
3. Using the layout shown in section 15.15, work out what kind of offspring would be produced if the heterozygous fly in question 2 mated with one which was homozygous for normal wings.
4. In humans, the gene for red hair, c, is recessive to the gene for brown hair, C. A man and his wife both have brown hair. They have five children, three of whom have red hair, while two have brown hair. Explain how this may happen.
5. In Dalmatian dogs, the gene for black spots is dominant to the gene for liver spots. If a breeder has a black-spotted dog, how can she find out whether it is homozygous or heterozygous for this characteristic?

15.18 Sex is determined by X and Y chromosomes.

If you look carefully at Figs 15.2(b) and 15.3 you will see that the last pair of chromosomes is not the same in each case. In the second photograph, of a woman's chromosomes, the last pair are alike. In the first photograph which is of a man's chromosomes, the last pair are not alike. One is much smaller than the other.

This last pair of chromosomes is responsible for determining what sex a person will be. They are called the **sex chromosomes.** A woman's chromosomes are both alike and are called X chromosomes. She has the genotype XX.

A man, though, only has one X chromosome. The other, smaller one, is a Y chromosome. He has the genotype XY (see Fig 15.9).

Fact! An isolated group of people living in the Zambesi river valley in Africa have only two toes on each foot. This mutation occurred several generations ago, and there are now dozens of these two-toed people. They can walk perfectly well, and can even use their two toes to pick up bottles, and so on.

15.19 Sex is inherited

You can work out sex inheritance in just the same way as for any other characteristic, but using the letter symbols to describe whole chromosomes, rather than individual genes.

parents' phenotypes	male	female
parents' genotypes	XY	XX
gametes	X or Y	X
F_1 genotypes	XX	XY
F_1 phenotypes	female	male

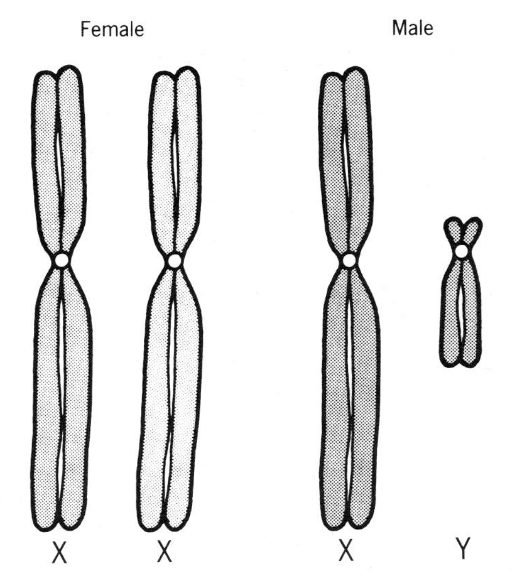

15.9 The sex chromosomes

Half the children will probably be male, and half female.

15.20 Some genes are sex linked.

The X and Y chromosomes do not only determine sex. They have other genes on them as well.

If you look back to section 15.4, you will remember that a pair of homologous chromosomes always carry genes for the same characteristic at the same place on the chromosome. This means that you have two genes for every characteristic in your cells.

However, most parts of the X and Y chromosomes do not carry the same genes. So a man has only one of most of the genes which are carried on the X chromosome. A woman, though, has two, just as for any other kind of gene.

We know quite a few of the genes which are carried on the X chromosome. One of them is a gene for blood clotting. The dominant gene, *H*, allows your blood to clot normally. But the recessive gene, *h*, causes **haemophilia,** a disease where even a bruise or small scratch will go on bleeding for a very long time.

There are three possible genotypes that a woman might have for the haemophilia characteristic (see Fig 15.10).

genotype	phenotype
X^HX^H	normal
X^HX^h	carrier
X^hX^h	haemophiliac

A **carrier** is someone who has a recessive gene in their cells, but has a normal phenotype. A woman with the genotype X^HX^h seems perfectly normal. Her blood clots in the usual way.

Fact! If a colour blind man married a normal woman none of his children would be colour blind. But if one of his daughters marries a normal man the chances are that half their sons will be colour blind. Can you explain this?

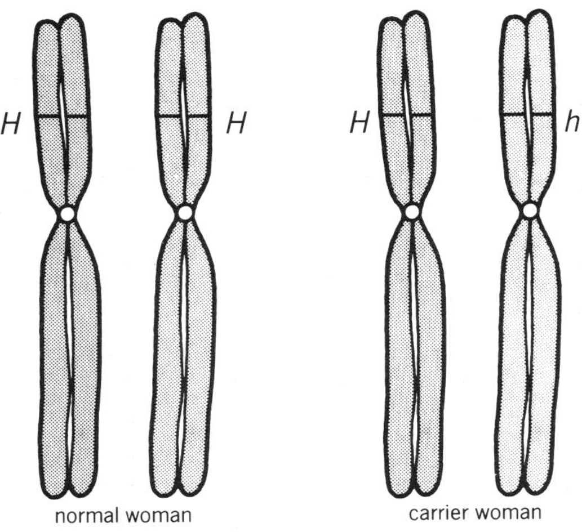

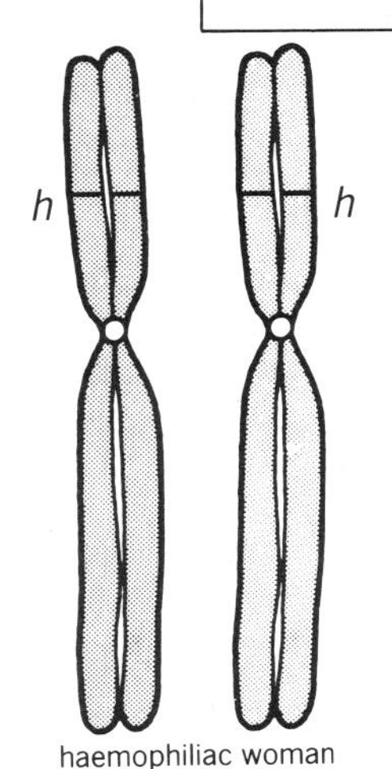

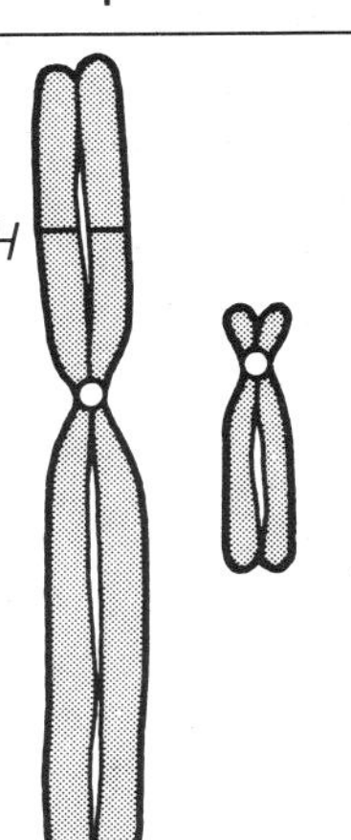

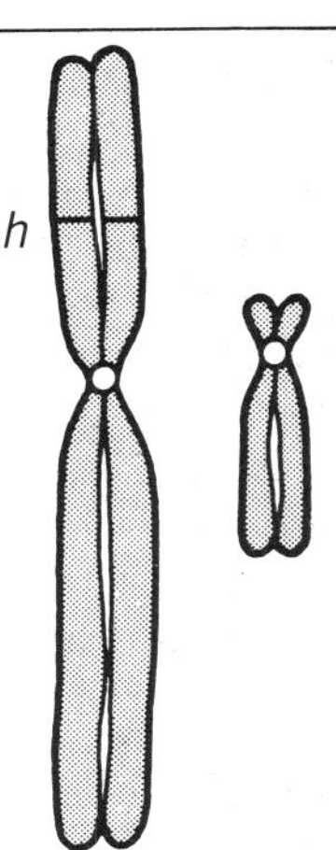

15.10 Haemophilia genotypes and phenotypes

There are only two possible genotypes for a man. This is because the Y chromosome does not have a haemophilia or blood clotting gene of any kind.

genotype	phenotype
X^HY	normal
X^hY	haemophiliac

A gene such as this, which is carried on the non-homologous part of a sex chromosome, is called a **sex-linked gene.** This is because the way it is inherited is linked with a person's sex. You will see how this happens in the next section.

15.21 Women can pass sex linked genes to their sons.

What would happen if a carrier woman married a normal man and produced children?

parents' phenotypes	normal man		carrier woman	
parents' genotypes	X^HY		X^HX^h	
gametes	X^H	Y	X^H	X^h
F_1 genotypes	X^HX^H	X^HX^h	X^HY	X^hY
F_1 phenotypes	normal female	carrier female	normal male	male haemophiliac

About half of the male children will probably be haemophiliacs.

Queen Victoria was a carrier for haemophilia. Her husband, Prince Albert, was normal. Fig 15.13 shows Queen Victoria's family tree.

15.22 Fruit flies are good for breeding experiments.

To find out how characteristics are inherited, breeding experiments must be done. Some organisms are much better than others for this kind of experiment. The best kind of organism to choose is one with the following features.

It is easy to keep in a laboratory, and breeds readily.

It produces large numbers of offspring.

It does not take too long to become adult, and breeds quickly, so that you can obtain a large number of generations in a short time.

It has several variations which are easy to see.

One organism which fits all these criteria is the fruit fly *Drosphila melanogaster*. Many geneticists have worked with this fly, and some still do. Fig 15.12 shows some of its varieties.

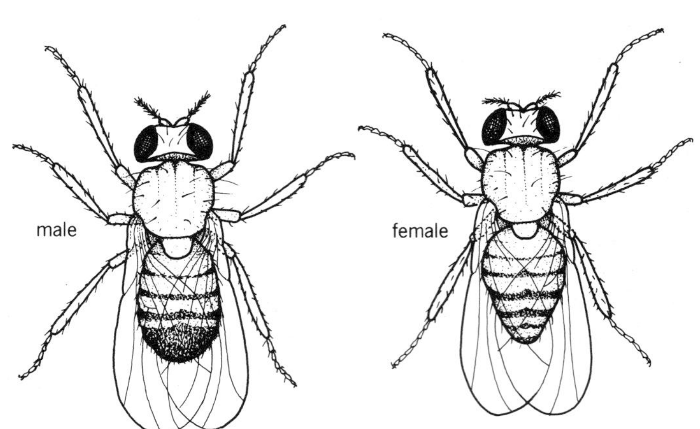

15.11 Male and female Drosophila. How do their abdomens differ?

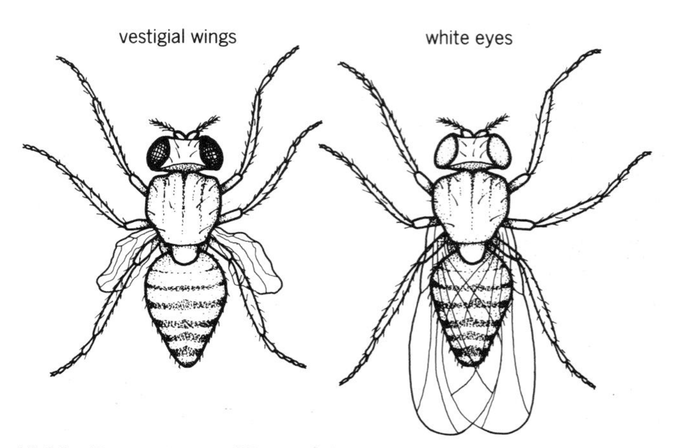

15.12 Two varieties of Drosophila

Questions

1. On which chromosome are most sex-linked genes carried?
2. What is a carrier?
3. The gene for colour-blindness, *c*, is a recessive gene, carried on the non-homologous part of the X chromosome.
   (a) Write down the genotype of a colour-blind man.
   (b) A man and his wife both have normal vision. They have three sons, two of whom are colour-blind. Explain how this may happen.
4. Why is the fruit fly, *Drosophila melanogaster*, often used in breeding experiments?

Investigation 15.1 Breeding beads

In this experiment, you will use two containers of beads. Each container represents a parent. The beads represent the gametes they make. The colour of a bead represents the genotype of the gamete. For example, a red bead might represent a gamete with genotype *R*, for tongue rolling. A yellow bead might represent a gamete with the genotype *r*, for non tongue rolling.

1 Put 100 red beads into the first beaker. These represent the gametes of a person who is homozygous for tongue rolling, *RR*.
2 Put 50 red beads and 50 yellow beads into the second beaker. These represent the gametes of a heterozygous person with the genotype *Rr*.
3 Close your eyes, and pick out one bead from the first beaker, and one from the second. Write down the genotype of the 'offspring' they produce. Put the two beads back.
4 Repeat step 3 one hundred times.
5 Now try a different cross, for example *Rr* crossed with *Rr*.

continued

Questions

1 In the first cross, what kind of offspring were produced, and in what ratios?
2 Is this what you would have expected? Explain your answer.
3 Why must you close your eyes when choosing the beads?
4 Why must you put the beads back into the beakers after they have 'mated'?

Fact!

There are 1115 known mutant forms of the fruit fly, *Drosophila melanogaster*.

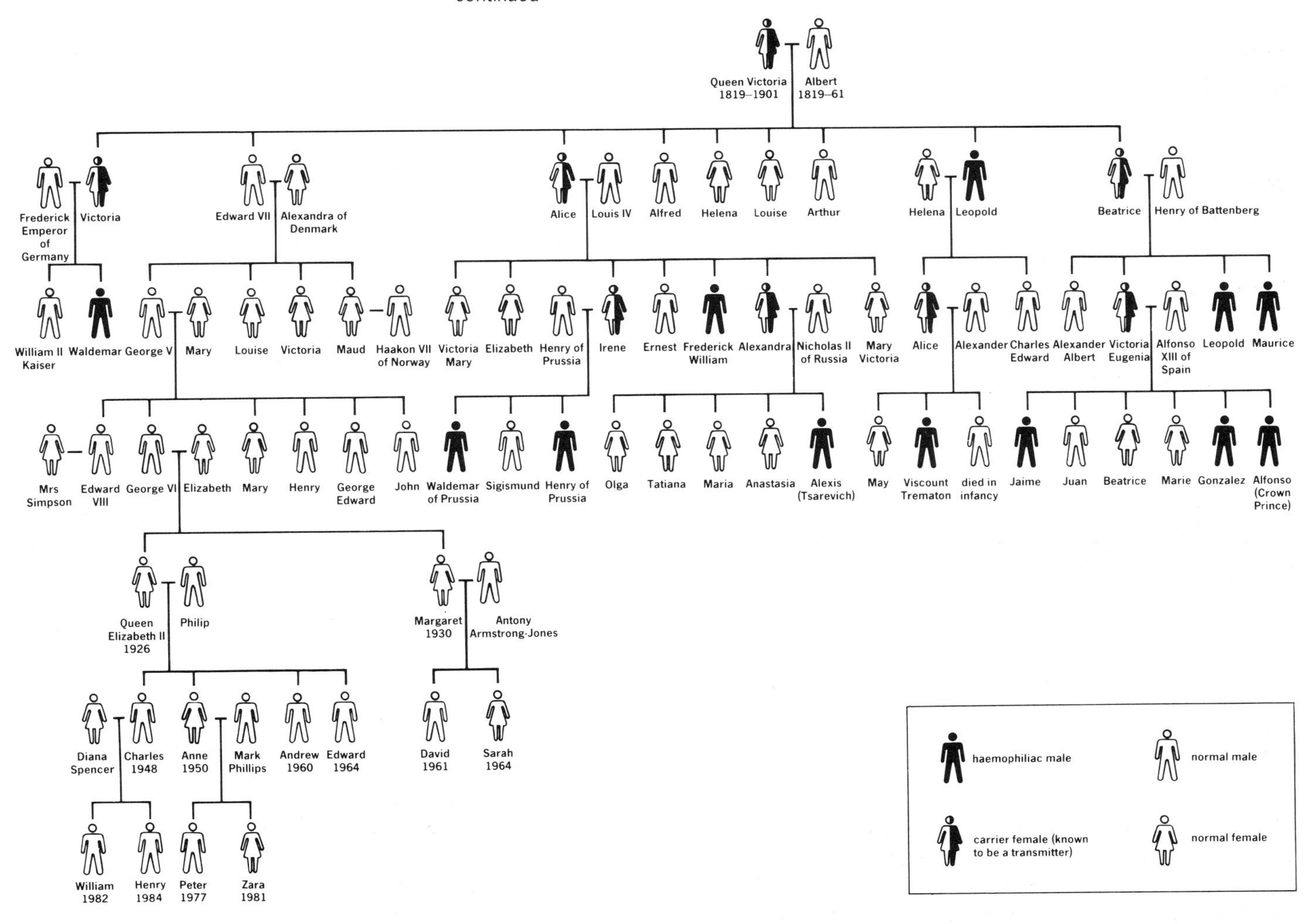

15.13 Queen Victoria's family tree

DNA and protein synthesis

15.23 A DNA molecule is a double helix.

Chromosomes are made of protein, and a substance called **DNA.** It is the DNA which carries the instructions for the proteins that the cell is to make.

DNA is short for **deoxyribonucleic acid.** Fig 15.14 shows a short length of DNA. It is made of two strands, twisted together into a spiral or helix. The two strands are linked together through the bases.

There are four kinds of base in DNA. They are adenine (**A**), thymine (**T**), cytosine (**C**) and guanine (**G**). The bases are different sizes and shapes, so that **A** will only fit next to **T**, and **C** will only fit next to **G.**

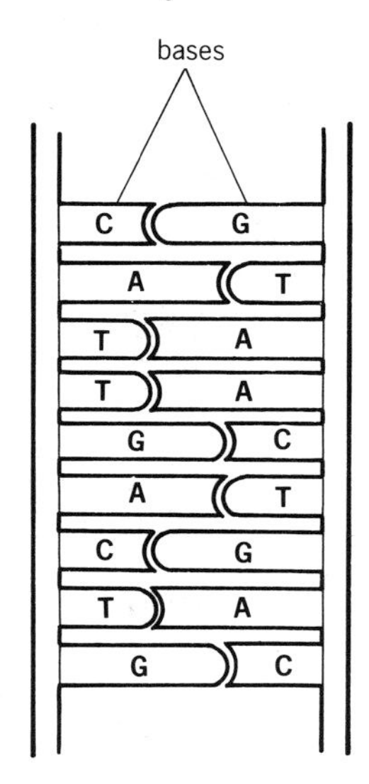

A DNA molecule is made of two strands, linked through the bases.

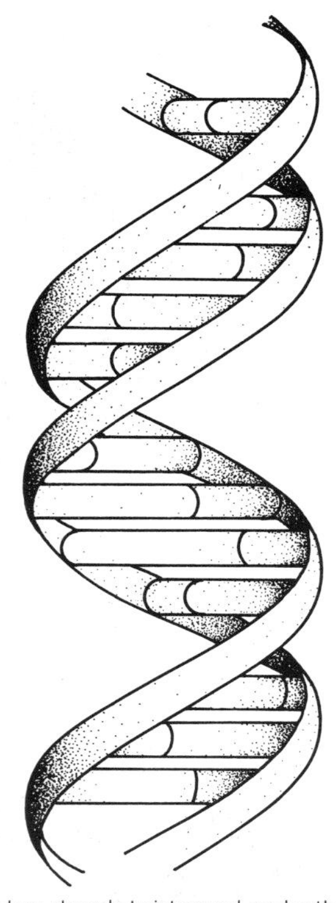

The two strands twist round each other, forming a double helix.

15.14 Part of a DNA molecule

15.24 Each protein has its own amino acid sequence.

A protein is a long chain of amino acids. There are about twenty different amino acids, and there may be hundreds or thousands of them in one protein. The order in which the different amino acids are linked together determines what kind of protein is made. There are usually plenty of amino acids of each kind, in solution in the cytoplasm. They are linked together, one by one, to make protein molecules. Proteins are made in the cytoplasm of every cell, on the ribosomes (see section 1.11).

15.25 A codon names an amino acid.

The amino acids are not joined together haphazardly. The order in which they are joined, and so the kind of protein they make, is very carefully organized. The instructions for this are kept in the DNA, in the nucleus.

The sequence of the bases in the DNA is a code for the sequence of amino acids in the proteins to be made. A row of three bases, called a **codon,** codes for one amino acid. Each amino acid has a different code 'word' of three 'letters'. For example, the sequence **GGC** on the DNA codes for the amino acid glycine. **GUC** means valine. So if a section of a DNA molecule runs **GGCGUC** then glycine will be joined to valine when a protein is made.

There is a codon for every amino acid, and also for beginning and ending a protein molecule. A length of DNA which codes for one protein is called a **gene.**

15.26 Messenger RNA copies the genetic code.

Although DNA is found in the nucleus, proteins are made in the cytoplasm. A messenger molecule is used to carry the information from the nucleus to the ribosomes in the cytoplasm.

The messenger is a molecule called messenger RNA, or **mRNA.** It copies the instructions from DNA, and then takes them out to the ribosomes (see Fig 15.15).

15.27 Accidental changes in DNA produce mutations.

DNA is a very important material. Almost everything that happens in a living organism is controlled by it. So all of the processes that DNA is involved in – mitosis, meiosis and protein synthesis – are very carefully controlled. The working of the cell is designed to ensure

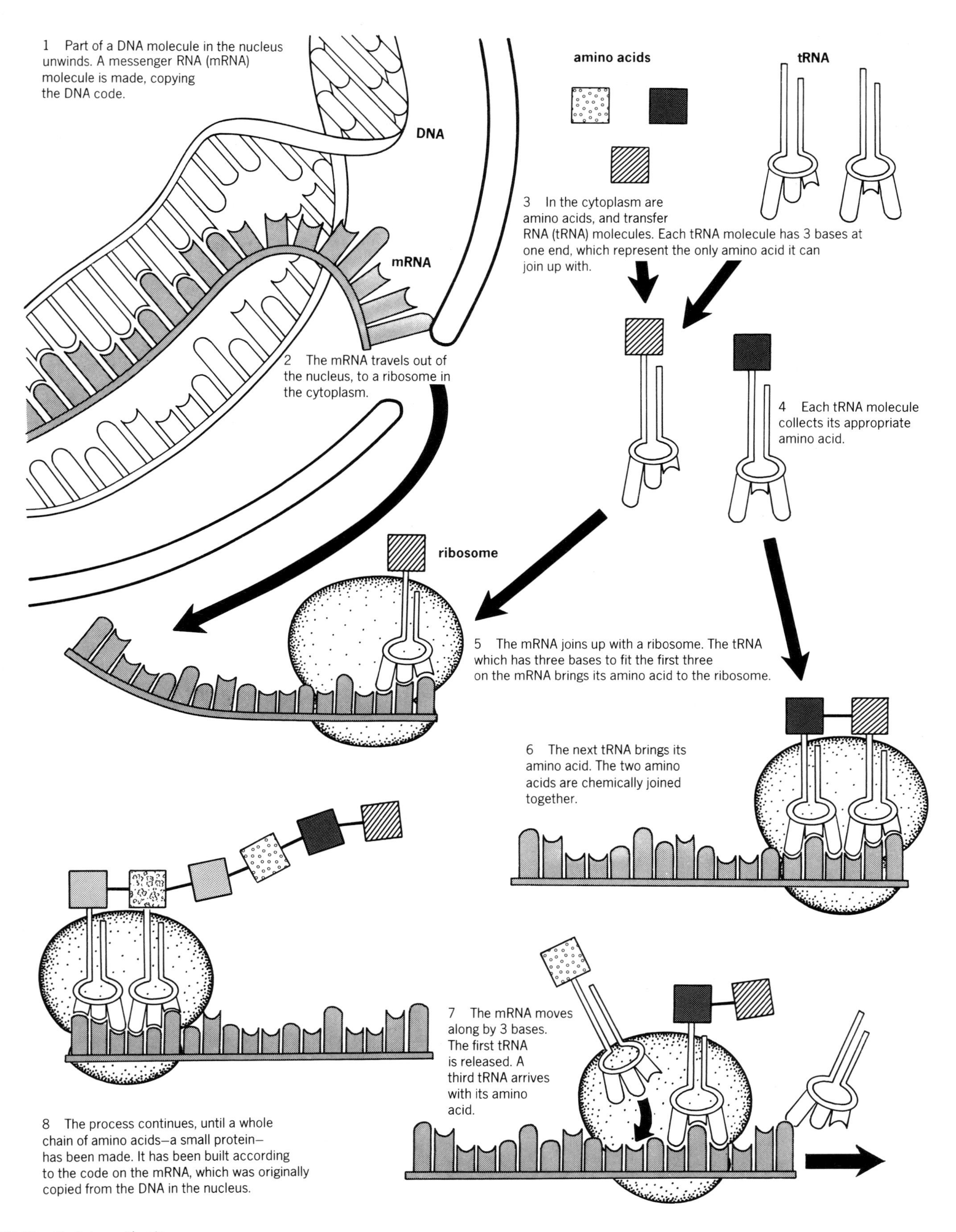

15.15 Protein synthesis

that the instructions carried on the DNA molecules are never damaged.

But occasionally things do go wrong. One time that this can happen is during meiosis. Sometimes, instead of homologous chromosomes separating perfectly, one may go the wrong way. Gametes will then be formed with the wrong number of chromosomes.

This sometimes happens when eggs are being made in a woman's ovaries. The chromosome 21's may fail to separate. Eggs are made with two chromosome 21's instead of one. If such an egg is fertilised by a normal sperm, the child which results will have three chromosome 21's in every cell. It will have **Down's syndrome,** sometimes called mongolism.

Another type of change that may occur is in the DNA molecule itself. Normally, the sequence of bases in DNA never changes. It is copied very carefully, and passed on unchanged from parent to offspring. But sometimes, one or more bases in the DNA may be altered, or moved out of sequence. This changes the sequence of amino acids in the protein molecule that is made, often with unfortunate effects (see Fig 15.16).

Changes like this are called **mutations**. Most mutations are harmful, but just occasionally a mutation may turn out to produce a better characteristic than the original. An example of this is the mutation of the pale form of the peppered moth to a dark form in some parts of Great Britain. This is described in Chapter 16.

15.16 A mutation has caused the normal black skin colour to be missing in this African boy

15.28 Radiation can increase mutation rate.

Mutations often happen for no apparent reason. However we do know of many factors which make mutation more likely. One of the most important of these is **radiation.** Radiation can damage the bases in DNA molecules. If this happens in the ovaries or testes, then the altered DNA may be passed on to the offspring.

15.29 Genetic engineering makes genes work for man.

The understanding of how DNA codes for protein synthesis in a cell has opened up tremendous possibilities. A new kind of technology is now developing, called **genetic engineering**.

Genetic engineering involves the introduction of an extra piece of DNA into a cell. If this is done in a particular way, the cell will make the kind of protein which is coded for by that DNA.

For example, one protein which many doctors would like to have good supplies of is **interferon.** This is a substance which human cells make to stop viruses multiplying inside the body. Interferon may prove to be very useful in treating some diseases, possibly even some kinds of cancer. We do not as yet know how useful it could be, because it has been very expensive to produce and is in very short supply.

Genetic engineering may make it possible to produce large quantities of interferon quite cheaply. A piece of DNA which codes for the production of interferon could be introduced into some bacteria. The bacteria would then be bred in very large numbers. They would produce interferon, which could be purified and marketed.

Genetic engineering is still a very young field of technology, and there are many problems to be overcome in the large scale production of substances in this way. However the possibilities are almost endless. In theory, it would be possible to 'design' a new kind of organism, by introducing a carefully worked out package of DNA into a cell. Should biologists be allowed to do this? It is a difficult question, which has no easy answer.

Chapter revision questions

1. One strand of a DNA molecule has bases in the sequence **ACCGATAG**. What will be the sequence on the other strand?
2. Where is DNA found?
3. Where are proteins made?
4. What is a codon?
5. What is a gene?
6. How is the information on DNA transferred to the ribosomes?

16 Evolution

16.1 Organisms have changed through time.

One of the questions which has always interested people is 'Where did all the different kinds of living organism come from?' Were they all created at the same time, or have they gradually changed, or evolved, to become what they are now?

Most cultures, all around the world, have some kind of creation story, which describes how the world was made, and all the animals and plants created. Up until the nineteenth century, most people in Europe and America believed that the creation story described in Genesis, the first book of the Bible, was literally true. Some people still believe this today.

Today, the generally accepted idea is that the forms of life that now exist have gradually developed from much simpler ones. We think that life began on Earth about 4000 million years ago. Since then, more complex and varied organisms have developed – and are still developing. This is the process of evolution.

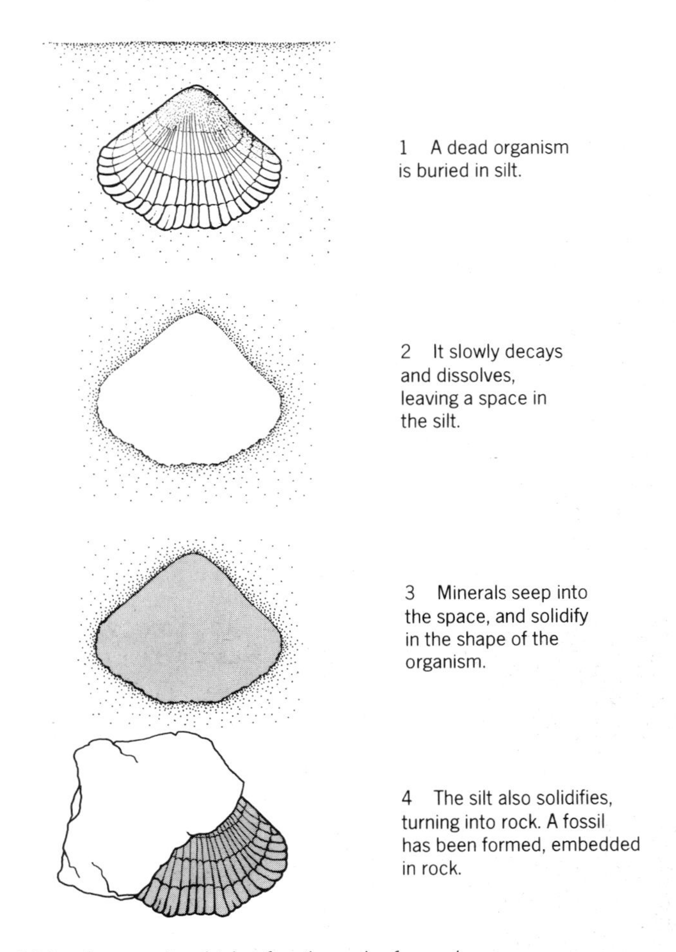

16.1 One way in which a fossil may be formed

Evidence for evolution – fossils

16.2 Fossils provide evidence for evolution.

A **fossil** is the remains or impression of a living organism which have become preserved in rock. Fig 16.1 illustrates how a fossil can be formed.

It is usually possible to work out the age of the rock, and so we know roughly how long ago the fossil was formed. Some of the oldest fossils that have been found are about 3000 million years old. They are fossils of very simple organisms, rather like bacteria. No other kinds of living things seem to have existed at this time. It is not until 1200 million years ago that simple protistans like *Amoeba* came into existence.

As we look at fossils from more recently formed younger rocks, more complex kinds of organism begin to appear. Fig 16.2 shows the age of the rocks in which the fossils of various kinds of plants and animals have first been found.

One way of explaining this sequence is to suggest that the more recent organisms such as mammals have developed from the earlier forms such as reptiles. We can find even more convincing evidence for this by looking at a small section of the fossil record in detail (see Fig 16.5).

In fact, it is only very rarely that enough fossils have been found for us to be able to 'see' one kind of organism evolving from another. Usually, there are big gaps in the fossil record, and we can only guess how the changes took place. The formation of a fossil is a rare event, and so it is not surprising that not enough fossils have yet been found for us to see exactly what happened to each kind of organism that lived in the past.

16.2 The ages of some of the earliest fossils of some types of organism

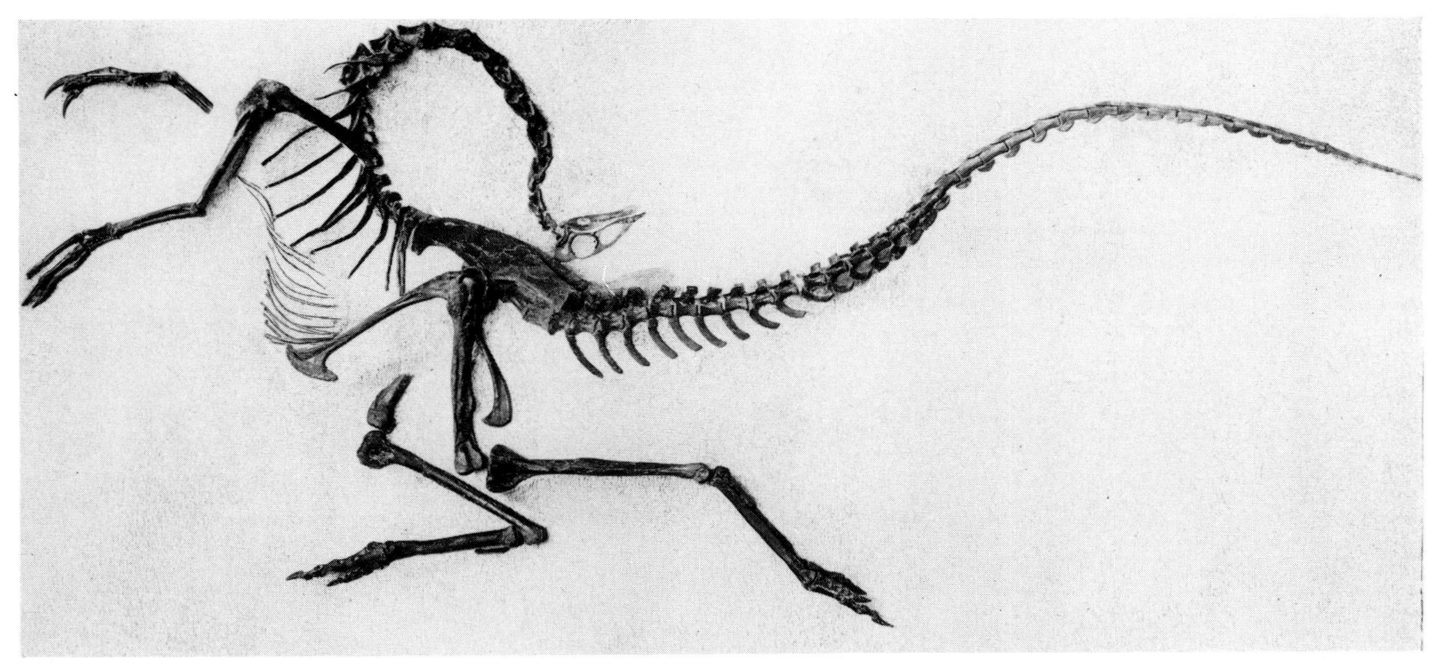

16.3 The fossil bones of a dinosaur, Struthiomimus, which lived about 100 million years ago. It ran fast on its long hind legs.

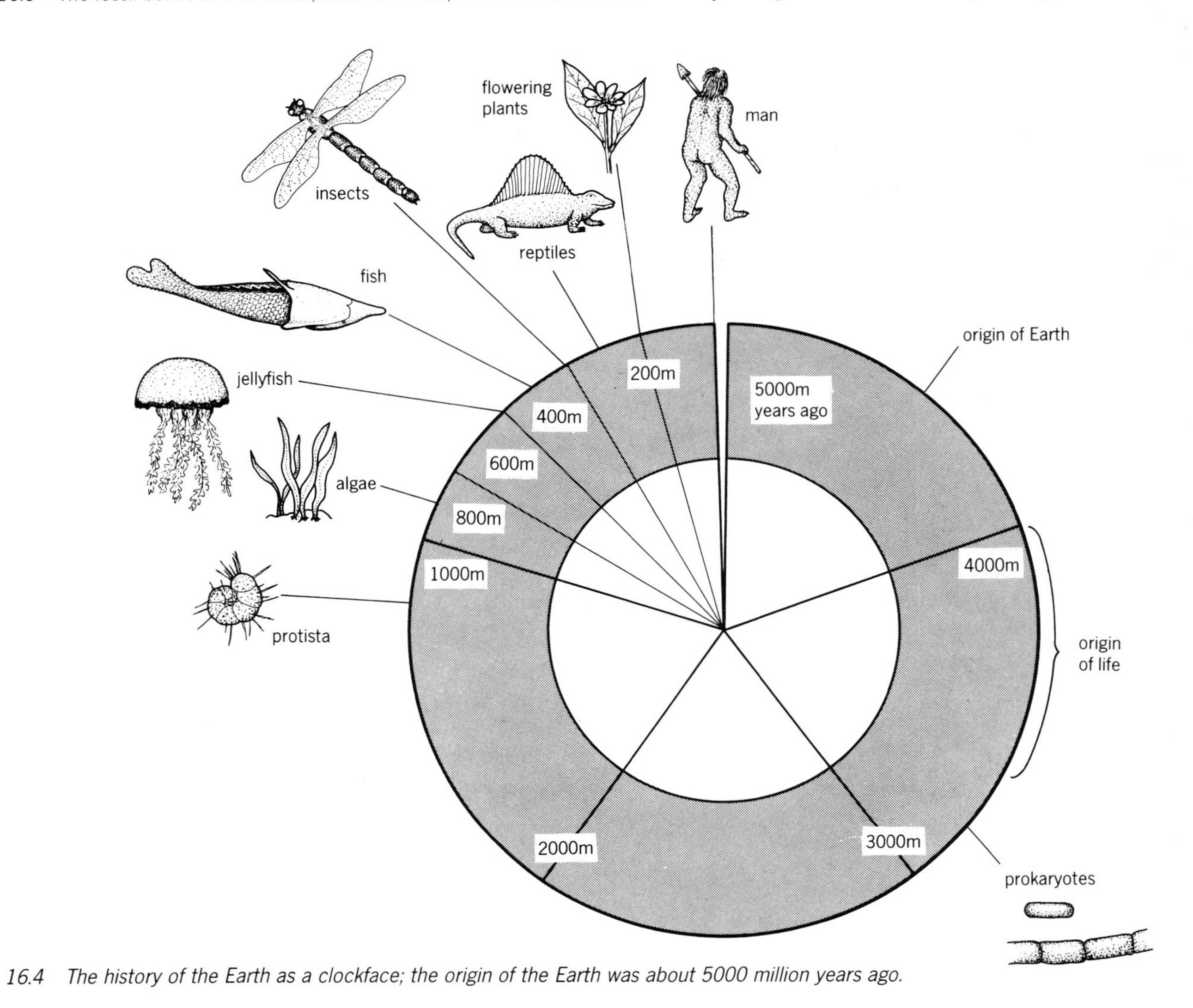

16.4 The history of the Earth as a clockface; the origin of the Earth was about 5000 million years ago.

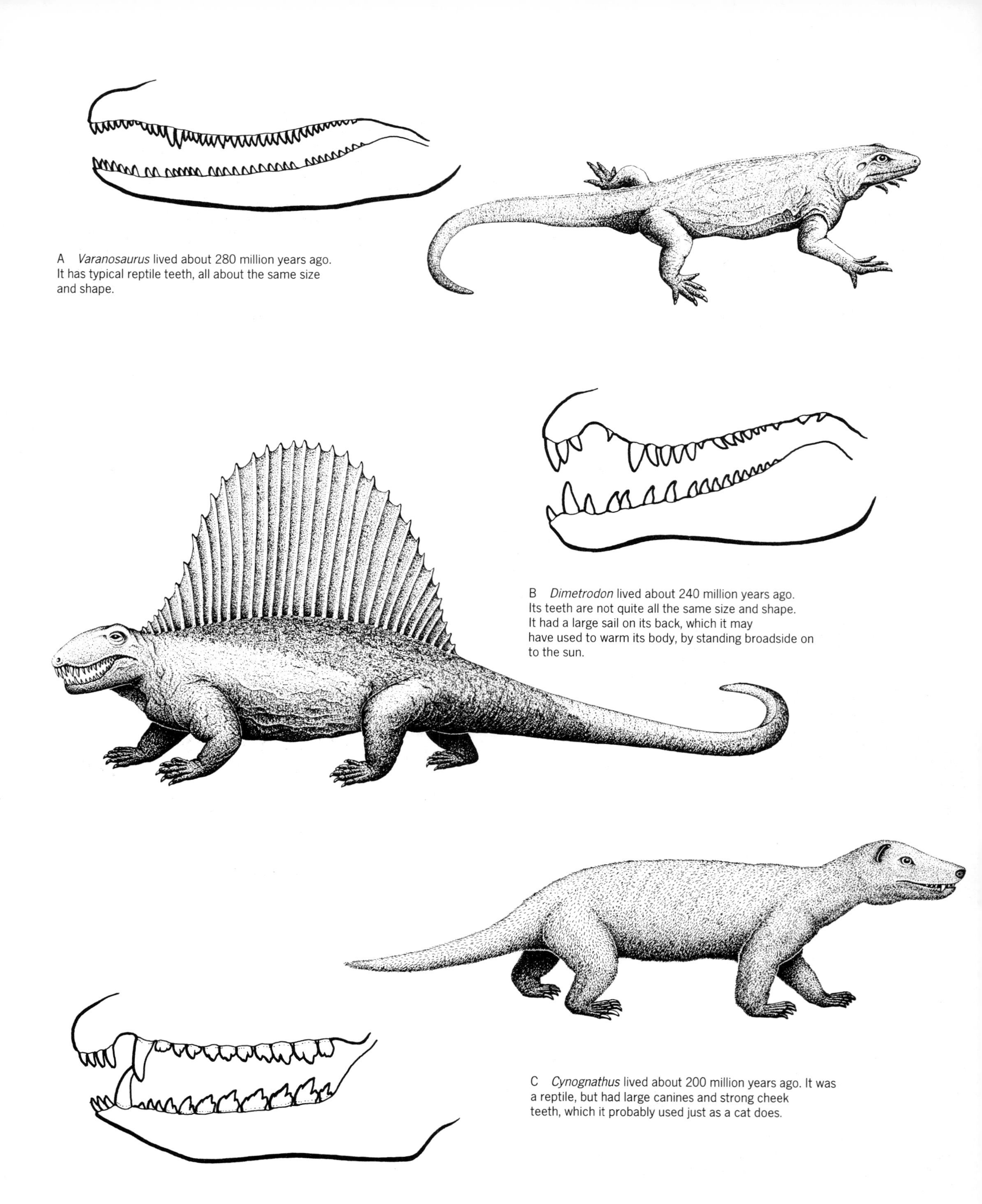

A *Varanosaurus* lived about 280 million years ago. It has typical reptile teeth, all about the same size and shape.

B *Dimetrodon* lived about 240 million years ago. Its teeth are not quite all the same size and shape. It had a large sail on its back, which it may have used to warm its body, by standing broadside on to the sun.

C *Cynognathus* lived about 200 million years ago. It was a reptile, but had large canines and strong cheek teeth, which it probably used just as a cat does.

D *Priacodon* lived about 180 million years ago. It was an early mammal. Its teeth are well differentiated into incisors, canines and cheek teeth.

E Modern cats have teeth which are very like *Priacodon's*.

16.5 The gradual development of mammalian teeth. Mammals evolved from reptiles. Skulls A, B and C are of reptiles belonging to the groups which were the ancestors of mammals. In the first two, the teeth were probably only used for catching and holding their prey, which was swallowed almost whole. Cynognathus, though, probably chopped up its food before swallowing it.

Skulls D and E are of mammals. Mammals need specialised teeth to help them to digest their food really efficiently. This is because they are homeothermic. They need a lot of food to produce heat energy to keep their bodies warm.

Evidence for evolution – homologous structures

16.3 Vertebrate limb bones show homology.

Although the fossil record strongly suggests that evolution has happened, it does not actually prove it. What other evidence is there to support the theory of evolution?

If we look at the way in which living organisms are made, we can often see quite striking similarities in their construction. Fig 16.6 shows the limb bones of several different kinds of vertebrate. Although their limbs are used for many different purposes, they all seem to be built to the same basic design. Structures like this are called **homologous** structures.

One way to explain the existence of these homologous bones is to suggest that all of these animals have evolved from an ancestral animal which had a 'basic design' limb.

The most likely ancestors for the amphibians, birds, reptiles and mammals are fish. One group of fish seem particularly likely candidates. These are the lobe-finned fishes, which first appear in the fossil record about 350 million years ago. In fact, one species of these fish is still alive today. This is the coelacanth (see Fig 16.7), which lives in deep water off Madagascar.

The arrangement of the bones in the pectoral fin of a lobe-finned fish is very similar to the arrangement of the bones in Fig 16.6. It is quite easy to imagine that a fish like this could have been the ancestor of the amphibians, reptiles, birds and mammals.

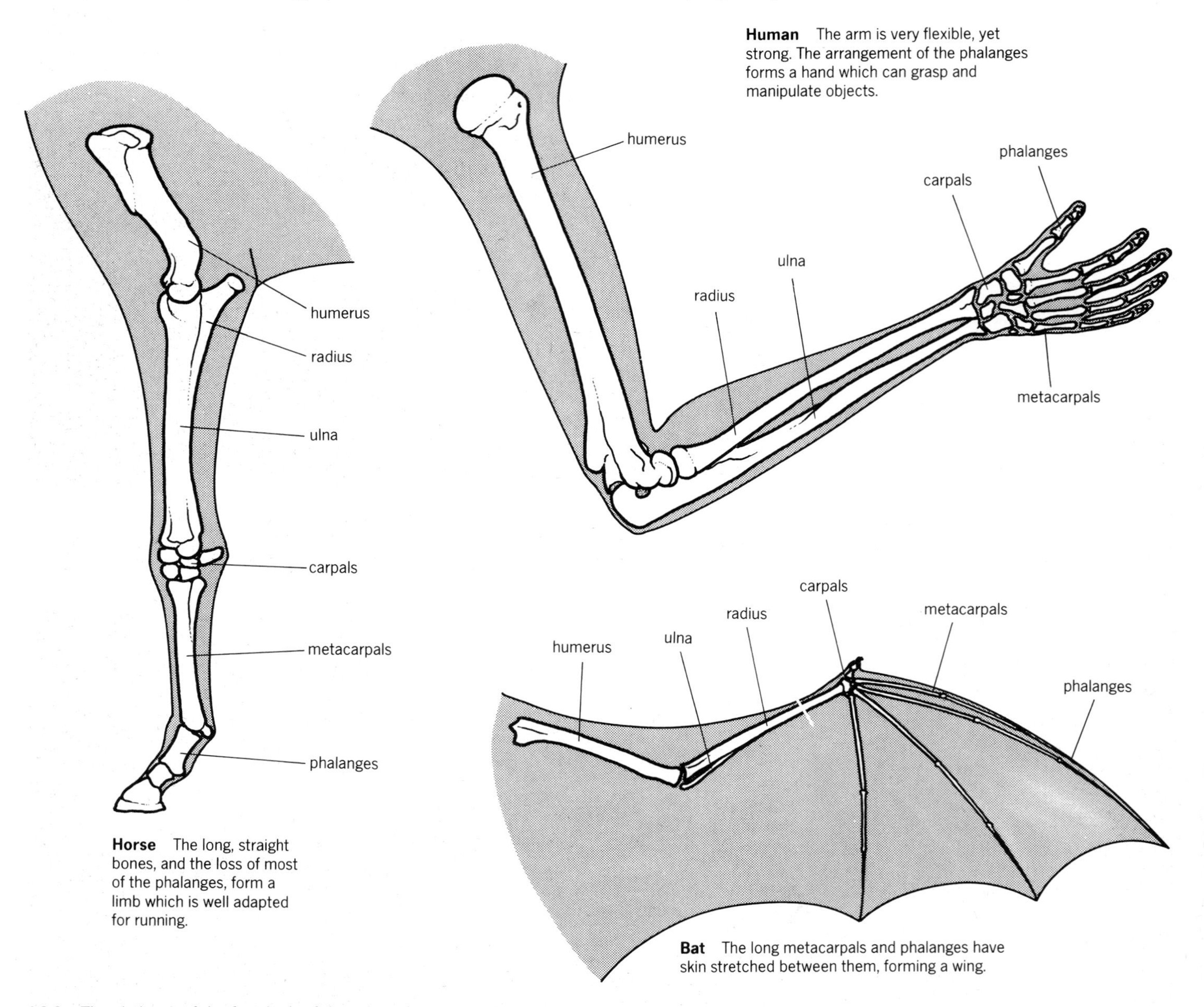

16.6 The skeleton of the fore-limb of three vertebrates; an example of homology

16.4 Vestigial structures have no obvious function.

Another example of the way in which homologous structures provide evidence for evolution is given by animals or plants which possess structures which do not seem to be used for anything. Structures like this are called **vestigial.** One example is the small limb bones of some snakes, such as pythons (see Fig 16.8).

Snakes are reptiles. A logical way of explaining the presence of these bones is to suggest that snakes have evolved from other reptiles which used their limbs for walking. The limb bones have gradually become smaller. In most snakes, there is no longer any trace of them.

Fact! There are about 1 million species of insects (Class Insecta) living in the world today, and about 8000 new species are discovered each year.

Questions

1 About how long ago did life probably begin on Earth?
2 In what way do fossils suggest that evolution has happened?
3 How long ago did the first of each of these organisms appear?
(a) bacteria
(b) land plants
(c) land animals
(d) mammals
(e) birds
4 What is meant by homologous structures?
5 How are (a) a horse's forelimb, and
(b) a bat's forelimb adapted for their way of life?
6 What is meant by vestigial structures?
7 In what way are vestigial structures evidence for evolution?

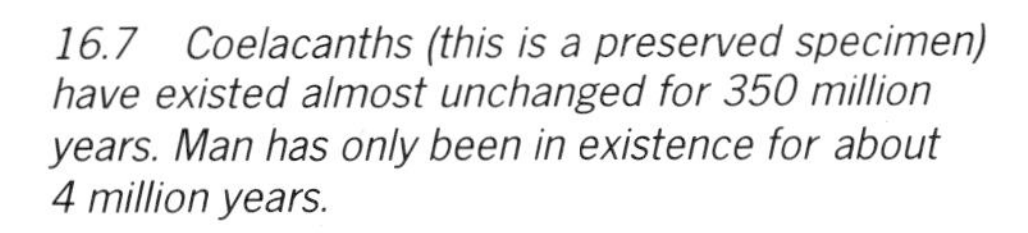

16.7 Coelacanths (this is a preserved specimen) have existed almost unchanged for 350 million years. Man has only been in existence for about 4 million years.

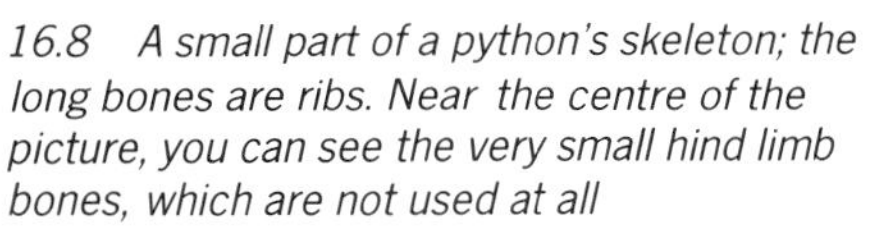

16.8 A small part of a python's skeleton; the long bones are ribs. Near the centre of the picture, you can see the very small hind limb bones, which are not used at all

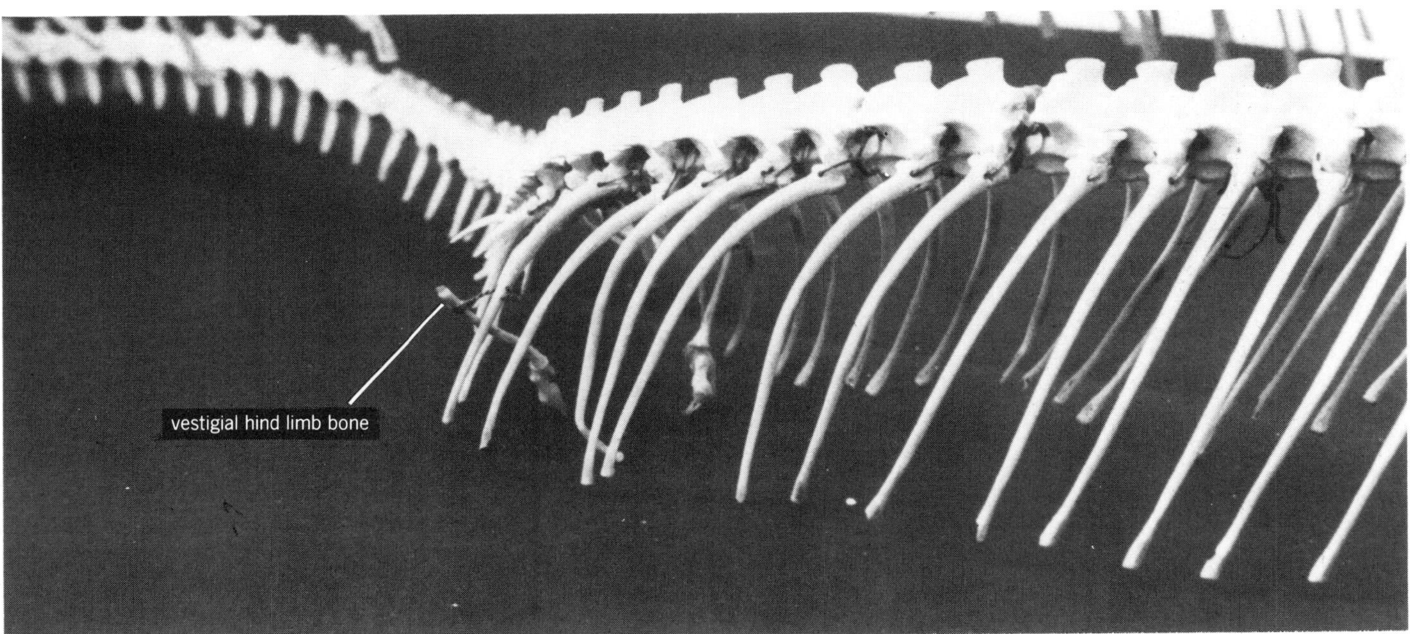

The theory of natural selection

16.5 Natural selection can cause gradual change.

The evidence for evolution has been known for a very long time. Yet until the second half of the nineteenth century, most people believed that all the different species of living organism had been created at the same time, when the world began. Some people still believe this.

One of the reasons why many people were reluctant to accept the idea of evolution was that they could not see why or how it could happen. In the nineteenth century, several ideas were put forward. One, still widely accepted today, was suggested by Charles Darwin. He put forward his theory in a book called *The Origin of Species*, which was published in 1859.

Darwin's theory of how evolution could have happened can be summarised like this.

Variation Most populations of organisms contain individuals which vary slightly from one to another. Some slight variations may better adapt some organisms to their environment than others.

Over-production Most organisms produce more young than will survive to adulthood.

16.9(a) A portrait of Charles Darwin at the age of 76

16.9(b) When large numbers of organisms, such as these wildebeest of the East African plains, live together, there is competition for food, and a tendancy for the weaker ones to be killed by predators. The fittest organisms survive.

Struggle for existence Because populations do not generally increase rapidly in size there must therefore be considerable competition for survival between the organisms.

Survival of the fittest Only the organisms which are really well adapted to their environment will survive.

Advantageous characteristics passed on to offspring Only these well adapted organisms will be able to reproduce successfully, and will pass on their advantageous characteristics to their offspring.

Gradual change In this way, over a period of time, the population will lose all the poorly adapted individuals. The population will gradually become better adapted to its environment.

This theory is often called the **theory of natural selection,** because it suggests that the best adapted organisms are selected to pass on their characteristics to the next generation.

Darwin proposed his theory before anyone understood how characteristics were inherited. Now that we know something about genetics, his theory can be stated slightly differently. We can say that natural selection results in the genes producing advantageous phenotypes being passed on to the next generation more frequently than the genes which produce less advantageous phenotypes.

Variation

16.6 Variation can be continuous or discontinuous.

Variation is the raw material for natural selection to act on. To understand how natural selection might work, we must try to understand how and why organisms vary.

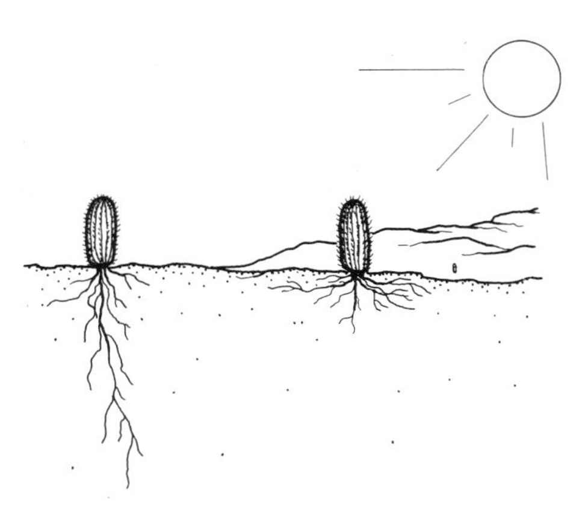

1 Genetic variation In a population of cacti, some have longer roots than others.

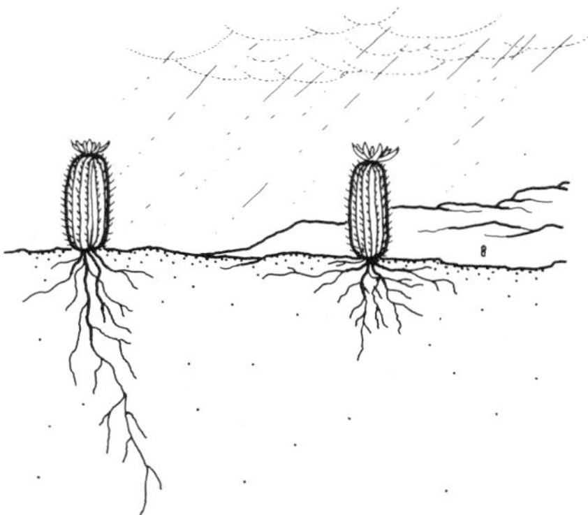

In the wet season they flower.

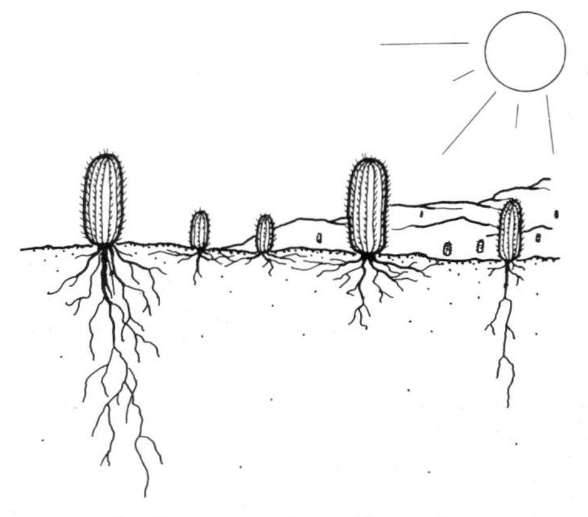

2 Overproduction The cacti produce large numbers of offspring.

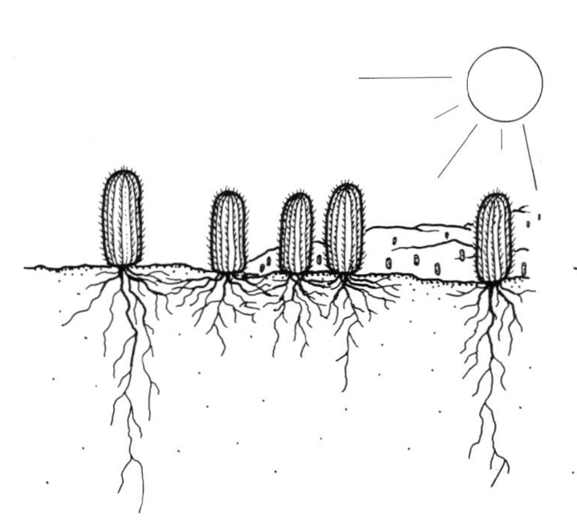

3 Struggle for existence During the dry season, there is competition for water.

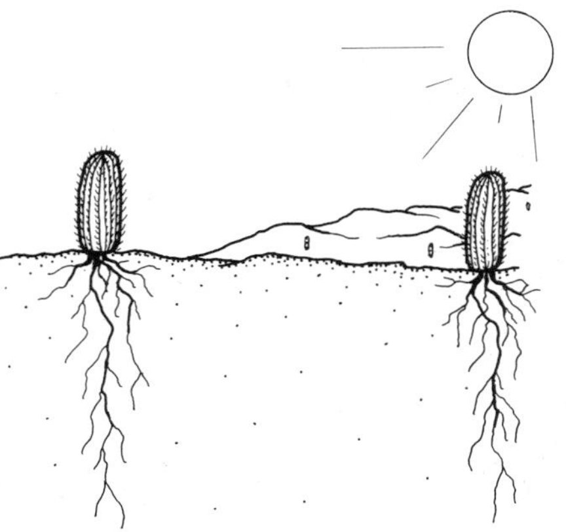

4 Survival of the fittest The cacti with the longest roots are able to obtain water, while the others die from dehydration.

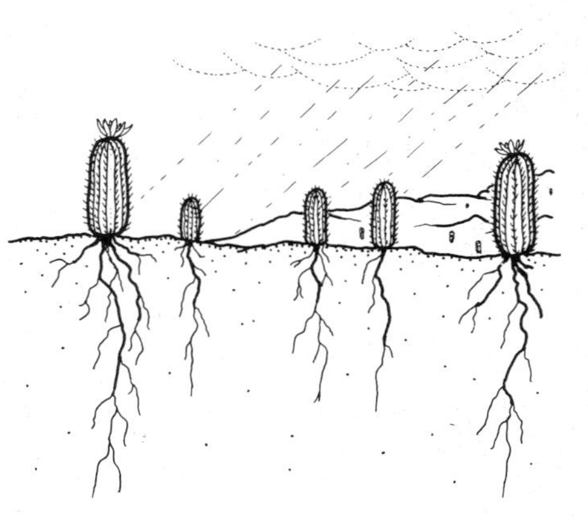

5 Advantageous characteristics passed on to offspring When conditions are suitable, the long-rooted cacti reproduce, producing long-rooted offspring.

16.10 An example of how natural selection might occur

You have only to look around a group of people to see that they are different from one another. Some of the more obvious differences are in height or hair colour. We also vary in intelligence, blood groups, whether we can roll our tongues or not, and in many other ways.

There are two basic kinds of variation. One kind is **discontinuous variation**. Tongue rolling is an example of discontinuous variation. Everyone fits into one of two definite categories – they either can or cannot roll their tongue. There is no in between category.

The other kind is **continuous variation.** Height is an example of continuous variation. There are no definite heights that a person must be. People vary in height, between the very lowest and highest extremes.

You can try measuring and recording discontinuous and continuous variation in Investigation 16.1. Your results for continuous variation will probably look similar to Fig 16.12. This is called a **normal distribution**. Most people come in the middle of the range, with fewer at the lower or upper ends.

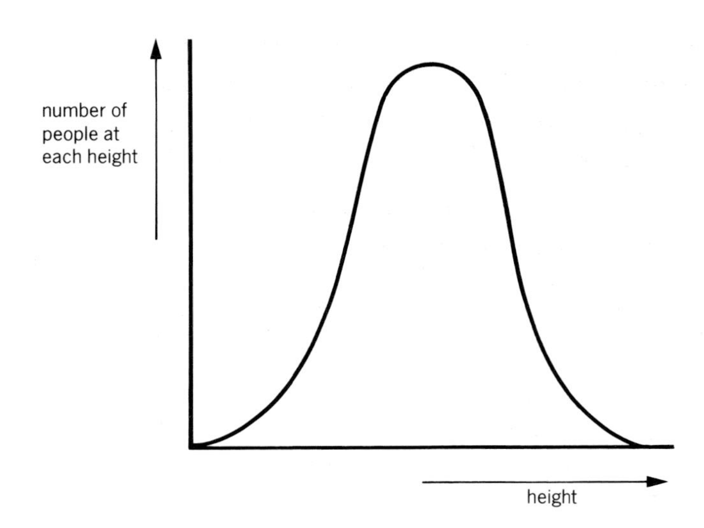

16.12 A normal distribution curve–a graph showing the numbers of people of different heights

16.7 What causes variation?

By describing variation as continuous or discontinuous, we can begin to explain how organisms vary. But the cause of the variation is another question altogether.

Genetic variation One reason for the differences between individuals is that their **genotypes** are different. Tongue rolling, for example, is controlled by genes. There are also genes for hair colour, eye colour, blood groups, height and many other characteristics.

Environmental variation Another important reason for variation is the difference between the **environments** of individuals. Scots pine trees possess genes which enable them to grow to a height of about 35 m. But if a Scots pine tree is grown in a very small pot, and has its roots regularly pruned, it will be permanently stunted (see Fig 16.13). The tree's genotype gives it the potential to grow tall, but it will not realise this potential unless its roots are given plenty of space and it is allowed to grow freely.

Characteristics caused by an organism's environment are sometimes called acquired characteristics. They are not caused by genes, and so they cannot be handed on to the next generation.

16.11 Human height shows continuous variation. What characteristic here shows discontinuous variation?

Investigation 16.1 Measuring variation

1. Make a survey of at least 30 people, to find out whether or not they can roll their tongue. Record your results.
2. Measure the length of the third finger of the left hand of 30 people. Take the measurement from the knuckle to the finger tip, not including the nail.
3. Divide the finger lengths into suitable categories, and record the numbers in each category, like this.

length	number
8.0–8.4 cm	2
8.5–8.9 cm	4 and so on

4. Draw a histogram of your results.

Questions

1. Which of these characteristics is an example of continuous variation, and which shows discontinuous variation?
2. Your histogram may be a similar shape to the curve in Fig 16.12. This is called a **normal distribution.** The class which has the largest number of individuals in it is called the **modal class.** What is the modal class for the finger lengths of your sample?
3. The **mean** or average finger length is the total of all the finger lengths, divided by the number of people in your sample. What is the mean finger length for your sample?

A bonsai pine tree is dwarfed by being grown in a very small pot, and continually pruned.

A Shetland pony's genes are responsible for its small size.

A cutting from a bonsai pine would grow into a full size tree, if given sufficient space.

Variation caused by the environment is not inherited

The offspring of Shetland ponies are small like their parents, no matter how well they are fed and cared for.

Variation caused by genes is inherited

16.13 The inheritance of variation

16.8 Genetic variation arises in several ways.

Meiosis During sexual reproduction, gametes are formed by meiosis. In meiosis, homologous chromosomes exchange genes, and separate from one another, so the gametes which are formed are not all exactly the same.

Fertilisation Any two gametes of opposite types can fuse together at fertilisation, so there are many possible combinations of genes which may be produced in the zygote. In an organism with a large number of genes the possibility of two offspring having identical genotypes is so small that it can be considered almost impossible.

Mutation Sometimes, a gene may suddenly change. This is called mutation. Most mutations are harmful, but occasionally one may happen which gives the mutant organism an advantage in the struggle for existence. It will then survive to pass its new characteristic on to the next generation. The mutant may even replace the normal form over a period of time.

Questions

1. When was the idea of natural selection first suggested?
2. Using the six points listed in section 16.5, explain how giraffes may have evolved from a short-necked ancestor by natural selection.
3. Give one example of discontinuous variation.
4. Give one example of continuous variation.
5. What is a normal distribution?
6. Explain the difference between genetic variation and environmental variation, giving examples of each.

Evidence for natural selection

16.9 Melanic moths are selected near cities.

Darwin's theory of natural selection provides a good explanation for our observations of the many types of animals and plants. It could explain what we see in the fossil record, and it could explain the presence of homologous structures. But it is almost impossible to prove that these were produced by natural selection. The only way we can really be sure that natural selection works is if we can watch it happening.

The peppered moth, *Biston betularia,* lives in most parts of Britain. It flies by night, and spends the daytime resting on tree trunks. It has speckled wings, which camouflage it very effectively on lichen covered tree trunks (see Fig 16.14).

People have collected moths for many years, so we know that up until 1849 all the moths in collections were speckled as in Fig 16.14(a). But in 1849, a black or **melanic** form of the moth was caught near Manchester. By 1900, 98% of the moths near Manchester were black.

16.14a Lichen-covered bark hides a speckled moth perfectly

16.14b Dark moths are better camouflaged on lichen-free trees

The distribution of the black and speckled forms today is shown in Fig 16.15.

How can we explain the sudden rise in numbers of the dark moths, and their distribution today?

We know that the black colour of the moth is caused by a single dominant gene. The mutation from a normal to a black gene happens fairly often, so it is reasonable to assume that there have always been a few black moths around, as well as pale speckled ones. Up until the beginning of the Industrial Revolution, the pale moths had the advantage, as they were better camouflaged on lichen covered tree trunks.

But in the middle of the nineteenth century, some areas became polluted by smoke. Because the prevailing winds in Britain blow from the west, the worst affected areas are to the east of industrial cities like Manchester and Birmingham. The polluted air prevents lichens from growing. Dark moths are better camouflaged

The proportion of dark/light areas in each circle shows the proportion of dark/light moths in that part of the country.

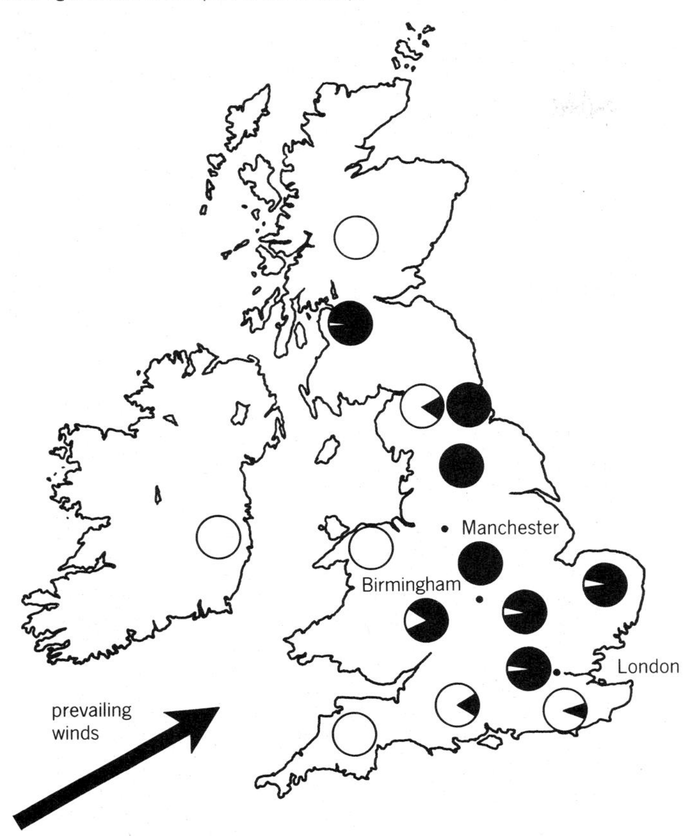

16.15 The distribution of the pale and dark forms of the peppered moth, Biston betularia

than pale moths on trees with no lichens on them.

Proof that the dark moths do have an advantage in polluted areas has been supplied by several experiments. Fig 16.16 summarises one of them.

The factor which confers an advantage on the dark moths, and a disadvantage on the light moths in polluted areas, is predation by birds. This is called a **selection pressure**, because it 'selects' the dark moths for survival. In unpolluted areas, the pale moths are more likely to survive.

16.10 Antibiotic resistance in bacteria is selected.

Another example of natural selection can be seen in the way that bacteria may become resistant to antibiotics, such as penicillin. Penicillin works by stopping bacteria from forming cell walls. When a person infected with bacteria is treated with penicillin, the bacteria are unable to grow new cell walls, and they burst open.

However, the population of bacteria in the person's body may be several million. The chances of any one of them mutating to a form which is not affected by penicillin is quite low, but because there are so many bacteria, it could well happen. If it does, the mutant bacterium will have a tremendous advantage. It will be able to go on reproducing while all the others cannot. Soon, its descendants may form a huge population of penicillin resistant bacteria.

This does, in fact, happen quite frequently. This is one reason why there are so many different antibiotics available – if some bacteria become resistant to one, they may be treated with another.

The more we use an antibiotic, the more we are exerting a selection pressure which favours the resistant forms. If antibiotics are used too often, we may end up with resistant strains of bacteria which are very difficult to control.

16.11 Natural selection does not always cause change.

Natural selection does not always produce change. Natural selection ensures that the organisms which are best adapted to their environment will survive. Change will only occur if the environment changes, or if a new mutation appears which adapts the organism better to the existing environment.

For example, in the south-west of Britain the environment of the peppered moth has never changed very much. The air has not become polluted, so lichens have continued to grow on trees. The best camouflaged moths have always been the pale ones. So selection has always favoured the pale moths in this part of Britain. Any mutant dark moths which do appear are at a disadvantage, and are unlikely to survive.

Most of the time, natural selection tends to keep populations very much the same from generation to generation. It is sometimes called **stabilising selection.** If an organism is well adapted to its environment, and if that environment stays the same, then the organism will not evolve. Coelacanths, for example, have remained virtually unchanged for 350 million years. They live deep in the Indian Ocean which is a very stable environment.

Questions

1. Using the six points listed in section 16.5, explain why the proportion of dark peppered moths near Manchester increased at the end of the nineteenth century.
2. Why is it unwise to use antibiotics unnecessarily?
3. What is meant by stabilising selection? Give one example.

Fact! Research is taking place all the time to produce better varieties of domestic animals and plants. This is called artificial selection. All the modern varieties of wheat and cattle, for example have been produced by selection by man.

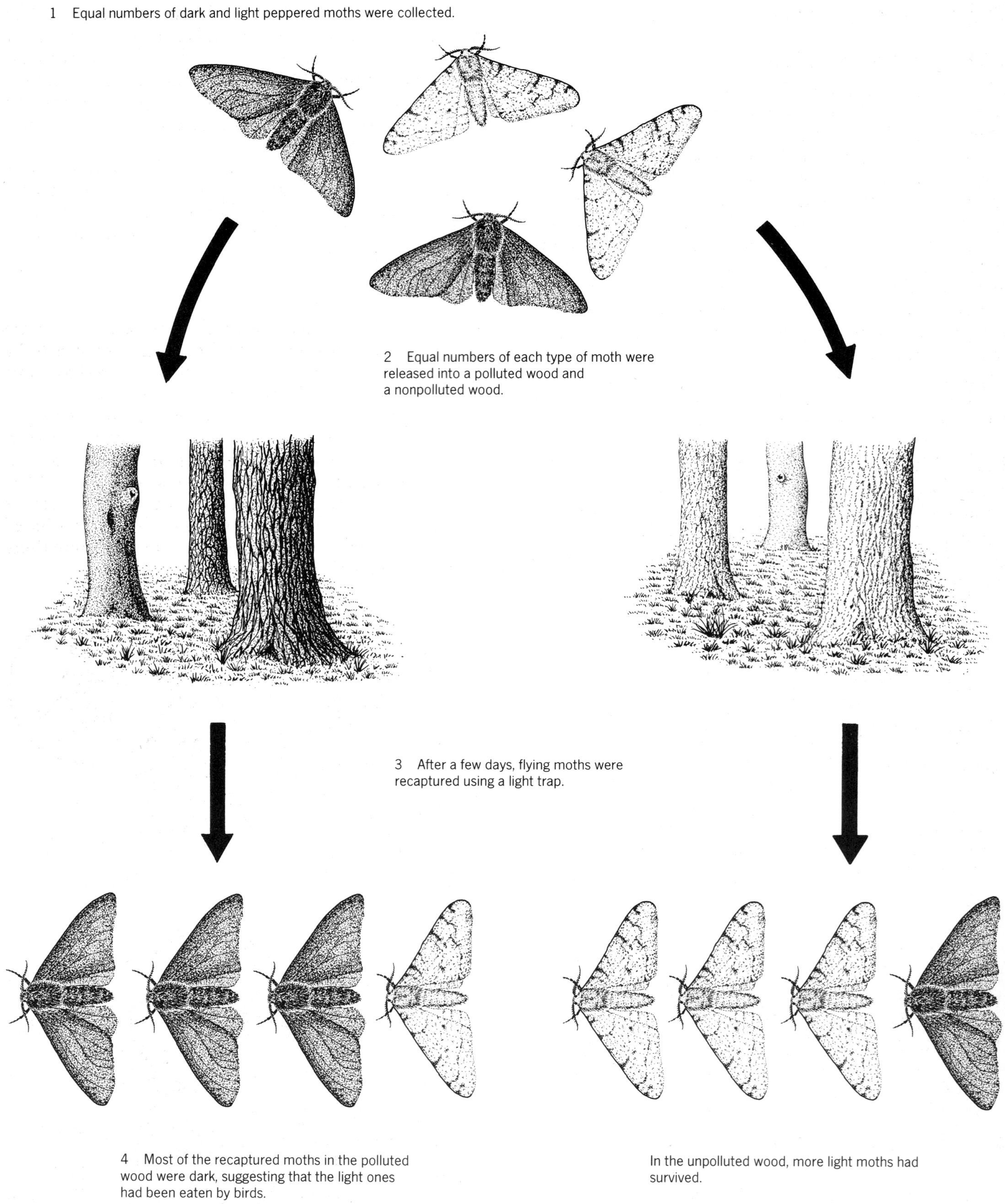

1 Equal numbers of dark and light peppered moths were collected.

2 Equal numbers of each type of moth were released into a polluted wood and a nonpolluted wood.

3 After a few days, flying moths were recaptured using a light trap.

4 Most of the recaptured moths in the polluted wood were dark, suggesting that the light ones had been eaten by birds.

In the unpolluted wood, more light moths had survived.

16.16 An experiment to measure the survival of dark and light peppered moths in polluted and unpolluted environments

The origin of species

16.12 What is a species?

All living organisms may be classified into groups. You will find more about this in Chapter 17. The smallest groups are called **species.**

A species is a group of living organisms which are all very similar to one another, and which can interbreed successfully with one another. Members of different species cannot interbreed to produce fertile offspring.

Sometimes, it is not possible to tell whether organisms can interbreed or not. Garlic, for example, only reproduces asexually, and never produces seeds. We can never know anything about the breeding behaviour of extinct organisms, such as dinosaurs. In cases like this, the decision about whether a group of organisms is a species or not has to be made according to how similar they are to each other, and how different they are from other groups of organisms.

16.13 New species from old.

According to the theory of evolution, the millions of different species that exist today have all evolved from other species which existed in the past. One of the big questions in evolution is 'How can a new species be formed?'

Darwin's theory of natural selection can explain how new varieties of an organism can become more common. The appearance of the dark form of the peppered moth since 1849 can be explained very neatly in this way. But this moth is not a new species. It is still a peppered moth; it can still interbreed with the pale peppered moth.

The problem to be solved is this. How can one species split into two groups, which can no longer interbreed with one another? Although Darwin's book is called *The Origin of Species,* he did not succeed in answering this question.

16.14 Isolation can produce new species.

Fig 16.17 shows how we think that new species are formed. The first stage is for the existing species to be split into two groups. They must be separated by some kind of barrier which they cannot cross.

Each group continues to live and breed in its environment. If the two environments are different, then the selection pressures on the organisms will be different. They will gradually become less and less alike.

After many years, the two populations may become so different that they will no longer be able to breed with one another. Two species now exist, where there was only one before.

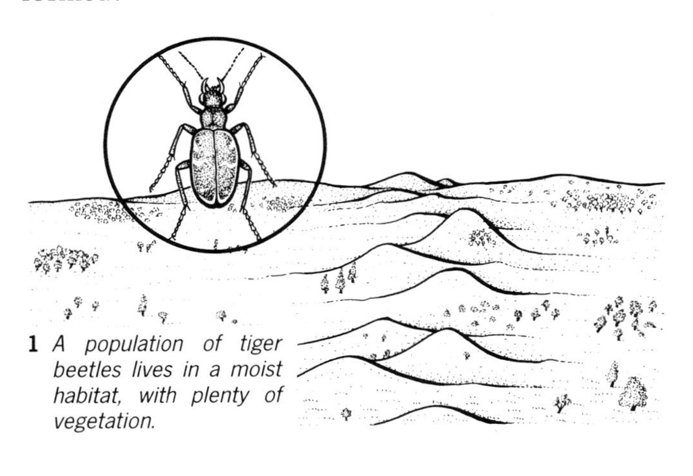

1 *A population of tiger beetles lives in a moist habitat, with plenty of vegetation.*

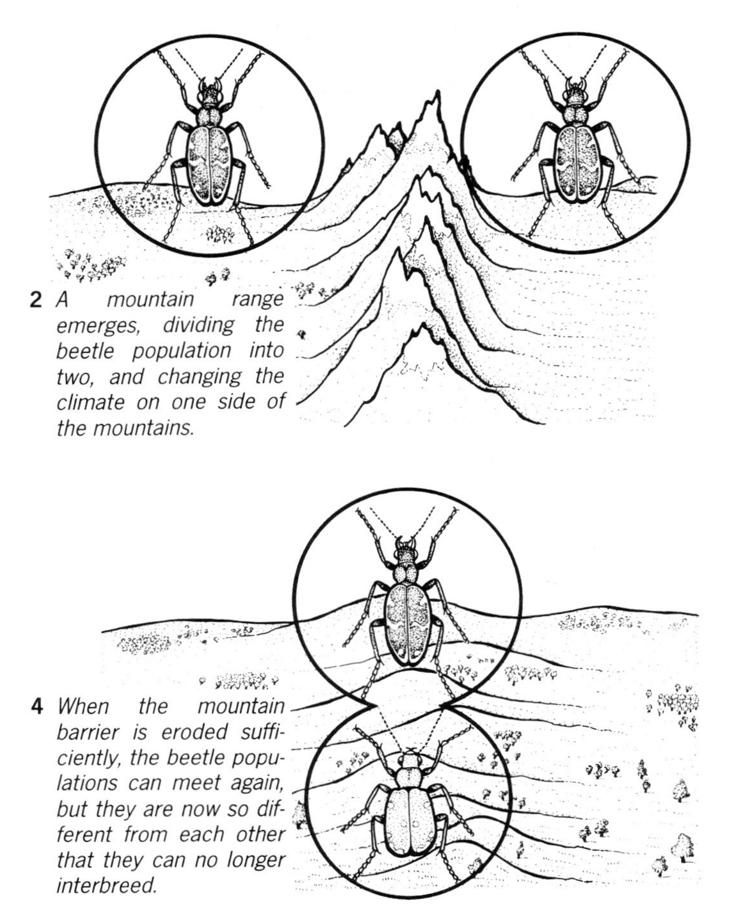

2 *A mountain range emerges, dividing the beetle population into two, and changing the climate on one side of the mountains.*

4 *When the mountain barrier is eroded sufficiently, the beetle populations can meet again, but they are now so different from each other that they can no longer interbreed.*

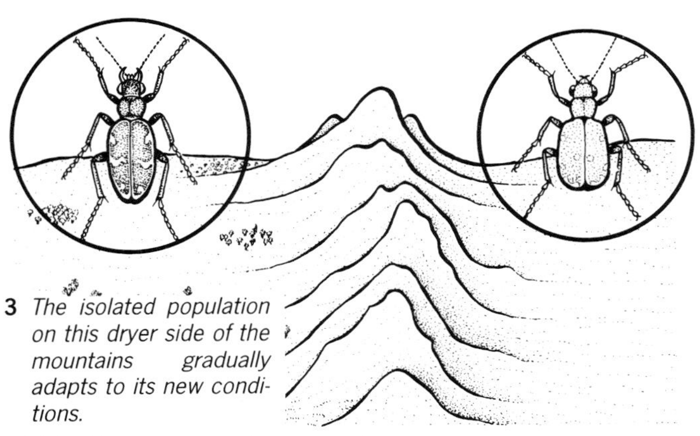

3 *The isolated population on this dryer side of the mountains gradually adapts to its new conditions.*

16.17 How a new species may evolve

Like many of the theories concerned with evolution, this one sounds very convincing, but is difficult to prove. The main difficulty is that it takes so long to happen. We can find examples where we think it has happened in the past, and other examples where it is beginning to happen now. But the formation of new species in this way is such a slow process that no-one has yet been able to watch it happening from beginning to end.

16.15 Apple and hawthorn flies are evolving.

One case where the beginning of this process can be seen is happening now in North America, involving a species of fruit fly (see Fig 16.18). The maggots of these flies feed on fruit. In the nineteenth century, they all fed on hawthorn berries, which ripen in October. Then, about one hundred years ago, apple trees began to be grown. Some of the flies laid their eggs on apple trees, and the maggots fed on apples instead of hawthorn berries.

Apples ripen in September, quite a bit earlier than hawthorn berries. Only the maggots hatching early were able to survive successfully on apple trees. Late hatching maggots would not survive on apple trees. Early hatching maggots would not survive on hawthorn trees.

16.18 A preserved specimen of a hawthorn fly

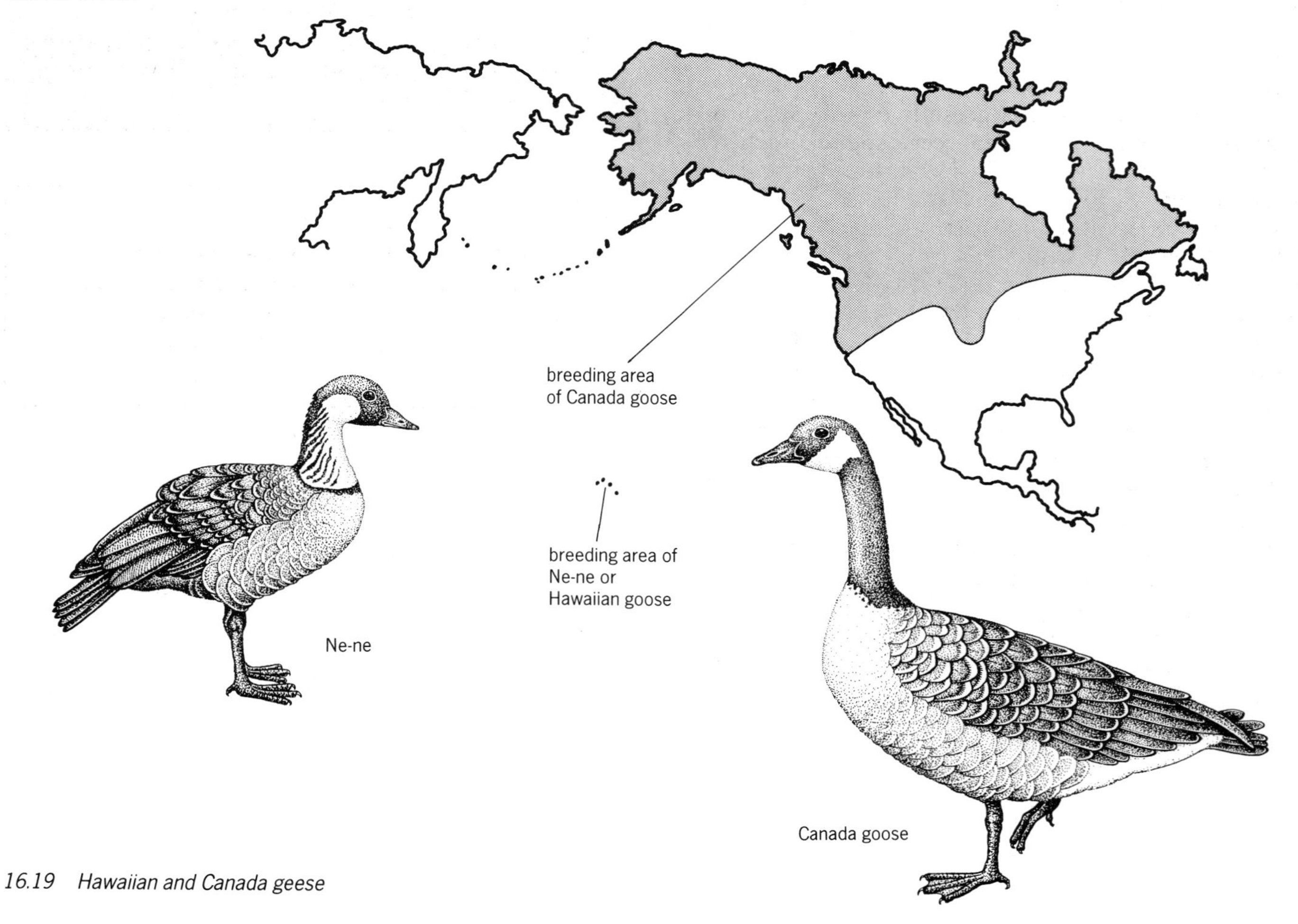

16.19 Hawaiian and Canada geese

Over the past hundred years, natural selection has divided the flies into two groups. One group lays its eggs early, on apple trees. The other group lays its eggs later, on hawthorn trees. Because they breed at different times, they never interbreed with one another. The two groups are separated by an ecological barrier.

At the moment, the two groups of flies are still capable of interbreeding, although they never actually do so. They still belong to the same species. Will they ever become two separate species? We will have to wait a long time to find out.

Questions

1 What is a species?
2 Why is it sometimes difficult to decide whether two organisms belong to the same or different species?
3 Briefly explain how new species are thought to evolve.

16.16 Did Hawaiian Geese evolve from Canada Geese?

An example of **speciation** (the formation of new species) which happened a long time ago is the formation of the Né-Né or Hawaiian Goose. The Né-Né lives on the Hawaiian islands, in the Pacific Ocean (see Fig 16.19). It is very similar to the Canada Goose, which breeds in Canada and some parts of the Northern United States.

The places where Canada geese live have plentiful vegetation and water. Canada geese spend much of their time in or near water. The habitat of the Né-Né, however, is very different. They live on solidified lava flows, where vegetation and water are scarce.

Although the Né-Né is basically very similar to the Canada goose, it differs in several ways. These differences seem to be adaptations which help it to live successfully in its environment. It is smaller than the Canada goose, and lighter in colour. Its legs, toes and beak are longer in proportion to its overall body size. The webs between its toes are smaller than those of the Canada goose. Large webs would be a handicap, as it spends so little time in water.

It seems likely that, thousands of years ago, a few Canada geese arrived in Hawaii. There, separated by thousands of miles from the nearest breeding grounds of Canada geese, they gradually evolved to suit their new environment. Now they are so different from Canada geese that they are classed as a separate species.

Chapter revision questions

1 Match each word with its definition.
evolution speciation homologous structures
natural selection species
(a) parts of different kinds of organisms, which seem to have a similar basic design
(b) a group of organisms which are very similar to one another, and can breed with each other
(c) gradual changes in the types of living things over a long period of time
(d) a process which selects only the fittest organisms to survive and reproduce
(e) the way in which new species are formed
2 Make a short summary of the evidence that suggests that
(a) evolution has happened in the past,
(b) evolution is happening now,
(c) evolution happens by means of natural selection.
3 Natural selection causes living organisms to become well adapted to their environment. The best adapted organisms are the most successful. Explain each of the following.
(a) A population of organisms which can reproduce sexually can often become adapted to a new environment faster than a population of organisms which can only reproduce asexually.
(b) Evolution does not come to a halt once organisms have become well adapted to their environment.

Fact!

It has been calculated that the dinosaur *Brachiosaurus* probably weighed as much as 80 tonnes when fully grown. This is double the weight of an average bull sperm whale.

17 Classification

17.1 Classification.

Classification means putting things into groups. Biologists now classify living organisms according to how closely they think they are related to one another.

Fact! It has been calculated that human beings share the Earth with 3 000 000 000 000 000 000 000 000 000 000 other living things.

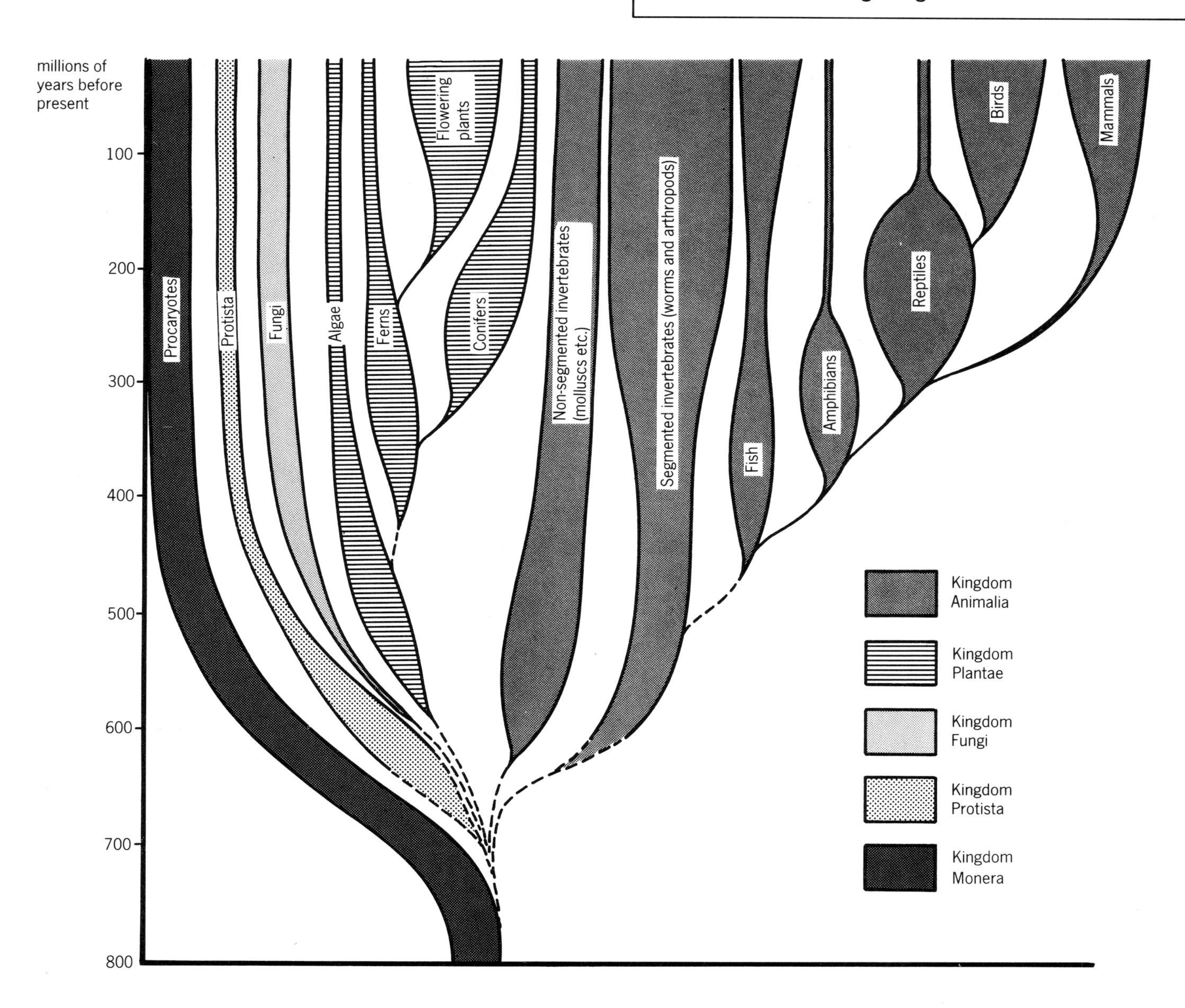

17.1 The probable evolutionary relationships between the main groups of organisms

The animals can be sorted like this:–

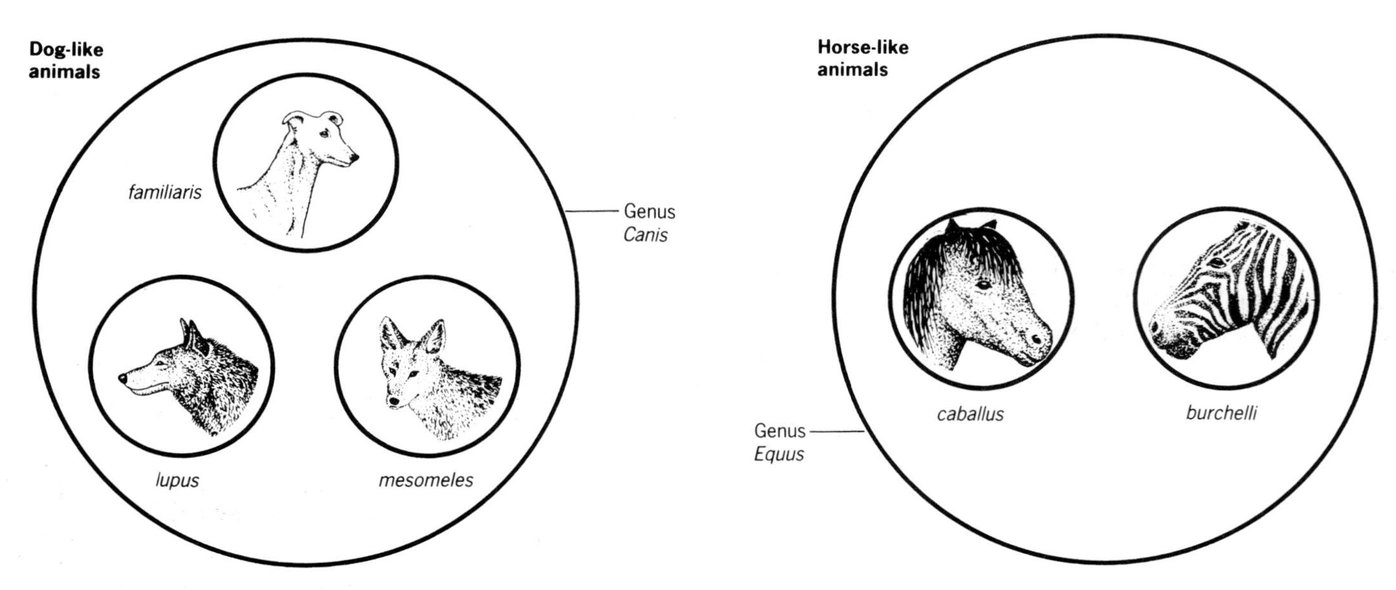

Each large group is a genus. Each small group is a species.

To name an animal, you write the name of its genus and the name of its species. A wolf, for example, is called *Canis lupus*.

17.2 Homologies help to classify organisms.

The first person to try to classify living things in a scientific way was a Swedish naturalist called **Linnaeus**. He introduced his system of classification in 1735.

Linnaeus grouped living organisms according to how similar they are. He looked for **homologies** (see section 16.3). He lived a century before Darwin, so he did not realise that the different groups of living organisms he studied were actually related to one another. Today, we know that homologies are a result of evolution. The more homologies different kinds of living things share, the more closely they are related to one another.

17.3 The species is the basic group in classification.

Linnaeus divided all the different kinds of living things into groups called **species**. He recognised 12 000 species.

Linnaeus' species were groups of organisms which had a lot of features in common. Today we look even more closely at a group of organisms to decide whether or not it is a species. The modern way of deciding what is a species is described in section 16.12.

17.4 The classification system.

Species are grouped into **genera** (singular **genus**). Each genus contains several species with similar characteristics (see Fig 17.2).

Several genera are then grouped into a **family**, families into **orders**, orders into **classes**, classes into **phyla** and finally phyla into **kingdoms**. Some of the more important ones are described in this chapter.

17.5 Each species has two Latin names.

Linnaeus gave every kind of living organism two names. The first name is the name of the genus it belongs to, and always has a capital letter. The second name is the name of its species, and always has a small letter.

For example, a wolf belongs to the genus *Canis* and the species *lupus*. Its Latin name is *Canis lupus*.

When Linnaeus was alive, Latin was a language that every scientist used and understood. He therefore chose Latin names, not Swedish ones, because everyone who was interested would understand them. We still use Latin names today. Although Latin is no longer used as a language, it is very useful if all scientists use exactly the same names for a particular kind of living organism. Any language would do, but as Linnaeus began with Latin, we continue to use it today. Many of the scientific names for animals and plants that are used today are the same ones that Linnaeus gave them 250 years ago.

Kingdom monera

These are the bacteria and blue-green algae. The oldest fossils belong to this kingdom, so we think that they were the first kind of organisms to evolve.

Characteristics unicellular (single-celled) organisms, have no nucleus.

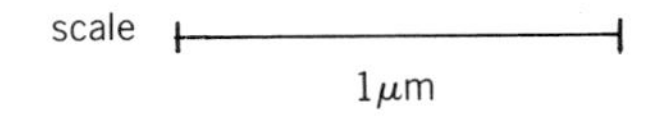

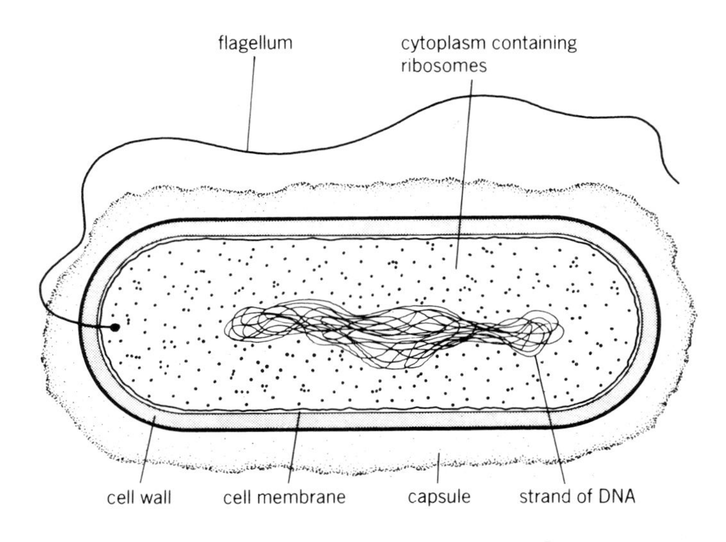

17.3 A bacterium, Escherichia coli

Kingdom protista

Like the bacteria, these organisms are unicellular. However, they do have a nucleus. They evolved later than the bacteria.

Almost all the protista live in water, because they have no protection against drying out. Some, like *Chlorella*, are plant-like, and feed by photosynthesis. Others, like *Amoeba*, are animal-like, feeding on other living things.

Characteristics unicellular organisms, have a nucleus.

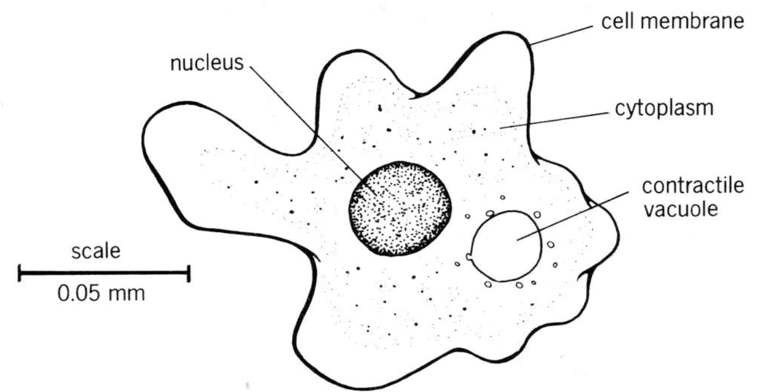

17.4 *Amoeba proteus*

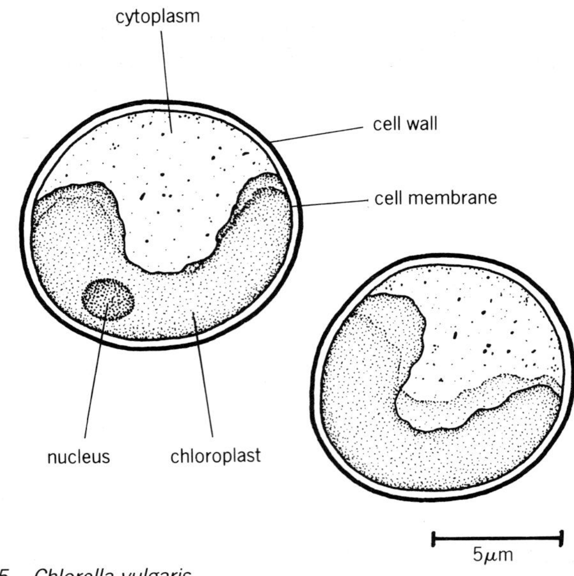

17.5 *Chlorella vulgaris*

Kingdom fungi

In many ways, the fungi are like plants, and are often put into the same kingdom. However, they do not have chlorophyll, and do not photosynthesise. Instead, they feed saprophytically, or parasitically, on organic material like faeces and dead plants and animals.

Characteristics multicellular (many-celled), have cell walls, do not have chlorophyll, feed saprophytically.

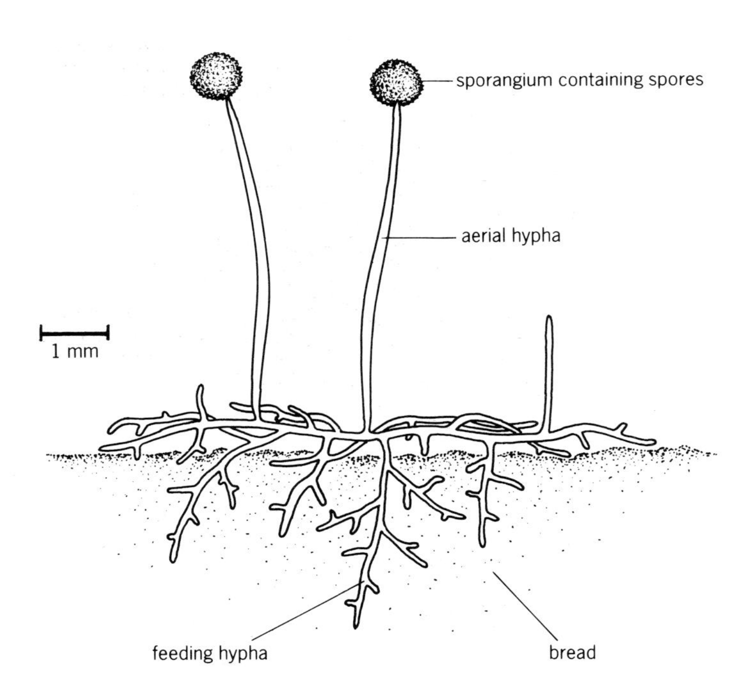

17.6 *Mucor haemalis*

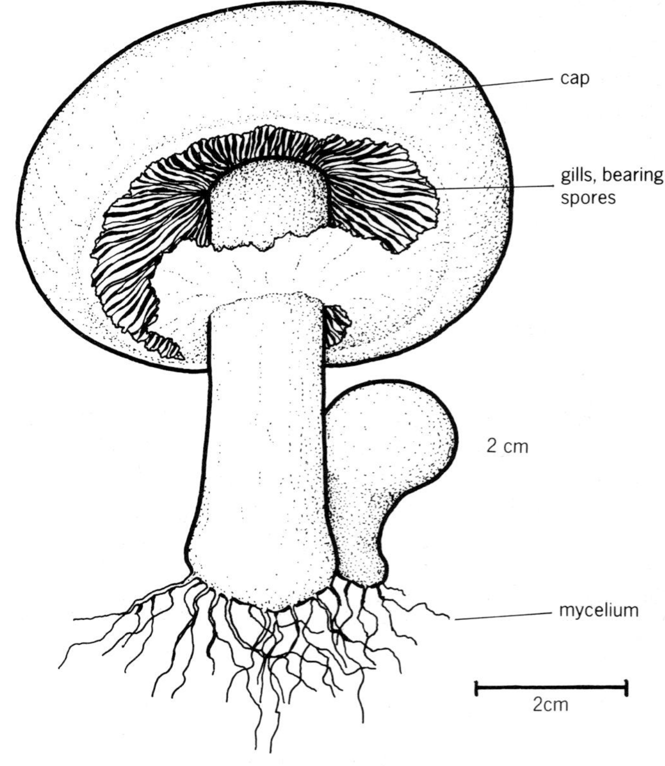

17.7 Field mushroom, *Agaricus campestris*

Kingdom plantae

Almost all plants are green, because they contain chlorophyll which they use for photosynthesis.

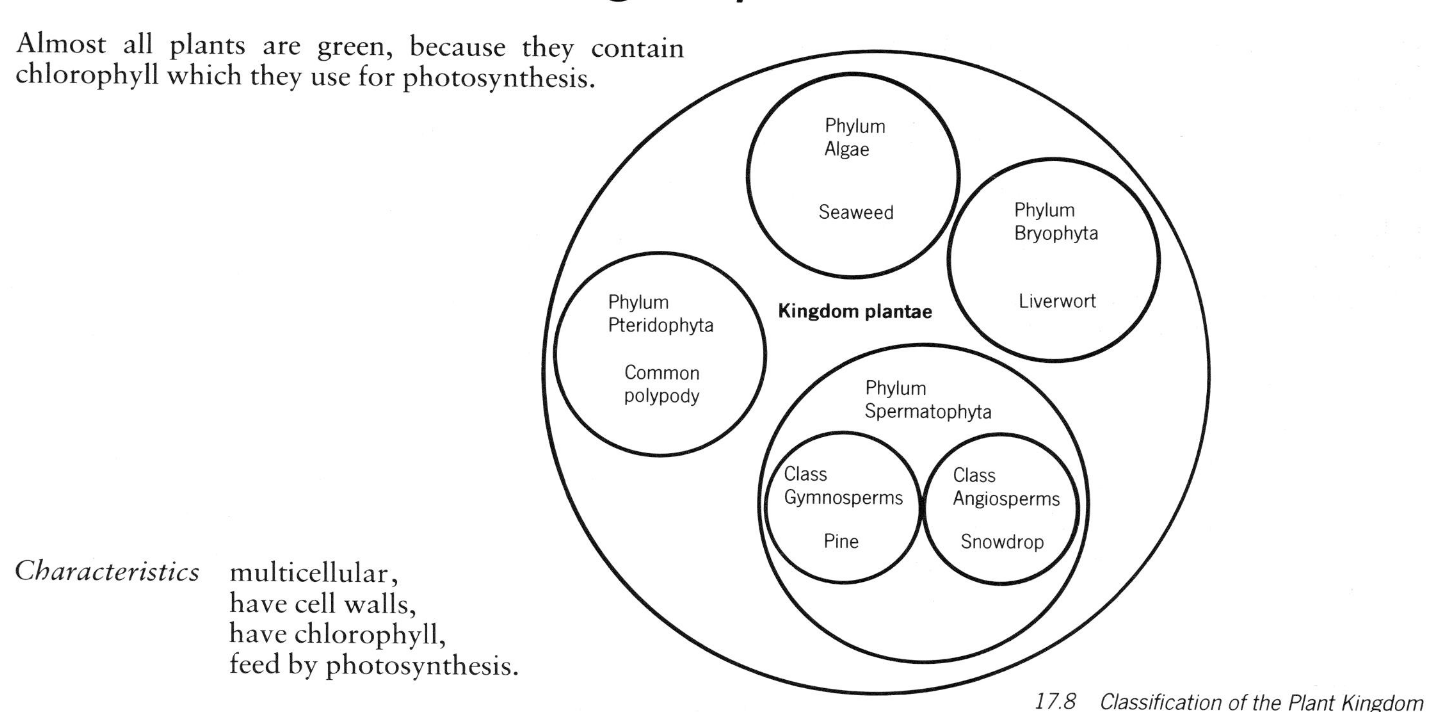

Characteristics multicellular,
have cell walls,
have chlorophyll,
feed by photosynthesis.

17.8 Classification of the Plant Kingdom

Phylum algae

These are the most simple kinds of plants. They have no real roots, stems or leaves, just a simple body called a **thallus**. They all live in water or very wet places.

Characteristics no roots, stems or leaves,
body called a thallus,
no xylem or phloem.

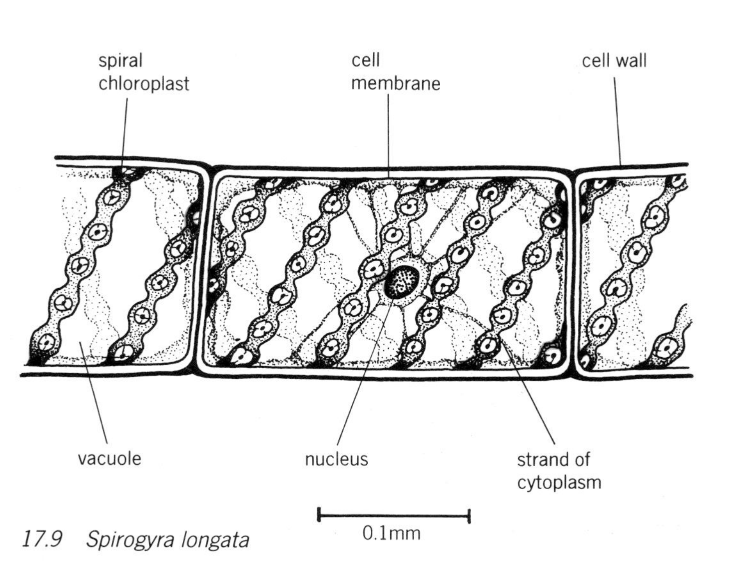

17.9 Spirogyra longata

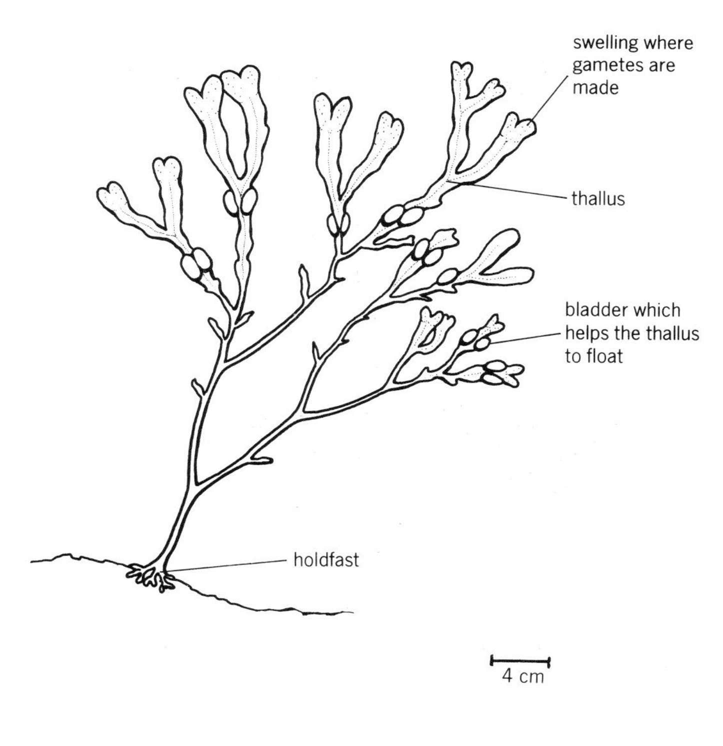

17.10 Bladder-wrack, Fucus vesiculosus

Phylum bryophyta

These are the mosses and liverworts. Although they live on land, they can only grow successfully in wet places. One reason for this is that they have no xylem to carry water. Another reason is that the male gametes have to swim in water to fertilise the female gametes.

Characteristics simple stems and leaves,
single-celled rhizoids (rootlets),
no xylem or phloem,
reproduce by means of spores.

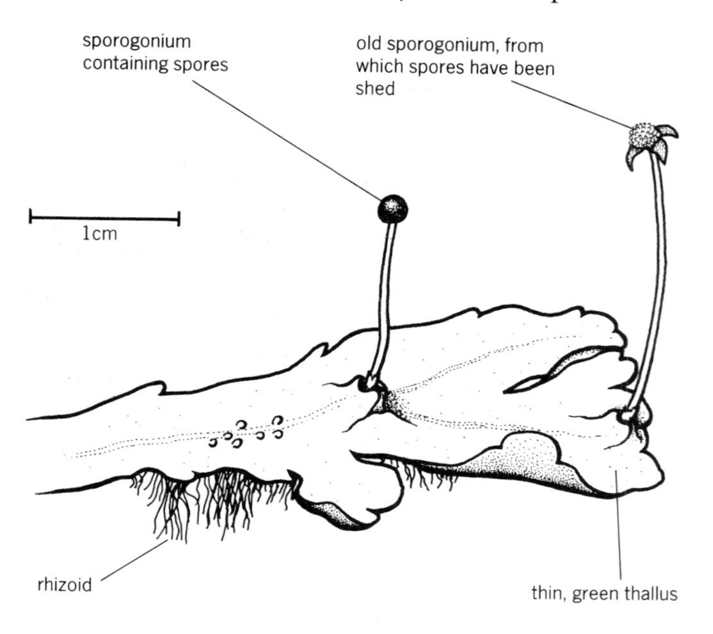

17.11 Liverwort, Pellia epiphylla

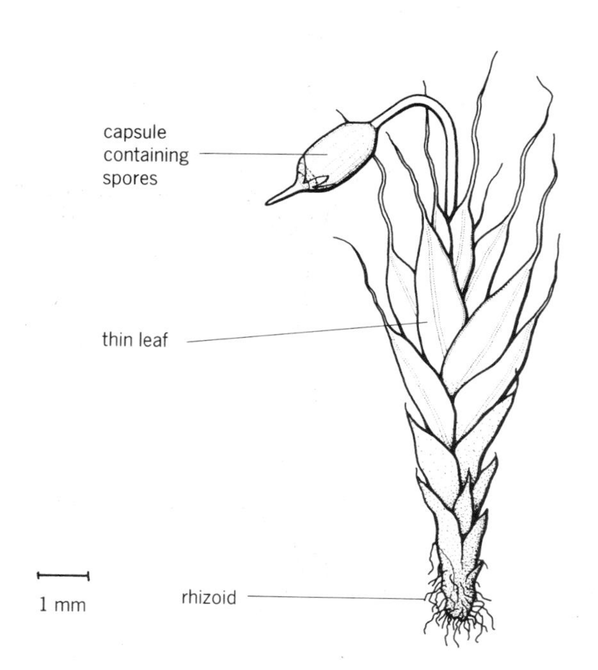

17.12 Moss, Grimmia pulvinata

Phylum pteridophyta

These are the ferns and horsetails. Like bryophytes, they need water for fertilisation. However, they do have xylem and phloem, so they can live in slightly drier places than the bryophytes.

Characteristics have roots, stems and leaves,
have xylem and phloem,
reproduce by means of spores, which grow on the backs of the leaves.

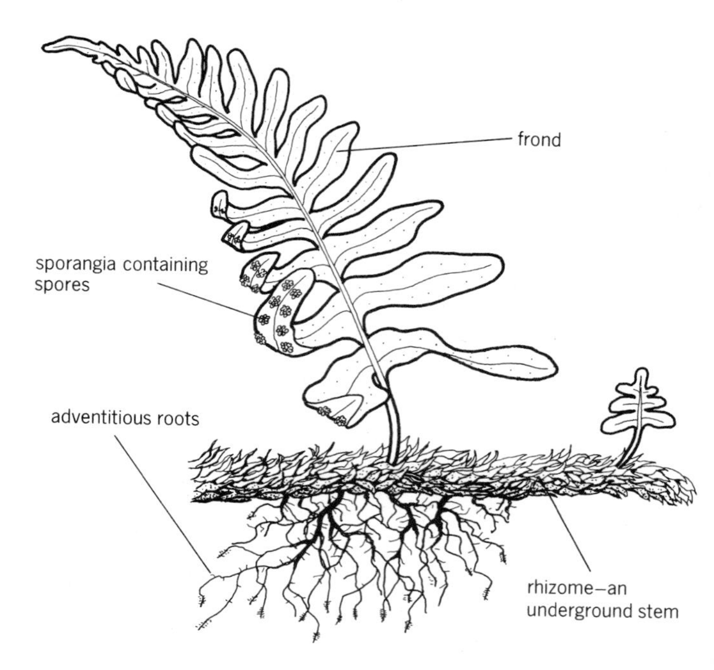

17.13 Common polypody, Polypodium vulgare

17.14 Bracken is a common type of fern

Phylum spermatophyta

These are the seed plants. Their male gametes are contained inside pollen grains, which can be carried to the female gametes by insects or the wind. This means that they do not need water for the male gamete to swim in, so they are able to live even in very dry places like deserts.

Characteristics have roots, stems and leaves, have xylem and phloem, reproduce by seeds.

Class gymnosperms

This group of seed plants does not have real flowers. Instead, the seeds grow inside cones.

Characteristics seeds grow inside cones.

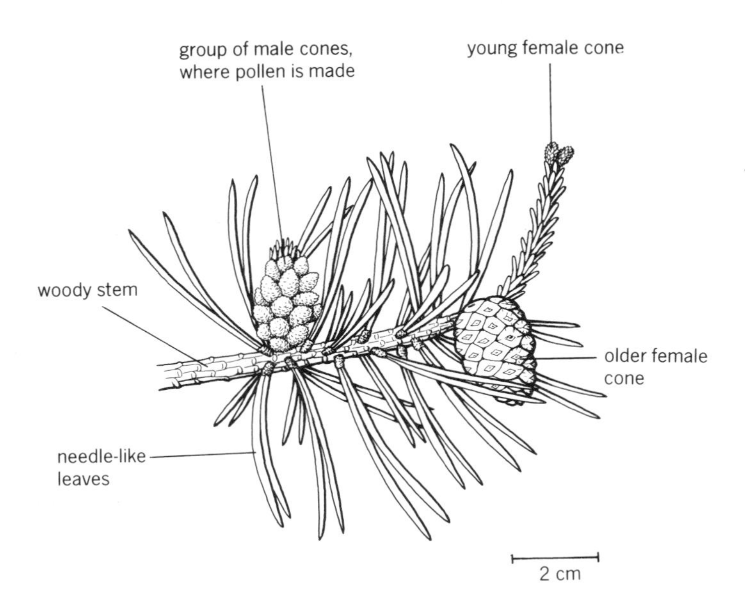

17.15 A branch of Scots pine, Pinus sylvestris

17.16 A group of Scots pine

Class angiosperms

These are the flowering plants. The seeds grow inside a fruit which developed from an ovary, inside a flower.

Characteristics seeds produced inside ovary, inside flower.

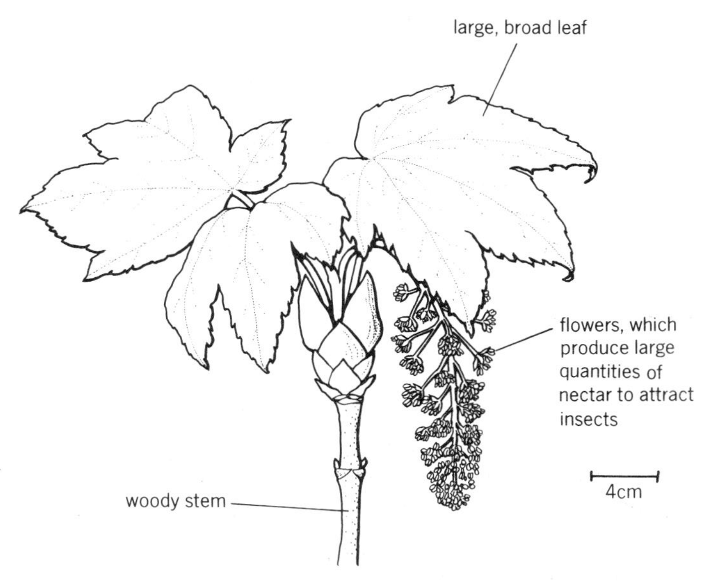

17.17 Twig of sycamore, Acer pseudoplatanus

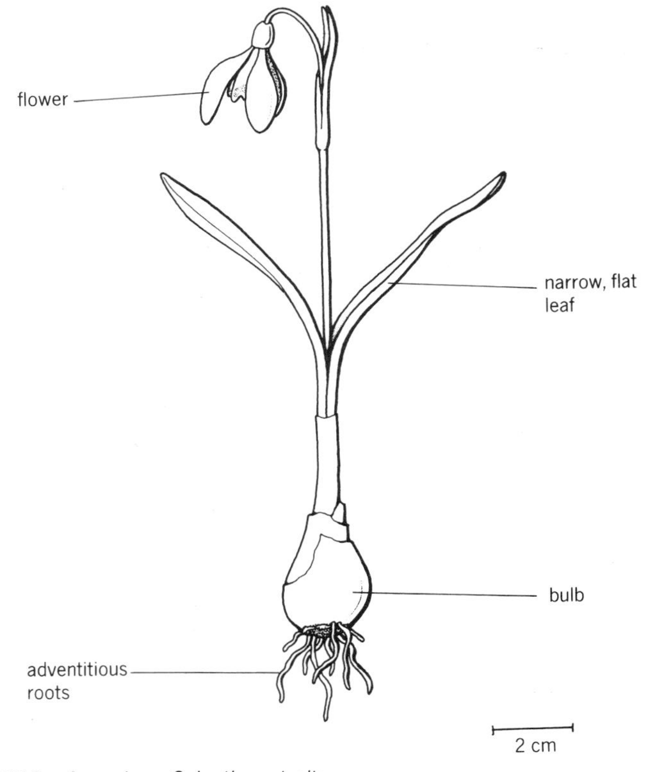

17.18 Snowdrop, Galanthus nivalis

Kingdom animalia

Animals do not photosynthesise, so they never have chlorophyll. They eat other living organisms, so they are usually able to move, to find their food. They do not have cell walls as this would make it difficult for them to move easily.

Only a few of the most important groups are shown.

Kingdom animalia

Phylum Arthropoda
- Class Crustacea — Crab
- Class Arachnida — Spider
- Class Insecta — Locust

Phylum Coelenterata — *Hydra*

Phylum Annelida — Earthworm

Phylum Chordata
- Class Pisces — Herring
- Class Reptilia — Snake
- Class Amphibia — Frog
- Class Mammalia — Cat
- Class Aves — Robin

Characteristics multicellular,
do not have cell walls,
do not have chlorophyll,
feed heterotrophically.

17.19 Classification of the Animal Kingdom

Phylum coelenterata

These are the jellyfish and sea anemones. They all live in water, because their soft bodies would dry out very quickly on land. They have a ring of tentacles surrounding a mouth. The mouth is the only opening in their digestive system – they have no anus.

Characteristics animals made of only two layers of cells,
have a ring of tentacles, with a mouth in the centre,
only one opening to gut.

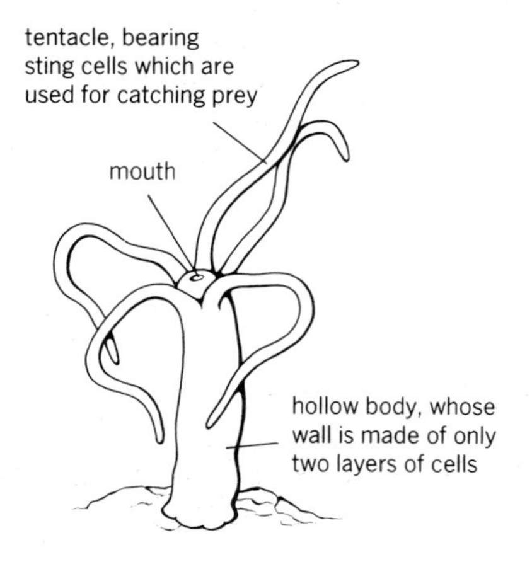

17.20 Hydra viridis

Phylum annelida

These are worms, with bodies made up of ring-like segments. Most of them live in water, though some, like the earthworm, live in moist soil.

Characteristics animals with bodies made up of ring-like segments,
no legs,
have chaetae (bristles).

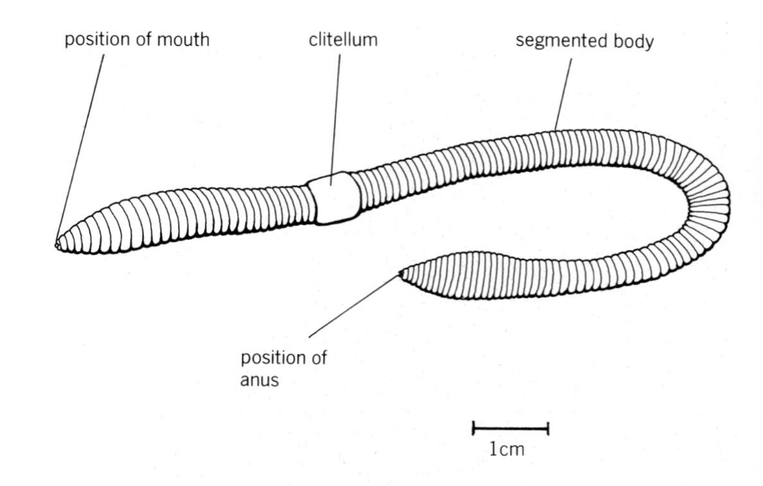

17.21 Earthworm, Lumbricus terrestris

Phylum arthropoda

Arthropods are animals with jointed legs, but no backbone. They are a very successful group, because they have a waterproof exoskeleton which has allowed them to live on dry land. There are more kinds of arthropod in the world than all the other kinds of animals put together.

Characteristics animals with several pairs of jointed legs,
have an exoskeleton.

Class crustacea

These are the crabs, lobsters and woodlice. They breathe through gills, so most of them live in wet places.

Characteristics Arthropods with more than four pairs of jointed legs,
breathe through gills.

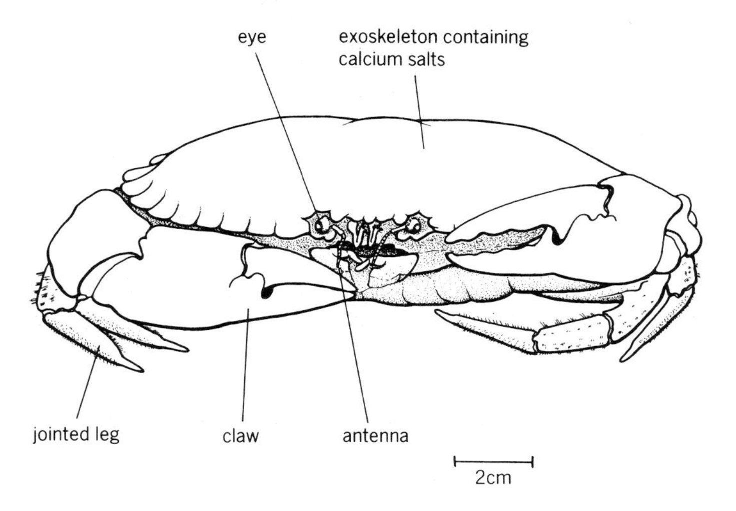

17.22 Edible crab, Cancer pagurus

17.23 Lobster, Homarus vulgaris

Class arachnida

These are the spiders, ticks and scorpions.

Characteristics Arthropods with four pairs of jointed legs,
breathe through gills called book lungs.

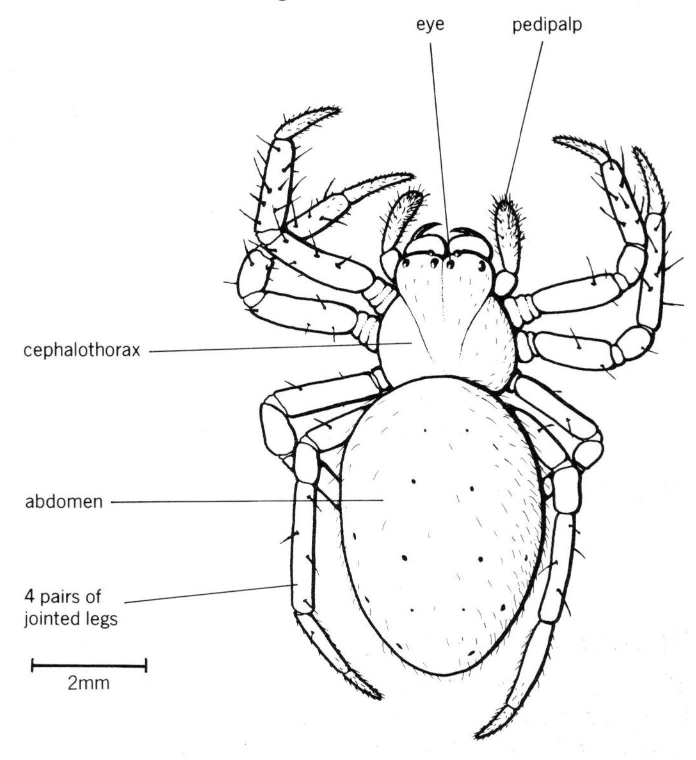

17.24 Spider, Areneus diadema

Class insecta

Insects are a very successful group of animals. This is mostly because their exoskeleton and tracheae are very good at stopping water from evaporating from the insects' bodies, so they can live even in very dry places.

Characteristics Arthropods with three pairs of jointed legs,
two pairs of wings (one or both may be vestigial)
breathe through tracheae.

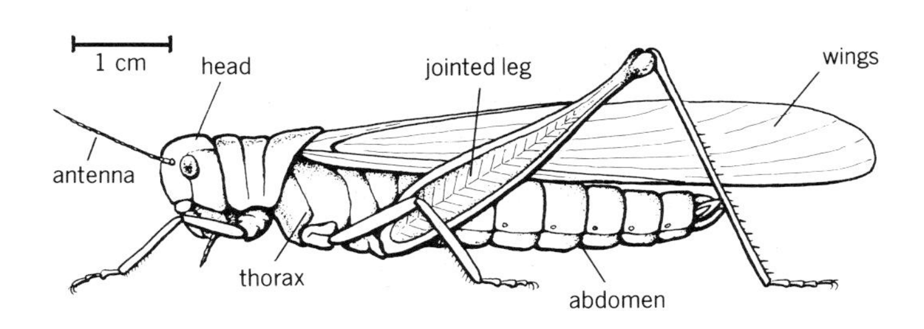

17.25 Desert locust, Locusta migratoria

Phylum chordata

These are animals with a supporting rod running along the length of the body. The most familiar ones have a backbone, and are called vertebrates.

Class pisces

The fish all live in water, except for one or two like the mud skipper, which can spend short periods of time breathing air.

Characteristics vertebrates with scaly skin,
have gills,
have fins.

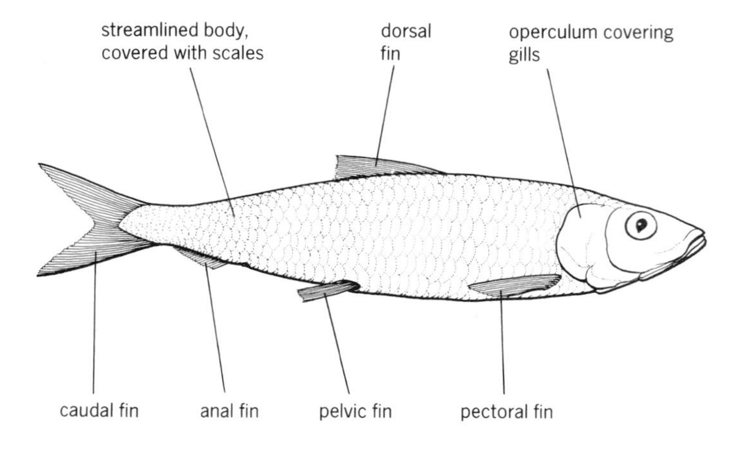

17.26 Herring, *Clupea harengus*

Class reptilia

Reptiles do not need to go back to the water to breed, because their eggs have a waterproof shell which stops them from drying out.

Characteristics vertebrates with scaly skin,
lay eggs with rubbery shells,

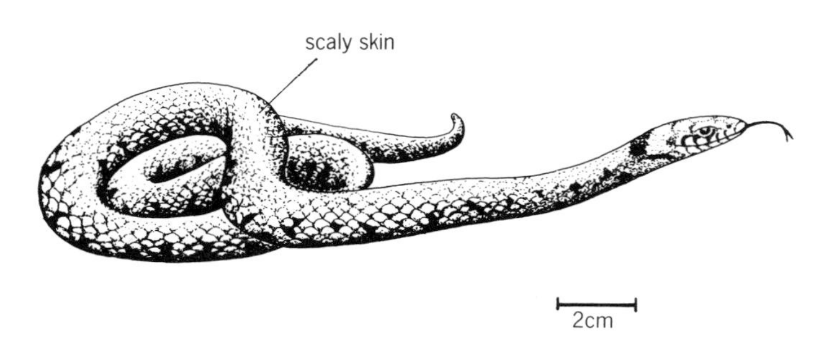

17.27 Grass snake, *Natrix natrix*

Class amphibia

Although most adult amphibians live on land, they always go back to the water to breed.

Characteristics vertebrates with moist, scale-less skin,
eggs laid in water,
larva (tadpole) lives in water, but adult often lives on land,
larva has gills, adult has lungs.

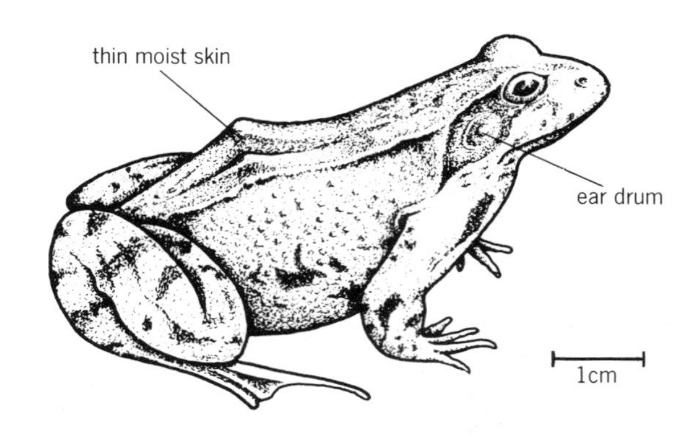

17.28 Frog, *Rana temporaria*

Class aves

The birds, like reptiles, lay eggs with waterproof shells.

Characteristics vertebrates with feathers,
forelimbs have become wings,
lay eggs with hard shells,
homeothermic,
have a beak.

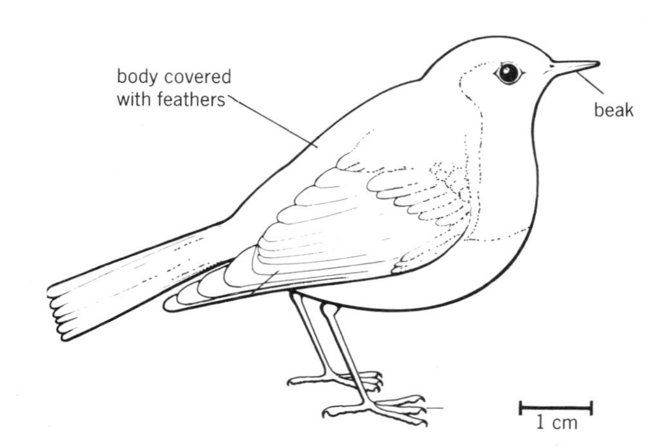

17.29 Robin, *Erithacus rubecula*

Class mammalia

This is the group which humans belong to.

Characteristics vertebrates with hair,
have a placenta,
young fed on milk from mammary glands,
homeothermic,
have a diaphragm,
heart has four chambers,
have different types of teeth (incisors, canines, premolars and molars)
cerebral hemispheres very well developed.

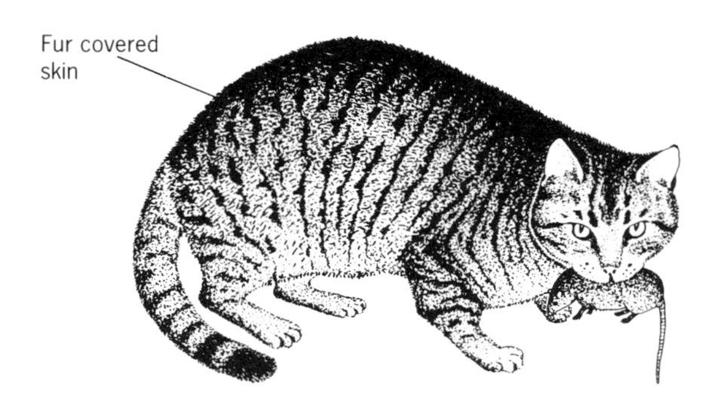

17.30 Cat, Felis catus

Chapter revision questions

1 Put these words into the correct order, beginning with the largest group and ending with the smallest:
species phylum genus kingdom
order class family
2 What is the Latin name for a horse? (See Fig 17.2).
3 List three similarities and two differences between Spermatophyta and Algae.
4 List three similarites and two differences between Annelids and Coelenterates.
5 List five differences between birds and mammals.
6 Classify as many of the organisms shown on the cover of this book as you can.

Revision questions

To find the answers to these questions, you will have to use many different parts of the book. The index will help you to find the information you need. For example, in question 2 you should look up *saprophytic nutrition, bacteria,* and *pathogens*.

1 Copy out and fill in the gaps in this passage.
All living organisms need energy. They obtain their energy from food, usually by combining it with .... This is called ...., and it takes place in organelles called ...., which are found in almost all animal and plant cells.
The type of food normally used in this process is .... Plants make theirs, by combining .... and ...., in the process of .... This happens in organelles called ...., in the mesophyll cells of green leaves.
Animals obtain their carbohydrates in foods such as .... If the carbohydrate is in the form of starch, it must first be digested. Saliva contains the enzyme ...., which breaks down starch to .... Further on in the alimentary canal, this complex sugar is broken down to the simple sugar ...., which can then be absorbed into the blood. The absorption takes place in the ...., whose inner surface is covered with .... to increase its surface area.

2 (a) What is meant by saprophytic nutrition?
(b) Some bacteria feed saprophytically. Describe one instance where this is useful to humans, and one instance where it is harmful to humans.
(c) What other group of organisms feeds saprophytically?
(d) Briefly describe three ways in which pathogenic bacteria may enter the human body, and the natural body defences against each of these methods of entry.

3 (a) What proportion of the atmosphere is carbon dioxide gas?
(b) Describe how the carbon from this carbon dioxide becomes part of a carbohydrate molecule in a plant.
(c) Briefly describe the ways in which the carbon dioxide may subsequently be returned to the atmosphere.

4 (a) What are the important features of the phylum arthropoda?
(b) What characteristics distinguish insects from other members of this phylum?
(c) Make a large, labelled diagram of a typical adult insect.
(d) Briefly describe three features of insects which have allowed them to live so successfully on dry land.
(e) (i) What is metamorphosis?
(ii) What is the advantage of metamorphosis to insects?

5 (a) Which of these processes require use of energy by a living organisms?
(i) diffusion of oxygen out of a stoma
(ii) movement of blood around a mammal's body
(iii) movement of phosphate ions into a plant's root hairs
(iv) movement of water up xylem vessels
(v) osmoregulation in *Amoeba*
(b) Describe how living organisms release energy from food.
(c) From where does the energy in food originate?
(d) Briefly explain how energy may be passed along a food chain in a named ecosystem.

6 (a) Why do living organisms need water?
(b) How is water obtained by a mesophyll cell in a leaf of a green plant?
(c) Describe an experiment you could perform to measure the rate of uptake of water by a shoot.
(d) List the ways in which water may be lost from the body of a mammal.
(e) Describe the processes by which the kidneys can help to regulate the amount of water in the blood of a mammal.

7 A wallflower with normal green leaves was self-pollinated, and 50 of the resulting seeds were germinated and allowed to grow.
(a) Draw a large, labelled diagram to show the events which would have taken place in one of the ovaries of the plant between pollination and fertilisation.
(b) Describe how the fertilised ovules would develop into seeds.
Of the young plants, 33 had green leaves like their parent, but 17 had yellow leaves.
(c) Using the symbols *G* for the gene for green leaves and *g* for the gene for yellow leaves, explain how this occurred.
(d) It was not possible to breed from the yellow-leaved plants. Why not?

Apparatus required for investigations

The apparatus listed is that required for each group performing the experiment.

1.1 Looking at animal cells

section lifter
slide
cover slip
pipette
very small amount of methylene blue (diluted)
filter paper or blotting paper
microscope

1.2 Looking at plant cells

sections of onion bulb; these can be cut beforehand, and kept in a beaker of water
slide
cover slip
pipette
filter paper or blotting paper
microscope
seeker or needle

2.1 Testing foods for carbohydrates

variety of foods
glucose solution (any strength)
sucrose solution (any strength)
starch powder
test tube rack
bunsen burner
test tube holder
boiling tubes
pipette
tile
scalpel
Benedict's solution
iodine solution
dilute hydrochloric acid
sodium bicarbonate solution

2.2 Testing food for proteins

variety of foods
albumen solution
test tube rack
test tubes
pipettes
tile
scalpel
potassium hydroxide solution
1% copper sulphate solution

2.3 Testing food for fats

variety of foods
cooking oil
test tube rack
clean, dry test tubes
pipette
tile
scalpel
absolute alcohol
distilled water
filter paper

3.1 Looking at the epidermis of a leaf

variety of leaves – ivy-leaved toadflax is good
forceps
slides
coverslips
pipette
clear nail varnish

3.2 Testing a leaf for starch

geranium plant which has been photosynthesising boiling water-bath, or beaker etc. as in Fig 3.12
boiling tube
methylated spirits
glass rod
iodine solution
forceps
white tile
pipette

3.3 To see if light is necessary for photosynthesis

Session 1 geranium plant
Session 2 destarched plant
apparatus as for 3.2
black paper or aluminium foil
scissors
paperclips
Session 3 apparatus as for 3.2

3.4 To see if carbon dioxide is necessary for photosynthesis

Session 1 geranium plant
Session 2 destarched plant
apparatus as for 3.2
two conical flasks, fitted with split corks
potassium hydroxide solution
distilled water
Vaseline
clamp stands or other means of support for flasks
Session 3 apparatus as for 3.2

3.5 To see if chlorophyll is necessary for photosynthesis

Session 1 plant with variegated leaves
Session 2 destarched plant with variegated leaves
apparatus as for 3.2
Session 3 apparatus as for 3.2

3.6 To show that oxygen is produced during photosynthesis

Session 1 large beaker
funnel which fits entirely inside beaker
test tube
Canadian pondweed (*Elodea*) or other water plant
Session 2 bunsen burner
splint

4.1 To show diffusion in a solution

gas jars
crystals of potassium permanganate, copper sulphate, potassium dichromate

4.2 To find the effects of different solutions on animal cells

solution A – distilled water	3 cover slips
solution B – 8.5% saline (isotonic)	labels for slides
solution C – 20% saline	filter paper

pipette for each solution
blood (either from pupils, or from a blood pack)
microscope
3 slides

N.B. Care must be taken when using blood, to prevent hepatitis infection, unless the blood has been screened for hepatitis. If using samples of blood from pupils, make sure that the skin is swabbed with 70% alcohol before a sample is taken; that each lancet is disposed of immediately after use, and is only used once; and that a sterile dressing is immediately applied to the wound. If using blood from a blood pack, prevent this from coming into contact with open wounds on pupils' skin, and mop up any spillages immediately.

4.3 To find the effects of different solutions on plant cells

solution A – distilled water	3 microscope slides
solution B – 0.3M sucrose solution	3 cover slips
solution C – 1.0M sucrose solution	labels for slides
red rhubarb petioles	filter paper or
forceps	blotting paper
scalpel	microscope

4.4 To demonstrate osmosis using eggs

Session 1	2 fresh eggs 2 large beakers containing enough dilute HC1 to cover eggs	
Session 2	deshelled eggs two large beakers Chinagraph pencils	distilled water 20% salt solution

4.5 To demonstrate osmosis using potatoes

4 petri dishes (lids not needed)	kitchen knife
Chinagraph pencil	distilled water
2 potatoes	20% salt solution
apparatus for cooking potatoes	2 pipettes

5.1 To show that carbon dioxide is produced in respiration

see Fig 5.2

5.2 To show the uptake of oxygen during respiration

see Fig 5.3

5.3 To show that heat is produced in respiration

Session 1 pea seeds
beaker
Session 2 boiled peas
soaked peas from Session 1
mild disinfectant solution
two Thermos flasks
two cotton wool plugs to fit flasks tightly
two thermometers
clamp stands to support flasks

5.4 To show that carbon dioxide is produced when yeast respires anaerobically

four boiling tubes, fitted with bungs and glass tubing as in Fig 5.6
boiled, cooled water, or apparatus for boiling it

sucrose or glucose	lime water or bicarbonate indicator solution
fresh or dried yeast	
boiled yeast solution	two beakers to support boiling tubes
glass rod	Chinagraph pencil
pipette	liquid paraffin

5.5 Examining lungs

set of sheep's or cow's lungs   burette tube

5.6 Using a model to show the action of the diaphragm

see Fig 5.18

5.7 Comparing the carbon dioxide content of inspired and expired air

see Fig 5.19

5.8 Investigating how breathing rate changes with exercise

stop watch
A good exercise in a confined space is stepping on and off a chair.

5.9 Investigating the structure of gills

small fish, e.g. sprat
seeker
scissors
Petri dish

5.10 To investigate the effect that plants and animals have on the carbon dioxide concentration of water

four boiling or specimen tubes, fitted with bungs
Elodea or other pond weed
snails or other pond animals
bicarbonate indicator solution
(This experiment may also be performed with terrestrial animals such as woodlice, or leaves of a terrestrial plant, held above the indicator solution on a gauze platform.)

6.1 To find the effect of exercise on the rate of heart beat

stop watch

6.2 To see which part of a root transports water and solutes

Session 1 freshly-pulled groundsel plant

	eosin solution	
	beaker	
Session 2	plant from Session 1	tile
	slide	paint brush or section lifter
	cover slip	pipette
	razor blade	microscope

6.3 To see which surface of a leaf loses most water

potted plant with smooth leaves
forceps
cobalt chloride paper in desiccator
self-adhesive book-covering film
scissors

6.4 To measure the rate of transpiration of a potted plant

two plants of similar size, in pots of the same size

two large polythene bags	Vaseline
rubber bands	balance

6.5 Using a potometer to compare rates of transpiration under different conditions

a potometer (not necessarily of the type in Fig 6.32)
plant with firm stems, such as geranium, which will fit tightly into the apparatus

wire	stop watch
pliers	electric fan
vaseline	

7.1 To find the conditions necessary for the germination of mustard seeds

five test tubes, fitted with gauze or perforated zinc platforms
pyrogallol in NaOH solution (take care; this is very caustic)

cotton wool	mustard seeds
one rubber bung to fit test tube	
test tube racks	Chinagraph pencil

7.2 To find which part of a root is the growing region

Session 1	soaked broad bean seeds
	gas jar
	blotting paper
Session 2	blotting paper
	Indian ink
	pen or fine paint brush
	ruler

8.1 Examining the structure of a hen's egg

hard boiled egg
fresh unfertilised egg
one fertilised egg for demonstration, with a window cut in the top of the shell.

two Petri dishes	binocular microscope
paper towel	tile
blunt forceps	kitchen knife

8.2 Investigating the structure of a wallflower

flower stalk of wallflower (*Cheiranthus*) or other simple insect-pollinated flower, with flowers in various stages of development

hand lens	microscope slide
razor blade	microscope
tile	seeker

8.3 Growing pollen tubes

four cavity slides	four types of flowers, with ripe pollen
vaseline	seeker
four cover slips	microscope
labels for slides	incubator at 20°C
variety of sugar solutions, e.g. distilled water	5% sucrose 10% sucrose 15% sucrose, each with a very small amount of boric acid added.

9.1 Investigating the effect of temperature on enzyme activity

four test tubes	spotting tile
beakers	four glass rods
water baths at 35°C and 80°C	
thermometer	stop watch
boiling tube	Chinagraph pencil
distilled water	pipette
10% starch solution	two syringes to measure
iodine solution	$5cm^3$

11.1 Using a model arm to investigate the action of the biceps muscle

see Fig 11.12
variety of weights
spring balance

12.1 To find which part of the skin contains the most touch receptors

two pins
Plasticene or piece of polystyrene, through which pins may be pushed to support them in position
ruler

12.2 To see which parts of the tongue can taste which flavours

solutions of salt, sugar, quinine and lemon juice
4 straws or cotton wool sticks

12.3 To measure reaction time

stop watch

12.4 To find out how shoots respond to light

3 Petri dishes
Chinagraph pencil
cotton wool or filter paper
mustard seeds
2 light-proof boxes, one with a slit in one end
clinostat

12.5 To find out how roots respond to gravity

Session 1	broad bean seeds	
	blotting paper	gas jars
Session 2	2 clinostats	pins
	blotting paper	

12.6 To find which part of a shoot is sensitive to light

Session 1	pots containing seed compost	oat seeds
Session 2	germinated coleoptiles	foil
	razor	ruler
	3 light-proof boxes with a slit in one end	
Session 3	coleoptiles from Session 2	ruler

12.7 To find how auxin affects shoots

Three pots of germinated oat coleoptiles.
Lanolin, warmed slightly.
IAA solution. Make this by dissolving 0.1g IAA in 1 litre of distilled water to make a solution of 100 parts per million. This can then be diluted to 1 ppm for use in the experiment. The solution should be used within a few hours of being made up.
Labels.
Three clinostats.

13.1 Estimating the size of a bead population, using the mark, release, recapture technique

large tray or bucket
about 1,000 beads of one colour and size
about 50 more beads of the same size, but a different colour

14.1 Making a rough estimate of the proportions of particles of different sizes in a soil sample

sample of soil, enough to ¼ fill a gas jar
gas jar
large stirring rod

14.2 To estimate the percentage of water in a soil sample

Session 1	evaporating dish
	spatula
	balance
	soil sample
	oven set at about 50°C
Session 2	soil sample from Session 1
	balance

14.3 To estimate the percentage of humus in a soil sample

Session 1	dried soil sample from Investigation 14.2
	balance
	bunsen etc. and crucible, or oven at very high temperature
Session 2	cooled sample from Session 1
	balance

14.4 To find the effect of lime on clay particles

small sample of powdered clay
boiling tube
glass rod
small amount of calcium hydroxide or oxide
spatula

15.1 Breeding beads

two containers
150 beads of one colour
100 beads of a second colour

Glossary

abiotic factor: an influence on an organism caused by a non-living feature of its environment.

absorption: the uptake of a substance into the cells of an organism's body.

accommodation: the adjustment of the shape of the lens and eyeball, so that light is focussed accurately onto the retina.

active transport: the movement of substances through cell membranes, using energy. The energy is in the form of ATP, which is first made by respiration. The substances are often moved against their concentration gradient.

adaptation: a feature of an organism which enables it to live successfully in its environment.

adolescence: the time between childhood and adulthood.

ADP: adenosine diphosphate – a substance found in all living cells, which is converted to ATP during respiration.

adventitious root: a root growing out of a stem.

aerobic respiration: the release of energy from glucose, by combining it with oxygen.

afterbirth: the placenta, which leaves the mother's body through the vagina, just after the baby is born.

albumen: a protein which forms a jelly-like substance when dissolved in water, and which is found in egg-white.

alga: a simple plant, with no stems or roots.

alleles: a pair (or more) of genes which code for the same characteristic, and are found on the same locus of homologous chromosomes.

amino acids: molecules containing carbon, hydrogen, oxygen, nitrogen and sometimes sulphur. A long chain of amino acids forms a protein molecule.

amnion: a membrane surrounding a developing foetus.

amniotic fluid: the liquid contained within the amnion, which supports the foetus and protects it.

ampulla: a swelling at one end of each of the semi-circular canals in the ear, containing cells which detect movements of the head.

amylase: an enzyme which digests starch to maltose.

anaemia: a disease caused by lack of haemoglobin, often because of a shortage of iron.

anaerobic respiration: the release of energy from glucose, without combining it with oxygen.

androgen: a hormone which produces male characteristics.

annual: a plant which completes its life cycle within one year or less.

antagonistic: antagonistic muscles work in pairs, one causing a joint to bend, and the other causing it to straighten.

anther: the part of a flower where male gametes are produced.

antibiotic: a drug which kills microorganisms without harming other cells.

antibodies: proteins made by white cells which attach to specific foreign cells or other substances (antigens) and help to destroy them.

antigen: a cell or other substance which is recognised as foreign by the body's white cells.

aquatic: living in water.

arachnid: a spider or scorpion; an arthropod with eight or more legs, which breathes by means of book lungs.

arthropod: an invertebrate with jointed legs.

articulation: the movement of two bones at a joint.

assimilation: the incorporation of absorbed food into various parts of the body.

ATP: adenosine triphosphate; a high-energy compound, found in all living cells, in which the energy released from glucose in respiration is stored.

atrio-ventricular: between the atria and ventricles of the heart.

atrium: one of the upper chambers of the heart, which receive blood from the veins and pass it on to the ventricles.

autotrophic nutrition: feeding by converting inorganic materials into organic ones.

auxin: a hormone produced in the meristems of plants, which affects cell elongation.

axon: a long process stretching from the cell body of a neuron, which carries impulses away from the cell body.

back-cross: breeding an organism showing a dominant characteristic in its phenotype, with one which is homozygous for the recesive characteristic. The offspring from this cross will show you whether the first organism is homozygous or heterozygous.

balanced diet: a daily intake of food containing all types of food in the correct proportions.

ball and socket joint: a joint where a 'ball' on one bone fits into a 'socket' on another, allowing a circular movement.

beri-beri: a disease caused by lack of vitamin B_1. The symptoms are weak muscles and tiredness.

bicuspid valve: the valve between the left atrium and ventricle of the heart.

bile: a liquid made by the liver, stored in the gall bladder, and passed into the duodenum along the bile duct.

bile salts: substances found in bile, which emulsify fats in the duodenum.

binary fission: a form of asexual reproduction in which one cell splits into two.

biotic factor: an influence on an organism, caused by other organisms.

blind spot: the part of the retina where the optic nerve leaves; it has no receptor cells, and so light falling onto it cannot be sensed.

bolus: a ball of food formed after chewing, which is swallowed.

bone marrow: a cavity of some bones where blood cells are made.

bronchiole: a small tube carrying air to and from the alveoli in the lungs.

bronchus: a large tube connecting the trachea to the bronchioles.

bulb: an underground bud, consisting of a short stem and many fleshy leaves tightly packed together.

caecum: part of the alimentary canal next to the appendix, used for cellulose digestion in some herbivores.

cambium: a tissue found in the stems and roots of plants, made of cells which can divide.

canine: a pointed tooth between the incisors and premolars, used by carnivores for killing prey and tearing meat.

carbohydrase: an enzyme which breaks down carbohydrate molecules.

carbohydrate: sugars and starches; a substance made of carbon, hydrogen and oxygen, where the ratio of hydrogen to oxygen atoms is 2:1.

carpal: one of the bones in the wrist.

carpel: the female part of a flower, made up of ovary, style and stigma.

carnassial teeth: the largest premolars of a carnivore, which slice past each other to crush bones.

carnivore: an animal which feeds on other animals which it kills.

carrier: a heterozygous organism, possessing a recessive gene which does not show in its phenotype.

cartilage: a tissue made of living cells in a matrix of collagen. It is found in several places in the mammalian skeleton, and makes up the entire skeleton of sharks.

catalyst: a substance which alters the rate of a reaction, without being changed itself.

cell membrane: a very thin layer of protein and fat, which surrounds the protoplasm of every living cell. Membranes are also found inside cells.

cell sap: a solution of sugars, amino acids and many other substances, found in the vacuoles of plant cells.

cellulase: an enzyme which breaks down cellulose molecules.

cellulose: a polysaccharide (long-chain carbohydrate) with molecules made from glucose molecules linked together in very long chains. Cellulose forms fibres, which make up the cell walls of plants.

cerebellum: an area of the brain which controls muscular co-ordination.

cerebrum: the part of the brain responsible for conscious thought, language and personality. In mammals it is very large and folded, and forms two cerebral hemispheres.

chalaza: a coil of albumen which supports the yolk in the centre of a bird's egg.

chitin: a carbohydrate-like substance which makes up the exoskeleton of insects.

chlorophyll: a green pigment found in all plants and some bacteria and protistas, which absorbs energy from sunlight to be used in photosynthesis.

chloroplasts: organelles found in many plant cells, which contain chlorophyll, and where photosynthesis takes place.

chordate: an animal with a supporting rod running inside its dorsal surface. Vertebrates are the most familiar chordates.

choroid: a black layer lining the eye, which absorbs light and so cuts down reflections inside the eye.

chromosome: a coiled thread of DNA and protein, found in the nucleus of cells.

chyme: a mixture of partly digested food, enzymes and hydrochloric acid – the result of digestion in the stomach.

cilia: small hair-like structures which project from some cells, and perform waving movements in synchrony with each other.

ciliary muscle: a ring of muscle surrounding the lens in the eye, which can adjust the size of the lens.

climax community: the mixture of species of plants and animals which will finally exist in an environment.

cochlea: a coiled tube in the inner ear, containing cells which are sensitive to sound waves.

codominance: the existence of two alleles for a characteristic where neither is dominant over the other.

codon: a row of three bases in a DNA molecule, which codes for one amino acid.

coelomic fluid: the fluid which fills up the space, or coelom, in an earthworm, and also in vertebrates.

coleoptile: a protective sheath which covers the plumule of a seedling of oats, barley, grasses etc.

collagen: a protein which is found in bone and many other tissues.

colon: the part of the alimentary canal between the ileum and the rectum, where water is absorbed.

community: all the organisms which live in a particular habitat.

companion cell: a cell found next to a sieve tube element in phloem tissue.

connective tissue: any tissue which fills in spaces, or connects various parts of the body, e.g. adipose tissue.

consumer: an organism which consumes other organisms for food; all animals are consumers.

continuous variation: variations in which organisms do not belong to definite categories, but may fit an anywhere within a wide range.

contractile vacuole: a vacuole found in *Amoeba* and many other protista, in which excess water is collected before being emptied out of the cell.

control: a piece of apparatus identical in every way to the experimental apparatus, except for the one thing whose effect you are investigating.

cork cambium: a layer of meristematic (dividing) cells which produce cork cells.

cornified layer: a layer of dead cells, containing keratin, on the surface of skin. It protects and waterproofs the layers underneath it.

corpus luteum: a structure in a mammalian ovary, formed from a follicle, which secretes the hormone oestrogen.

cortex: (a) the part of a stem or root between and around the vascular bundles or stele.
(b) the outer part of a kidney.

cretin: a child whose thyroid gland does not function properly, resulting in retarded mental and physical development.

cupula: a structure inside a semi-circular canal, which moves as the head moves, and stimulates sensory cells embedded in it.

cuticle: a waxy, waterproof covering, found on various parts of various organisms, e.g. leaves of plants and the exoskeleton of insects.

dark reaction: the stage in photosynthesis when hydrogen is combined with carbon dioxide, to make glucose.

deamination: a reaction which takes place in the liver, where amino acids are converted to urea and carbohydrate.

deciduous tree: a tree which sheds all its leaves in autumn.

denitrifying bacteria: bacteria which often live in damp soil, and which convert nitrates into nitrogen gas.

dermis: the inner layer of skin, made of connective tissue and containing capillaries, nerve endings etc.

destarching: keeping a plant in the dark, so that it cannot photosynthesise and will use up its starch stores.

diastema: a toothless gap found in many herbivores, between the incisors and premolars.

diastole: the stage in heart beat when muscles are relaxed.

dichotomous key: a way of identifying an organism, in which you are given successive pairs of descriptions to choose between.

diffusion: the movement of particles of gas, solvent or solute, from an area of high concentration to an area of low concentration.

diploid cell: a cell which has two of each kind of chromosome.

disaccharide: a sugar made of molecules which consist of two monosaccharide units joined together, with no intermediates.

DNA: deoxyribonucleic acid; a substance found in chromosomes, which carries a code used by the cell when making proteins.

dominant gene: a gene which has the same effect on the phenotype of an organism, whether the organism is homozygous or heterozygous for that gene.

Down's syndrome: a condition caused by having an extra chromosome, where full mental development does not take place; sometimes called mongolism.

duodenum: a short length of alimentary canal between the stomach and the jejunum, into which the bile duct and the pancreatic duct empty bile and pancreatic juice.

ecdysis: moulting; the shedding of the exoskeleton of an insect or other arthropod.

ecosystem: the living organisms and their environment, in a certain area – e.g. a wood, or a pond.

effector: a part of an organism which carries out an action, often in response to a stimulus, e.g. muscles and glands.

egestion: the removal of indigestible food from the body.

embryo: a plant or animal as it develops from a fertilised egg.

emulsification: the breaking up of large droplets of fat into small ones, which can then disperse in water.

endolymph: the fluid contained in the central chamber of the cochlea.

endoplasmic reticulum: a network of membranes in the cytoplasm of most cells, where large molecules are built up from small ones.

enzymes: biological catalysts; proteins made by living organisms, which speed up chemical reactions.

epidermis: an outer covering made of one or more layers of cells, found in many parts of many organisms.

epithelium: a tissue which covers surfaces either inside or outside the body.

etiolated plant: a plant which has grown in insufficient light, and is yellow, thin, and taller than normal.

Eustachian tube: an air-filled tube leading from middle ear to the back of the throat.

extensor: a muscle which straightens a limb when it contracts.

extracellular: outside cells.

faeces: remains of indegestible food, bacteria, mucus etc., which are egested from the alimentary canal.

family: a group of genera with similar characteristics.

fatty acids: molecules made of long chains of carbon atoms with many hydrogen atoms and some oxygen atoms attached to them. Fatty acids can combine with glycerol to make fats.

fermentation: the conversion of sugar into alcohol and carbon dioxide, often by yeast, by means of anaerobic respiration.

fertilisation: the joining together of the nucleus of a male gamete with the nucleus of a female gamete.

flaccid: a flaccid plant cell is one which has lost water, so that the cytoplasm does not push outward on the cell wall.

flexor: a muscle which bends a limb when it contracts.

flocculation: the clumping of clay particles into larger crumbs, which can be caused by the addition of lime.

foetus: a mammalian embryo in a fairly advanced stage of development.

fovea: the part of the retina where receptor cells are most densely packed, and onto which light is normally focussed.

fruit: a plant's ovary after fertilisation; it contains seeds, and usually helps in their dispersal.

gamete: a sex cell, containing only one of each kind of chromosome (i.e. haploid). Eggs and sperm are gametes.

gene: a length of DNA which codes for the making of a particular protein.

genotype: the genes possessed by an organism.

genus: a group of species with similar characteristics.

geotropism: the directional response of a plant, by growth, to gravity.

gestation period: the time between fertilisation and birth.

glomerulus: a tangle of blood capillaries inside the cup of a Bowman's capsule in the kidney.

glycerol: an organic molecule containing carbon, hydrogen and oxygen, which forms the backbone of many kinds of fat molecules.

graft: a piece of one organism which is attached to another in such a way that it will become part of it.

habitat: the place where an organism lives.

haemoglobin: a protein containing iron, found in red blood cells, which carries oxygen.

haploid: having only one of each kind of chromosome.

hepatic: of the liver.

herbaceous: a plant with little or no wood, normally dying back in the winter.

herbivore: an animal which eats plants.

hermaphrodite: able to make both male and female gametes.

heterotrophic: using food made by other organisms; all animals and fungi are heterotrophic.

heterozygous: possessing two different genes for a certain characteristic.

hinge joint: a joint such as the elbow or knee, where movement is possible in one plane only.

holozoic nutrition: feeding by taking in pieces of food which are digested inside the alimentary canal, as mammals do.

homeostasis: the maintenance of a constant internal environment.

homeothermic: able to maintain a constant body temperature.

homologous chromosomes: chromosomes which carry genes for the same characteristics in the same positions.

homologous structures: structures which are used for different purposes, but which appear to be built to the same basic design, e.g. the limbs of a bat and a horse.

homozygous: possessing two identical genes for a certain characteristic.

hormone: a chemical which is made in one part of an organism, and travels through it to affect another part.

hypothalamus: a part of the brain to which the pituitary gland is joined, and which is responsible for several aspects of homeostasis, such as temperature regulation.

ileum: the part of the alimentary canal between the duodenum and colon; it is very long, and is lined with villi to help with absorption of digested food.

immunity: the possession of antibodies against a particular disease.

incisor: a tooth at the front of a mammal's mouth, normally used for biting off pieces of food for chewing.

ingestion: taking food into the alimentary canal.

insoluble: not able to dissolve in water.

instar: a stage between moults of an insect which has incomplete metamorphosis (e.g. locust), where each stage becomes progressively more like the adult.

integument: a covering.

intracellular: inside cells.

islets of Langerhans: patches of cells in the pancreas which secrete insulin.

karyotype: the shapes, sizes and numbers of the chromosomes in a cell.

keratin: the protein which makes up hair, nails, horn and the outer layer of skin.

lactase: an enzyme which breaks down the disaccharide lactose into monosaccharides.

lactation: the secretion of milk by a female mammal.

lacteal: a lymphatic vessel inside a villus, which looks milky because it contains absorbed fat.

lactic acid: a chemical produced during anaerobic respiration in animals.

lactose: milk sugar; a disaccharide found in milk.

lamina: the blade of a leaf.

larva: a young organism which looks very unlike its parent, e.g. a caterpillar or tadpole.

leaching: the loss of soluble substances from soil, as they are washed out by rain water.

lenticel: an area in a woody stem where the cells are loosely packed, allowing gas exchange to take place.

light reaction: the stage in photosynthesis where light energy, absorbed by chlorophyll, is used to split water molecules into hydrogen and oxygen.

lignin: the substance present in the walls of xylem vessels; wood contains lignin.

limiting factor: a factor whose supply limits the rate of a metabolic reaction; e.g. low light intensity may limit the rate at which photosynthesis takes place.

lipase: an enzyme which digests fats.

lumen: the space in the middle of a tube.

lymph node: an organ through which lymph flows, containing many white cells, and where antibodies are made.

lysozyme: an enzyme found in tears, which can destroy bacteria.

Malpighian layer: a layer of cells at the base of the epidermis in the skin, which divide to provide new cells, and which contain the pigment melanin.

maltase: an enzyme which digests the disaccharide maltose.

maltose: a disaccharide found in germinating seeds, formed from the breakdown of starch.

mechanical digestion: the breakdown of large particles of food into small ones, by teeth and the churning movements produced by muscles.

medulla: the part of a kidney between the cortex and the pelvis.

medulla oblongata: the part of the brain nearest to the spinal cord, responsible for the control of heart-beat, breathing movements etc.

meiosis: a type of cell division in which homologous chromosomes separate, resulting in four haploid cells being produced by one diploid cell.

meninges: membranes surrounding the brain and spinal cord.

menstruation: the breakdown and loss of the soft lining of the uterus.

meristem: a part of a plant which contains cells which can divide.

mesophyll: the central layers of a leaf, where photosynthesis takes place.

metabolism: the chemical reactions taking place in a living organism.

metamorphosis: a change from a larva to an adult organism.

microclimate: the climate in a small area, such as under a log.

micropyle: a small gap in the integuments of an ovule, through which the pollen tube grows. Later, when the ovule becomes a seed, the micropyle remains as a hole through which water enters at germination.

mitosis: a type of cell division in which two identical cells are formed from a parent cell.

mitral valve: the valve between the left atrium and ventricle.

molar: a large tooth near the back of a mammal's mouth, used for chewing, grinding or slicing.

monosaccharide: a simple sugar: a sugar whose molecules are made of a single sugar unit.

mucus: a slimy liquid, secreted by goblet cells, used in many parts of the body for lubrication.

mutation: an unpredictable change in an organism's genes or chromosomes.

mycelium: the tangle of threads (hyphae) which makes up the body of a fungus.

natural selection: the selection of only the best adapted organisms for survival and reproduction, by natural factors such as predators or shortage of food supply.

nectar: a sugary liquid secreted by flowers to attract insects for pollination.

nephron: a kidney tubule, where urine is formed.

nerve: a group of nerve fibres, surrounded by connective tissue.

nerve fibre: an axon or dendron; a strand of cytoplasm extending from a nerve cell body.

niche: the role of a living organism in a community.

nitrifying bacteria: bacteria which convert proteins and urea into nitrates.

nitrogen fixation: the conversion of nitrogen gas into some compound of nitrogen, such as ammonia, nitrates or proteins.

nymph: the young stage of an insect, which resembles its parent rather more than a larva does.

oesophagus: the part of the alimentary canal between the mouth and the stomach.

oestrogen: a hormone secreted by the ovaries which produces female secondary sexual characteristics.

operculum: the covering over the gill openings of a fish.

organelle: a membrane-bound structure inside a cell.

organ: part of an organism; a structure made of several tissues, which performs a particular function, e.g. heart.

osmoregulation: the control of the water content of the body.

osmosis: the movement of water molecules from a dilute solution to a concentrated solution, through a selectively permeable membrane.

ovulation: the release of an egg from the ovary.

ovule: a structure inside a plant's ovary which contains a female gamete, and which develops into a seed after fertilisation.

ovum: a female gamete; an egg.

oxidation: the combination of a substance with oxygen.

oxyhaemoglobin: haemoglobin combined with oxygen; it is a brighter red than haemoglobin.

palisade layer: a layer of rectangular cells near the upper surface of a leaf, where photosynthesis takes place.

pancreatic juice: watery fluid secreted by the pancreas, containing various digestive enzymes, which flows into the duodenum along the pancreatic duct.

parasite: an organism which lives in very close association with another, and feeds on it.

parthenogenesis: the production of young from unfertilised eggs.

pathogen: an organism which causes disease.

pepsin: an enzyme secreted by glands in the wall of the stomach, which digests proteins.

perennial: a plant which survives one or more winters.

pericarp: the outer layers of a fruit, developed from the overy wall.

perilymph: a fluid found in the semicircular canals in the ear, and also the outer part of the cochlea.

peristalsis: rhythmic contractions of the muscles in the walls of tubes, such as the alimentary canal or oviduct, which squeeze the contents along.

permeable: allowing substances to pass through.

pH: a measure of the acidity of a solution; pH 7 is neutral, below 7 acidic, and above 7 alkaline.

phagocytosis: 'cell feeding'; the intake of particles of food by a cell.

phenotype: characteristics shown by an organism, a result of interaction between its genotype and its environment.

phloem: a plant tissue in which substances made by the plant are carried from one part to another.

photolysis: the splitting of water molecules into hydrogen and oxygen, using energy from light, in photosynthesis.

phylum: a major group of organisms; a subdivision of a kingdom.

pioneer species: a species which colonises new areas while conditions are still not suitable for many other species to live there.

placenta: the organ through which a mammalian embryo is connected to its mother, and through which it obtains food, oxygen etc.

plankton: microscopic organisms which float in water.

plasma: the liquid part of blood.

plasmolysis: shrinkage of the cytoplasm of a plant cell, so that the cell membrane begins to tear away from the cell wall; caused by loss of water.

pleural membranes: membranes surrounding the lungs and lining the thoracic cavity.

plumule: part of an embryo plant which will develop into the shoot.

poikilothermic: unable to control body temperature accurately.

pollination: the transfer of pollen from an anther to a stigma.

pollution: the addition of some factor to an environment, such as sewage, toxic gases or noise, which has an adverse effect on the organisms in that environment.

polysaccharide: a carbohydrate such as starch or cellulose, whose molecules are made of many sugar units joined together.

population: all the organisms of a particular species living in a certain area.

predator: an animal which hunts and kills other animals (known as its prey) for food.

premolar: large tooth between the canines and molars of mammals; unlike molars, premolars are present in the milk dentition as well as the permanent dentition.

primary consumer: the first consumer in a food chain; a herbivore.

producer: the first organism in a food chain, which produces food, i.e. a green plant.

progesterone: a hormone secreted by the ovary and later the placenta, which maintains the uterus lining during pregnancy.

prokaryotes: organisms such as bacteria, whose cells have no nucleus or other organelles with a membrane round them.

prostate gland: a gland near the junction of the two sperm ducts with the urethra, which secretes a fluid in which sperm swim.

protease: an enzyme which digests protein.

protein: a substance whose molecules are made of long chains of amino acids.

protista: unicellular organisms with nuclei.

pseudopodium: 'false foot'; a projection from a moving cell such as *Amoeba* or a white blood cell.

puberty: the age at which secondary sexual characteristics appear, and gametes begin to be produced.

pulmonary embolism: a blockage in a capillary or small artery in the lungs, which may be caused by a blood clot.

pure breeding: producing offspring like themselves; homozygous.

quadrat: a square which is placed over an area so that the numbers of organisms within it may be estimated.

radicle: a young root.

receptor: part of an organism which receives stimuli,

recessive gene: a gene which only shows in the phenotype in a homozygous organism.

rectum: the last part of the alimentary canal, in which faeces are formed before being egested through the anus.

reflex action: an automatic, unchanging response to a stimulus.

reflex arc: the series of neurones and synapses by which an impulse passes from a receptor to an effector in a reflex action.

refraction: the bending of light rays as they pass through materials of different densities.

relay neurone: a neurone in the central nervous system, which passes impulses from one neurone to another.

respiration: the release of energy from carbohydrates; it happens in every living cell.

respiratory surface: the surface of an organism's body across which gas exchange takes place.

rhizome: an undergound stem.

ribosomes: very small particles found in all cells, where protein molecules are assembled from amino acids.

rickets: a disease of the bones caused by lack of vitamin D.

RNA: ribonucleic acid; a substance found in all cells, one type of which copies instructions from the DNA in the nucleus and carries them to the ribosomes.

root hair: part of a root cell which projects into the soil, where it absorbs water and mineral salts.

roughage: fibrous, indigestible food, which stimulates the muscles of the alimentary canal to perform peristalsis.

saliva: watery fluid containing salivary amylase and mucus, secreted into the mouth by salivary glands.

saprophyte: an organism which feeds on dead organic material, by secreting enzymes onto it and absorbing it in liquid form.

secondary growth: growth in diameter of stems and roots.

secretion: the production and release of a useful substance.

selection pressure: a factor acting on a population which favours certain varieties for survival.

selectivly permeable: allowing some substances, but not others, to pass through.

septum: a structure which divides one part of an organism from another; a partition.

sex chromosomes: chromosomes which determine the sex of an organism.

sex linked: sex linked characteristics are determined by genes carried on the sex chromosomes, so that their inheritance is linked with the inheritance of sex.

sieve plate: the perforated end wall of a sieve-tube element in phloem tissue.

smooth muscle: muscle found in the walls of the alimentary canal, bladder etc., which contracts slowly and smoothly over long periods of time.

solute: a substance which dissolves in water or another solvent, to form a solution.

solvent: a liquid such as water, in which other substances can dissolve.

species: a group of organisms with similar characteristics, which can breed with each other, but not with organisms of different species.

spermatozoa: sperm; the male gametes of animals.

spermatophyte: a plant which reproduces by means of seeds; conifers and flowering plants.

sphincter muscle: a muscle round a tube, which can close the tube when it contracts.

sphygmomanometer: an instrument used to measure blood pressure.

spinal nerve: a nerve entering or leaving the spinal cord.

spiracle: a hole in the side of an insect, through which air enters the tracheal system.

spongy mesophyll: layer of cells near the underside of a leaf where photosynthesis takes place; they have large air spaces between them.

stabilizing selection: natural selection which acts on a population to keep it very much as it is; the most common form of natural selection.

starch: the polysaccharide storage material of plants, made from molecules of hundreds of glucose units linked together.

stele: xylem and phloem tissue in the centre of a root.

stimulus: a change in an organism's environment which is detected by a receptor.

striated muscle: muscle attached to the skeleton, which contracts when stimulated by a nerve.

stylets: mouthparts of an insect such as an aphid, adapted for piercing and sucking.

suberin: a waterproof substance which forms the cell walls of cork cells.

substrate: (a) a substance which is converted to another during a chemical reaction.
(b) the material on which a bacterium or fungus lives and feeds.

succession: a gradual change in the numbers and variety of organisms living in a habitat, beginning with colonisation and ending with a climax community.

sucrase: an enzyme which digests sucrose.

sucrose: a non-reducing disaccharide sugar.

sugar: a type of carbohydrate made of molecules consisting of one (monosaccharide) or two (disaccharide) sugar units; it tastes sweet, and is soluble.

suture: a join, e.g. between the bones in the cranium.

sweat gland: coiled gland in the dermis, which extracts water, salt and urea from the blood and secretes it as sweat.

swim bladder: an air-filled sac lying just under the backbone of bony fish, which aids buoyancy.

symbiosis: organisms of two different species living together for mutual benefit.

synapse: very small gap between two nerve fibres.

synovial joint: a joint between two bones, where free movement can occur.

synovial membrane: a membrane enclosing a synovial joint, attached to the bones on each side, which secretes and encloses synovial fluid.

systole: the stage in heart-beat when muscle contracts.

tap root: a root system with a main root which grows vertically downwards.

taste bud: a group of cells on the tongue which are sensitive to one or more of the four tastes sweet, sour, salty or bitter.

tendon: tough band of fibres which joins a muscle to a bone.

testa: hard outer covering of a seed.

testosterone: a hormone secreted by the testes, responsible for male secondary sexual characteristics.

thorax: chest; the part of the body containing heart and lungs, separated from the abdomen by the diaphragm.

thrombosis: a blood clot in a vein or artery.

thyroid gland: an endocrine gland in the neck, which secretes thyroxine.

thyroxine: a hormone containing iodine, which speeds up metabolic rate.

tissue: a group of similar cells which together perform a particular function.

tissue fluid: fluid which fills in spaces between cells in the body, formed by plasma which leaks from blood capillaries.

toxin: a poison, especially one produced by pathogens inside the body.

trachea: a strengthened tube in vertebrates or insects, through which air passes on its way to and from the respiratory surface.

tracheole: a small tube branching from a trachea in an insect, with thin walls where gas exchange occurs.

transect: a line along which vegetation or animal life is recorded, usually to investigate changes from one habitat to another.

translocation: the movement of materials within a plant, particularly ones which the plant itself has made, such as sugars.

transmitter substance: a chemical which diffuses across a synapse to transmit an impulse to the other side.

transpiration: the evaporation of water from a plant, mostly from the leaves.

tricuspid valve:: the valve between the right atrium and ventricle.

trophic level: the level in a food chain at which an organism feeds.

tropism: a directional growth response of a plant to a stimulus.

turgid: a turgid plant cell contains plenty of water, so that the cytoplasm pushes outwards on the cell wall.

trypsin: an enzyme secreted by the pancreas, which digests proteins in the duodenum.

umbilical cord: cord containing an artery and vein, which connects a foetus to its placenta.

urea: substance containing nitrogen, made in the liver by the deamination of excess amino acids, and excreted in urine.

ureter: a tube carrying urine from the kidney to the bladder.

urethra: a tube leading from the bladder to the outside; it carries urine, and also semen in males.

uric acid: a semi-solid, nitrogenous excretory product of birds and many insects.

urine: a watery liquid containing urea and other excretory substances, produced by the kidneys.

uterus: womb; muscular organ with a soft lining, where a foetus develops.

vacuole: an organelle containing liquid, and surrounded by a membrane.

vagina: tube leading from the uterus to the outside.

vascular bundle: a group of xylem vessels and phloem tubes.

vas deferens: sperm duct; the tube carrying sperm from the testis to the urethra.

ventricle: one of the thick-walled lower chambers of the heart, which pumps blood into the arteries.

vestigial: small and useless.

villus: a small 'finger' of tissue, which increases surface area; found e.g. in the ileum and the placenta.

xylem vessel: a long, narrow tube made of many dead, lignified cells arranged end to end; conducts water in plants, and supports them.

yolk: a store of fat and protein food in an egg.

zygote: a cell formed at fertilisation, which will develop into an embryo.

Historical notes

Some landmarks in the study of cells.

Anton van Leeuwenhoek (1632–1723) was a Dutchman, whose hobby was making lenses. Van Leeuwenhoek improved the design of the microscope and made many of them in his lifetime. His microscopes could magnify about 240 times. Van Leeuwenhoek saw things which had never been seen before – tiny living creatures which he called 'little animals'.

Robert Hooke (1635–1703) used a more advanced kind of microscope than van Leeuwenhoek, called a compound microscope – rather like the ones in school laboratories today. He was able to see much smaller things and his drawings and descriptions of a piece of cork are especially famous, because he called the shapes he saw 'cells'. The word was soon being widely used to describe the small units from which all living organisms are made.

Lazaro Spallanzani (1729–1799) was an Italian scientist who was very interested in the 'little animals' which van Leeuwenhoek had described, and in where they came from. Many people thought that they just appeared, by spontaneous generation. Spallanzani carried out many experiments which showed that this was not true – these tiny organisms always appeared from eggs or spores, just as larger ones did.

The electron microscope was invented in 1930. It was not until 1952, however, that it could be used easily with living tissues, because it was so difficult to get suitably thin slices of them. The electron microscope has enabled us to see things 500 times smaller than with the light microscope.

On its own, the electron microscope can only show us what the different parts of a cell probably look like. But by comparing what can be seen with an electron microscope with what can be found out in other ways, a better picture can be built up about what goes on inside cells.

Some landmarks in the history of medicine.

Edward Jenner (1749–1823) was an English country doctor who investigated smallpox, a terrible disease, which often killed the sufferer. In the eighteenth century, it was common practice in England to inoculate children with smallpox germs from a person who had had a mild attack of the disease. The hope was that the child would get a mild form of smallpox, survive, and be immune to the disease for the rest of their lives. Sometimes it worked; sometimes the child died.

Edward Jenner carried out many such inoculations. But he noticed that patients who had had a disease called cowpox – a mild disease caught from cows by people who milked them – did not get smallpox when he inoculated them with smallpox pus.

He tried an experiment. He inoculated a boy with cowpox pus, and later with smallpox pus. The boy did not get smallpox.

This was a much safer method of protection against smallpox. It was called 'vaccination' (Latin 'vacca' means 'cow'). Within a few years, it was being used all over Europe.

Louis Pasteur (1822–1895) lived and worked in France. He was a great thinker, who had a major influence on other scientists of his time. Much of his work was concerned with the causes and prevention of disease.

An industrialist was having trouble with the alcohol he was producing from fermented beet juice; it sometimes went sour. Pasteur proved that the yeast which caused fermentation was a living organism, and that microscopic rod-shaped organisms were responsible for making the alcohol go sour. The science of microbiology had begun.

Several years later Pasteur showed that the organisms which turned wine sour could be killed by heating the wine gently. The method was called pasteurisation, and is now used all over the world as a way of treating milk to help it to keep longer.

Pasteur's first investigation into disease was with silkworms. He was able to show that a disease of silkworms was caused by a microbe; the first time microbes had been identified as the cause of a disease.

Pasteur firmly believed that each kind of infectious disease was caused by a particular kind of microbe, and he tried, unsuccessfully, to isolate the micro-organism that caused cholera in humans. In 1880, Pasteur was working on a disease of chickens, called chicken cholera. He discovered that if his cultures of the microbe were left exposed to the air for some time, they were weakened. If they were then injected into chickens, the chickens became immune to the disease. Like Jenner, Pasteur had discovered a vaccine – and he went on to work out how to make other vaccines. Within a year, he had produced a vaccine for anthrax, and by 1885 a vaccine for rabies.

Pasteur was a great scientist. He worked very hard, was capable of logical thought and deduction, and loved argument. He always liked to show that his intellect was greater than that of the other scientists of his day – and his efforts to prove that his ideas were right and theirs were wrong often led him to great discoveries.

Robert Koch (1843–1910) was a German doctor. Early in his career he began to study anthrax. This was a disease which affected sheep, cattle, horses and sometimes humans. French scientists had shown that the blood of animals with anthrax contained a micro-organism. Koch showed that the micro-organism actually caused the disease. This was the first time that a particular germ had been proved to cause a particular disease. Koch later went on to identify the micro-organisms responsible for many other diseases.

In the course of his experiments, Koch developed many techniques which are now widely used in microbiology – for example, growing micro-organisms on a jelly made from agar.

Joseph Lister (1827–1912) was a surgeon in Scotland, at a time when surgery was done with no anaesthetics, and one in three people died from infection after major operations. Lister read of Pasteur's work, and realised that the infection was probably caused by micro-organisms. In 1865, he tried applying carbolic acid to wounds to kill the microbes, with great success. Gradually, he came to realise the importance of washing his hands and surgical instruments thoroughly before an operation. It was many years, however, before his techniques became widely used by other surgeons.

Elie Metchnikoff (1845–1916) was a Russian scientist. In 1882, he inoculated starfish larvae with microbes and could see, in their transparent bodies, how cells moved towards the infected place and attacked the microbes. He called the cells phagocytes, and suggested that a similar defence system operated in the human body. His discoveries began studies into how immunity works – the science of immunology.

Paul Ehrlich (1854-1915) was interested in the use of chemicals to attack disease. Working in Berlin, he tried out a substance called atoxyl, which contained arsenic, as a treatment for sleeping sickness. The results were hopeful, and he tried making other chemicals from atoxyl. Number 606, which he called salversan, proved particularly effective against syphilis and relapsing fever. Ehrlich's work began a new branch of medicine – chemotherapy.

Alexander Fleming (1881–1955) was born in Scotland, but worked as a bacteriologist in a London hospital. In 1928, while working with *Staphylococci* which were growing on an agar plate, Fleming saw that a mould had got in by mistake. Around the mould, there was a clear patch, where no bacteria had grown. Fleming suggested that the mould, *Penicillium*, had made a substance which diffused through the agar, and which killed the bacteria.

Howard Florey (1898–1968) isolated the substance made by the *Penicillium* mould, ten years after Fleming's discovery. He called it penicillin. It was the first antibiotic – a substance which kills bacteria but not human cells. Initially there were problems with producing penicillin in large quantities, but by the end of the Second World War it was being manufactured in enormous quantities. Since then, many other antibiotics have been discovered.

Steps in the understanding of evolution and genetics.

Carl Linnaeus (1707–1778) a Swedish naturalist, introduced the idea of classifying living things. He worked with plants, giving each one a Latin name, as described in Chapter 17. His work helped to remove a lot of confusion which had resulted from all the different common names which could be given to one kind of plant.

Jean Baptiste de Lamarck (1744–1829) was a French naturalist and philosopher. He was the first person to put forward a consistent theory of evolution, in 1809. He realised that the earth was extremely old, and that organisms had evolved, so that they became better adapted to their environment. However, Lamarck believed that each organism had a kind of inbuilt 'drive towards perfection', and that characteristics that an organism acquired during its lifetime could be passed on to its offspring. He also believed in spontaneous generation. For Lamarck, each group of organisms began as something created by spontaneous generation, which then gradually became more 'perfect' over a long period of time.

Charles Darwin (1809–1882), like Lamarck, thought that the earth must be very old, and that organisms evolved to become better adapted to their environment. His theory of natural selection, however, which is explained in Chapter 16, has turned out to be much nearer the truth than Lamarck's ideas.

Darwin was a great naturalist, who worked on many biological subjects during his lifetime. His ideas about evolution and natural selection were inspired during a voyage on the *Beagle* to South America from 1831 to 1836. However, it took him many years to sort these ideas out, and he published his most famous book, *The Origin of Species*, in 1859.

Gregor Mendel (1822–1884), an Austrian monk, worked on the inheritance of various characteristics in the garden pea. Other people had previously done experiments like Mendel's, but they did not take as much trouble as he did over accurate recording of results, and were not able to interpret them correctly. Mendel's hypothesis to explain inheritance was that there were particles (we now call them genes) responsible for the appearance of certain characters; that each parent had two of these genes for each characteristic, but only passed on one to its offspring; and that the genes stayed separate in the offspring, so that they would separate again to be passed on to their offspring.

Mendel's work was overlooked for 34 years. In 1900, it was rediscovered by several scientists also working on genetics. Once his results were published, they sparked off a tremendous amount of research on the mechanism of heredity, which continues today.

Index

Page numbers in **bold** indicate illustrations.